578.9786   164370
Per   Perkins, Eric John

The biology of
estuaries and
coastal waters

The Biology of Estuaries and Coastal Waters

The Biology of Estuaries and Coastal Waters

E. J. PERKINS

Department of Biology, University of Strathclyde, Scotland

1974

ACADEMIC PRESS · LONDON · NEW YORK

A Subsidiary of Harcourt Brace Jovanovich, Publishers

ACADEMIC PRESS INC. (LONDON) LTD.
24/28 Oval Road,
London NW1

United States Edition published by
ACADEMIC PRESS INC.
111 Fifth Avenue
New York, New York 10003

Library of Congress Catalog Card Number: 73 9473
ISBN: 0 12 550750 X

Printed in Great Britain by
Page Bros (Norwich) Ltd, Norwich

Preface

Within recent years, the world has seen a remarkable development of interest in all things marine: an interest which seems to have arisen as a result of many influences—commercial, military and no doubt the many wonderful films which bring the sea and its life into our homes. All have given a greater appreciation of the fact that we know extraordinarily little about the larger part of our planet, and, in common with the rest of the world's salt water, estuaries and coastal waters have become a focus for increasing attention.

During most of my adult life, I have been happily involved with a number of British estuaries. However, during the past ten years these interests have become formalized, and yet more generalized, because of an association with a number of waste disposal problems. In these cases, it has been necessary to try to visualize an estuary as a whole, not just as a collection of plants, animals, water movements, or sediment movements, but rather as a completed jigsaw puzzle: a fascinating and obsessive puzzle which has often seemed to have only differently shaped, but not varicoloured, pieces as a clue to its completion.

There are, of course, many excellent works dealing with individual facets of seaboard life. I am not attempting to imitate or supplant these works, which remain a source of inspiration, but because it seems to be necessary, I have tried to present an overall view of these areas and their interaction with the rest of the planetary environment. The purist may feel that certain topics have been dealt with inadequately and no doubt this is true. However, I have tried to present information which a worker faced with the need to answer questions of industrial and economic interest will find a useful starting point.

Succeed or fail in this object, it still seems to me, as it did in 1959, that a biologist can make a valuable contribution to man's well-being by working in this field. We, in Britain, enjoy a high standard of living because of the fruits of industrial development, but I see no reason why we should foul our habitat by poor management practices. There must surely be a rational way of receiving the benefits of our industrial society without inhabiting a midden. Indeed, it seems to me that much of our aquatic environment is in a squalid state today largely because so many biologists in the past, despite all warning signs, turned their back on the less pleasant consequences of life. Having chosen this path, it has led me to an appreciation of the estuarine and coastal environment which I could have gained in no other way.

The award of a Winston Churchill Travelling Fellowship, in 1969, did much to enlarge my knowledge of aquatic biology, and I am deeply indebted to the Trustees and staff of the Trust for providing me with this opportunity and for the superlative organization which supported it. I am also very grateful for the many kindnesses shown to me by my hosts in Canada and the United States; after so universal a display of kindness it would be invidious to make an individual distinction.

I am indebted to Dr. J. E. Forrest and Mr. R. S. A. Beauchamp of the Central Electricity Research Laboratory, Leatherhead, for permission to publish hitherto unpublished work on the River Blackwater, Essex. I am similarly indebted to Dr. C. E. Lucas, F.R.S., and Dr. J. H. Fraser, D.A.F.S., Marine Laboratory, Aberdeen, for permission to publish hitherto unpublished work on the Firth of Clyde.

Much of the work carried on in the Solway Firth during the period 1965–71, and referred to in the text, was conducted in conjunction with the Cumberland Sea Fisheries Committee, Thames Board Mills Ltd., Workington, and Marchon Products Ltd., Whitehaven. To these bodies an expression of appreciation is due.

I have been greatly helped by Mr. J. R. S. Gilchrist and Mr. J. W. M. Logan of the University of Strathclyde, Marine Laboratory, Garelochhead, and to them my thanks are due.

I am especially indebted to Professor D. J. Crisp, F.R.S., for his kindness in preparing Table 7.4, and to Professor W. MacNae for his kindness in preparing Table 11.5 and for making additional comments which have been incorporated in the text.

Figure 1.4 is reproduced by permission of U.N.E.S.C.O.

To Catherine, John and Elizabeth, who know too well that a trip to the seaside is just a busman's holiday, a special thank you for their tolerance and understanding.

January, 1974   E. J. PERKINS

Contents

For MARY

1

Introduction

"I must go down to the seas again, for the call of the
running tide,
Is a wild call and a clear call that may not be denied;"

John Masefield "Sea-Fever"

In his poem, John Masefield gave expression to the feelings of all who are associated with the sea, but these two lines seem to express most vividly the attraction felt by those whose life is spent beside, or upon, estuaries and coastal waters. These waters of the seaboard exert an influence upon the affairs of mankind far out of proportion to their size, for it is here that land and sea meet and the fresh and salt waters of the earth intermingle. Still more, however, they offer the first impression that the seafaring man and the jet airline traveller alike experience of the new land: for, if you are a sailor, what can compare with the smells borne by the breeze off the land; or, if you are in an airliner, the sight of the glacier tumbling into the coastal waters of Greenland as you fly over at 34 000 ft. To the fisherman, these waters present an individual way of fishing and of life. To the engineer there is the promise of undreamed of opportunities. To the teeming multitudes of the big cities a place of relaxation and peace is there for the seeking.

In the widest sense, the coastal or neritic waters are those which lie to the landwards of the 200 m depth contour. However, in the present work, it is proposed to use the term coastal in a more local context, viz. estuaries and that part of the inshore waters with which estuaries interact more directly, and which are under the influence of, or influence, those processes which occur at the margin between land and sea.

What then is an estuary? It is a semi-enclosed body of water connected freely with the open sea and within which measurable dilution of sea water, by fresh water, occurs[1]. Evidently, this definition covers a wide range of situations from the bustling Thames and Elbe, so important to commerce, to the overwhelming and lonely grandeur of the Norwegian fjords; from the beautiful, strange estuaries of Sutherland, to the equally beautiful, no less strange, but claustrophobic tropical estuaries colonized by mangroves.

Estuaries may be classified by a variety of criteria, although no one system of classification is universally used. Within given criteria, there are marginal cases, or cases where an estuary may change its classification depending upon the prevailing conditions. The balance between freshwater inflow and loss due to evaporation provides the basis of one such system. Where the freshwater inflow exceeds evaporation the estuary is defined as a *positive estuary*; where evaporation exceeds the freshwater inflow the estuary is defined as *negative*; and in a neutral estuary, evaporation and freshwater inflow are in approximate equilibrium.

In many ways, use of geomorphological structure gives a classification which is the most satisfactory. *Bar-built* estuaries may arise from the development of an offshore bar on a shoreline of low relief; such estuaries normally have narrow connections with the sea, e.g. Chincoteague Bay area, Maryland. The fiords resulting from glacial action on the coasts of Scotland, Norway, New Zealand and Canada represent a second class called *deep basin estuaries*. Such estuaries are normally long, narrow indentations of the coastline, with a shallow sill at the mouth and to the seawards of a deep basin; Loch Fyne, Loch Long and the Gareloch which form a part of the Firth of Clyde complex on the Scottish west coast are excellent examples of this kind of estuary. The *coastal plain estuaries* which represent the third class are formed by the lower reaches of river valleys and drowned river mouths. Normally, estuaries of this type are elongated and shallow, branched and irregular in outline and at the upstream end all receive a river. The Thames, Mersey, Solway Firth, Elbe, and Chesapeake Bay are all of this type. Although, these coastal plain estuaries are regarded as the true estuaries, and have received greater attention than the bar and deep basin estuaries, they each have many features in common and this distinction should be treated with caution.

General Nature of the Sea

Some 71% of the earth's surface is ocean; of this, 90% lies at a depth greater than 200 m, is known as the abyss, and covers about 64% of the earth's surface (Fig. 1.1). Indeed, if the earth were a perfectly rounded sphere, the entire globe would be covered by the oceans to a depth of 8000 ft or approximately 1300 fathoms.

The area of shallow seas within the 200 m (100 fm) contour is known as the *continental shelf*, and its seaward margin is known as the *continental edge*. The continental shelf forms some 3% of the earth's surface and is the site of the major commercial fisheries and most other exploitation of the sea. Where young mountain ranges border the sea the continental shelf may be very narrow, as along the western coasts of the Americas, but

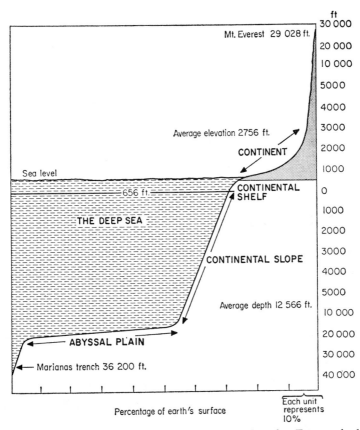

Fig. 1.1. The principal features of the land and sea. Note that Everest the highest mountain at 29 028 ft would be submerged by the depths of the Marianas Trench at 36 200 ft. The mean height of the land is 2756 ft and the mean depth of the sea is 12 566 ft (after Idyll[2]).

elsewhere, as around our coasts, it is much wider and off the coast of Asia it is 800 miles wide[2].

Away from the rich continental shelf and areas of upwelling of water off continental shores, e.g. Humboldt Current, the fertility of the sea decreases. In comparison with these rich waters, the true "blue-water" or central water masses are impoverished and are characterized by a greater number of species which occur at a low density of population. Some large forms exist in these waters, but in general there is a reduction in size; a phenomenon which is particularly evident when species from the same genus, occur in both types of water. The reduction in size is accompanied by a similar reduction in organ development. Indeed, many forms previously thought to be juvenile, because of their small size, are now known to be adult. Few birds are associated with blue-water, and any flying fish present may be strays[3].

Composition of Sea Water

Sea water is a very complex solution, but a relatively few constituents make up more than 99·9% of the solute (Table 1.1).

Table 1.1. Major constituents of an ocean water. $S‰ = 35·00$ (from H. W. Harvey[16])

Constituent	g/kg
Sodium	10·77
Magnesium	1·30
Calcium	0·409
Potassium	0·388
Strontium	0·010
Chloride	19·37
Sulphate as SO_4	2·71
Bromide	0·065
Carbon, present as bicarbonate, carbonate and molecular carbon dioxide	0·023 at pH 8·4 to 0·027 at pH 7·8

Salinity

The amount of inorganic matter dissolved in sea water, expressed as grams per kilogram of sea water, is known as the *salinity* (S). In the open ocean, where this value approximates to 35 g/kg, or 35‰, the proportions of the inorganic constituents remain constant; a property which was discovered by Dittmar who analysed the samples taken by H.M.S. Challenger, 1872–76. This proportionality and the presence of the chloride ion can be used in the determination of salinity by titration with a silver nitrate solution. This method determines the amount of chloride, plus a chloride equivalent of bromide and iodide, present in a sample. The total weight of these elements per kilogram of sea water is known as the *chlorinity* (Cl). To derive the salinity from the chlorinity, an empirical relationship, known as Knudsen's Formula[4], i.e. $S = 0·030 + 1·8050$ Cl, was widely used. However, in strict terms, this formula applies to the Baltic only, for the constant 0·030 refers to the salinity of the river waters contributed to the Baltic. Clearly, this constant is not generally applicable to inshore waters away from this area, or to the open oceans. The implications of this problem were considered by a Joint Panel of Experts sponsored by U.N.E.S.C.O. and they concluded that the relationship between salinity and chlorinity could be expressed more correctly thus:

$$S = 1·80655 \times \text{chlorinity.}$$

This expression eliminates the constant, 0·030, without greatly affecting salinity in the range 30–40‰[5]. To ensure accurate, reproducable results, the silver nitrate solution is standardized with a sample of *Normal Seawater* (*Eau de Mer Normale*) of a given chlorinity which approximates to 19·38 g/kg. The titration is performed at a temperature as close to 20°C, as possible; the results for the chlorinity are then corrected to that at 20°C, i.e. the $Cl/litre_{(20)}$. Normal sea water may be obtained from the Depot d'Eau Normale, Laboratoire Hydrographique, Charlottenlund Slot, Denmark[6]. In practice, this method may be used in conjunction with Knudsen's Hydrographical Tables which also define the relationship between salinity, specific gravity and temperature[7, 8]. Determinations of salinity are now more frequently made by use of a conductivity bridge or *salinometer*.

The amount of salt present in sea water, at N.T.P., influences its density (σ_0) according to the relationship:

$$\sigma_0 = -0·069 + 1·4708\ Cl - 0·001570\ Cl^2 + 0·0000398\ Cl^3.$$

The maximum density of fresh water occurs at 4°C, but for sea water the temperature of maximum density decreases with increased salinity, until at a salinity of 24·7‰, the temperature of maximum density coincides with the freezing point at $-1·332°C$[4]. Use can be made of the salinity/density relationship to determine the salinity by means of hydrometry. This method is of particular interest in estuarine work, for it is here that the relationship found by Dittmar breaks down and the salinity/chlorinity relationship becomes unreliable. Clearly, the land can, and indeed often does, supply large amounts of mineral salts to inshore waters. Some of this contribution is due to the chemical composition of the land itself, but some arises from the fertilizers and other chemicals added to the land by man and washed out in run-off. Under these conditions the conductivity bridge will indicate correctly the amount of inorganic salt present, but the titration method will not. Furthermore, the variations of salinity are so great in estuaries, that the degree of precision required for oceanic work is unnecessary, and the use of an accurate hydrometer/thermometer system is justified. The use of nomograms facilitates the conversion of hydrometer measurements into salinity values[9].

The specific heat (c_p) of sea water at 0°C and atmospheric pressure is also dependent upon the salinity[4], according to the relationship:

$$c_p = 1·005 - 0·004136\ S + 0·0001098\ S^2 - 0·000001324\ S^3.$$

It changes with salinity at 17·5°C and atmospheric pressure as follows[4]:

$S‰$	0	10	20	30	40
c_p	1·00	0·968	0·951	0·939	0·926

The surface tension and salinity of sea water are related thus:

$$\text{Surface tension} = 0\cdot3185\ S + 73\cdot6\ \text{ergs/cm}^2.$$

However, this relationship is affected by the presence of organic matter and pollutants[5].

Because the salinity of inshore waters is so variable, particularly at lower values, advantages to interpretation can be gained by use of the concept of Defined Coastal Influence (D.C.I.)[10]. This value is defined thus:

$$\text{D.C.I.} = 1 + \log_{10}(35\cdot5 - S)$$

where S is the salinity in parts per thousand.

Dissolved Gases

All the atmospheric gases are dissolved in sea water at concentrations which depend primarily upon the salinity and temperature.

Oxygen

Theoretically, the concentration of oxygen in sea water can range between 0 and 14 ml/litre at 30‰, but normally lies within the range 1–6 ml/litre. It is generally accepted that changes in the solubility of oxygen with salinity are linear. However, the solubility of oxygen in saline water can be expressed by the equation,

$$C_S = 14\cdot161 - 0\cdot3943\,T + 0\cdot007714\,T^2 - 0\cdot0000646\,T^3$$
$$- S(0\cdot0841 - 0\cdot00256\,T + 0\cdot0000374\,T^2)\ \text{mg/litre}$$

where C_S is the solubility, in p.p.m., of oxygen in water of salinity $S‰$, T is the temperature in °C and the air pressure is 760 mm Hg [11, 12]. A rather less accurate result is given by the equation

$$C_S = \frac{475 - (2\cdot83 - 0\cdot011\ T)S}{33\cdot5 + T}\ \text{mg/litre}.$$

The results given by these equations are identical for freshwater, but at a salinity of 35‰ differ by not more than 0·04 mg/litre from −1·5 to 34°C. A useful nomograph is available also[13].

In temperate latitudes, and at a given salinity, the oxygen content of sea water varies with the season. This is due not only to differences in temperature, but to differences in production and utilization also. It has been estimated that about 30×10^4 cm^3 of oxygen per m^2 leaves the surface of the Gulf of Maine, between the months of October and March, and a like amount re-enters from the atmosphere in the alternating period. Consumption and production by living organisms accounts for some two fifths of this change. The rate of exchange of oxygen across the sea surface can be defined by the equation

$$dQ/dt = E.s. (P-p)$$

where dQ/dt is the rate at which oxygen enters the sea surface, s is the surface area, P and p are the partial pressure of oxygen in the atmosphere and the oxygen tension in the water, respectively; E is a constant, the *exchange coefficient*, in cm^3O_2 per m^2 per atmosphere per month. In the Gulf of Maine, values of E obtained for the summer and winter are $2\cdot8 \times 10^6$ and 13×10^6 cm^3O_2 per m^2 per atmosphere per month, respectively. The difference, in these values, has been attributed to errors arising from conditions at the sea surface[14]. A similar value of E was obtained, in the summer months, off the coast of Oregon[15]. It was formerly suggested that the phytoplankton of the oceans contributed some 70% of the world's production of oxygen[15a]; this is no longer considered to be true[15b].

Carbon Dioxide

Sea water is slightly alkaline in reaction. Values of pH normally lie within the range 7·5–8·4, with a mean of 8·2[16, 17]. In the open sea, the pH may be stable over wide areas, but latitudinal differences may be observed[18]. However, in sunlit rock pools, it may rise to 9·6, while it may fall to 7·0 in isolated basins where decomposition is taking place[16, 17] and occasionally at the margin of an estuary[X]. Normally, the pH is at a maximum within the surface 100 m, falling to a minimum within a depth of 200 to 1200 m at which depth an oxygen minimum occurs; deeper still the pH rises to a deep sea maximum and then declines once more[19].

Table 1.2. Anion–cation balance in average sea water (from Fearon[19a])

	Na^+	K^+	Ca^{2+}	Mg^{2+}	Cl^-	CO_2^{2-}	SO_4^{2-}	PO_4^{3-}
g/kg	11·0	0·4	0·4	1·3	19·0	0·09	2·7	0·015
milli-equivalent/kg	480	10	20	108	535	18	56	0·16

Total cation charge	618	Total anion charge	609

The slight alkalinity arises from the excess of strong cations, viz. Na^+, K^+ and Ca^{2+}, over anions derived from strong acids (Table 1.2). Consequently, a considerable amount of carbon dioxide can be carried in solution as bicarbonate and carbonate[16], thus:

$$CO_2 \underset{\text{In air}}{\rightleftharpoons} CO_2 \overset{+H_2O}{\underset{\text{In water}}{\rightleftharpoons}} H_2CO_3 \rightleftharpoons H^+ + HCO_3^- \rightleftharpoons H^+ + CO_3^{2-}.$$

Whereas the minimum of pH observed in waters of 200–1200 m depth is probably due to biological processes, effects at greater depth, and especially the deep sea maximum, are probably due to the effects of hydrostatic pressure upon the dissociation constants of carbonic acid[19, 20]; sea water at high pressure can dissolve substantially more calcium carbonate than surface waters[21].

Evidently, the oceans act as a mechanism to regulate the amount of CO_2 in the atmosphere. This poses an interesting problem, for at the present time we are burning fossil fuels and releasing CO_2 to the atmosphere at a colossal and increasing rate per annum. Theoretically, two consequences are possible: (1) that the pH of the oceans will fall; and (2) when the sea can no longer absorb the excessive amounts of CO_2 produced, the concentration in the atmosphere will rise. As we shall see, marine animals are adapted to live at a pH lower than 8·2, but such a lowering of pH could have significant consequences in waste disposal problems (see Chapter 16). It is known that the amount of CO_2 in the atmosphere is increasing, but effects upon the sea are thought to be small[21a].

At the present time however, many waters are not saturated with calcium carbonate[22]. On the other hand, calcium carbonate deposition can occur by biological and inorganic processes, but the latter do not generally occur because the Mg^{2+} ion has an inhibiting effect upon the nucleation of calcium carbonate[23, 24]. Nevertheless, the current trend is for the deposition of calcium carbonate on the stable deep ocean floor; a trend which could lead to atmospheric depletion of CO_2, despite additions from the burning of fossil fuels. Should such a situation develop plant life could be diminished[25].

Because the excess base is equivalent to the ions of bicarbonate, carbonate and borate, sea waters possess a limited buffering power. Of these two buffering systems, the carbonate is more important than the borate at normal pH of sea water[6].

The role of silicates in the buffer system of the oceans was ignored until recently. However, it seems unlikely that silicates play any significant role in problems whose duration does not exceed 1000 years, and that for shorter periods the CO_2–carbonate system controls the pH of oceanic waters[24].

Just as the addition of CO_2 to the system can depress the pH, so its removal can raise it. Such a removal is normally the result of photosynthesis, hence the high values found in sunlit rock pools. Formerly, it was considered that carbon for photosynthesis was obtained primarily from molecular CO_2, but recent work has shown that the algae *Platymonas* sp., *Nitzschia closterium* and *Porphyridium cruentum* utilize the carbamino complex of alanine or of sea water concentrates rather than inorganic components of the CO_2 system[26], while the bicarbonate ion may be utilized by *Platymonas* sp., *Nitzschia closterium* and a marine *Chlorella* sp.[27]. The freshwater species *Scenedesmus quadricauda* has a similar ability under appropriate conditions[28], whereas *Chlorella pyrenoidosa* lacks it[27].

It should be appreciated that changes in the calcium carbonate-bicarbonate system which arise as a consequence of photosynthesis affect the conductivity of sea water[29, 30]: yet another reason why conductivity measurements of salinity in waters which lie close to the shore, in rock pools, and in estuaries may be very unreliable. Because of the relationship between CO_2 concentration and the pH, the latter can be used experimentally to monitor changes in the former[31, 32].

Minor Inorganic Constituents

In addition to the major constituents, there is a wide range (Table 1.3) of minor constituents. Their combined weight does not exceed 2 mg/kg (2 p.p.m.), but nevertheless, they are of great importance to marine life

Table 1.3. Geochemical parameters of sea water (from Goldberg[32a])

Element	Abundance, mg/litre	Principal species	Residence time, years
H	108 000	H_2O	
He	0·000005	He (g)	
Li	0·17	Li^+	$2·0 \times 10^7$
Be	0·0000006		$1·5 \times 10^2$
B	4·6	$B(OH)_3$; $B(OH)_2O^-$	
C	28	HCO_3^-; H_2CO_3; CO_3^{2-}; organic compounds	
N	0·5	NO_3^-; NO_2^-; NH_4^+; N_2 (g); organic compounds	
O	857 000	H_2O; O_2 (g); SO_4^{2-} and other anions	
F	1·3	F^-	
Ne	0·0001	Ne (g)	
Na	10 500	Na^+	$2·6 \times 10^8$
Mg	1 350	Mg^{2+}; $MgSO_4$	$4·5 \times 10^7$
Al	0·01		$1·0 \times 10^2$
Si	3	$Si(OH)_4$; $Si(OH)_3O^-$	$8·0 \times 10^3$
P	0·07	HPO_4^{2-}; $H_2PO_4^-$; PO_4^{3-}; H_3PO_4	
S	885	SO_4^{2-}	
Cl	19 000	Cl^-	
A	0·6	A (g)	
K	380	K^+	$1·1 \times 10^7$
Ca	400	Ca^{2+}; $CaSO_4$	$8·0 \times 10^6$
Sc	0·00004		$5·6 \times 10^3$
Ti	0·001		$1·6 \times 10^2$
V	0·002	$VO_2(OH)_3^{2-}$	$1·0 \times 10^4$
Cr	0·00005		$3·5 \times 10^2$
Mn	0·002	Mn^{2+}; $MnSO_4$	$1·4 \times 10^3$
Fe	0·01	$Fe(OH)_3(s)$	$1·4 \times 10^2$
Co	0·0005	Co^{2+}; $CoSO_4$	$1·8 \times 10^4$
Ni	0·002	Ni^{2+}; $NiSO_4$	$1·8 \times 10^4$
Cu	0·003	Cu^{2+}; $CuSO_4$	$5·0 \times 10^4$
Zn	0·01	Zn^{2+}; $ZnSO_4$	$1·8 \times 10^5$

Table 1.3—(*Continued*)

Element	Abundance, mg/litre	Principal species	Residence time, years
Ga	0·00003		1·4 × 10³
Ge	0·00007	$Ge(OH)_4$; $Ge(OH)_3O^-$	7·0 × 10³
As	0·003	$HAsO_4^{2-}$; $H_2AsO_4^-$; H_3AsO_4; H_3AsO_3	
Se	0·004	SeO_4^{2-}	
Br	65	Br^-	
Kr	0·0003	Kr (g)	
Rb	0·12	Rb^+	2·7 × 10⁵
Sr	8	Sr^{2+}; $SrSO_4$	1·9 × 10⁷
Y	0·0003		7·5 × 10³
Nb	0·00001		3·0 × 10²
Mo	0·01	MoO_4^{2-}	5·0 × 10⁵
Ag	0·0003	$AgCl_2^-$; $AgCl_3^{2-}$	2·1 × 10⁶
Cd	0·00011	Cd^{2+}; $CdSO_4$	5·0 × 10⁵
In	<0·02		
Sn	0·003		5·0 × 10⁵
Sb	0·0005		3·5 × 10⁵
I	0·06	10_3^-; I^-	
Xe	0·0001	Xe (g)	
Cs	0·0005	Cs^+	4·0 × 10⁴
Ba	0·03	Ba^{2+}; $BaSO_4$	8·4 × 10⁴
La	0·0003		1·1 × 10⁴
Ce	0·0004		6·1 × 10³
W	0·0001	WO_4^{2-}	1·0 × 10³
Au	0·000004	$AuCl_4^-$	5·6 × 10⁵
Hg	0·00003	$HgCl_3^-$; $HgCl_4^{2-}$	4·2 × 10⁴
Tl	<0·00001	Tl^+	
Pb	0·00003	Pb^{2+}; $PbSO_4$	2·0 × 10³
Bi	0·00002		4·5 × 10⁵
Rn	0·6 × 10⁻¹⁵	Rn (g)	
Ra	1·0 × 10⁻¹⁰	Ra^{2+}; $RaSO_4$	
Th	0·00005		3·5 × 10²
Pa	2·0 × 10⁻⁹		
U	0·003	$UO_2(CO_3)_3^{4-}$	5·0 × 10⁵

as essential trace elements. For normal healthy growth, all plants have a requirement for phosphate, nitrate, iron, cobalt, manganese, copper and zinc; some require silicon, molybdenum and vanadium, also. All animals require iron, which may be essential for their respiratory pigments; other species require copper for this purpose. The ability of ascidians to concentrate, in their blood, vanadium from sea water in which it could not be detected, is a classic case, often quoted; here a concentration factor of 5×10^4 is known now. Because living organisms have an ability to accumulate these trace elements, some of them, e.g. Zn and Mn, have assumed a considerable importance in problems of radioactive waste disposal. In this instance, the oyster can accumulate zinc, principally in the gills and hepatopancreas[33, 34], with an accumulation factor of $2 \cdot 0$–$3 \cdot 1 \times 10^4$.

The cycle of phosphorus in the sea is summarized in Fig. 1.2[35]. It is possible to distinguish between different water masses by their phosphorus content, and the total amount of phosphorus in a water column determines its fertility. It should be noted that the phosphorus content is not related to the salinity. In the open sea dissolved inorganic phosphate is of considerable importance, but, in inshore water, as we shall see in Chapter 2,

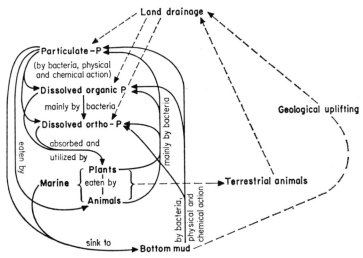

Fig. 1.2. The phosphate cycle in the sea (after Chu[35]).

the phosphate is associated particularly with the bottom sediments. As one might expect, the concentration of phosphate dissolved in sea water and the distribution of phosphorus in the environment and the biota undergoes marked annual changes which depend upon the annual cycle of production[36] (see Fig. 1.3). Conversely, the concentration of phosphate present may be expected to influence the productivity of sea water. It is of interest, therefore, that landings of the spurdog, *Squalus acanthias*, at Mevagissey are, broadly speaking, related to changes in the phosphate concentration at the nearby International Hydrographic Station El (Table 1.4[37]).

Silica is important to some marine organisms, notably the diatoms, radiolarian protozoa and sponges, in which it is a skeletal material, but other features of its chemistry are not so apparent. However, we have already seen that on a protracted time scale silicates may have an important role in controlling the buffering action of the water[24]. Although rivers are

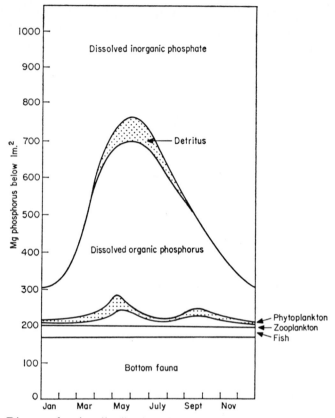

Fig. 1.3. Diagram showing distribution of phosphorus in water column 70 m deep, Plymouth area. (Phosphorus in solution and in detritus based on observations made in 1948 and 1949, in zooplankton based on observations made in 1935 farther inshore, in fish and bottom fauna on computations.) (After Armstrong and Harvey[36].)

continually adding silica, and very many other materials, to the sea, the concentrations of the salts in sea water seem to have remained much the same over long periods of geological time. In the short term, at least, the concentration of silica in sea water seems to be controlled by biological rather than physical processes[38].

In addition to these minor inorganic constituents, sea water contains organic compounds in small and varying quantities. In a way, sea water can be regarded almost as a living soup in which slight differences in composition can have profound effects upon the inhabitants. For example, from the same batch of fertilized eggs of the sea urchin, *Echinus esculentus*, larvae of different morphological types are produced depending on the

Table 1.4. The relationship between available phosphate and fish landed in the English Channel (after Cooper[37])

Year	Weight in fish landed (tons)	Edible phosphorus in fish landed (tons)	Phosphorus available for growth in Channel (tons)	Usable phosphorus landed as percentage of phosphorus available for growth (%)
1925	81 000	115	111 000	0·104
1926	79 000	103	137 000	0·075
1927	74 000	96	100 000	0·096
1928	71 000	92	126 000	0·073
1929	95 000	123	126 000	0·098
1930	58 000	75	—	—
1931	76 000	99	95 000	0·104
1932	88 000	114	86 000	0·133
1933	95 000	123	97 000	0·127
1934	93 000	121	88 000	0·137
1935	65 000	85	77 000	0·110
1936	79 000	103	86 000	0·120
1937	76 000	99	88 000	0·112

source of the sea water used in the cultures, viz. the Plymouth area or the Firth of Clyde. No one single factor, e.g. different copper content, is considered responsible, but it seems likely that the biological difference between sea waters from different sources is due to a number of factors[39].

Although earlier workers had postulated that the organic materials present in sea water might be of great significance, it remained for Lucas to formulate a theory of non-predatory relationships[40-43]. In essence, this theory states that in its life-time an organism, plant or animal, secretes organic materials which may be of advantage or disadvantage to associated organisms. Where these substances are of advantage they may be growth factors, or may act as a trigger which ensures that a particular succession of organisms can take place. Since that first expression of the significance of external metabolites, (which may be hormones, antibiotics or vitamins), the importance of these substances to life in the sea has become generally realized. Some species of Protista, for example, have a requirement for vitamins (Table 1.5)[44]; moreover, the yield of *Monochrysis lutheri*, a supra littoral chrysomonad, is directly proportional to the concentration of vitamin B_{12}, between 0·1 and 100 pg/ml[45].

Both vitamin B_{12} and divalent sulphur influence the development of the neritic diatom *Skeletonema costatum*. Absence of the former reduced all divisions significantly, and absence of the latter reduced divisions by 5% compared to fully enriched cultures[46]. The waters of fertile bays are

Table 1.5. Some vitamin requirements for some protista (after Droop[44])

	Requirement for thiamine	Portion of vitamin required	Requirement for vitamin B_{12}	Other heterotrophic tendencies
Chrysophyceae *Monochrysis* *Prymnesium* *Syracosphaera* *Microglena*	+	Pyrimidine	+	Can utilize some amino acids as N source
Chlorophyceae *Nannochloris* Bacillariophyceae *Phaeodactylum*	−	−	−	Can utilize some amino acids as N source
Skeletonema	−	−	+	Organic compounds sometimes stimulatory in an unspecific way
Cryptophyceae *Hemiselmis* Dinophyceae *Oxyrrhis*	+	Thiazole	+ ?	Amino N obligatory Amino N obligatory; acetate as C source; phagotrophic in Nature; other B vitamins
Glenodinium *Peridinium*	−	−	+	?

enriched with amino acids, e.g. tyrosine, phenylalanine, lysine, ornithine, histidine and arginine, when compared to the open sea[47].

Carbohydrates are also lost to the environment by phytoplankton, and a rise in diatom numbers is accompanied by increased amounts of carbohydrate and organic matter present in the supporting water[48-50]. Although it was once considered that the concentration of sugars in sea water could affect the pumping rate of oysters[51], it has not been confirmed[52].

External metabolites of a kind so far discussed are clearly beneficial, or at least neutral, in effect. However, the secretions produced by a "red tide" are far from beneficial; shellfish become tainted by toxin and enormous numbers of fish may die in the presence of such blooms[53-58].

Spray from a red tide carried inland by an onshore wind can cause considerable respiratory distress among the more susceptible members of a human population, e.g. bronchitics[59]. It is true that this phenomenon is observed mainly in the warm waters of the world; some truly spectacular red tides occur off the coasts of the Americas[2]. However, they are not confined to low latitudes. Red tides occur in European waters, also; the most recent occurred in the fine, dry summer of 1968 off the east coast of Britain, between Rosehearty and Flamborough Head. More than 80 people who ate mussels were poisoned, none fatally. In this outbreak, serious consequences were prevented by the British habit of boiling the mussels thoroughly and rejecting the supernatant liquor[60]. Generally, red tides are blooms of dinoflagellates in near monoculture; a variety of species is capable of producing this phenomenon, e.g. *Prymnesium parvum*, *Peridinium foliaceum*, *Gymnodinium breve*, *Gonyaulax polyedra*, *G. tamarensis*, *Exuviaella baltica*. All produce a neurotoxin which can be fatal to man and animals. Nutrient enrichment is an essential for their development[61-63]. It is interesting to note that in fresh water, toxic blooms of blue-green algae are induced by enrichment, and with the eutrophication of many lakes by sewage and other effluents, noxious blooms are becoming more wide-spread. It is a sobering thought that unbridled use of so large a volume of water as Lake Erie is causing serious concern.

Some Physical Parameters of the Sea

If we exclude the phenomena which occur under extreme conditions, the illumination and temperature structure of the ocean can be defined as in

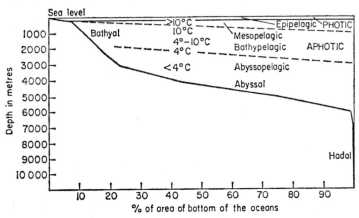

Fig. 1.4. Zones of the marine environment (after Bruun[64]).

Fig. 1.4[64]. Inshore waters reveal many variants of this general picture, and although these two parameters are closely related, it is convenient to consider them separately.

Temperature

In summer, in the middle latitudes, where there is a sufficient depth of water, the surface layers become heated more than the deeper water. Turbulence is not sufficient to carry all the heat below a given depth, and the temperature differences which result have the effect of inhibiting eddy motion at this depth. When this occurs a zone known as the *discontinuity layer* or *thermocline* is developed and the passage of heat from it to the waters below is slow. The amount of heat which does pass the thermocline can be carried downwards by the eddy motion, and consequently these layers are more or less isothermal. This seasonal thermocline normally develops at depths between 15 and 40 m. It may divide, very sharply, the upper and lower layers, and has a profound effect upon the life of the plankton. For example, in the warm, upper layer, the metabolic rate may be > 50% more rapid than in the water below[65].

The upper layer may be called the *epithalassa* (cf. epilimnion for lakes), while the layer below the thermocline may be called the *hypothalassa* (cf. hypolimnion for lakes).

In the English Channel, the temperature structure in summer conforms to a pattern similar to that in Table 1.6[65]. Clearly, the condition of the thermocline structure is dependent upon the weather, and may tend to breakdown under stormy conditions; however, the resumption of calm weather restores the structure. The thermocline system is so stable that it is only broken finally by the gales and lowered temperatures of autumn. The vertical mixing which takes place at this time is important because it returns to the surface nutrient salts which passed below the thermocline in the sinking bodies of dead plankton organisms[66]. Close inshore a turbulence due to tidal currents effectively transfers heat and prevents the formation of a thermocline, but in the open sea it occurs at a much greater depth than in the English Channel and in the tropics a thermocline is

Table 1.6. The temperature structure of the English Channel during summer (after Harvey[65])

Depth m	Temperature °C	
0	16·15	
5	16·08	
10	15·85	Epithalassa
12	15·82	
15	15·82	
		Discontinuity layer
17·5	12·09	
20	12·05	
30	12·05	Hypothalassa
60	12·05	

permanently present. The annual cycle of air and sea surface temperatures can be characterized by the mean, amplitude and first harmonic value of phase in an harmonic analysis. This first harmonic value of phase is expressed as a lag of temperature behind solar radiation. Coatal stations are late in phase and inland basins are earliest, while the western margins of continents lag substantially behind the eastern margins[67].

Not only do temperatures vary from place to place and from season to season, but long term changes also take place. In the English Channel, the annual mean surface temperature rose by about 0·5°C in the 50 years prior to 1960, while the bottom temperatures at International Hydrographic Station E1 rose by about 0·25°C in the period from the mid 1920s to 1960[68]. In Plymouth Sound, the mean sea temperature rose by 0·3°C in the 40 years prior to 1958[69]. The temperature, and its changes, are a vital factor in all metabolic processes. It is of considerable interest, therefore, to find that so small a change in the mean sea temperature affected the distribution of the fauna in the English Channel, viz. a change in the relative proportions of the intertidal barnacles *Chthamalus stellatus* and *Balanus balanoides*. As the *Balanus* retreated from its southern and western limits, so the *Chthamalus* advanced[70]. Since 1960, the mean sea temperatures around the British Isles have fallen slightly when compared with the previous 25 years; coincident with this fall in temperature *Balanus balanoides* has increased in abundance in S.W. England[71].

Before leaving the subject of temperature, it is essential to make reference to temperature measurements with respect to animals which dwell on the shore. It has become a custom in the marine biological literature to take temperatures as recorded by the Meteorological Office and use these temperatures as if experienced by the marine plants and animals concerned.

Table 1.7. Correction to be applied to the mean monthly temperature at 5 ft to obtain the corresponding temperature at 0·5 ft °F. Experimental data from exposed stations at Seabrook Farms, Bridgeton, New Jersey (after Baum[73])

	1947	1948	Average
January	—	−0·2	−0·2
February	—	−0·5	−0·5
March	—	0·4	0·4
April	—	0·9	0·9
May	—	0·9	0·9
June	—	1·0	1·0
July	—	1·4	1·4
August	—	1·5	1·5
September	2·0	1·6	1·8
October	1·3	—	1·3
November	0·3	—	0·3
December	−0·2	—	−0·2

Basically, this malpractice is due to a failure to appreciate the meaning of Meteorological Office measurements, moreover it disregards long standing knowledge of lapse rates. First of all, the Meteorological Office measures air temperatures at a height of 4 ft above a grass surface. Secondly, there is a gradient in temperature between the grass surface and that at a height of 4 ft[72], this difference is illustrated by Table 1.7[73], and, of course, shores are not grass surfaces. Again few meteorological stations are at or near the head of a shore, while most are situated at a significant distance from the site in question. Meteorological Office data, then, does not provide means of obtaining temperatures without effort, at best this information can only be used to show climatic trends which *may* be of significance to marine problems.

Illumination

Clearly, temperature as it is normally experienced in the environment is due to a balance between heat gained by radiation from the sun and heat lost by radiation from the earth. The heat gained from the sun is not directly proportional to the number of hours of sunshine, for the intensity of illumination is dependent upon the solar altitude, the amount, height and thickness of cloud cover, and also upon the reflectivity of the surface illuminated.

The reflection of light, by a surface, is expressed as the *albedo*, a, where

$$a = \frac{\text{Amount of reflected light}}{\text{Amount of incident light}} \times 100\%.$$

In a series of experiments, conducted from a blimp, the albedo of the sea surface was 14% at a solar altitude of 30–35°; at high solar altitudes, i.e. 50–70°, the albedo was lower and tended towards a more constant value of *ca.* 8%[74, 75].

Once that portion of the light not reflected by the sea surface has penetrated the water it is absorbed rapidly with depth, thus:

Depth (ft)	% Absorbed
6	50
14	75
25	90

In the clear waters of the Sargasso Sea, 490 ft of water is required to absorb 99% of the incident illumination, whereas a similar proportion is absorbed by 105 ft and 26 ft of water in the Gulf of Maine and the sediment laden waters at Wood's Hole, Massachusetts, respectively[65]. However, not all light is equally penetrative; red light is absorbed first, and blue light shows the greatest penetration. In his bathyscaphe descent, Beebe noted that from 50 to 150 ft the dominant colour was orange, while from 150 to 300 ft it was yellow.

The absorption of light in water is defined by the *extinction coefficient* (λ or μ), otherwise known as the *absorption coefficient* or *transmission exponent*[4].

A simple, early method of investigating the transparency of sea water employed the Secchi disc; it has been much used by biologists and is of considerable value in estuarine waters where rapid, but not terribly accurate, methods give valuable results. It is a disc, 30 cm in diameter, formerly painted white, but now more often used with alternating white and black quadrants to give greater visibility[76, 77]. In use it is lowered into the water to a depth at which it just disappears; the depths at which disappearance and reappearance occur is noted and the mean taken. A relationship exists between the mean maximum depth of Secchi disc visibility in metres (D) and the extinction coefficient of visible rays (λ). In the English Channel[78, 79]

$$\lambda = \frac{1 \cdot 7}{D}.$$

However, off the coast of southern California[80] a better fit is given by:

$$\lambda = \frac{1 \cdot 4}{D};$$

in very clear waters by[81]:

$$\lambda = \frac{1 \cdot 9}{D};$$

and in the turbid water of the Cochin Backwater by[82]:

$$\lambda \approx \frac{1 \cdot 5}{D}.$$

The Secchi disc method does not measure light intensity, but photo-electric cell techniques do. The values of light intensity measured by this means may be interpreted by use of the following relationship[83] or one of its variants[84, 85], thus:

$$\lambda = \frac{\log_e (I_0/I)}{d} \qquad \left(\text{or} = \frac{2 \cdot 3}{d} \log_{10} (I_0/I) \right)$$

where λ = extinction coefficient, I_0 = intensity of incident illumination and I = intensity at depth d.

We have already seen that the depth to which light penetrates is dependent upon the amount of sediment present in the water, however there is also an inverse relationship between the visibility of a Secchi disc and the amount of phytoplankton present[86].

In non-turbid offshore waters, the relationship between the extinction coefficient, λ, and the abundance of the plant population, expressed as the concentration of chlorophyll a in mg/m^3, C, is given by the formula[87]:

$$\lambda = 0{\cdot}04 + 0{\cdot}0088C + 0{\cdot}054C^{2/3}.$$

The degree of scattering of light by materials, either organic or inorganic, suspended in sea water is dependent upon the size of the particles[88].

Measurements of the absorption spectra of ultra-violet light have shown that natural sea water is twice as absorbent as artificial sea water. In shallow waters differences in absorbance have been ascribed to the presence of organic materials. In water from deep in the Atlantic Ocean, an enhanced absorption may be due to the presence of a high concentration of nitrate[89].

Hydrostatic Pressure

The pressure of the sea water increases by approximately 1 atmosphere for every 30 ft increase in depth (or 1 millibar $\equiv$ 1 cm sea water, or 1 decibar $\equiv$ 1 metre of sea water). The sensitivity of marine animals to small pressure changes was demonstrated first with decapod larvae[90]. While the physiological effects of pressure are not well understood, further work has shown that there are some animals which do not respond to pressure changes, e.g. *Tomopteris*, *Sagitta* and Phyllosoma lavae; however, other species do respond, e.g. *Corophium* to which a change in pressure is important to its emergence from the substratum, and *Aurelia* to which a response to changing pressure is essential to maintain its place in the plankton[91]. In common with responses to other phenomena, shallow water species of animals seem capable of withstanding, without harm, considerable increases in pressure. The crab *Pachygrapsus crassipes* survived pressure changes between 1 and 109 kg/cm^2, while the mussel, *Mytilus edulis diegensis*, survived pressure changes between 1 and 219 kg/cm^2[92].

It is not only the living organisms which are influenced by the effects of hydrostatic pressure. The dissociation constants of carbonic and boric acids are affected[93], and thus, as we have seen, the pH[19, 20]; however, an increased hydrostatic pressure influences proteins, enzyme systems and the ionization of water and other inorganic compounds[94, 95].

References

1. Pritchard, D. W. (1960). *In* "Encyclopedia of Science and Technology", McGraw-Hill, New York and London.
2. Idyll, C. P. (1971). "Abyss", 2nd edn. Crowell, New York.
3. Marshall, N. B. (1966). *Rep. Challenger Soc.* **3** (18), 40–41.
4. Sverdrup, H. U., Johnson, M. W. and Fleming, R. H. (1960). "The Oceans: Their Physics, Chemistry and General Biology", Prentice-Hall, New York.

5. Johnston, R. (1964). *Oceanogr. Mar. Biol. A. Rev.* **2**, 97–120.
6. Strickland, J. D. H. and Parsons, T. R. (1965). "A Manual of Sea Water Analysis", *Bull. Fish. Res. Bd. Can.* **125**, 1–203.
7. Knudsen, M. (1901). "Hydrographical Tables", G.E.C. Gad., Copenhagen.
8. Matthews, D. J. (1932). "Tables of the Determination of Density of Sea water Under Normal Pressure, Sigma-t", Cons. perm. int. Explor. Mer, Copenhagen, 55 pp.
9. Olson, F. C. W. "Nomograms for Hydrometer—Salinity and Seawater Density", Florida State Universtiy, Tallahassee Res. Council Studies, No. 22, pp. 13–18.
10. Craig, R. E. (1953). *Annls biol. Copenh.* **10**, 86–89.
11. Truesdale, G. A. and Knowles, G. (1956). *J. Cons. perm. int. Explor. Mer*, **25**, 263–267.
12. Truesdale, G. A. and Gameson, A. L. H. (1956). *J. Cons. perm. int. Explor. Mer* **25**, 163–166.
13. Gilbert, W. E., Pawley, W. M. and Park, K. (1967). *J. oceanogr. Soc. Japan* **23**, 252–255.
14. Redfield, A. C. (1948). *J. mar. Res.* **7**, 347–361.
15. Pytkowicz, R. M. (1964). *Deep Sea Res.* **11**, 381–389.
15a. Cole, L. C. (1968). *Bio Science* **18**, 679.
15b. Ryther, J. H. (1970). *Nature, Lond.* **227**, 374–375.
16. Harvey, H. W. (1955). "Chemistry and Fertility of Sea Water", University Press, Cambridge.
17. Nicol, J. A. C. (1967). "The Biology of Marine Animals", Pitman, London.
18. Park, K. (1966). *J. Oceanological Soc., Korea* **1**, 1–6.
19. Park, K. (1966). *Science, N.Y.* **154**, 1540–1542.
19a. Fearon, W. R. (1946). "An Introduction to Biochemistry", Heinemann, London.
20. Park, K. (1967). *Geol. Soc. Am. Annual Meetings* 1967, 171–172.
21. Pytkowicz, R. M. and Conners, D. N. (1964). *Science, N.Y.* **144**, 840–841.
21a. Dyrssen, D. (1972). *Ambio* **1**, 21–25.
22. Pytkowicz, R. M. (1965). *Limnol. Oceanogr.* **10**, 220–225.
23. Cloud, P. E. (1962). *U.S. Geol. Surv. Profess. Papers*, **350**, vi+138 pp.
24. Pytkowicz, R. M. (1967). *Geochimica et Cosmochimica Acta* **31**, 63–73.
25. Goldring, R. (1969). *New Scient.* **44**, 141–143.
26. Smith, J. B., Tatsumoto, M. and Hood, D. W. (1960). *Limnol. Oceanogr.* **5**, 425.
27. Hood, D. W. and Park, K. (1962). *Physiologia Pl.* **15**, 273–282.
28. Osterlind, S. (1949). *Symb. bot. upsal.* **10**, 1.
29. Park, K. (1964). *Science, N.Y.* **146**, 56–57.
30. Miyake, Y., Sugiara, Y. and Park, K. (1965). *La Mer.* **2**, 136–139.
31. Beyers, R. J., Larimer, J. L., Odum, H. T., Parker, R. B. and Armstrong, N. E. (1963). *Publ. Inst. mar. Sci., Texas* **9**, 454–489.
32. Beyers, R. J., Aiken, S. C. and Gillespie, B. (1964). *Am. Biol. Teacher* **26**, 499–510.
32a. Goldberg, E. D. (1963). *In* "The Sea: Ideas and Observations on the Progress in the Study of the Sea", Vol. I, Wiley-Interscience, New York.
33. Mauchline, J. (1961). H.M.S.O., U.K.A.E.A., P.G. Report 248 (W), 1–18.
34. Chipman, W. A., Rice, T. R. and Price, T. J. (1958). *Fishery Bull, Fish Wildl. Serv. U.S.* **135**, 279–292.

35. Chu, S. P. (1946). *J. mar. biol. Ass. U.K.* **26**, 285–295.
36. Armstrong, F. A. J. and Harvey, H. W. (1950). *J. mar. biol. Ass. U.K.* **29**, 145–163.
37. Cooper, L. H. N. (1948). *J. mar. biol. Ass. U.K.* **27**, 326–336.
38. Fanning, K. A. and Schink, D. R. (1969). *Limnol. Oceanogr.* **14**, 59–68.
39. Wilson, D. P. and Armstrong, F. A. J. (1961). *J. mar. biol. Ass. U.K.* **41**, 663–681.
40. Lucas, C. E. (1938). *J. Cons. perm. int. Explor. Mer* **11**, 343–362.
41. Lucas, C. E. (1955). *Deep Sea Res.* **3** (Suppl.), 139–148.
42. Lucas, C. E. (1961). *Oceanography, Am. Advanc. Sci.* 1961, 499–517.
43. Lucas, C. E. (1961). *Symposia Soc. exp. Biol.* No. XV, 190–206.
44. Droop, M. R. (1958). *J. mar. biol. Ass. U.K.* **37**, 323–329.
45. Droop, M. R. (1961). *J. mar. biol. Ass. U.K.* **41**, 69–76.
46. Curl, H. (1962). *Limnol. Oceanogr.* **7**, 422–424.
47. Park, K., Williams, W. T., Prescott, J. M. and Hood, D. W. (1963). *Publ. Inst. mar. Sci., Texas* **9**, 59–63.
48. Marshall, S. M. and Orr, A. P. (1962). *J. mar. biol. Ass. U.K.* **42**, 511–519.
49. Marshall, S. M. and Orr, A. P. (1964). *J. mar. biol. Ass. U.K.* **44**, 285–292.
50. Marker, A. F. H. (1965). *J. mar. biol. Ass. U.K.* **45**, 755–772.
51. Collier, A. (1953). *Trans 18th N. A. Wildl. Conf.*, 463–472.
52. Butler, P. A. (1962). "Biological Problems in Water Pollution, Third Seminar" U.S. Dept. of Health, Education and Welfare, Public Health Service Publication No. 999-WP-25: 92–104.
53. Nightingale, H. W. (1936). "Red Water Organisms: Their Occurrence and Influence Upon Marine Aquatic Animals, with Special Reference to Shellfish in Waters of the Pacific Coast", Argus Press, Seattle, Washington.
54. Sommer, H., Whedon, W. F., Kofoid, C. A. and Stoher, R. (1937). *Archs path (Lab. Med.).* **24**, 537–559.
55. Davis, C. C. (1948). *Bot. Gaz.* **109**, 358–360.
56. Needler, A. B. (1949). *J. Fish. Res. Bd. Can.* **7**, 490–504.
57. Pinto, J. Dos Santos and E. Silva, Esteca de Sousa. (1956). "The Toxicity of *Cardium edule* L. and its Possible Relation to the Dinoflagellate *Prorocentrum micans* Ehr", *Notos e Estudos do Instituto de Biologica Maritima*, No. 12.
58. Ray, S. M. and Wilson, W. B. (1957). *U.S. Fish Wildl. Serv. Spec. Sci. Rep. Fish.* **211**, 1–50.
59. Woodcock, A. H. (1948). *J. mar. Res.* **7**, 56–62.
60. Wood, P. C., Robinson, G. A., Coulson, J. C., Potts, G. R., Deans, I. R., Fraser, S. M., Adams, J. A., Seaton, D. D., Buchanan, J. B., Longbottom, M. R., Ingham, H. R. and Mason, J. (1968). *Nature, Lond.* **220**, 21–27.
61. Rae, B. B., Johnston, R. and Adams, J. A. (1965). *J. mar. biol. Ass. U.K.* **45**, 29–47.
62. Ketchum, B. H. and Keen, D. J. (1948). *J. mar. Res.* **7**, 17–21.
64. Bruun, A. F. (1957). Proc. U.N.E.S.C.O. Symposium on Physical Oceanography, Tokyo, 1955.
65. Harvey, H. W. (1945). "Recent Advances in the Chemistry and Physics of Sea Water", University Press, Cambridge.
66. Atkins, W. R. G. and Jenkins, P. G. (1952). *J. mar. biol. Ass. U.K.* **31**, 327–333.
67. Prescott, J. A. and Collins, J. A. (1951). *Q. Jl. R. met. Soc.* **77**, 121–126.
68. Southward, A. J. (1960). *J. mar. biol. Ass. U.K.* **39**, 449–458.
69. Cooper, L. H. N. (1958). *J. mar. biol. Ass. U.K.* **37**, 1–3.

70. Southward, A. J. and Crisp, D. J. (1954). *J. Anim. Ecol.* **23**, 163–177.
71. Southward, A. J. (1967). *J. mar. biol. Ass. U.K.* **47**, 81–95.
72. Pedgeley, D. E. (1962). "Elementary Meteorology", H.M.S.O., London.
73. Baum, W. A. (1949). *Ecology* **30**, 104–107.
74. Neiburger, M. (1948). *Trans. Am. geophys. Un.* **29**, 647–652.
75. Neiburger, M. (1954). *Trans. Am. geophys. Un.* **35**, 729–732.
76. Welch, P. S. (1948). "Limnological Methods", The Blakiston Co., Philadelphia.
77. Tyler, J. E. (1968). *Limnol. Oceanogr.* **13**, 1–6.
78. Poole, H. H. and Atkins, W. R. G. (1929). *J. mar. biol. Ass. U.K.* **16**, 297–324.
79. Clarke, G. L. (1941). *J. mar. Res.* **4**, 221–230.
80. Tibby, R. B. and Barnard, J. L. (1962). *In* "Proc. 1st International Conference on Water Pollution Research", Section 3, Paper 45: 1–32. Pergamon, Oxford.
81. Strickland, J. D. H. (1958). *J. Fish. Res. Bd. Can.* **15**, 453–493.
82. Qasim, S. Z., Bhattathiri, P. M. A. and Abidi, S. A. H. (1968). *J. exp. mar. Biol. Ecol.* **2**, 87–103.
83. Cooper, L. H. N. and Milne, A. (1938). *J. mar. biol. Ass. U.K.* **22**, 509–527.
84. Atkins, W. R. G., Poole, H. H. and Warren, F. J. (1949). *J. mar. biol. Ass. U.K.* **28**, 751–755.
85. Gall, M. H. W. (1949). *J. mar. biol. Ass. U.K.* **28**, 757–779.
86. Atkins, W. R. G., Jenkins, P. G. and Warren, F. J. (1954). *J. mar. biol. Ass. U.K.* **33**, 497–509.
87. Riley, G. A. (1956). *Bull. Bingham oceanogr. Coll.* **15**, 15–46.
88. Burt, W. V. (1956). *J. mar. Res.* **15**, 76–80.
89. Armstrong, F. A. J. and Boalch, G. T. (1961). *J. mar. biol. Ass. U.K.* **41**, 591–597.
90. Hardy, A. C. and Bainbridge, R. (1951). *Nature, Lond.* **169**, 354.
91. Knight-Jones, E. W. and Morgan, E. (1966). *Oceanogr. Mar. Biol. A. Rev.* **4**, 267–299.
92. Menzies, R. J. and Wilson, J. B. (1961). *Oikos* **12**, 302–309.
93. Culberson, C., Kester, D. R. and Pytkowicz, R. M. (1967). *Science, N.Y.* **157**, 59–61.
94. Morita, R. Y. (1967). *Oceanogr. Mar. Biol. A. Rev.* **5**, 187–203.
95. Morita, R. Y. (1967). *Seventh International Congress of Biochemistry*, Col. VII-6, 493–494.
X. Perkins, E. J. Previously unpublished work.

2
Physical and Chemical Features of Estuarine Waters

Estuaries represent a meeting place of fresh water, as run-off from the land, and the sea. Consequently, the estuarine environment is more extreme, and undergoes more violent fluctuations than open sea or freshwater habitats.

In general, estuaries are viewed too narrowly, in some cases as a closed "estuarine ecosystem" which is independent of outside forces. In others, they may be taken to refer only to that inner part of an estuary which either lies within the *ebb-flood channel system* (p. 154) if it is present, or occupies an equivalent position if this system is absent. However, the ebb-flood channel system when it is present is an integral part of an estuarine system. Indeed the seaward limit of an estuary may be difficult to define because it represents a gradual transformation from the mutability of the inner estuary to the more stable conditions of the open sea. Furthermore, it will become evident that an estuary may be influenced by physical and chemical processes induced at some relatively distant part of the sea.

The Composition of Estuarine Waters

Salinity

The range of salinity tends to be greater in the surface than in the bottom waters. In the Patuxent River, Chesapeake Bay, the salinity is lower in spring and summer, and greater in autumn and winter[1], but the regime of individual estuaries depends upon the climate of the estuary and the catchment area of its feeder rivers. The salinity tends to fall with increasing distance from the open sea, but this is not always true. This salinity gradient depends upon the relative balance of three factors: (1) run-off from the land, (2) rainfall to and (3) evaporation from the estuary itself[2, 3]. With a high run-off the last two may be unimportant. For example, after a period of prolonged drought, the St. Lucia estuary, Natal, becomes hypersaline in its upper reaches[2]; the Laguna Madre, a bar built estuary on the coast of Texas, lies in a semi-arid region and salinities exceeding 100‰ have been recorded[4, 5]: under such conditions a positive estuary becomes negative.

The estuarine environment with its wide, and often rapid, fluctuations of salt content, clearly presents would-be colonizers with such severe problems in adaptation, that the decrease in species number with increasing distance from the sea is not really surprising. Indeed, there is an interesting comparison which can be made with the effects of pollution: characteristic of both situations is a decline in the number of species, accompanied by an increase in the abundance of individual species: greater pressure due either to greater pollution or a much lowered salinity leads to the decline in abundance of these resistant species (Chapter 14).

Saline waters can be classified by the Venice system[6, 7], thus:

Zone	Salinity ‰
Hyperhaline	>40
Euhaline	40–30
Mixohaline	(40) 30–0·5
(Mixo)euhaline	>30 but < adjacent euhaline sea
(Mixo)polyhaline	30–18
(Mixo)mesohaline	18–5
α-mesohaline	18–10
β-mesohaline	10–5
(Mixo)oligohaline	5–0·5
α-oligohaline	5–3
β-oligohaline	3–0·5
Freshwater	<0·5

Clearly, the distribution of such salinity zones, in an estuary, depends upon variations in the amount of run-off from the land, tidal range and even wind direction, in a manner which is characteristic of the individual estuary (see Fig. 2.1). In an ideal estuary[2], the salinity regime in relation to currents and soil grades (i.e. soil particle sizes) would follow a pattern like that in Fig. 2.2. Because of tidal currents, a given situation will experience a range of salinity during the tidal cycle; it should be noted particularly that the salinity changes are greatest in and near the channel, and least at the higher levels of the banks[2, 8, 9]. However, when the effect of the tides is allowed for, the mixing of fresh and salt waters induces a non-tidal circulation which affects the salinity structure and the life of the estuary (p. 53).

Many biological investigations in estuaries require that the range and mean of salinity for a number of situations be plotted; the construction of isohalines may be of little value, instead a *haligraph* or diagram which combines both map and graphed salinity data is used[10].

In general, estuaries can be characterized by their salinity structure (p. 55), but some, notably Australian estuaries, experience wide fluctuations

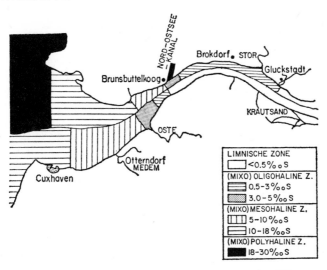

Fig. 2.1. Displacement of the brackish water region in the Elbe estuary (after Caspers[6]).

due to extreme climatic conditions, and are not easily classified except by the Venice system[6].

Oxygen Concentration

Fundamentally, the concentration of oxygen dissolved in sea water is inversely dependent upon salinity and temperature (p. 6), and is normally about 80% of the concentration in fresh water at the same temperature. In estuaries, which are polluted, the oxygen concentration falls and may become anaerobic under extreme conditions[11]. In an unpolluted estuary, the decomposition of organic matter can produce local oxygen deficiencies in the bottom water.

Oxygen concentrations may be affected significantly by the estuarine inhabitants: in *Zostera* beds and salt marsh pools, super-saturation created by photosynthesis during the day, can give way to oxygen deficiency at night when the plants no longer photosynthesize, but continue to respire[12-14]. In Australian estuaries, the concentration of dissolved oxygen is inversely related to the concentration of inorganic phosphate; in the bottom waters, the oxygen concentration varies directly with pH, in a manner considered to be related to the cycle of phosphate liberation and utilization[15]. In the Magothy River estuary, Chesapeake Bay, a direct relationship between pH and oxygen concentration exists for much of the year[3].

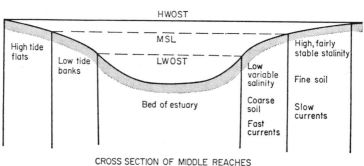

Fig. 2.2. The principal environmental characteristics of an estuary (after Day[2]).

Vertical gradients in oxygen concentration develop only in estuaries which have a vertical salinity stratification[1, 3]. Marked concentration gradients, with depletion of oxygen at the bottom, can occur in fiords with high sills[16].

Regeneration of oxygen, in estuaries, is brought about by mixing with well oxygenated water from rivers or the sea, direct re-aeration from the air, and by the photosynthetic activity of plants. Where the water is turbid, the last process is of minor importance, compared with re-aeration from the air. The exchange coefficient is lowered by salinity and synthetic detergents, but may be doubled during gales[11]. The exchange coefficient for the Thames estuary is about 5–6 cm/hr, compared to 10 cm/hr for an inland stream[17]; however, this is not representative of all estuaries for the Thames carries a heavy load of pollutants, including synthetic detergents. It is generally assumed that shallow waters affected by surf are fully oxygenated, but this is not always true. At Coos Bay, Oregon, investigation showed that water of low oxygen content from the adjacent estuary is not responsible for this deficiency[18].

The pH of Estuarine Waters

The pH of estuarine waters is more variable than that of the open sea (7·5–8·4, mean 8·2)[19]. Under normal, unpolluted conditions pH values ranging from 6·8–9·25 have been recorded[1, 3, 20, 21]. In vertically stratified estuaries, the surface waters generally have a higher pH than the bottom waters[1, 3, 21].

The pH values are generally highest in summer; a minimum value occurs early in spring[3, 20, 21]. Diurnal variations occur and are greatest in salt marsh pools. Here, oxygen deficiencies and a pH $< 7·0$ at dawn, give way to supersaturated oxygen concentrations ($> 200\%$ saturation) and pH values $> 9·0$ at, or slightly after, the time of maximum illumination[13, 14]. The diurnal changes in salt marsh and rock pools are greater than normal seasonal changes of open water. However, both arise from the effects of photosynthesis and respiration upon the amounts of CO_2 present and the effects noted in Australian estuaries (p. 27) are consistent with this. The release of acid and alkaline effluents to the marine environment may be expected to disrupt the buffer system and produce changes in the pH, at least locally.

Calcium

Although changes in pH may cause the precipitation or solution of calcium[14], this element may be removed actively from estuarine waters by biological processes. Where large populations of foraminifera exist, viz.

the Elbe and Christchurch Harbour, Hampshire, significant amounts of calcium are removed from the sea water at the time of the zooplankton maximum; some is added during the winter months. This phenomenon does not occur in estuaries which have small populations of calcareous foraminifera, viz. Orinoco; Gulf of Paria, Venezuela; Tamar, England[21].

Not only do the foraminifera remove calcium from sea water, but the algae *Nitzschia closterium, Amphidinium carteri, Cricosphaera elongata* and *Synechococcus* sp. all have a quantitative requirement for this element which they remove from solution[22]. Among the metazoan animals, the molluscs can obtain calcium directly from the water[23], whereas others such as the shrimp *Crangon* cannot, and food is the primary source of this element[24].

Iron, Manganese and Vanadium

Trace elements such as iron and manganese may occur in much greater concentration in estuaries than the open sea; sea waters contain 1–10 mg Mn/m^3, whereas river waters contain 500–1000 mg Mn/m^3 [25].

Vanadium, which was detected formerly only as a result of the ability of ascidians to concentrate this element, can now be measured and investigated by modern techniques. Such investigation has shown that whereas the Columbia River introduces considerable amounts of vanadium into the sea, it occurs there at concentrations of 3–5 μg/kg, and is due, at least in part, to biological mechanisms. Indeed it has now been found in algae, *Enteromorpha intestinalis*, zooplankton, *Eurytemora* sp. and mysid shrimps, polychaetes, *Nereis* sp., molluscs, *Mytilus* sp., sea cucumbers, *Parastichopus californicus*, as well as ascidians. The degree to which the element is found in living organisms is apparently related to trophic level, viz. from 37·5 to 85·8 μg/g ash in the organisms listed above, whereas in organisms from a high trophic level the element is below the limit of detection, e.g. coelenterate, *Chrysaora melanaster*, decapod crustacean, *Crangon franciscorum* and fish, *Psettichthys melanostictus* and *Microgadus proximus*. The biochemical function of this element declines with phylogenetic scale, and is vestigial in the more advanced ascidia[26].

Nutrients

When considering the nutrient salts of estuaries, it is well to bear in mind the comment "Turbidity, scouring and rapid tidal circulation together with a less stable environment preclude any direct correlation between inorganic nutrient salts and plankton production"[15]. Estuarine waters are generally richer in nutrient salts than are offshore waters[1, 27–29], (Tables 2.1, 2.2), but the concentrations recorded are characteristic of the

Table 2.1. Comparison of nutrient concentrations in Chesapeake Bay (Bay Group) and the Patuxent River Estuary (Groups 1–3) (after Nash[1])

Concentration (10^{-3} mg atoms/litre)	Bay group	Group 1	Group 2	Group 3
Surface silicate	17	49	84	112
Bottom silicate	14	46	73	110
Surface nitrite	0·2	0·2	0·1	0·1
Bottom nitrite	0·3	0·3	0·2	0·1
Surface nitrate	0·8	1·0	1·7	1·9
Bottom nitrate	1·4	1·7	1·9	2·0

Concentrations of phosphate are also higher in the Patuxent than in Chesapeake Bay.

Table 2.2. Comparison of average nutrient concentrations in the Dutch Waddensea with the North Sea, 15 km offshore (after Postma[27])

Nutrient	Concentration (μg—at/litre) North Sea	Waddensea
Phosphate, summer	0·05–traces	0·10–traces
Phosphate, winter	0·60	0·75
Phosphate, average	0·31	0·49
Dissolved organic P, summer	0·60	0·80
Dissolved organic P, winter	0·20	0·30
Dissolved organic P, average	0·38	0·53
Total dissolved P, average	0·69	1·02
Particulate P, average	0·25	0·60
Total P, average	0·94	1·62

individual estuary[1, 15, 16, 19, 20, 25, 27–30]. At present, there is no information on the manner in which nutrient concentrations may limit production in estuaries. Nitrate concentrations appear to vary inversely with plant production[1, 20], but nitrogen fixing algae such as the blue-green *Calothrix scopulorum* liberate to the environment nitrogen in the form of amino-acids which could be of importance to co-inhabitants[31]. Indeed it is of interest to note that of the 15 amino acids known to be liberated by *Calothrix*, 14 (the exception being isoleucine) have been recorded from a total of 20 in the waters of Redfish Bay, Texas. As we have seen, these waters are much enriched with amino acids when compared with the adjacent open sea[32].

In Australian estuaries, the phosphate cycles are a part of a complex relationship between the substratum and the overlying water, and are an effect of local turnover, (Fig. 2.3). Insoluble ferric phosphate is strongly bound to the sediment; under reducing conditions it becomes ferrous phosphate and may leach out to the overlying water. Evidently, an oxygen deficiency may be expected to facilitate the release of phosphate by the substratum. However, in Australian estuaries water movements permit deoxygenation of the overlying water only infrequently; consequently, such a release of ferrous phosphate is probably due to bacterial action in the sediment[15]. In Australian estuaries, too, there is an increase in phosphate coincident with the seasonal discharge of fresh water[15, 33].

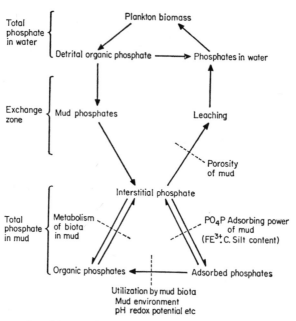

Fig. 2.3. Types of, and interactions between, phosphates occurring in the muds and overlying water masses of estuarine systems (after Rochford[15]).

In European estuaries, phosphate attached to the bottom sediments is released to the overlying water by gales. The largest amounts are found in the winter months[20, 27, 34, 35], when finely divided $FePO_4$ may be formed in the water[34]. Organic phosphorus compounds (dissolved and particulate) are found in largest amounts in late spring and summer[27, 34]. Estuarine muds may release or remove phosphate upon agitation with sea water. Release of phosphate is accompanied by a slight rise in the pH, while a rise in pH will facilitate this release[36-38].

Silicate concentrations may be raised at times of heavy run-off and immediately preceding the spring phytoplankton bloom, while low values coincide with spring and autumn blooms[1, 19]. When river water mixes with sea water, in estuaries, up to 20% of dissolved silicon may be removed from solution by non-biological processes[39].

It has long been appreciated that the land makes a contribution to the fertility of estuaries[40], but unlike many of the rivers of Europe, the Americas, India and Indonesia, the rivers and estuaries of Australia make only an intermittent contribution to, and have no persistent effect upon, the coastal (neritic) waters[14]. In the Menai Straits, accumulation of nitrate-nitrogen from the land occurs in winter, but phosphate concentrations are high in summer and low in winter. The concentration of silicate, nitrate and phosphate are affected significantly by occasional intrusions of high salinity water which is poor in nutrients[41]. In the River Blackwater estuary, Essex, nutrient levels are highest in winter and run-off makes a significant contribution[42]; in this estuary and others, nutrient levels tend to increase towards the head[43-45]. The transport of nutrient materials to lacustrine, estuarine and neritic waters, by rivers, is creating considerable problems in the "developed" countries. Lakes and estuaries are undergoing eutrophication, with dramatic effect[46, 47]. Further problems of unknown dimension are created by the removal and transport of agricultural pesticides by run off (p. 568).

Physical Properties

Temperature

With increasing distance from the open sea, the onset of seasonal changes in the water temperature takes place earlier, and the range of temperature increases. Surface layers tend to be more extreme than the bottom waters, and the greatest relative temperature stratification tends to occur in the middle third of an estuary[1, 3, 44, 48]. Although it is considered that estuarine water temperatures are controlled fundamentally by the temperatures of the sea and run-off water[2], this is true only for those estuaries which are short and have little development of sand and mud flats. Compared with insolation the heating effect of air temperature is unimportant[49]. The increased insolation in summer does not lead to increased estuarine temperatures in all cases: in the Yaquina estuary, Oregon, and other estuaries along the Pacific coast of the United States, an upwelling of coid water of a high salinity is carried several miles upstream with each flood tide. This effect, due to upwelling close inshore, may persist for several weeks[44].

The mean amount of solar radiation received at the earth's atmosphere, viz. *the solar constant*, is 1·94 g cal/cm^2/min. However, the proportion of this radiation which reaches the earth's surface depends on the time, of day and year, the length of air column and the amounts of water vapour and suspended material present. The earth is closer to the sun in January than in July, and at that time receives *ca.* 7% more radiation[50].

Evidently, warming by insolation occurs in two ways, viz. (1) the direct absorption of solar radiation by the water, and (2) the absorption of solar radiation by exposed sand and mud flats, which transfer accumulated heat, to the overlying waters, in the succeeding period of immersion. The amount of incoming radiation absorbed will depend upon the solar altitude and upon the albedo of the receiving surface; cloud amount is important, too.

The relationship between the amounts of radiation received from clear and cloudy skies at the same point may be defined by the formula[51]

$$Q_s = Q_0(1 - 0 \cdot 071 C) \text{ g cal/cm}^2$$

where Q_0 is the amount of insolation from a cloudless sky, and Q_s the insolation received with a cloud amount C (scale 0–10).

Because it can be shown that the amount of heat received by a large area of sand or mud flats is comparable with the output of heat from a sizeable power station, it seems likely that these areas may have a significant effect upon the rate and scale of temperature changes in adjacent waters. In the winter months, a microthermal environment develops on the tidal mud flats of some New South Wales estuaries[15]; while in Pipe Clay Lagoon, Tasmania, the tidal waters may be heated for some 50 ft from the advancing water's edge[52]. At Whitstable, Kent, pronounced gradients develop in the 400 ft seawards of the advancing tide's edge[53]; unlike the open sea, the waters of Whitstable Bay tend to be warmer than the air in summer, and cooler than the air in winter[54]. In New Zealand, the surface temperatures of Otago and Lyttleton harbours are higher than the offshore waters in summer. The reverse condition occurs in winter, and the lower water temperatures in these basins may be due to the combined effects of increased cold, fresh run-off, low air and ground temperatures, and general atmospheric cooling[55].

A direct loss of heat, by radiation, from sand and mud flats must have helped to lower the seawater temperatures during the severe winter of 1962–63. The River Blackwater, Essex, became choked with ice[56], and the Solway Firth, to the east of Annan, was covered by an ice-sheet to a depth of 6 in.; composite blocks to a height of 6 ft were stranded upon the shore[57].

In an interesting study of water temperatures in the Waddensea, it has been shown that temperatures taken at a greater interval than three days have no value in the compilation of monthly mean temperatures[58].

The Penetration of Light

The penetration of light into estuarine waters depends largely on the turbidity which is much greater than that of the open sea. This turbidity is due to sedimentary materials from three sources, viz. the rivers flowing into the estuary, transport into the estuary from the open sea, and a reworking within the estuary. In general, the turbidity decreases, and therefore the depth of light penetration increases as the open sea is approached. Estuarine determinations can be made by photoelectric cell methods[59], but it is generally more convenient to use the Secchi disc, whose readings can be related to the amount of sediment in suspension[2, 27, 49] (Table 2.3). If the Secchi disc can be seen at 1 m depth then an accurate direct determination of the quantity of silt cannot be made by direct methods[2]. Since the Secchi disc reading is related to turbidity which depends on the velocity of the tidal currents, it follows broadly, that with increasing distance from the sea, such readings will depend on time during the tidal cycle, height of tide and time of year.

Table 2.3. Variations in silt content and Secchi Disc reading of estuarine water during the period from low to high water (after Francis-Boeuf[49])

State of tide	Silt content (g/litre)		Secchi disc Reading (m)
	Surface	Bottom	
L.W.	0·27	0·39	0·35
H.W.−5 hr	0·26	0·36	0·35
H.W.−4 hr	0·22	0·32	0·45
H.W.−3 hr	0·13	0·24	0·60
H.W.−2 hr	0·11	0·19	1·00
H.W.−1 hr	0·18	0·26	1·00
H.W.	0·19	0·21	1·40

The suspended material present in most estuarine waters ensures that nearly all the light which penetrates is absorbed in the surface 1–2 m[59, 60]. Because of the effect of scattering by suspended sediment and absorption by the sea water, the maximum transmission of light occurred at 600 mμ, in the River Tamar estuary[59]. The absorption of light in sea and estuarine waters has been investigated by use of the attenuation coefficient μ. The attenuation coefficient is related to the percentage light transmission per half metre t[61], by

$$t = 100 \exp(-0.5\,\mu) \text{ per m.}$$

This coefficient of attenuation is *not* the same as the vertical extinction coefficient, λ. In some, but not all cases, there is a linear relationship between attenuation coefficient and the concentration of simple suspensions[60–62] (Fig. 2.4). Secchi disc visibility is inversely proportional to attenuation coefficient[61].

The effect of turbidity and rapid absorption of light upon the photosynthetic processes which can take place in an estuary, is considerable.

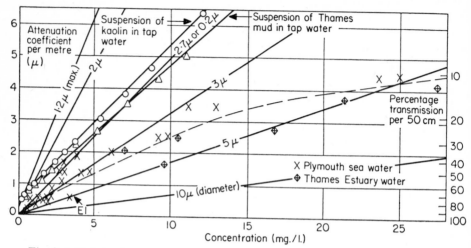

Fig. 2.4. Relation between attentuation coefficient per metre and concentration of suspended solid matter. The solid lines have been added to show theoretical attentuation due to scattering by small particles with diameters of 0·2, 1·2, 2, 2·7, 3, 5 and 10 μm. El refers to a hydrographic station south of the Eddystone (after Jones and Wills[61], Burt [62]).

Clearly, the phytoplankton will receive sufficient light for this vital process in the surface layers only, and will generally contribute little to primary production in turbid estuaries. Under these conditions, the shore-dwelling plants, particularly the microflora, take on a real importance as the primary source of food[27, 63].

References

1. Nash, C. B. (1947). *J. mar. Res.* **6**, 147–174.
2. Day, J. H. (1950). *Trans. R. Soc. S. Afr.* **33**, 53–91.
3. Pritchard, D. W. and Bunce, R. E. (1959). *Chesapeake Bay Institute Tech. Rept.* No. XVII, Ref. No. 59–2, pp. 1–87.
4. Hedgpeth, J. W. (1947). *Trans. 12th N. Am. Wildl. Conf.*, 364–380.
5. Hedgpeth, J. W. (1967). In "Estuaries" (G. H. Lauff, ed.), *Publs. Am Ass. Advmt. Sci.*, **83**, 408–419.
6. Caspers, H. (1959). *Estratta Dall'Archivo Di Oceanografia e Limnologia* **9**, Suppl. 153–169.
7. Segerstråle, S. G. (1964). *Oceanogr. Mar. Biol. A. Rev.* **2**, 373–392.
8. Milne, A. (1938). *J. mar. biol. Ass. U.K.* **22**, 529–542.
9. Newell, R. (1964). *Proc. zool. Soc., Lond.* **142**, 85–106.
10. Stauber, L. A. (1943). *J. mar. Res.* **5**, 165–167.
11. Klein, L. (1966). "River Pollution. **III.** Control", Butterworths, London.
12. Broekhuysen, G. J. (1935). *Archs. néerl. Zool.* **1**, 339–346.
13. Nicol, E. A. T. (1935). *J. mar. biol. Ass. U.K.* **20**, 203–261.
14. Bradshaw, J. S. (1968). *Limnol. Oceanogr.* **13**, 26–38.

15. Rochford, D. J. (1951). *Aust. J. mar. Freshwat. Res.* **2**, 1–116.
16. Braarud, T. and Ruud, J. T. (1937). *Hvalråd. Skr.* No. 15, 1–56.
17. Southgate, B. A. (1958). *J. Instn publ. Hlth Engrs.* **57**, 177–191.
18. Burt, W. V., McAlister, W. B. and Queen, J. (1959). *Ecology* **40**, 305–306.
19. Harvey, H. W. (1960). "The Chemistry and Fertility of Sea Water", University Press, Cambridge.
20. Perkins, E. J., Bailey, M. and Williams, B. R. H. (1964). H.M.S.O. U.K.A.E.A., P.G. Report 604 (CC), 58 pp.
21. Murray, J. W. (1966). *J. mar. biol. Ass. U.K.* **46**, 561–578.
22. Taylor, W. R. (1964). "Proceedings of Symposium on Experimental Marine Ecology" *Occasional Publication No.* 2, 1964, 17–24, Graduate School of Oceanography, University of Rhode Island.
23. Wilbur, K. M. (1964). *In* "Physiology of Mollusca" (K. M. Wilbur and C. M. Yonge, eds), Vol. 1, Academic Press, New York, London.
24. Plankemann, H. (1935–36). *Schr. naturw. Ver. Schlesw.-Holstein.* **21**, 195–216.
25. Harvey, H. W. (1949). *J. mar. biol. Ass. U.K.* **28**, 155–163.
26. Dyer, R., Forster, W. O. and Renfro, W. C. (1969). "Ecological Studies of Radioactivity in the Columbia River Estuary and Adjacent Pacific Ocean", Progress Report, Dept. of Oceanography, Oregon State Univ., Corvallis, Ref. No. 69–9: 82–84.
27. Postma, H. (1954). *Archs. néerl. Zool.* **10**, 1–106.
28. MacGinitie, G. E. (1935). *Am. Midl. Nat.* **16**, 147–174.
29. Hulbert, E. M. (1956). *J. mar. Res.* **15**, 181–192.
30. Howes, N. H. (1939). *J. Linn. Soc. (Zool.)* **40**, 383–445.
31. Jones, K. and Stewart, W. D. P. (1969). *J. mar. biol. Ass. U.K.* **49**, 475–488.
32. Park, K., Williams, W. T., Prescott, J. M. and Hood, D. W. (1963). *Publ. Inst. mar. Sci., Texas* **9**, 59–63.
33. Spencer, R. (1956). *Aust. J. mar. Freshwat. Res.* **7**, 193–253.
34. Oliver, J. H. (reported by L. H. N. Cooper) (1961). "Symposium on Marine Microbiology", *Bacteriological Proceedings D20*.
35. Bassindale, R. (1943). *J. Ecol.* **31**, 1–30.
36. Stephenson, W. (1949). *J. mar. biol. Ass. U.K.* **28**, 371–380.
37. Pomeroy, L. R., Smith, E. E. and Grant, C. M. (1965). *Limnol. Oceanogr.* **10**, 167–172.
38. Jitts, H. R. (1959). *Aust. J. mar. Freshwat. Res.* **10**, 7–21.
39. Burton, J. D. and Liss, P. S. (1969). *Proc. Challenger Soc.* **4** (1), 55.
40. Nelson, T. C. (1947). *Ecol. Monogr.* **17**, 337–346.
41. Ewins, P. A. and Spencer, C. P. (1967). *J. mar. biol. Ass. U.K.* **47**, 533–542.
42. Spencer, J. F. (1967). *In* "Hydrobiological Studies in the River Blackwater in Relation to Bradwell Nuclear Power Station", Central Electricity Generating Board, London.
43. Emery, K. O. and Stevenson, R. E. (1957). *Bull. geol. Soc. Am.* **67**, 673.
44. Frolander, H. F. (1964). *J. Wat. Pollut. Control Fed.* **36**, 1037–1048.
45. Mommaerts, J. P. (1969). *J. mar. biol. Ass. U.K.* **49**, 749–765.
46. Barlow, J. P., Lorenzen, C. J. and Myren, R. T. (1963). *Limnol. Oceanogr.* **8**, 251–262.
47. Stewart, K. M. and Rohlich, G. A. (1967). "Eutrophication—A Review", State of California Water Quality Control Board, Publication No. 34.
48. Perkins, E. J. (1968). *Trans. J. Proc. Dumfries. Galloway nat. Hist. Antiq. Soc.* Ser. 3, **45**, 15–43.

49. Francis-Boeuf, C. (1943). *C. r. somm. Séanc. Soc. Biogéogr.* Nos. 169–170, 19–26.
50. Allee, W. C., Emerson, A. E., Park, O., Park, T. and Schmidt, K. P. (1949). "Principles of Animal Ecology", Saunders, Philadelphia and London.
51. Neumann, J. (1953). *Bull. Res. Coun. Israel* 2, 337–357.
52. Guiler, E. R. (1951). *Pap. Proc. R. Soc. Tasm.* 1950, 29–52.
53. Perkins, E. J. (1964). H.M.S.O., U.K.A.E.A., P.G. Report 456 (CC): 18 pp.
54. Perkins, E. J. (1958). "The Microfauna of the Shore", Ph.D. thesis, University of London.
55. Skerman, T. M. (1958). *N.Z. Jl Geol. Geophys.* 1, 197–218.
56. Waugh, G. D. (1967). *In* "Hydrobiological Studies in the River Blackwater in Relation to Bradwell Nuclear Power Station", Central Electricity Generating Board, London.
57. Williams, B. R. H., Perkins, E. J. and Bailey, M. (1963). *Trans. J. Proc. Dumfries. Galloway Nat. hist. Antiq. Soc.* Ser. 3, 41, 30–44.
58. Meyer-Waarden, P. F. and Tiews, K. (1959). *I.C.E.S., Shellfish Committee,* C.M. 1959, No. 74: 14 pp.
59. Cooper, L. H. N. and Milne, A. (1938). *J. mar. biol. Ass. U.K.* 22, 509–527.
60. Qasim, S. Z., Bhattathiri, P. M. A. and Abidi S. A. H. (1968). *J. exp. mar. Biol. Ecol.* 2, 87–103.
61. Jones, D. S. and Wills, M. S. (1956). *J. mar. biol. Ass. U.K.* 35, 431–444.
62. Burt, W. V. (1957). *J. mar. biol. Ass. U.K.* 36, 223–226.
63. Williams, B. R. H., Perkins, E. J. and Hinde, A. (1965). H.M.S.O., U.K.A.E.A., P.G. Report 611 (CC): 84 pp.

3

Water Movements in Estuaries and Coastal Waters

Water movements in estuaries and coastal waters arise from three causes, viz. wave action, tides and non-tidal circulations due to interactions between fresh water introduced by rivers and fully saline water from the sea. Each of these phenomena will be considered in turn.

Waves

The wind is a potent factor in the economy of the oceans. Waves are relatively unimportant here, but *wind-drift currents* can transport water in the same direction over large areas if the prevailing wind blows from one direction, e.g. the *Trade Winds* generate the *Equatorial Currents*. The persistent movement of surface waters, by the wind, induces nutrient rich upwellings off the western continental coasts of Africa and America. These movements are better understood when it is appreciated that due to *Corioli's Force*, all bodies undertaking sustained movement appear to move to the right and left in the Northern and Southern Hemispheres respectively. Corioli's Force is not a force as such, but is a displacement which results from the rotation of the earth beneath a moving body. In shallow water, the relative importance of waves and wind drift currents is reversed.

The simplest water waves are oscillatory, e.g. an ocean swell passing through calm water. The wave-form, of regular undulations, moves steadily in the direction of propagation and perpendicular to the line of the crests. All are linear and parallel, and each is identical with its neighbour. In a wave train, each wave has four principal characteristics, viz. height, length, period and velocity (Fig. 3.1).

Wave height (H) is the distance of the crests above the troughs; if the wave height is variable, then it is assumed to be the height of the crest above the trough immediately preceding it.

Wave length (L) is the distance between crests, troughs, or other demonstrable features. The ratio $H:L$ is defined as the *wave steepness*.

Period (T) is the time, in seconds, between the passage of adjacent crests past a stationary observer.

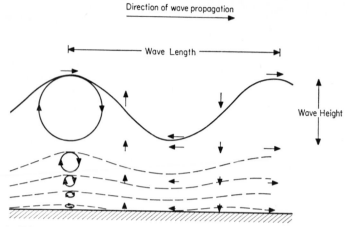

Fig. 3.1. Diagrammatic section through an oscillatory wave showing the rapid decrease in particle orbit with increasing depth and the tendency to elliptical particle orbits near the sea bed. The circles and arrows show the direction of movement of the water in different parts of the wave (after Johnson[2] and Russell and Macmillan[3]).

Velocity (*V*) is the speed, in ft/sec or knots, with which waves move past a stationary observer.

Only a swell moving over water, which is undisturbed by local wind, can be defined simply by these four parameters. Normally, wherever a wind is blowing over the sea, the wave crests are not straight and parallel, neither are successive waves precisely similar. Nevertheless, important generalizations can and have been made regarding the pattern of movement and transport which arises from wave action. The form of all oscillatory waves moves far over the sea's surface, but the individual water particles move a comparatively short distance (Fig. 3.1). The contrast between wave and water movement is demonstrated clearly by the advance of waves up an estuary during an ebbing tide.

In deep water, the particles in a typical oscillatory wave move in circular orbits: seaweed or refuse floating on the water performs a circular motion, but seems not to move forward. This is an important characteristic of wave motion: if it were not true, the velocity of a wave is such that the oceans would not be navigable. Normally, the limited orbital movement of the water particles is stressed, but the statement that in the open sea, the water particles have no progressive motion is not true. In the 1840s, Stokes[1] stated that in addition to the motion of oscillation, water particles undergo mass transport in the direction of wave propagation.

In shallow waters, the movement of water particles follows an ellipse with a horizontal axis. Reduction of the ellipse to a straight line occurs at the seabed where the water particles move backwards and forwards only. In deeper water, the surface particles move through circular orbits, the mid-water particles in ellipses, and those near the seabed in straight lines[2, 3]. The motion of normal oscillatory waves is characteristically progressive; consequently, when orbits are reduced to straight lines and opposing forces are absent, a slow, forward movement of sediment will occur in the direction of wave propagation.

The size of the water particle orbits decreases rapidly with depth (Fig. 3.1); reduction to an orbit 1/534·5 of that at the surface occurs at a depth of one wave length beneath the surface, or for every successive 1/9 of the wave length, beneath the still water level, the orbit diameters are reduced by one half[2].

In practical terms, if the sea is disturbed by waves of height 10 ft, and length 500 ft, then the water particles at the surface and 500 ft depth will move through orbits of 10 ft and ~0·22 in. diameter respectively. This principle is of fundamental importance to a consideration of the depths at which waves may erode the sea bottom and facilitate the transport of sedimentary materials (Chapter 6).

When an oscillatory wave passes into shallow water, i.e. depth $< \frac{1}{2}L$, it slows down and changes form, viz. it heightens and shortens, the front steepens until the height of the wave is so great compared with the depth of the water that it becomes unstable and breaks. This occurs when the ratio of depth to wave height is about $4 : 3$[3]; such a wave is known either as a *breaker* or much less frequently as a *combing wave*[2]; this latter term is applied also to a deep water wave whose crest is pushed forward by the wind. It is important to realize that the breaker does *not* result from the frictional drag of seabed upon the wave, but is a function of decreasing water depth making insufficient water available for the completion of the orbits of the water particles.

The velocity of an oscillatory wave may be calculated approximately by a variety of simple formulae[2, 3]

$$v = \sqrt{2\cdot25 \times L} \qquad \text{m.p.h.} \qquad (a)$$

where L is in feet.

$$v = 2\cdot26 \sqrt{L} \qquad \text{ft/sec} \qquad (b)$$

where L is in feet.

$$v = \sqrt{5\tfrac{1}{8} \times L} \qquad \text{ft/sec} \qquad (c)$$

where L is in feet.

$$v = 3 \times T \qquad \text{knots} \qquad (d)$$

where T is measured in seconds.

$$v = L \times T \qquad \text{ft/min} \qquad (e)$$

from (b) it follows that because the period is more easily measured than length or velocity, the advantage of these relationships is evident; however, the shape of sea waves is different to those for which these relationships strictly apply[3]. Provided that the depth of water is greater than about half a wave length, the wave velocity is independent of precise depth, and depends only on wave length. A wave of height and length 4 ft and 600 ft respectively has a steepness ($H:L$) of 1/150 and is a *low wave*, whereas a wave of height and length 4 ft and 60 ft respectively has a steepness of 1/15 and is a *steep wave* and is of the same order of steepness as the maximum possible in deep water. It can be shown by $v=2\cdot26\sqrt{L}$ ft/sec that the velocity of the long, low wave is 55 ft/sec and the shorter, steeper wave is 17·5 ft/sec. However, the latter is a steep wave and the calculated velocity is about 3% too small[3].

Initially, wave height depends upon three parameters: (1) strength of the wind; (2) its duration; and (3) the extent of the open water over which it blows, i.e. *the fetch*. Casual estimates of wave height vary considerably; accurate measurements require that the observer shall be on a fixed platform, and the movement of the waves recorded by use of fixed graduated posts[3]; modern techniques use transmitting buoys.

The average wave height, measured in feet, can be related to the wind velocity, in statute miles per hour[2], thus:

$$H=\frac{\text{Wind velocity}}{2\cdot05}\text{ ft.}$$

Winds below a critical velocity (about 1·1 m/sec or 2·1 knots) do not build up waves on a still water surface. At the critical value, which is difficult to measure precisely because winds blow in gusts, small waves only are generated: they have a length of 8 cm and velocity of one-third of that of the wind. Once the critical value is markedly exceeded, the waves grow in height, until a particular wave is predominant; eventually such a wave will move as fast as the wind, and its length, if the wind blows long enough, can be calculated from the formulae on p. 41.

The relationship between wave height, wind velocity and duration is given in Fig. 3.2[4]. In the most severe storms, the highest waves may not coincide with the maximum wind velocity, but occur when the wind begins to die away. This may be due either to the more violent wind blowing the tops of the waves off into the troughs, thereby diminishing wave height, or, with the moderation of the storm, waves which were kept irregular and independent by the gusty winds at the height of the storm, combine into larger ones upon which the dying wind has small effect[3].

In the open sea, waves grow to a particular size under the influence of a particular wind strength. If a breeze springs up over a circumscribed

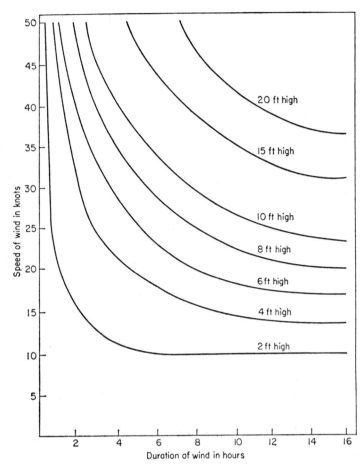

Fig. 3.2. The variation in wave height with wind strength and duration. The distance from the windward shore is assumed to be sufficiently great as to have no influence (after Sverdrup and Munk[4]).

body of water, e.g. a pond, ripples appear over the whole water surface, but increase in size towards the leeward side of the pond. On the windward side the ripples remain small whatever the duration of the wind. On the leeward side, however, where the fetch of the wind is greatest, waves of some size develop; here the waves are generated by the direct effect of the wind combined with waves which originate on the opposite side of the pond. This development of waves in a pond illustrates a principle which is of considerable importance in, and to, estuaries viz. waves are limited by fetch in places close to the windward shore and by wind duration at great distances from it. For example, in a westerly gale, the waves arriving on the

West Coast of Scotland are limited by duration only; the American coast is so distant that the fetch is uninfluential. Conversely, the same wind blowing in the Irish Sea will produce waves influenced by the fetch only: here waves of a maximum height, related to the particular fetch, will grow no higher whatever the duration. For ordinary gales and distances there is an empirical relationship between wave height and fetch, thus:

$$H = 1.5 \sqrt{F} \text{ ft}$$

where H is the greatest height of wave generated by winds blowing over F, the fetch in sea miles. This simple relationship which ignores wind strengths, has been found accurate for long and short fetches[2].

The total energy of a wave varies approximately as the square of the height and the first power of the length, hence the ratio of height to length is of great interest; storm waves are usually higher and steeper than those in a moderate sea, the ratio $H:L$ ranges from $1:39$ to $1:13$ for a light sea and storm conditions respectively. The ratio $H:L$ also becomes smaller with increasing wave length[2]. At sea, the period of a wave train can be measured by watching the rise and fall of patches of foam, and on the beach by counting the arrival of the number of breakers in a given time. It should be appreciated that while wave height, length and velocity change when waves enter shallow waters the wave period remains constant. The period of locally generated waves breaking on English Channel coasts is ~6 sec; waves of longer period require more room for development. If gale force winds blow consistently for a day in the same direction, waves with a period of *ca.* 20 sec develop; this seems to be the maximum for wind generated waves[2].

An oscillatory wave which enters very shallow water, i.e. water with a depth of less than 1/50 of the wave length, undergoes a metamorphosis from a shallow water oscillatory wave. Under such conditions, a very low ground swell, just before breaking, passes onto a flat, wide beach of low gradient and is characterized by a series of individual crests separated by long, flat troughs: the distance between adjacent crests is such that the velocity of the individual crest is unaffected by its neighbours. Such waves are *solitary waves* or *waves of translation*; their velocity is primarily a function of depth, and not of wave length[2, 3].

The velocity (v) of solitary waves, of small height, is given by,

$$v = \sqrt{gd} \text{ ft/sec}$$

where d is the depth of water, in feet, measured at the wave crest; when the wave height exceeds 1/1.28 of the water depth, the velocity of the crest exceeds that of the wave, and the expression becomes invalid[3].

The name wave of translation arose because such waves are accompanied by a forward transport of water. In each wave, the particles move as in Fig. 3.3, and lack the recoil of oscillatory waves. Movement of the water particle occurs on the passage of the wave only. At this time, the surface particles lift forwards, but with increasing depth the vertical rise decreases,

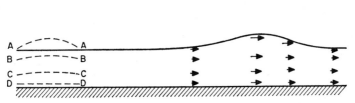

Fig. 3.3. Section through a solitary wave showing the paths followed by particles at several depths. All particles are displaced forwards to a new position at the same elevation as they had initially, and there they stay. The velocity and direction of particles are shown by the arrows (after Russell and Macmillan[3]).

although the horizontal movement of all the particles is the same[2]: the horizontal distance moved (S) is given approximately by the expression

$$S = 4 \sqrt{\frac{dH}{3}}$$

where d is the depth to still water level (trough) and H the height of the crest above the trough[3]. The resemblance to a solitary wave is good[5], when the distance between crests exceeds

$$2\pi d \sqrt{\frac{d}{3H}}.$$

Upon entry into shallow water, only the steepest waves do not measure up to this requirement, e.g. a wave 5 ft high and 100 ft long is a solitary wave in less than 16 ft of water according to this criterion.

A sudden wind acting upon a surface layer of low density which overlies water of a greater density may induce *boundary waves*; however, these waves have little effect upon shore-line processes[2].

In the open sea, the crest of an oscillatory wave is always normal to the direction of propagation. Near the coast, however, contact with shoaling ground off the headlands occurs before that off the bays. The velocity of the wave off the headlands decreases relative to that part of the wave off the bays, consequently the wave begins to follow the shape of the coast. Where the coastline is not too exaggerated, by the time the shoreline is reached, the wave crest will have adjusted itself to become parallel with all major features of the shoreline[2, 3], i.e. the phenomenon of *wave refraction*.

By definition, this is change in direction due to a change in wave length, and the law relating to this change is the same as that for light refraction[3], viz.

$$\frac{\sin r}{\sin i} = \frac{L_2}{L_1}$$

where i = the angle of incidence, r = the angle of refraction, L_1 and L_2 are the incident and refracted wave lengths respectively. Because, in general, coastal contours are curved, the morphology of the seabed ultimately influences the direction of the waves which can only be deduced, stepwise, by means of a *wave refraction diagram*[3]. Evidently, concentration of wave energy occurs at the headlands, and the reverse occurs in the bays, i.e. convergence and divergence occur respectively. Consequently, waves gain height at a convergence, but lose it at a divergence[3].

An oscillatory wave, moving inshore, breaks when the water becomes too shallow, but may reform, break and reform again before its dissolution upon the beach: a sand *bar* and, just inshore, a trough are formed at each of the breakers. If the approach of a wave train is normal to the shoreline, then mass transport by the waves, particularly those which are higher than average, accumulates water within the bars[1, 6]. The vigorous seaward

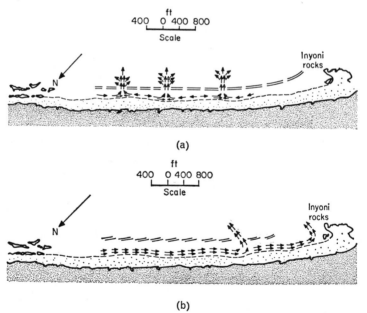

Fig. 3.4. (a) Schematic diagram of typical circulation pattern for normal breaker approach on test beach (after Harris *et al.*[8]). (b) Schematic diagram of typical circulation pattern for oblique breaker approach on test beach (after Harris *et al.*[8]).

return of this water creates a *rip-current*, which breaches a *rip-channel* where the bar narrows[3, 7] (Fig. 3.4a)[8]. Characteristically, a rip-current is an intermittent, narrow body of water which diverges beyond the breakers: a velocity of up to 4 knots, and lengths between 200 ft and 2500 ft have been recorded. Rip-channels may be formed at a pier or breakwater, but generally occur at random. Rip-currents are easily seen from a cliff, aeroplane or other vantage point. Rip-currents appear in folk lore as *under-tow*, but experiments using dyes, e.g. Rhodamine B, have shown that this phenomenon does not exist: danger to the swimmer arises from the high velocity of the rip-current. Fortunately, they are so narrow that by swimming a short distance to one side, the swimmer can be freed of its influence[3, 7, 8]. On exposed coasts, rip-current systems may influence waste disposal[8], but in estuaries they are important only in those with wide mouths.

A different situation results when a wave-train impinges obliquely upon the shore line. Water which flowed obliquely up the shore, can only under the influence of gravity, flow directly downwards. Consequently, i, has undergone a net movement along the shore. Evidently, the persistenct of such a wave-train will give rise to a current, the *longshore* or *littora current*: where the prevailing waves arrive thus, then an unidirectional, longshore current flows constantly[3–8]. In these conditions, rip-currents occur infrequently, and are associated with some topographic feature, e.g. a headland, where they may reach a width of 70–100 ft (Fig. 3.4b)[8]. A miniature of this system occurs in Brodick Bay, Firth of Clyde; here primary and secondary bars and a north to south littoral current develop. Bar formations occur in the Moray Firth also. On open coasts, the transport of sediment by longshore currents is very important; in estuaries, the effect of longshore currents and drift is diminished, but real nevertheless.

Tides

The gravitational pull of all the celestial bodies, particularly the sun and moon, induces periodic movements of the waters of the earth. These periodic movements are the *tides*. The worlds oceans are separated incompletely into a series of natural basins, in which the bodies of water have a natural period of oscillation of approximately twelve hours, which corresponds to the diurnal rhythm of gravitational forces exerted by the sun and moon. These forces are considered to maintain this oscillation in the sea. Formerly, the tides of the world were thought to originate in a tidal wave which passed around the Southern Ocean. However, this concept is inadequate, for the tides observed are much greater than could thus be expected. Now each basin is considered to have standing oscillations which resonate with the attractive forces. This principle can be illustrated by a consideration in which water oscillates in a bowl, with sea-saw movement. There is a large rise and fall at the two ends and the middle, but none along the nodal line. In the sea, the nodal line is reduced to a nodal point, due to the earth's rotation. In the Northern Hemisphere the tidal oscillation rotates in an anticlockwise direction round these points, whereas the rotation is clockwise in the Southern Hemisphere. Points which have a high water simultaneously may be joined by *cotidal lines*, which radiate from the nodal point. This nodal point is called an *amphidrome* or *amphidromic point*. In the North Sea, there are three *amphidromic points* viz. east of Lowestoft, west of Esbjerg and near Stavanger, all related to a triple oscillation[9].

The global tidal pattern is complex and the oscillations observed are by no means uniform. In the open sea, the amplitude of the tidal oscillation or the *tidal range* rarely exceeds 3 ft. However, tides of extreme range may be found on the eastern coast of Asia; the Gulf of Cambay, India; the western coast of Europe; and in the Bay of Fundy, North America. Conversely, tides in the South Pacific may not exceed 20 in., and in the Gulf of Mexico tides of 1–2 ft recorded generally are reduced to 0·5 ft in the Mississippi estuary. However, such marked contrasts in height may occur in relatively close proximity: in the Gulf of Maine, Minas Basin at the head of the Bay of Fundy has a spring tide of 50 ft in height, whereas at Cape Cod on the south side of the Gulf the range is reduced to less than 3 ft. In the English Channel, the range of the spring tides at Portland is 7 ft, whereas at St. Malo on the French coast the range is 38 ft.

All tidal oscillations have semi-diurnal and diurnal components. Around the coast of Britain, the former are dominant, there are therefore two high and two low waters per day; the former are almost, but not quite, equal in height, indeed it is usual for tides which occur at approximately the same time on succeeding days to be more nearly similar in range than succeeding tides on the same day. On the Pacific coast of North America, the tides like those of N.W. Europe, are semi-diurnal, but are more markedly unequal in height. In the Gulf of Mexico the diurnal component is dominant and here the tidal cycle is restricted to one high and one low water per lunar day[9a, 9b, 36]. As the tidal waves enter the shallow waters near the coast their height is increased. Around the British Isles, the average tidal height is about 15 ft and 11½ ft at *springs* and *neaps* respectively. In the Bristol Channel, the range is 42 ft and 21 ft at springs and neaps respectively. Furthermore, it should be noted that the very high tides of the Bay of Fundy and the Gulf of Cambay result from this cause,

A phenomenon of a double high water occurs near Southampton.

whereas a double low water occurs to the west of Portland. This is not due to a difference in the time of arrival of the tidal wave around the Isle of Wight, but results from the interaction of harmonics of the main tidal period, i.e. shallow water tides. Double tides can only form where the amplitudes of the harmonics are appreciable compared with that of the main tidal period. Generally this will occur only near an *amphidromic point*[9].

Meteorological phenomena may have an effect upon tidal height. A high barometric pressure and an offshore wind may result in a lower tide than predicted, while there are few marine biologists who have not at some time or other been unable to reach a desired location on the shore because an onshore wind has prevented the tide from retiring as expected. The disastrous floods experienced along the east coast of England and in the Low Countries during the night of 31 January 1953, resulted from a surge coincident with a violent storm in the northern North Sea. The intensity of the wind action, and thus of waves, upon a sea-level raised several feet above normal resulted in disastrous floods. In addition to the widespread inundation by salt water, marked erosion occurred[10].

Thus far the terms spring and neap tide have been used without definition. It will be recalled that not only has the earth a daily rotation, but it moves in orbit around the sun also; simultaneously the moon moves in orbit around the earth. At new and full moon, the sun, earth and moon lie approximately in a straight line, i.e. full opposition (Fig. 3.5a); consequently the combined gravitational pull of the sun and moon is greatest. At these times the vertical range of the tides is at a maximum, i.e. the *spring tides*. It is essential to appreciate that, in this case, spring has nothing to do with the season of the year. For seven days after the new or full moon, the sun and moon move progressively out of a straight line, until the sun and moon are situated at right angles relative to the earth, i.e. in quadrature (Fig. 3.5b), when their combined gravitational pull is at a minimum; the tidal range is at a minimum, i.e. the *neap tides*. After the neap tide, the range increases once more to the succeeding spring tide and so on; the cycle being repeated twice in the 28-day period of the lunar cycle. It will be appreciated that in consequence the tidal range changes daily, and that even in the same day successive tides are not of the same height.

Because the earth moves in orbit around the sun, an annual rhythm is superimposed on the 28-day lunar cycle. As the equinoxes (about 21 March and 21 September) are approached, the spring tides become progressively larger, for it is only at these times that the earth, sun and moon achieve most nearly a linear alignment. At the time of the summer and winter solstice (about 21 June and 21 December), the sun, earth, and moon are most out of linear alignment, and the springs, although greater than the

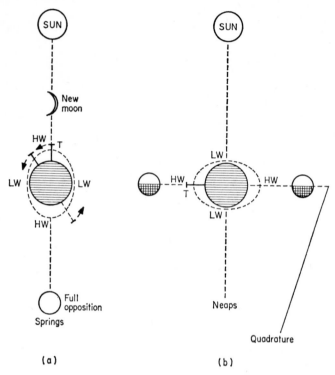

(a)   (b)

Fig. 3.5. Astronomical forces which produce spring and neap tides (after Russell and Macmillan[3]).

neaps, are much smaller tides than those which occur about the equinoxes. In Essex, where the lowest tides of the year occur at the end of May and the beginning of June, the saltings, i.e. salt marshes, are not inundated by the tides. It is at these times that the birds nest and hatch their young. For this reason these tides are known as "*bird tides*"[11].

The oscillatory movement of the tides is a vertical movement basically, and in the open sea comparatively little horizontal movement results. However, such movements do occur and are known as *tidal currents*. The period which occurs from the time of high water to the time of low water is known as the ebb tide, while the reverse is known as the flood tide; the corresponding horizontal water movements are known as ebb and flood currents respectively.

In the open sea, the orbital path of the oscillatory tidal wave is undisturbed, hence flood and ebb movements persist for some time after high and low water[2]. At the same time, it has been shown by measurements at light-ships, for example, that the tidal currents in the open sea normally rotate. The velocities of these currents are low and the velocity of the maximum flood current is approximately equal to the velocity of the maximum ebb current[3, 12].

In coastal and estuarine waters, the orbit of the tidal wave becomes a much flattened ellipse. There is a shoreward movement of water particles until the time of high water, after which a movement seaward takes place. The modification of the tidal wave as it passes into the shallow water of estuaries and bays is such that the period of the flood tends to become shorter and that of the ebb longer[2]. Where a river flows directly into a tidal sea, the tides are modified compared with oceanic tides in three ways: (1) the speed of the tide as it travels upstream depends on the depth of the channel, thus

$$v \approx \sqrt{gd}$$

where v is the speed, g the acceleration due to gravity and d the depth; (2) with increasing distance upstream the duration of the ebb is increased while that of the flood is decreased; and (3) the range of the tide tends to decrease with increasing distance from the open sea[13]. Tidal currents can be classified according to estuarine geomorphology. If there is no obstruction at the entrance to the estuary, then the tidal wave length can be estimated thus:

$$L \approx 48 \cdot 1 \sqrt{d}$$

where L is the wave length in miles and d is the mean depth in feet. L is an approximate value because a mean depth is not easily determined. This parameter L, controls the scale and duration of the estuarine tidal currents; where the length of the tidal reach exceeds $\frac{1}{4}L$ then the fastest currents and slack water may be expected at high tide and mid-tide respectively. However, when the length of the tidal reach does not exceed $\frac{1}{4}L$, then the flood currents will accumulate water at its head, and will themselves be diminished before high water; slack water occurs at high water and the fastest currents around mid-tide. Estuaries with narrow mouths constitute a third group: the tidal flow is restricted and the tidal range reduced. Here the water movements are essentially hydraulic in nature[14].

Evidently, estuaries are very varied. In the Hudson River, which is long and narrow, with a slightly constricted entrance, the difference in the depths of water at higher high tides and lower high tides significantly affects the speed at which these tides move; at Tarrytown, 24 miles from the mouth, this difference is 10 minutes, whereas at Albany, 101 miles further upstream, it is 60 minutes. In this same estuary, the difference in depth at different points affects the duration of the flood and ebb, furthermore the river flow helps to prolong the ebb (6 hr 22 min) and shorten the flood (6 hr 3 min); at Tarrytown the ebb and flood durations are 6 hr 33 min and 5 hr 52 min respectively, whereas at Albany they are 7 hr 21 min and 5 hr 4 min respectively[13]. In the Solway Firth, which has a wide, unrestricted mouth, the run-off from the rivers is negligible in comparison with the total volume of the estuary as a whole. Here the length of the ebb and flood periods increases and decreases with increasing distance upstream until at Newbie, 40 miles from the mouth, the ebb and flood periods are 8 and 4 hours respectively. In estuaries of this kind the change in ebb/flood duration is accompanied by a marked disparity between the maximum ebb tide current velocity and the maximum flood tide current velocity. In the Solway, at Newbie the maximum flood tide velocity exceeds that of the maximum ebb tide velocity by 2·5 knots[15]; this asymmetry in the tidal flow has significant consequences with respect to the accumulation of sediment[16].

A decrease in tidal range upstream may result from a loss of energy due to friction within the containing channel. In the mouth of the Hudson the average tidal range is 4·4 ft, whereas at Troy, 131 nautical miles upstream, it is 3·0 ft. However, the loss of energy, and hence the reduction in tidal range, may be modified by a decrease in the size of the channel; the Bristol Channel is 40 miles wide at its mouth, where the average tidal range is 20 ft, however 80 miles upstream where the channel has narrowed to 5 miles, the range is 33 ft.

The curious Lake Maracaibo system of Venezuela is connected by a constricted entrance, the Strait of Maracaibo, with the Gulf of Venezuela. The tides of the Lake Maracaibo system have been attributed to a standing wave produced by tidal waves which have entered from the Gulf and have undergone reflection from the head of the "lake". An antinode for the semi-diurnal constituents of this system occurs in Tablazo Bay at the seaward end of the Strait. This antinode has had marked effects on the sedimentation regime and is an important factor in preventing the entry of salt water into the lake[17]. Lough Ine in Eire has an entrance so narrow and shallow that tidal conditions in the Lough are modified significantly. Here, although tides outside follow a symmetrical sine curve, within the Lough the flood lasts 4 hr and the ebb 8½ hr. Furthermore the tidal range within is ca. 1 m compared with 2½–3½ m outside[18]. Clearly this is a case in which hydraulic currents are important.

In estuaries which are characterized by a large tidal range, shoals and a rapid decrease in width upstream, a phenomenon known as a *tidal bore* may develop. A tidal bore occurs when the tidal rise is so rapid that the water advances as a wall which may be several feet in height. In essence, a shoaling channel steepens the tidal curve and a bore results if it becomes vertical. The decrease in width increases the tidal range which is further increased by spring tides and it is at these times that bores are more generally seen. Although a bore is a very striking feature, the tide continues to rise once it has passed, and this rise is normally the greater[19]. The conditions under which a bore forms are critical and they may be eliminated by a change in channel shape or depth[19]; however, bores can develop in an individual part of a larger system[20]. Bores have been observed in many parts of the world, e.g. Bay of Fundy, the head of the Gulf of California, the head of Cook Inlet, Alaska[19], the Solway Firth and its tributary estuaries[20], and the Severn estuary[21]. The largest known bore is observed with spring tides in the Tsientang Kiang, China, where a wall of water 15 ft high moves upstream at 25 ft/sec[19].

It is of interest that until a model of the Forth estuary was constructed by the Department of Civil Engineering, University of Strathclyde, the presence of a bore in the headwaters of the Forth was not appreciated. However, inspection of the estuary, subsequent to its recognition in the model, confirmed the development of the bore.

In estuaries which have no complicating hydraulic factors, it is generally true that the duration of the flood tide is shorter than that of the ebb, and accompanying this condition, the tidal currents lose the rotational behaviour characteristic of the open sea, and become rectilinear. With a significant shallowing of depth, a marked lengthening of the *tidal excursion* and an increase in current velocity can occur. The tidal excursion, E, is the distance which a particle of water would travel in moving from low water to high water along the axis of the estuary; it may be calculated thus:

$$E = \frac{2}{\pi} \times \text{average maximum velocity} \times 6 \cdot 2 \text{ hr}$$

the average maximum velocity is obtained from the multiplication of the maximum spring-tide velocity by $0 \cdot 85$[22].

In the Solway Firth area, the generally slack water, characterized by low velocity currents off St. Bees Head, gives way to increasingly rectilinear streams and increased velocity with increasing distance upstream. Under estuarine conditions where a quick change of direction of the tidal current takes place at the time of high and low water, the ebb and flood currents not only tend to, but in fact do, follow different courses. In those estuaries which lack an easily worked sediment, no effects can be observed, but where such sediment is present effects of considerable importance result (p. 154).

Non-Tidal Circulation

In estuaries, wave and tide induced water movement are the result of extrinsic forces. However if compensation is made for tidal movement, then water movements (i.e. non-tidal circulations) characteristic of the individual estuary, and induced by the interaction of fresh and saline waters, become

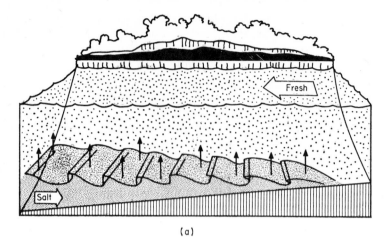

(a)

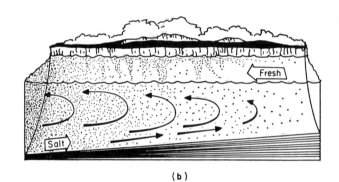

(b)

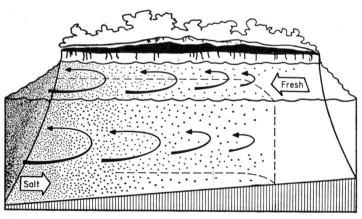

(c)

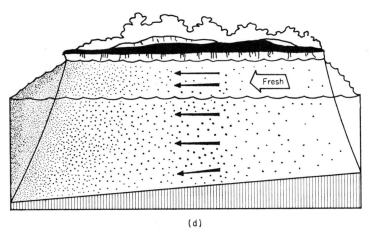

(d)

Fig. 3.6. Estuarine circulation systems. (a) Two-layered, salt wedge, Type A estuary; (b) partially mixed, Type B estuary; (c) vertically homogeneous, Type C estuary, and (d) vertically homogeneous, Type D estuary (after Pritchard[34]).

manifest[23]. First observed by Ekman in 1876[24], these currents were neglected generally, although considered in a few waste disposal problems[25]. The movement of salt water upstream at the seabed, noted by Ekman and called *reaction currents*, may be called *induction currents* also.

In 1949, interest in estuarine circulations was renewed[26], a spate of investigation followed and continues to the present day[22, 27–47]. This information, and especially a classification of the coastal plain estuaries of the U.S.A.[34, 36], has improved significantly our understanding of estuarine processes.

The *Type A estuary* is characterized by a dominant freshwater inflow, a small tidal range and a large depth to width ratio (Fig. 3.6, 3.7). A salt wedge protrudes beneath the surface waters, to a distance which is inversely dependent upon the river flow. Salt water passes into the upper layers by advection, but the mixing of fresh water downwards into the salt wedge is minimal. Evidently, the seaward flow in the surface layers increases with the addition of salt water from the wedge: an upstream reaction current of salt water results in compensation for this loss[34]. In the Umpqua estuary, Oregon, this current flows at an average rate of 0·5 kt below a depth of 35 ft and at a point 3·3 miles upstream (Fig. 3.8). Suspended material, living or inanimate, can be transported upstream to the limit of the salt wedge[36]. The estuary of the River Mississippi belongs to this type.

There is little tidal mixing in a Type A estuary. However, an increase in tidal range, which facilitates mixing, coupled with a decline in river flow, becomes of greater importance in succeeding estuarine types. The *Type B*

3

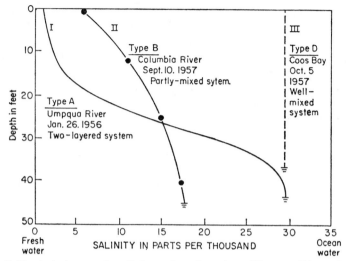

Fig. 3.7. Vertical changes in salinity, selected stations, Umpqua River, Columbia River and Coos Bay, illustrating typical changes in different types of estuaries (after Burt and McAlister[36]).

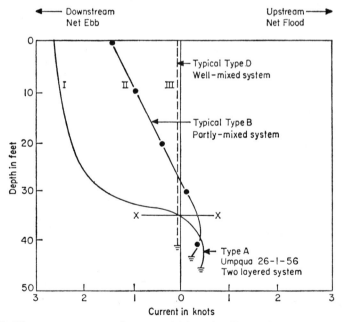

Fig. 3.8. Net current averaged over a tidal cycle, illustrating type average flow expected in three classes of estuaries. Line X–X shows depth of no net motion where average current is zero over a tidal cycle for line 1 (after Burt and McAlister[36]).

estuary (Fig. 3.6b, 3.7) is characterized by vertical mixing between the low saline upper seaward-flowing and the saline upstream-flowing lower layers: mixing prevents the formation of the distinct boundary characteristic of the Type A estuary. The maximum ebb and flood currents are found near the surface and bottom respectively. The volume of the freshwater inflow is small compared with the total volumes engaged in the net circulation pattern, e.g. the discharge from the James River estuary is about twenty times that of the river inflow, and the upstream flow about nineteen times this value[34]. A net upstream flow of 0·3 kt takes place below a depth of 28 ft in the Columbia River[36] (Fig. 3.8). Vertical and horizontal salinity changes are gradual, but because of the earth's rotation, i.e. Corioli's Force, there is a slight lateral gradient in this type of estuary. In the Northern Hemisphere, the boundary between the surface and lower layers is tilted, so that the upper layer flows in a greater volume and to a greater depth on the right looking downstream, whereas the lower, saline layer is nearer the surface and flows more strongly on the left side of the estuary looking downstream[34] (Fig. 3.9a). Because this effect is a consequence of Corioli's Force, one would expect the reverse situation to pertain in the Southern Hemisphere, and this was demonstrated in the Hawkesbury River by D. J. Rochford (Fig. 3.9b). A pollutant released near the bottom, at the mouth of such an estuary may be carried upstream a considerable distance before mixing completely with the upper layers: therefore the whole could become polluted[36].

The *Type C estuary* is characterized by an increase of width, and tidal velocities to the point at which vertical stratification is abolished. There is, however, a lateral gradient of salinity (Fig. 3.6c), with a near vertical boundary between the more saline water flowing upstream and the less saline water flowing downstream, on the left and right sides of the estuary respectively. Examples are the lower, relatively wide portions of Delaware Bay and of the Raritan estuary[34], New Jersey. Curiously, the Solway Firth, superficially of this type, fits the Type D estuary more nearly[40] and is presumably a consequence of the constraint placed upon it by the underlying geological structures[20].

The vertically homogeneous *Type D estuary* is narrower than the Type C: there is no lateral salinity gradient (Figs. 3.6d, 3.7). An equilibrium between the advective longitudinal flux of salt out of the estuary and longitudinal flux into the estuary by mixing, maintains the salt balance in this type of estuary. The length of such an estuary is related to the tidal excursion and cannot exceed the sum of a small number of excursions. Unlike the other types of estuaries, there is a slow net drift seawards at all depths[34] (Fig. 3.8). The intense mixing experienced in a Type D estuary may be enhanced by *tidal overmixing* which is an increase of turbulence

(a) SURFACE ISOHALINES IN THE JAMES RIVER ESTUARY

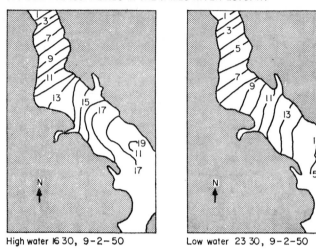

High water 16 30, 9-2-50   Low water 23 30, 9-2-50

(b) PROFILE OF ISOHALINES ACROSS CHANNEL OF HAWKESBURY RIVER,
(Facing direction of stream flow)   AUSTRALIA.

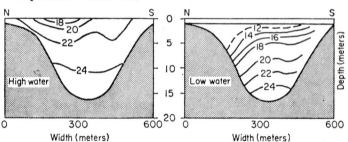

Fig. 3.9. Asymmetric distribution of salinity in estuaries of the northern and southern hemispheres caused by Coriolis force (after Pritchard[32], Bates and Freeman[87]).

and mixing, due to the incoming tide flooding more strongly at the surface and overriding less saline water[48]. A contaminant introduced near the mouth of this estuary will move upstream in significant amounts and the estuary will be contaminated to the limit reached by salt from the sea[49]. Examples of this type are Netarts Bay and Coos Bay, Oregon[36], and the Solway Firth[40].

In summary, an increase in width and tidal velocities, and a decrease in depth and river flow tend to shift an estuary from Type A, to Type B and ultimately to Type C or D. Subsidiary factors, which act directly upon the estuary and influence its type, are wind direction and velocity, solar radiation, bottom roughness, evaporation and rainfall[34]. A change in the weather conditions may influence type, e.g. during heavy run-out the Umpqua, Oregon, is a Type A estuary, but under more moderate flows becomes a Type B[36]; the Savannah River shows Types A and B; characteristics at river flows of 60 000 to 70 000 sec/ft and 5000 to 10 000 sec/ft respectively[34].

A determinant of the estuarine mixing pattern is the *flow ratio*, i.e. the ratio of the volumes of the run-off in the 12·4 hr half tidal cycle and the *tidal prism*. The tidal prism is the volume contained between mean high and low water marks. With a high run-off and a flow ratio $\geqslant 1$, a distinct vertical salinity gradient, i.e. Type A estuary, is maintained; with a flow ratio of 0·2–0·5, an estuary is probably partially mixed, i.e. Type B; and when the flow ratio falls below 0·1, the estuary is probably vertically homogeneous, i.e. Type D[36]. However, estuarine type is not always easy to define[45]. In the Mersey estuary, long regarded as vertically homogeneous, a 1% difference between the salinity of surface and bottom waters is sufficient to induce net flows seawards and upstream in the upper and lower halves of the water column respectively[38]: a system which extends at least 12 miles offshore from the mouth of the Mersey[50].

The residual movements of estuarine waters may be determined from current measurements or deduced from salinity data. An early concept proposed that the non-tidal drift (N.T.D.) in an estuary could be derived thus:

$$\text{N.T.D.} = \frac{\text{River flow}}{F \times \text{Cross sectional area}}$$

where F is the average proportion of river water in a sample, and the formula gives the net distance per day which the water must move seaward in order to carry the water from the river out of the area,

$$F = \frac{S - S_0}{S},$$

where S is the salinity of the open sea and S_0 the salinity of the estuarine water sample. A knowledge of F permits the calculation of the volume of fresh water in a given segment of estuary, hence the *flushing time* is the time required for the river flow to supply this volume. The flushing time for the whole estuary is the sum of the flushing times of the individual segments in which the estuary may be considered[53]. Similarly, an *exchange ratio* (R_s) can be defined for each segment of the estuary, thus

$$R_s = \frac{P_s}{P_s + V_s}$$

where P_s = the intertidal volume, i.e. prism, and V_s = the low tide volume of each segment[28].

Evidently, the river cannot move independently of the sea water, but, as the rising salinity indicates, must carry admixed salt water in its seaward flow. Assuming that all such mixed water moves seawards, then its cross-sectional area times the non-tidal drift gives the total volume of water moving downstream. All the saline water entrained in this flow was introduced by a reaction current. The magnitude of this current may be derived from the ratios of the volumes of river flow and the total flow seawards[27, 33]. An empirical expression of exchanges, of fresh and salt waters in tidal estuaries[28], has been replaced by powerful mathematical techniques.

Where knowledge of an estuary is confined to the river inflow and the vertical distribution of salinity a valuable appreciation of exchange may be obtained, assuming that a steady state pertains[35], thus: at a station, A, the downstream flow, F_1, of mean salinity S_1, and the upstream flow, F_2, of mean salinity S_2, is induced by a freshwater inflow F (Fig. 3.10). Then,

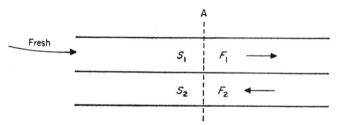

Fig. 3.10. Water exchange across a section A, laid across a narrow estuary (after Craig[35]).

at equilibrium, the total quantity of water and of salt above A is constant, and

$$F_1 = F_2 + F$$

hence

$$F_2 = F\left(\frac{r}{1-r}\right) \text{ and } F_1 = F\left(\frac{1}{1-r}\right)$$

where

$$r = \frac{S_1}{S_2}.$$

All the foregoing considerations depend upon a knowledge of river inflow, and ignore the effect of extrinsic factors. The river inflow may be derived from its velocity thus:

$$\text{Flow} = 538\ 168 \times A.v. \text{ gal/day}$$

where A = cross-sectional area of the channel in ft^2 and v = mean river velocity in ft/sec[51, 52]. Lacking this information use can be made of river flow and standard meteorological records[51, 51a, 51b].

A precise measure of the exchange of water between an estuary and the open sea is difficult to obtain because all practical methods so far devised depend on the use of river inflow as a measure of fresh water added to the estuary[53]. It is difficult to account for water gained by direct rainfall, and as run-off from land in other than monitored streams; loss by evaporation from the surface presents problems also, but it may be defined by the expression,

$$E = (0.26 + 0.77v) (0.98 \, e_w - e_a)$$

where E = Evaporation in mm per cm^2 per 24 hr; v = wind speed in m/sec; e_w = saturated vapour pressure of water at the temperature of the water surface; e_a = partial pressure of water vapour in the atmosphere (mb)[45]. In the River Blackwater, Essex, it was appreciated in 1960, that these two factors were significant[X], a result confirmed by later studies[45]: here, in the summer, evaporation exceeds rainfall and the salinity tends to exceed that of the open sea[45]. Similar results were obtained on the Magothy River estuary Chesapeake Bay[54]. Wind can induce numerous short and long term salinity variations[55]; a marked increase in the abundance of the chaetognath, *S. setosa*, indicating that the introduction of water from the open sea occurred in the River Blackwater under the influence of prolonged easterly winds[X].

There is a further problem with respect to estuarine water movements. For obvious reasons a hydrographer is looking for an expression of the average situation, essentially with respect to salinity. Such information, although very valuable, does not give the detail required by a biologist. In the Blackwater estuary, Essex, the hydrographers appreciation, although very elegant[45], did not offer an adequate explanation for the occurrence of the known lack of homogeneity in the distribution of the planktonic larvae of the oyster, *Ostrea*. In 1960, an analysis of all the chemical and physical data relating to the individual day revealed an inhomogeneity of the water which was not apparent in average data, and that water derived from the different creeks, e.g. Thirslet, Goldhanger, tended to remain in discrete cells as it moved into and down the estuary[X]. Since then further work on the Blackwater has revealed the existence of sub-surface patches of warm water which could be derived from water warmed by the effects of insolation of the shore and/or the power station. The sediment load of these patches may be important to their formation, and by an influence upon specific gravity affect the time for which they can persist. However, their existence is consistent with the early view that the River Blackwater is not entirely well mixed (J. F. Spencer, private communication). Such conclusions would seem to be consistent with recent work in the open sea where it has been shown that a pattern of interleaved sheets and layers may be present, and that an introduced dye may drift along in a lamina undisturbed by turbulence[56]. Clearly, the appreciation that a marked inhomogeneity may exist in the waters of estuaries and of the open sea has important consequences. The association of oyster larvae with particular cells of water implies that it is important to understand the nature, behaviour and fate of such patches if the underlying causes for the success or failure of these or any other larvae are to be understood. Moreover, it is implicit from such a consideration that information regarding the condition and behaviour of estuarine and coastal waters as a whole may be different from that of the individual cells and may therefore be misleading. A second consequence follows from this conclusion, viz. that if a cell or cells of water containing planktonic larvae meet with adverse conditions, since the cell or cells as a whole will be affected totally this is much more serious than if an adverse influence was imposed upon a part only of a large homogeneous system. Since it has been widely assumed in relation to waste disposal and dumping that inshore waters are homogeneous, such a conclusion implies that the basis of these practices would benefit from further scrutiny.

The consequences of Corioli's Force can only be ignored in estuaries less than a quarter of a mile wide. With a greater width, as in the lower reaches of most estuaries, the upper, outflowing layer will form a wedge against the right-hand shore (looking downstream) in the Northern Hemisphere, and on the left-hand shore in the Southern Hemisphere; the inward flowing, more saline water will occupy the remainder of the estuary, e.g. Firth of Forth and Moray Firth[34]. The situation is more complex in the Solway Firth and cannot be observed within the seaward limit of the sand banks[15]. However, to the seawards there are indications of a movement to the north, i.e. to the right, but its interpretation remains uncertain[43].

The outer areas of estuaries may be subject intermittently to powerful influences generated a considerable distance away. Water from the Coral Sea may pass southwards along the eastern coast of Australia, and influence life in the estuaries of New South Wales[57], some 1500 miles distant. The Solway Firth is influenced by currents which sweep into its mouth from the west, and may be derived from the North Channel and hence from changes in the flow of the Gulf Stream Drift[42]; a south-westerly wind may induce the accumulation of water in this estuary, the release of which may lead to irregular flows in the North Channel. This flow was demonstrated by use of *plankton indicator species*[58]. It is possible that populations of the chaetognath, *Sagitta setosa*, found in the Firth of Clyde and off the west coast of Scotland, north to the Minch, have the Irish Sea, some 280 miles away, as their source[59, 60]. The circulation of the Firth of Clyde and the distribution of its plankton may be influenced dramatically by irregular flows of water south through the North Channel[61, Y]. The Yaquina estuary, Oregon, is influenced by the summer upwelling of cold water offshore[62]. The plume from the Columbia River characterized both by its alkalinity[63] and its content of ^{51}Cr[64] can be detected up to 200 miles south of the river mouth, and up to 100 miles offshore. It may be pushed offshore by coastal upwelling, which therefore controls its influence on the coast of Oregon, however ^{65}Zn, from this source, can be detected in mussels, *Mytilus californianus*, south from Tillamook Head, Oregon, into northern California, including Coos Bay estuary[65]. By the use of measurements of both kinds this outflow can be detected from southern Alaska to northern California, a distance of some 1600 miles. To summarize, the inner estuarine circulation is largely a result of the interaction between fresh and salt water. However, at its downstream end, extrinsic forces may modify the circulation and have far reaching consequences upon the estuary and coastal waters as a whole.

The essential difference between coastal plain and *Fiord type* estuaries is that the latter are deep, elongated and have a U-shaped cross section; furthermore they are separated from the open sea by a submarine elevation, or sill[34]. Fiords may be simple, e.g. Pacific coast of Canada[26], or branched, e.g. Norway[34] and Scotland. A freshwater inflow at the landward end, is normally small in comparison to the volumes transported by the tide; moreover, the degree of mixing is dependent upon the tidal velocities.

If the depth of the sill exceeds 50 ft, then the circulation and mixing between fresh and salt water occurs in the surface 30 ft, and resembles a Type B coastal plain estuary. The basin below sill depth is filled with water of a salinity comparable to that outside: an upstream current flows between this layer and the surface[34].

In summer, a thermocline forms at approximately sill depth; below it oxygen depletion and decomposition product accumulation are continuous while it persists. Complete deoxygenation occurs in some parts of the Oslo Fiord system, particularly under the pressure of sewage effluents[66]; it is uncertain whether this has occurred in the depths of the fiordic Scottish sea lochs[35]. Phosphate values are usually high in these deep waters[35, 66].

The sea lochs of the complex Firth of Clyde system have the normal fiordic vertical circulation. However, the deep waters of these lochs reflect a stagnation imposed by an outer sill upon the main body of the Firth water. The degree of stagnation experienced in all fiordic systems may be a function of sill width[35].

3*

Methods of Current Measurement

Generalized information upon water movements in estuaries may be deduced from salinity data, however many problems require more detailed information.

Short term studies, i.e. up to a few days, may be performed by use of a Kelvin Hughes Direct Reading Current Meter which revolutionized this aspect of hydrography. Further developments in design resulted in recording current meters which give a long series of observations in the absence of a ship; very recently, the Nereus Corporation of Seattle has devised an exciting new current meter which lacks moving parts exposed to the sea, and depends upon electric currents induced as the sea crosses lines of magnetic force.

A less expensive method depends on the fact that a body submerged in a flow of water is subject to a force, F, of magnitude

$$F = C_D A_p v_0^2 / 2$$

where C_D = the coefficient of drag, A the cross sectional area of the body in a plane perpendicular to the direction of flow, p and v_0 = the density and velocity of the fluid, respectively. If the body is attached by a wire, and suspended freely in the current, it hangs in an equilibrium position, so that θ, the angle through which the suspending wire is transposed from the vertical, is related to v_0, thus

$$v_0 = k \sqrt{\tan \theta}$$

where $k = \sqrt{2m.g./C_D.A.\ \rho}$, and m = the mass of the body, and g = acceleration due to gravity. In practice, the body is a weighted, wooden drogue formed of two planes intersecting at right angles in the long, central axes; each plane is 4 ft × 3 ft. Current measurements to 50 ft depth may be made easily and rapidly[67].

The Carruthers Jelly Bottle is simple, cheap and elegant. A bottle containing a compass floating on melted gelatine, is tethered by its cap and allowed to swing freely. Suspended at an appropriate depth, the gelatine sets at an angle related to current velocity. The direction to which the current flows is given by the compass skirt graduation nearest the highest setting point of the gelatine: the same bottle may be used repeatedly[68].

The foregoing methods for fixed stations do not indicate the movements of a particular body of water throughout a tidal cycle. Such studies may be performed with the drogue or logship[69]. Individual preferences dictate type, but variants of the Iroquois drogue are used widely, and perform well. This drogue consists of two, right angled and bolted cross frames joined by strips of hessian to form four vanes which do not meet at the axis. The length of each cross piece, and the separation between the two, is 1 m.

The bottom is weighted so that the whole just sinks. The top is attached to a float by rope: variation of rope length permits study of the currents at any depth desired[25]. Tracking by means of the distinguishing float, fixes, at desired intervals, can be taken by sextant or prismatic compass: similarly theodolite fixes of the tracking boat's position may be taken from a suitable vantage point.

Use of aerial photography was proposed in 1952[70] and although photogrammetry may be used[71], this technique has been more immediately valuable in relation to the movement of large lateral floats, e.g. sheets of plywood or polythene film[72]. Recent studies, performed by Oregon State University, Corvallis, of the relative movement of fixed and free-floating markers which release dye, show great promise. Use of an infra-red film in an ordinary 35 mm camera has a value in the study of water movements[73].

All the preceding methods are concerned with short term studies, i.e. up to 24 or 48 hr, but some problems have a duration of days or weeks. Classic methods employed the drift bottle, suitably weighted with sand and carrying a questionnaire: it was broken by the finder who gave details of date and position found, receiving a small reward in return. The drift card, in a polythene envelope, was a replacement used in the late 1940s and early 1950s in oil pollution studies: this interesting device was abandoned because it was blown from wave to wave so easily. Until the late 1950s seabed currents were investigated with drift bottles, additionally weighted with wire tails; recovery of such bottles was poor, variable (ca. 2–30%), and required a protracted time lapse[74, 75]. The first real improvement was made by Craig who used polythene discs and wire weights. However, Woodhead added a touch of genius in the development of the seabed drifter, which is named after him and used the world over. Each is a red polythene disc, $7\frac{1}{4}$ in. diameter, pierced by four holes, 1 in. diameter, at right angles on four mid-radii, and having a 21 in. white polythene stalk with a cylindrical copper weight crimped onto its distal end. Returns are good, i.e. 48–100% within one year of release, and consequently provide good data for statistical analysis. Many variants of it are used to investigate surface currents. All have instructions for return stamped on them, and, unlike drift bottles, do not have to be destroyed: because polythene is almost indestructable, they may be used repeatedly[41–43, 76–83].

While the effects of friction at the sea bed can lead to peculiar movements in the water above it[84], an asymmetrical tidal wave, and hence ebb and flood currents of widely differing maximum velocities may have a significant effect on seabed drifter movement. For example, if rectilinear tidal currents flowed in opposite directions for 9 and 3 hr at constant velocities of 10 cm/sec and 30 cm/sec, respectively, then in this 12 hr period the residual movement is zero. However, the velocity of a seabed drifter (x cm/sec) is related to that of the current (y cm/sec) thus:

$$y = 1{\cdot}099x + 3{\cdot}90$$

and a drifter in the currents above would have a residual movement of 1·8 cm/sec in the direction of the stronger current[82, 83]. Such a concept is vital to an understanding of the rapid upstream movement of sea bed drifters in the Solway Firth at the time of the equinoxes[42, 82, 83, 85] and to an understanding of sediment movement in estuaries generally. For sediments, like sea-bed drifters, may be transported by the tidal asymmetry in the opposite direction to that taken by the residual movement of the sea water[85].

Estuarine systems, which consist of a series of interconnected channels, are complex and little investigated[86].

References

1. Stokes, G. G. (1847). *Camb. phil. Trans.* **4**, 8th part, 441–445.
2. Johnson, D. W. (1919). "Shore Processes and Shoreline Development", Wiley, New York.
3. Russell, R. C. and Macmillan, D. H. (1952). "Waves and Tides", Hutchinson, London.
4. Sverdrup, H. U. and Munk, W. H. (1947). *United States Navy, Hydrographic Office, Technical Report in Oceanography* No. 1, H.O. Pub. 601.
5. Bagnold, R. A. (1946). *Proc. R. Soc.* Series A. **187**, 1–15.
6. Grant, U. S. (1943). *Am. J. Sci.* **241**, 117–123.
7. Bascom, W. M. (1960). *Scient. Am.* **203**, 81–94.
8. Harris, T. F. W., Jordaan, J. M., McMurray, W. R., Verwey, C. J. and Anderson, F. P. (1962). *In* "Proc. 1st International Conference on Water Pollution Research", Section 3, Paper 34: 1–25. Pergamon, Oxford.
9. Deacon, G. E. R. (1947). *Sci. News* **4**, 77–103.
9a. Bowden, K. F. (1962). *In* "Oceans" (G. E. R. Deacon, ed.), Hamlyn, London.
9b. Figuier, L. (1872). "The Ocean World", Cassell, Petter and Galpin, London, Paris, New York.
10. Steers, J. A. (1962). "The Sea Coast", Collins, London. New Naturalist Series No. 25.
11. Day, J. Wentworth. (1969). *Essex Countryside* **17** (146), 21–22.
12. Sverdrup, H. U., Johnson, M. W. and Fleming, R. H. (1960). "The Oceans: Their Physics, Chemistry and General Biology", Prentice-Hall, New York.
13. Pritchard, D. W. (1960). *In* "Encyclopedia of Science and Technology", McGraw-Hill, New York and London.
14. Caldwell, J. M. (1955). *Proc. Am. Soc. civ. Engrs.* **81**, 1–12.
15. Perkins, E. J., Bailey, M. and Williams, B. R. H. (1964). H.M.S.O., U.K.A.E.A., P.G. Report 604 (CC).
16. Perkins, E. J. and Williams, B. R. H. (1966). H.M.S.O., U.K.A.E.A., P.G. Report 587 (CC).
17. Redfield, A. C. (1961). *Limnol. Oceanogr.* **6**, 1–12.

18. Bassindale, R., Davenport, E., Ebling, F. J., Kitching, J. A., Sleigh, M. A. and Sloane, J. F. (1957). *J. Ecol.* **45**, 879–900.
19. Kinsman, B. (1960). *In* "Encyclopedia of Science and Technology", McGraw-Hill, New York and London.
20. Perkins, E. J. and Williams, B. R. H. (1963). H.M.S.O., U.K.A.E.A., P.G. Report 500 (CC).
21. Bassindale, R. (1943). *J. Ecol.* **31**, 1–28.
22. Ketchum, B. H. and Keen, D. J. (1952). *J. Fish. Res. Bd. Can.* **10**, 97–124.
23. Defant, A. (1961). "Physical Oceanography", 2 vols, Pergamon, Oxford, London, New York, Paris.
24. Ekman, F. L. (1876). *Nova Acta R. Soc. Scient. upsal.* Ser. 3, **10**, 23.
25. Alexander, W. B., Southgate, B. A. and Bassindale, R. (1935). *D.S.I.R. Water Pollution Research Technical Paper No.* 5, pp. 1–171.
26. Tully, J. P. (1949). *Bull. Fish. Res. Bd. Can.* **83**, 169 pp.
27. Ketchum, B. H. (1950). *J. Boston Soc. civ. Engrs.* **37**, 296–314.
28. Ketchum, B. H. (1951). *J. mar. Res.* **10**. 18–38.
29. Pritchard, D. W. (1951). *Trans. 16th N. Am. Wildl. Conf.* 368–376.
30. Arons, A. B. and Stommel, H. (1951). *Trans. Am. geophys. Un.* **32**, 419–421.
31. Stommel, H. and Farmer, H. G. (1952). *Woods Hole Oceanographic Institution, Tech. Rept. Ref.* No. 52–51.
32. Pritchard, D. W. (1952). *J. mar. Res.* **11**, 106–123.
33. Ketchum, B. H. (1952). *In* "Proc. 3rd Conf. Coastal Engineering held at Cambridge, Mass., Oct. 1952", 65–76.
34. Pritchard, D. W. (1955). *Proc. Am. Soc. civ. Engrs.* **81**, 1–11.
35. Craig, R. E. (1959). *Mar. Res.* 1959. No. 2, 1–30.
36. Burt, W. V. and McAlister, W. B. (1959). *Res. Brief, Fish Commission of Oregon.* **7**, 14–27.
37. Pritchard, D. W. (1960). *Chesapeake Sci.* **1**, 48–57.
38. Bowden, K. F. (1962). *In* "Proc. 1st International Conference on Water Pollution Research", Section 3, Paper 33: 15 pp. Pergamon, Oxford.
39. Allen, J. H. (1963). *I.A.H.R. Congress.* London, 1963: 53–60.
40. Perkins, E. J., Bailey, M. and Williams, B. R. H. (1964). H.M.S.O., U.K.A.E.A., P.G. Report 604 (CC).
41. Perkins, E. J., Bailey, M. and Williams, B. R. H. (1964). H.M.S.O., U.K.A.E.A., P.G. Report 550 (CC).
42. Perkins, E. J., Bailey, M. and Williams, B. R. H. (1964). H.M.S.O., U.K.A.E.A., P.G. Report 577 (CC).
43. Perkins, E. J., Bailey, M. and Williams, B. R. H. (1964). H.M.S.O., U.K.A.E.A., P.G. Report 605 (CC).
44. Nelson-Smith, A. (1965). *Fld. Stud.* **2**, 155–188.
45. Talbot, J. W. (1966). *Fishery Invest., Lond.* Series II. **20** (6).
46. Rochford, D. J. (1951). *Aust. J. mar. Freshwat. Res.* **2**, 1–116.
47. Dyer, K. R. and Ramamoorthy, K. (1969). *Limnol. Oceanogr.* **14**, 4–15.
48. Burt, W. V. and Queen, J. (1957). *Science, N.Y.* **126**, 973–974.
49. Pritchard, D. W. (1958). *In* "Peaceful Uses of Atomic Energy. Proc. 2nd International Conference held at Geneva, Sept. 1958", U.N. Publication, **18**, 410–413.
50. Bowden, K. F. and El Din, S. H. S. (1966). *Geophys. J. R. astr. Soc.* **11**, 279–292.
51. H.M.S.O. "Surface Water Year Book of Great Britain", London.

51a. Meteorological Office. "British Rainfall", H.M.S.O., London.
51b. Klein, L. (1966). "River Pollution. III. Control", Butterworths, London.
52. Troskolanski, A. T. (1961). "Hydrometry. The Theory and Practice of Hydraulic Measurements", Pergamon, Oxford, London, New York, Paris.
53. Neal, V. T. (1966). In "Proc. 10th Conf. Coastal Eng. (Tokyo)" Vol. II, pp. 1463–1480.
54. Pritchard, D. W. and Bunce, R. E. (1959). Chesapeake Bay Institute, Johns Hopkins University, Technical Report No. 17, Ref. No. 59–2: 1–87.
55. Barlow, J. P. (1956). J. mar. Res. 15, 193–203.
56. Stubbs, P. (1969). New Scient. 41, 562–564.
57. Crosby, L. H. and Wood, E. J. F. (1958). Trans. R. Soc. New Z. 85, 483–530.
58. Williamson, D. I. (1956). J. mar. biol. Ass. U.K. 35, 461–466.
59. Fraser, J. H. (1952). Mar. Res. 1952. No. 2, 52 pp.
60. Barnes, H. (1950). Nature, Lond. 166, 447.
61. Vannucci, M. (1956). Glasg. Nat. 17, 243–249.
62. Frolander, H. F. (1964). J. Wat. Pollut. Control. Admin. 36, 1037–1048.
63. Park, K. (1966). Limnol. Oceanogr. 11, 118–120.
64. Osterberg, C., Cutshall, N. and Cronin, J. (1965). Science, N.Y. 150, 1585–1587.
65. Osterberg, C. (1965). Trans. Oc. Sci. and Oc. Engr. Symposium. 2, 968–979.
66. Braarud, T. and Ruud, J. T. (1937). Hvalråd. Skr. No. 15, 56 pp.
67. Pritchard, D. W. and Burt, W. V. (1951). J. mar. Res. 10, 180–189.
68. Carruthers, J. N. (1958). Fishg. News No. 2362, 25 July, pp. 6–7.
69. Hydrographic Department of the Admiralty. (1948). "Manual of Hydrographic Surveying."
70. Cameron, H. L. (1952). Photogramm. Engng. 18, 99.
71. Cameron, H. L. (1962). Photogramm. Engng. 28, 158.
72. Waldichuk, M. (1966). "Proc. 3rd International Conference on Water Pollution Research", Section 3, Paper 13: 1–22. Pergamon, Oxford.
73. Scherz, J. P., Graff, D. R. and Boyle, W. C. (1969). Photogramm. Engng., Jan., 1969, 38–43.
74. Brown, C. H. (1914). Fisheries, Scotland, Sci. Invest., 1913, II. (August 1914): 22 pp.
75. Brown, C. H. (1916). Fisheries, Scotland, Sci. Invest., 1916. II. (March 1916): 10 pp.
76. Woodhead, P. M. J. and Lee, A. J. (1960). I.C.E.S. Hydrographical Committee (No. 12, Annual Meeting, 1960).
77. Craig, R. E. (1962). Scott. Fish. Bull. No. 17, June 1962, 14–15.
78. Robinson, A. H. W. (1964). Dock Harb. Auth. 44 (52): 3 pp.
79. Ramster, J. W. (1965). Fisheries Laboratory, Lowestoft. Laboratory Leaflet (New Series) No. 6: 8 pp.
80. Lee, A. J., Bumpus, D. F. and Lauzier, L. M. (1965). International Commission for Northwest Atlantic Fisheries. Research Bulletin No. 2, 1965: 42–47.
81. Harvey, J. G. and Gould, W. J. (1966). J. Cons. perm. int. Explor. Mer 30, 358–360.
82. Harvey, J. G. (1967). "Drifter Studies in the Irish Sea", Liverpool Essays in Geography. A Jubilee Collection. 137–156. Longmans, Green, Co., London.
83. Harvey, J. G. (1968). Sarsia 34, 227–242.
84. Hunt, J. N. and Johns, B. (1963). Tellus 15, 343–351.

85. Perkins, E. J. and Williams, B. R. H. (1965). *Trans. J. Proc. Dumfries Galloway nat. Hist. and Antiq. Soc.*, Ser. 3, **42**, 1–5.
86. Swain, F. E. (1951). *U.S. Department of the Interior, Bureau of Reclamation*: *Tech. Memo.* **640**, 1–41.
87. Bates, C. C. and Freeman, F. (1952). *In* "Proc. 3rd Conference Coastal Engineering", Cambridge Mass. 1952 (12), 165–175.
X. Perkins, E. J., Lewis, B. G., Williams, B. R. H.,Howard, T. E., Davis, D. S. and Burdett, J. R. Unpublished work on the C.E.G.B. survey of the River Blackwater estuary, Essex.
Y. Perkins, E. J. Unpublished work.

4

Plankton

All swimming and floating organisms belong to the *pelagic* division: active swimmers constitute the *nekton*, e.g. herring and mackerel, whereas weak swimmers and the floating drifting life form the *plankton*, individually referred to as *plankters*. Most planktonic organisms are small or microscopic in size, capable of localized movement, but lead an essentially passive life. They may be plant, i.e., *phytoplankton*, or animal, i.e., *zooplankton*.

The phytoplankton includes three principal groups, viz. diatoms (e.g. *Coscinodiscus, Skeletonema*), dinoflagellates (e.g. *Noctiluca, Peridinium*), and the nanoplankton or μ-flagellates (e.g. *Isochrysis, Monochrysis*). In contrast, the zooplankton is derived from many phyla, viz. protozoans (e.g. Foraminifera, Radiolaria), coelenterates (e.g. the medusae *Aurelia, Sarsia*), ctenophores (e.g. *Pleurobrachia, Beroë*), polychaetes (e.g. *Tomopteris*), rotifers (e.g. *Synchaeta*), copepods (e.g. *Calanus, Eurytemora*), mysids (e.g. *Neomysis, Praunus*), euphausiids (e.g. *Meganychtiphanes, Thysanoessa*), chaetognaths (e.g. *Sagitta*), molluscs (e.g. *Limacina, Clione*). The term *neuston* refers to those species which dwell in the surface 10 cm of the sea, and includes *Halobates*, a water measurer and oceanic wanderer, *Ianthina*, a beautiful violet-shelled mollusc supported by a raft of mucus bubbles, the pontellid copepods and siphonophore coelenterates, e.g. the notorious Portuguese man-of-war.

Species with an entirely planktonic life history are referred to as *holoplankton* or *holopelagic* forms; those which dwell within the 200 m contour (i.e. the continental edge) are coastal or *neritic*, in contrast to the *oceanic* species which live seawards of this limit. *Tychopelagic* species are those inshore or estuarine forms which live in the plankton, or as bottom dwellers, i.e. *benthos*, with equal facility. In the spring and early summer, the burgeoning holoplankton is augmented by vast numbers of a temporary plankton, i.e. *meroplankton*, the larvae of fish and benthic invertebrates seeking food and dispersal.

All estuaries, which drain incompletely as the tide ebbs, have a characteristic plankton. If this condition is not fulfilled such populations cannot be sustained, and benthic invertebrate populations which have planktonic larvae depend upon external water movements to bring recruits from outside colonies. Even when this condition is satisfied a more variable salinity and temperature, varying degrees of turbidity and, negative estuaries apart, a net loss of water to the open sea result in a special challenge to the inhabitants of an estuary.

Phytoplankton

The phytoplankton requires sunlight for photosynthesis and, therefore, is confined to the euphotic zone. Consequently, the depth of light penetration controls the volume of sea water in which photosynthesis can occur.

At a given depth, viz. the *compensation depth*, the amounts of oxygen which the phytoplankton produces by photosynthesis and requires for respiration are exactly equal (the light has a *compensation intensity*): it varies with latitude, season, water transparency, wave action and turbidity. The compensation depth ranges from 100 m in the Sargasso Sea, to 25–30 m in the Gulf of Maine, to $2\frac{1}{2}$–12 m for *Skeletonema* in Oslo Fjord[1]. Photosynthesis does not cease at the compensation depth, but below it production is ineffective: the concept can be applied to the soil-dwelling littoral diatoms, also (p. 112).

In the turbid waters of many coastal plain estuaries, the compensation depth is exiguous and, the phytoplankton contribution to primary production, is minimal. Conversely, in the deep, clear waters of the mouths of some coastal plain, and in many fiordic, estuaries the compensation depth is considerable and the contribution of the phytoplankton to primary production is dominant.

Estuarine plankton, at any given situation, may be derived from three components: (1) *autochthonous* populations, the permanent residents; (2) *temporary autochthonous* populations, introduced from an outside area by water movements, are capable of limited proliferation only, and are dependent upon reinforcement from the parent populations; and (3) *allochthonous* populations, recently introduced from fresh water or the open sea, are unable to propagate and have a limited survival potential[2].

The effect of salinity upon estuarine plankton populations is considerable. In the Elbe estuary, below Cuxhaven, where the average salinity is *ca.* 16‰ the holoplankton is a typical neritic community which consists of diatoms, *Biddulphia* spp. and *Chaetoceros* spp.; dinoflagellates, *Ceratium* spp.; *Noctiluca miliaris*; rotifers, *Trichocerca marina*, *Synchaeta littoralis*; hydromedusae, *Nemopsis bachei*; diverse copepods; *Sagitta* and *Oikopleura*. The meroplankton is composed largely of polychaete and barnacle larvae. In the reach above Cuxhaven, where the salinity ranges from 16–18‰, the diatoms consist of *Chaetoceros subtilis*, *Keratella cruciata* var. *eichwaldi* together with a massive development of *Coscinodiscus commutatus*; the copepods are made up of *Acartia* spp. In the reach between Ottendorf and Gluckstadt where the salinity ranges from 8–0·5‰ the typical oligohaline autochthonous species are the diatoms *Coscinodiscus fluviatilis*, *Cyclotella striata*, and *Stephanodiscus lucens* and the rotifer *Synchaeta bicornis*. In the reach where the salinity has fallen to 0·2–1·0‰ populations consist of the diatom *Actinocyclus normanni* and the copepod, *Eurytemora hirundoides*[2a].

Table 4.1. Ecological categories of phytoplankton in the Tamar estuary[3]

Autochthonous	Temporary autochthonous	Allochthonous
	Marine species	
	Melosira jurgensii (?)	*Asterionella japonica*
	Nitzschia closterium	*Biddulphia* spp.
	Skeletonema costatum	*Chaetoceros densum*
	Thalassiosira decipiens	*Corethron criophilum*
	T. gravida	*Coscinodiscus concinnus*
	Gyrodinium aureolum (?)	*Rhizosolenia* spp.
	Heterocapsa triquetra	*Dinophysis* spp.
	Peridinium trochoideum	*Diplopsalis lenticula*
	Prorocentrum micans	*Peridinium* spp.
		Phaeocystis pouchetii
	Brackish-water species	
Chaetoceros danicum		
C. wighami		
Cryptomonas spp.		
Gonyaulax tamarensis		
Katodinium rotundatum		
	Freshwater species	
	Synedra ulna	*Anabaena* sp.
		Euastrum sp.
		Scenedesmus obliquus
Littoral species		
Achnanthes brevipes		
Cocconeis scutellum		
Diploneis crabro		
Navicula distans		
Nitzschia angularis		
Paralia sulcata		
Pleurosigma spp.		
Podosira stelliger		
Surirella gemma		

In the Tamar estuary, England, where the population was derived from a range of ecological categories (Table 4.1), the phytoplankton species could be ascribed to two types viz. (1) those with optima in the euhaline zone, e.g. *Thalassiosira decipiens*, *Skeletonema costatum* and *Gymnodinium* sp., and (2) those with optima in the mesohaline zone, e.g. *Katodinium rotundatum*, *Nitzschia closterium*, *Cryptomonas* spp. and flagellates (Chrysophyta)[3].

This is comparable to the Navesink estuary, U.S.A., where the open sea population dominated by the diatom *Skeletonema costatum*, was replaced successively by dominant populations of dinoflagellates, including *Peridinium trochoides*, *P. conicoides* and *Glenodinium danicum*, of diatoms: *Cerataulina bergonii* and *Rhizosolenia* sp., and finally at salinities below 20‰ by euglenoids[4]. Dinoflagellates may dominate in some estuaries, e.g. the Niantic[5, 6]. Some mangrove swamps, e.g. in Puerto Rico, may have characteristic populations of dinoflagellates, including red tide and lumines-

Table 4.2. (A), predominant species in the 0–2 m layer, and (B), in the deepest sample for each date of observation, June 1952–May 1953. (C), the degree of water exchange within the 0–2 m layer between successive dates of sampling, as indicated by the composition of the plankton of the Hunnebunnen[2]

Month	A 0–2 m layer	B Deepest sample (4 or 6 m)	C Water exchange in the 0–2 m layer
June	*Cyclotella caspia*	*Ceratium lineatum* *Dinophysis borealis* Euglenaceae	
July	*Cyclotella caspia*	*Ceratium lineatum* *Dinophysis borealis* Euglenaceae	Small if any
August	*Chaetoceros socialis* (*Cyclotella caspia*)	*Skeletonema costatum* *Dinophysis borealis* *Exuviaella baltica*	Very extensive
September	*Chaetoceros socialis*	*Dinophysis borealis* *Exuviaella baltica* *Chilomonas marina*	Small if any
October	*Skeletonema costatum*	*Skeletonema costatum* (*Dinophysis borealis*)	Very extensive
November	*Coscinodiscus excentricus* *Exuviaella baltica*	*Nitzschia delicatissima* *Exuviaella baltica* (*Dinophysis borealis*)	Extensive
January	*Chilomonas marina*	No samples	
February	Euglenaceae	No samples	No
March	Monads Pennate diatoms	No samples	No
April	*Chaetoceros wighami*	Euglenaceae	Very extensive
May	*Skeletonema costatum*	*Skeletonema costatum* Euglenaceae	Extensive

cent species, e.g. *Pyrodinium bahamense*[7], in other areas, e.g. S. Florida, few may occur[8].

The highly stratified Norwegian fiords support a phytoplankton which resembles that of stratified coastal plain estuaries. Surface waters may support a freshwater, brackish or marine plankton depending upon the amount of mixing between the surface and deeper layers which have a more typically marine population (Table 4.2). In the brackish waters of the oyster polls, diatoms are unimportant when compared to the nano-plankton and the permanent autochthonous species are represented by Euglenaceae; temporary autochthonous species include diatoms, e.g. *Skeletonema costatum* and *Chaetoceros* spp., and dinoflagellates, e.g. *Ceratium* spp., *Dinophysis borealis*, *Exuviaella baltica* and *Peridinium triquetrum*. The allochthonous element is difficult to define but it probably represented by coccolithophores, and diatoms and green algae derived from fresh and marine waters[2, 9]. Under conditions of heavy rainfall Lake Macquarie, New South Wales, shows a similar intense salinity stratification accompanied by characteristic phytoplankton blooms[6].

Similarities exist between the phytoplankton population structure of all estuaries, especially those with a stable salinity structure[4], and in general there is a decrease in species diversity with increasing distance moved upstream[3, 10]. Each estuary, in common with all inshore waters, has a characteristic phytoplankton (and zooplankton), the ultimate survival of

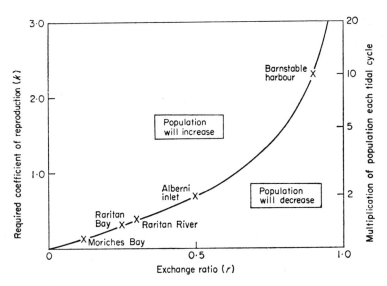

Fig. 4.1. The coefficient of reproduction required to maintain steady state, endemic populations in estuaries as a function of the exchange ratio (after Ketchum[11]).

which depends on individual reproduction rates and the water circu-lation[11] (Fig. 4.1), i.e. degree of mixing, exchange ratio and interactions with outside waters, which may include oceanic incursions[6, 12]. These interactions with oceanic waters are very well illustrated by events described for Port Hacking, Australia, in 1943 "In this year, there was only a slight diatom peak, at the entrance in January, the phytoplankton consisting mainly of the dinoflagellate *Ceratium buceros*. In April, *Rhizosolenia alata* and *R. styliformis* (oceanic forms) were numerous, and in August, the oceanic *Stephanopyxis turris* (East Australian current) appeared with *Chaetoceros secundum*, which although ubiquitous, appears also to be autochthonous in Port Hacking. In November–December, Coral Sea influence is suggested by a peak of *Rhizosolenia calcar-avis*, *R. alata*, *R. styliformis*, *R. stolterfothii* and *Streptotheca thamensis*"[12] (see also p. 98 *et seq.*).

Wind and tide induced turbulence is another important factor. For, in addition to indirect effects on compensation depth, such turbulence is responsible for raising benthic and tychopelagic species into planktonic suspension[8, 12–14]. At Aberystwyth, surface hauls were characterized by *Rhizosolenia*, *Ditylium*, *Guinardia*, *Bacillaria*, *Chaetoceros* and *Thalas-siothrix*; sub-surface hauls by *Coscinodiscus*, *Actinoptychus*, *Biddulphia*, *Paralia*, *Asterionella* and *Lauderia*; and bottom hauls by *Pleurosigma*, *Surirella*, and *Nitzschia*. Motile benthic species, other than *Nitzschia closterium*, are not truly planktonic and disappear after periods of pro-longed, calm weather[13]. Such differences in habit may be generic or specific, e.g. within the genus *Biddulphia*, *B. mobiliensis*, *B. granulata* and *B. rhombus* are true planktonic species. The first two are compressed, have elaborately developed frustules, and live in the open sea, whereas *B. rhombus* with a simpler structure is rarely found more than half a mile seawards from the coast. *B. antediluviana*, which has no spines or protuber-ances, normally forms epiphytic filaments and rarely occurs in the plank-ton[15]. *B. aurita* is tychopelagic[14, 16], and at Whitstable, Kent, where it is the dominant species, has been found to live at depth in the substratum[17, X]: it can withstand desiccation[17].

For much of the year, benthic diatoms contribute to the plankton[13, 14], although the reverse does not generally hold[14]. However, a well-defined group of genera, viz. *Synedra*, *Asterionella*, *Bacillaria*, *Fragilaria* and *Nitzschia*, after assumption of the benthic habit, show an interesting reversion to the plankton where they occur widely, but not abundantly[15].

In general, diatoms are most abundant where a reduced salinity is accompanied by high phosphate, nitrate and organic concentrations[13]. Some species, viz. *Skeletonema costatum* and *Nitzschia closterium* are euryhaline and eurythermal in habit, others, viz. *Guinardia flaccida*,

Lauderia borealis and *Rhizosolenia stolterfothii* tolerate a low salinity, whereas *Rhizosolenia styliformis, Biddulphia regia* and *Chaetoceros teres* do not[3]. Off the Castellon coast significant diatom production does not take place when temperatures exceed 18°C[18].

In neritic areas, salinity changes affect dinoflagellate distribution to an uncertain degree. In the Baltic, *Ceratium tripos* will grow at a salinity of 10‰, whereas *C. fusus, C. furca* and *C. lineatum* will not: below 10‰ *Ceratium* spp. are absent[19, 20]. In Oslo Fjord, a low salinity accompanied by a high temperature may lead to the development of a red tide by *Ceratium furca* and *Gonyaulax polyedra*. Apparently, temperature plays a critical role in controlling dinoflagellate distribution, opinions differ with respect to nutrients[20], but for the development of a red tide high nutrient concentrations are essential (Chapter 1).

The importance of flagellates, in the Norwegian fiords, has led to a greater understanding of the effect of salinity and temperature upon their growth and distribution (Figs. 4.2, 4.3). The coccolithophore *Syracosphaera carterae* is an euryhaline, oceanic species which has been recorded from brackish water in England and Norway; it has a high rate of reproduction within the salinity range 25–45‰ with an optimum at 30‰. On the other hand, the armoured dinoflagellate *Exuviaella baltica*, is an euryhaline, neritic species with a range from the Baltic to the open sea at a salinity

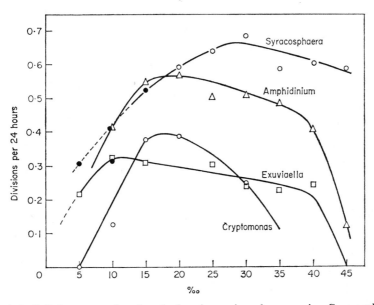

Fig. 4.2. Salinity-curves for the planktonic marine algae species *Syracosphaera carterae, Cryptomonas* sp., *Amphidinium* sp. and *Exuviaella baltica* (after Braarud[21]).

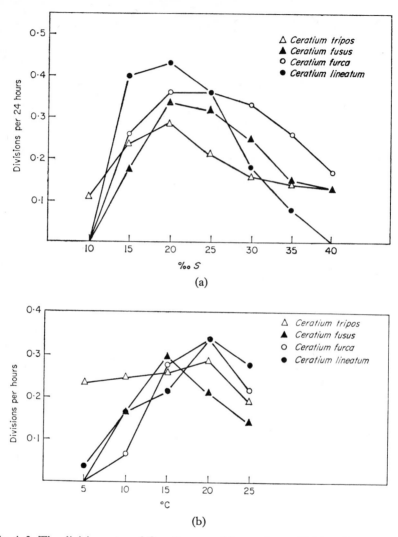

Fig. 4.3. The division rates of *Ceratium* spp. (a) at various salinities, (b) at various temperatures (after Nordli[20]).

of 35‰: growth is good in the salinity range 10–40‰, with an optimum at 10‰. The salinity optima of 15–20‰ for *Cryptomonas* sp. and *Amphidinium* sp. corresponds with the summer salinity of Oslo Fjord[21]. Within a wide tolerance range, *Ceratium tripos*, *C. fusus*, *C. furca* and *C. lineatum* show an optimal growth at a salinity of 15–25‰, and a temperature of 15–20°C, thus corresponding with conditions when intense blooms develop in Oslo Fjord. The ability of *C. tripos* to grow rather better at a lower salinity than the other species is consistent with the distribution found in the Baltic. Unlike the other species *C. lineatum* is affected by a high salinity; indeed *C. fusus* and *C. furca* are found in the Gulf of Oman at a salinity of 39·14‰, and in the Eastern Mediterranean *C. tripos* and *C. furca* occur at a salinity of 38·64 and 39·22‰ respectively[20].

Difficult to investigate because of their small size, the naked flagellates of less than 60 μm diameter constitute the *nanoplankton* (or *μ-flagellates*): those less than 5 μm diameter form the *ultraplankton*. This group of minute, fragile organisms occurs abundantly in the sea and probably contributes the major portion of phytoplankton biomass. Some, the coccolithophores may produce a condition, similar to a red tide, known to herring fisherman as "white water", other species are of particular importance as food for planktonic larvae. Because of the size of its mouth, the early larva of the oyster, *Ostrea*, can only take flagellates up to 10 μm diameter, and grows best on the naked *Isochrysis galbana*, *Monochrysis lutheri*, *Dunaliella euchlora* and *Dunaliella* sp.; older larvae can take forms with thicker cell walls viz. *Chlorella*, *Platymonas* sp., *Chlorococcum* sp., and *Phaeodactylum tricornutum*. Unlike the oyster, the larvae of *Mercenaria mercenaria*, hard shell clam and *Mytilus edulis*, the mussel, can, from the earliest larval stages, utilize species which have a cell wall. However, the food value of a particular flagellate is not determined by the thickness of the cell wall. A close relative of *Isochrysis galbana*, viz. *Prymnesium parvum*, is naked, but produces metabolites which are toxic to larvae[22–26].

Primary Production and Seasonal Changes

In the open ocean, in temperate latitudes, diatom blooms occur in spring and autumn, while dinoflagellate blooms occur in summer[15]. In tropical seas the relationship is modified by the development of a permanent thermocline and upwellings of deep, nutrient rich water: nevertheless within these limits the primary production of phyoplankton is related to the supply of nutrients. The essential factors controlling the development of phytoplankton blooms are summarized in Table 4.3. Estimates of primary production may be calculated from the utilization of nutrients, e.g. phosphates[28], by the uptake of ^{14}C or from the concentration of chlorophyll[29, 30]. With the nutrient methods difficulties may arise because of the phenomenon of "luxury consumption", in which the nutrient is absorbed rapidly from the medium, but is then utilized over a much longer period[31].

In estuaries, the same parameters are fundamental to primary production. However, thermoclines do not develop and the effects of turbidity,

Table 4.3. Summary of the main physical factors affecting the growth of
phytoplankton[27]

Season	Nutrients	Temperature	Light	Phytoplankton
Winter	Abundant	Cold	Weak	Scarce
Spring	Abundant	Moderate	Moderate	Major bloom
Summer	Scarce	Warm	Strong	Scarce
Autumn	Moderate	Moderate	Moderate	Minor bloom

scouring and rapid tidal circulation in this more labile and extreme environment obscure the relationship which can be demonstrated in the open sea[32, 33]. Although difficult to estimate precisely, much of the primary productivity is due to the littoral microbenthos[22, 33–38]. Although the distribution of benthic diatoms may be controlled by tidal currents[39], the mucilage which they secrete helps to stabilize the sediment. Unlike the phytoplankton, the microbenthos is equipped thus to withstand scouring[33], moreover turbidity can affect the former continuously, while the latter can photosynthesize actively in the exposure period. Furthermore, benthic diatoms are equipped for survival in the plankton, whereas the reverse does not apparently hold.

Evidently, the contribution of the phytoplankton to primary productivity depends upon the individual estuary. In the deep Hood Canal, off Puget Sound, effects due to nutrients, e.g. phosphate, temperature and salinity are not limiting, and the development of phytoplankton blooms are related to seasonal changes in solar radiation and water transparency[40]. In Sennix Creek, Long Island Sound, the significant factor is length of day[41]. Australian estuaries are characterized by an addition of nutrient in the short, annual flood period; this transient effect is considered to be responsible for the poor development of phytoplankton at Port Hacking where nutrient uptake, and therefore primary production, is dominated by benthic algae and plants[33]. In Southampton Water, chlorophyll determinations indicate a persistently high productivity with a major peak from March to May and a lesser one from July to September. Here primary productivity is dependent, not on high phosphate values, but on nutrients introduced by the River Test[42–44]. In the impoverished Solway Firth, there is no clear relationship between phytoplankton blooms, nutrient concentrations and run-out[45]. In all estuaries, the situation is confused by the release of phosphate when the seabed is agitated (Chapter 2) and under these conditions chlorophyll determinations become difficult to interpret because of the contribution of tychopelagic and benthic elements. In New Zealand waters, phytoplankton blooms in the moderately turbid waters of Hauraki Gulf are generally similar to those of the open sea; in the more turbid waters of Wellington Harbour, the dominant diatom population has a peak of abundance in autumn. Both populations disappear abruptly with the onset of winter[46].

In the inner Drams Fjord, large allochthonous diatom populations occur in the spring influx; at other times, few are found. Between the final thaw and the onset of river floods, dinoflagellates, derived from an outside source, bloom briefly; summer populations are very sparse because of the transport outwards due to run-off, but from August onwards the influx introduces new allochthonous populations. In the outer Drams Fjord and Oslo Fjord large diatom populations develop in the spring and continue through the summer into the autumn. The dinoflagellates increase during the late spring, summer and early autumn, declining late in the autumn. During the ice-free months, an abundant phytoplankton develops in the Hunnebunnen despite water exchange; when ice is present, sparse diatom and dinoflagellate populations develop in a thin layer between the ice and the anaerobic layers containing H_2S[9]. The success or failure of an estuarine phytoplankton bloom is not determined solely by nutrients and physical factors. Trace elements, e.g. Fe, Mn, may have their origin in the estuary or its tributaries (sea water contains 1–10 mg Mn/m^3 whereas river water contains 50–1000 mg Mn/m^3). Deficiency of these elements, in colloidal or particulate form, can affect the growth and production of *Chlorella*, *Chlamydomonas* and *Skeletonema*; vitamin B_{12} and divalent sulphur have important effects also[47–53]. Important though they are, trace elements apparently do not control the *succession of phytoplankton species*.

The annual cycle of phytoplankton blooms is normally due to a succession of species in which each depends upon its precursors to provide the external metabolites (Chapter 1) which ensure a suitable environment for its development[27, 54]. Diatoms and dinoflagellates bloom at different times which depend upon differing temperature optima, furthermore the dinoflagellates, especially *Ceratium* spp., can survive at lower nutrient concentrations than the diatoms, but they too grow better in neritic than oceanic water[20].

For much of the year, the succession in the 0–2 m and 4–6 m layers of the Hunnebunnen is due to different species (Table 4.2); the diatoms and dinoflagellates predominate in the surface and deeper waters respectively. *Skeletonema costatum* is common to both layers in May and September only[9]. In the Drams Fjord and Oslo Fjord, the diatom population, resembling that of the coastal waters, has a succession which is dependent upon the peculiar hydrographic system. The dinoflagellates develop continuously, and without marked quantitative changes, from spring to autumn: consistent with their known temperature and salinity tolerances the *Ceratium* spp. have a marked succession, viz. *C. longipes* in early spring, *C. fusus* in summer, *C. furca* in later summer and autumn, and *C. tripos* from May to September.

Like the Norwegian fjords, diatoms, especially *Thalassiothrix nitzschioides* and *Th. hyalina*, predominate in the brackish surface waters of New Zealand. Generally, *Th. nitzschioides* and *Nitzschia seriata* bloom in late summer in Wellington Harbour compared with early summer at Hauraki Gulf, some 5° to the north. In sub-surface water mixtures of diatoms and dinoflagellates are found, the former composed of mixtures of *Coscinodiscus* spp., *Thalassiosira decipiens*, *Biddulphia simensis* and *Chaetoceros* spp., the latter composed of *Ceratium* spp., *Peridinium pellucidum* and *Phalacroma ovum* in proportions depending upon location and time of year[46]. At Port Hacking, New South Wales, the phytoplankton succession is complicated by incursions of oceanic water[12] (see p. 98).

At Whitstable, Kent, many species, including *Coscinodiscus* spp., *Skeletonema costatum*, *Biddulphia* spp., *Chaetoceros* spp., *Rhizosolenia* spp., *Asterionella* spp. and *Nitzschia* spp., contribute to the diatom population. An interesting succession is shown by the genus *Biddulphia*: *Biddulphia aurita* the dominant species is most abundant in March, *B. granulata* and *B. regia* are common always, with more than one marked peak; *B. rhombus* of common occurrence, also, is most abundant in autumn and winter; *B. sinensis*, is very abundant from late August to January, and has a maximum in September; *B. alternans*, *B. antediluviana*, *B. mobiliensis* and *B. pulchella* occur sporadically in autumn and winter[14]. *Skeletonema costatum* is the most important species in March/April at Whitstable[14], in April at Southampton Water[55], and July in Loch Striven (some 400 miles to the north)[56].

At Scoresby Sound, East Greenland, there is some evidence of spring and autumn phytoplankton blooms comparable to those of temperate seas. An early phase of *Nitzschia* spp., *Navicula* spp. and *Amphora* spp., associated with the ice, is succeeded by *Fragilaria* spp. and *Thalassiosira* spp., and then by a summer plankton of *Chaetoceros* spp., *Ceratium arcticum* and other peridinians—a stage which persists through the winter[57].

Zooplankton

In both estuaries and the open sea, the zooplankton is more diverse than the phytoplankton; a difference made more pronounced by the larval meroplankton of spring and early summer. Like the phytoplankton, the quality of the zooplankton population depends on the relationship between individual rates of reproduction or development and the flushing time. In Barnstaple Harbour, Cape Cod, clam populations declined, and attempts

at restocking failed, because of an adverse relationship between these parameters with respect to the larvae[11]. In the Gulf of St. Lawrence, the
Margaree River which is drained completely at low water has no characteristic plankton. Instead, at high water, a marine assembly, characterized
by *Podon, Evadne, Calanus*, ctenophores, medusae and polychaete larvae,
is replaced at low water with a freshwater assemblage, characterized by
Bosmina, Cyclops, Daphnia, Diaptomus and *Holopedium*[58]. Of course, such
an estuary is an extreme case, most positive estuaries have a relatively long
flushing time, and consequently develop a stable system with a gradation
from a freshwater to a brackish to a fully marine plankton. In the Columbia
River, *Cyclops vernalis, Daphnia longispina* and *Bosmina* spp. are characteristic of fresh water (salinity $\leqslant 0.1‰$), *Eurytemora hirundoides* brackish
water and *Acartia clausi, A. longiremis* and *Pseudocalanus minutus* the salt
water intrusion[59].

Although, the estuarine zooplankton, other than meroplankton, is
drawn from many groups, copepods are a dominant element, and most
show the broad relationships found in the Columbia River estuary. There
are individual variations, but in general, the brackish, upstream end is
characterized by *Eurytemora hirundoides*, the middle reaches by *Acartia*
spp. (including *A. clausi, A. bifilosa, A. discaudata* and *A. tonsa*) and the
seaward end by *Centropages hamatus, Temora longicornis, Paracalanus
parvus* and *Pseudocalanus minutus*. The occurrence of harpacticoids is
variable, and depends upon the individual estuary, but the holoplanktonic
Euterpina acutifrons is generally the most important. Cyclopoid species
occur from time to time[42, 60–67, Y].

Although cyclopoid copepods of the genus *Oithona* viz. *O. similis* and
O. nana are associated with *Temora longicornis, Paracalanus parvus* and
Pseudocalanus minutus at the mouth of the Columbia, the Solent, Plymouth
Sound, and in the River Blackwater estuary[42, 59, 66, Y], they are notable
absentees from many coastal plain estuaries[42]. However, *O. similis* is
present with these species, together with *Acartia clausi* and the characteristically deep-water species *Euchaeta norvegica* in the fiordic Loch
Striven and Loch Fyne[56, 69]. In the Black Sea, *Oithona similis* and *O.
nana* are found in association with *Pseudocalanus minutus, Paracalanus
parvus* and *Acartia clausi* in waters with a surface salinity of $17–18‰$[70, 71].
It seems possible, therefore, that the absence of *Oithona* spp. from many
coastal plain estuaries is due to their relative lack of depth.

The estuaries of Australia are characterized by complex communities
of centropagid copepods. Like other estuaries, populations are dependent
upon salinity. In the Brisbane River estuary, the euryhaline *Gladioferens
pectinatus* and *Sulcanus conflictus* inhabiting the upstream end are replaced
successively by *Isias uncipes, Acartia* sp. and at the seaward end by

Pseudodiaptomus spp. Some species also inhabit saline waters of the interior, e.g. *Calamoecia salina* and *Boeckella triarticulata*. The genus *Gladioferens* ranges from fresh water viz. *G. spinosus* to the seaward limit: *G. pectinatus* is widely distributed, but its occupation of the more saline reaches depends upon a seasonal reduction of temperature. *Sulcanus conflictus*, another euryhaline species, is characteristic of shallow estuaries and lagoons[71–74].

Like the copepods, the distribution of Cladocera is linked to the salinity structure of the estuary. In Chesapeake Bay, apparently, the upstream penetration of *Podon polyphemoides*, *Penilia avirostris* and *Evadne tergestina* is controlled by minimum salinities of 3·00, 18·12, and 15·75‰ respectively[75].

Fundamentally, the proportion of the estuary occupied by individual species depends upon run-off from the rivers and the exchange ratio. However, this system may be modified by external forces which not only affect the estuary, but also its influence upon offshore waters. A build up of water in the Solway Firth, due to a prolonged south-westerly wind leads to an intermittent flow through the North Channel when the winds moderate: this water has a characteristic plankton[76]. Irregularities in the flow of the Gulf Stream Drift can induce pronounced short term changes in the character of the zooplankton of the Firth of Clyde[77, X]: some species, viz. *Cosmetira pilosella*, *Neoturris pileata* and *Laodicea undulata*, have become resident in consequence[77]. The phytoplankter *Phaeocystis* may develop large blooms in the southern North Sea, during the summer, but this organism penetrates the Essex estuaries to a variable degree: in 1959 and 1960 spectacular blooms developed outside the River Blackwater, but penetrated in 1959, only. Here too, the chaetognath *Sagitta setosa* penetrates in large numbers only after a period of prolonged easterly winds. A similar situation to that of *Phaeocystis* may arise with *Noctiluca scintillans*, here, however, the situation is complicated by the upstream development of blooms of which portions are transported to the rest of the estuary by cells of water[Y]. The phenomenon of widespread colonization of an estuary by *Acartia* spp. and *Eurytemora* spp. from circumscribed, upstream loci is also dependent upon the circulation pattern and its relationship to outside water movements[66, 67]. The plankton of the Irish Sea, Firth of Clyde, west coast Scottish sea-lochs and the Minches are all linked by extrinsic forces (Chapter 3).

Long term salinity changes may have a marked effect; the Baltic has become more saline since the early decades of this century. *Limnocalanus grimaldii*, a copepod species with a low salinity tolerance, has had its range reduced, on the other hand, the range of many species, viz. the jellyfish *Aurelia aurita* and *Cyanea capillata*, the ctenophore *Pleurobrachia pileus*, the copepods *Acartia longiremis*, *Temora longicornis*, *Centropages hamatus*,

Pseudocalanus minutus and *Oithona similis*, and the tunicate *Oikopleura dioica*, has extended. The medusa *Melicertum octocostatum* was resident in the southern Baltic from 1949–55[78].

Of course, individual plankton populations are transported by tidal currents, so that at any one situation marked differences are observed during the tidal cycle[63, 79] (Table 4.4). At Whitstable, Kent, the plankton at high and low waters is very different. At high water, the plankton is characterized by the harpacticoid copepods *Canuella perplexa*, *Alteutha oblonga*, *Thalestris longimana* and *Cleta lamellifera*; the amphipod *Hyperia galba* may be taken at this time also. At low water the plankton is characterized by the forminiferan *Elphidium* sp., the ostracod *Loxoconcha impressa*, the harpacticoid copepods *Alteutha interrupta*, *Amphiascus tenellus*, *Laophonte* sp. and *Asellopsis intermedia*, the amphipods *Microprotopus longimanus* and *M. maculatus* and the isopod *Gnathia maxillaris*[61, 62].

Table 4.4. Three major copepod species at three stations in Yaquina Bay at high, mid, and low tide[63]

| Station | Species | Copepod Concentration (no/m³) | | |
		High tide	Mid tide	Low tide
15	*Acartia tonsa*	—	10	94
	Acartia clausi	3 426	3 947	33 394
	Pseudocalanus minutus	12 331	2 621	113
21	*Acartia tonsa*	—	122	8 993
	Acartia clausi	6 080	12 497	3 767
	Pseudocalanus minutus	12 560	70	—
39	*Acartia tonsa*	38 227	26 509	27 014
	Acartia clausi	4 937	545	429
	Pseudocalanus minutus	—	—	—

N.B. Station 15 = Zone of strong oceanic influence
Station 21 = Transition area
Station 39 = Region removed from strong oceanic influence

The vertical migrations of zooplankton, both seasonal and diurnal, were first recognized in 1876[80], but it remained for Russell, in his classic work conducted between 1925 and 1934, to demonstrate the importance of this phenomenon in the open sea[81, 82]. The diurnal migration is primarily a response to changes in light intensity, but changes in temperature, salinity and the concentration of dissolved gases, the addition of some chemicals, or a sudden change in the light intensity may reverse this positive phototropism[80].

The estuarine zooplankton also exhibits some form of diurnal migration, although it may be modified significantly by the turbidity of the waters.

Many animals react positively to low light intensities only, and are consequently found at the surface around sunset. These include copepods, e.g. *Acartia* spp.[83], Cladocera, chaetognaths, and brachyuran zoea[80]. Other species including the cladoceran *Evadne* sp., and some of the dinoflagellates (claimed equally by botanists and zoologists) viz. *Ceratium tripos, C. fusus, C. furca, Peridinium triquetrum, Prorocentrum micans* and *Gonyaulax polyedra*, migrate to the surface during the day and away from it at night[80, 84, 85]. The behaviour of the harpacticoid copepods at Whitstable, Kent shows an interesting gradation related to occurrence in the plankton and on the shore: the holoplanktonic *Euterpina acutifrons* has photopositive responses, the tychopelagic *Tachidius discipes* and wholly benthonic *Stenhelia palustris* have strong positive responses to light also, but *Nitocra typica, Laophonte setosa* and *L. foxi* are benthonic and exhibit a strong photonegative response[86]. Although vertical migration may be a powerful maintenance mechanism for some species, it is not so for *Eurytemora hirundoides* in the Columbia[59].

While the copepods *Acartia tonsa* and *A. clausi* may react strongly to cold water, the species *Calanus finmarchicus* and *Metridia lucens* which occur in outer estuarine areas react positively to light at temperatures $\leqslant 10°C$, and negatively to all light intensities at higher temperatures[80]; the harpacticoid copepods *Tachidius discipes, Microarthridion fallax* and *Parathalestris intermedia* show no temperature reversal of phototactic response[86]. Except for the copepod *Oithona similis* and the chaetognath *Sagitta elegans baltica* which are confined to deep saline waters and do not undergo vertical migration, the distribution and abundance of zooplankton in the Baltic is controlled primarily by temperature which controls the vertical distribution and migration of many species[87]. In a stratified estuary vertical migration may be reduced because those animals living in the more saline layers cannot ascend into the reduced salinity of the upper layers: the copepod species *Acartia tonsa, A. bifilosa, A. discaudata, A. clausi, Centropages hamatus* and *Temora longicornis* are all limited thus[88, 89].

Although not all the species show the same degree of movement, an increase in temperature or of light intensity may be responsible for the migration of the copepods, *Pseudocalanus minutus, Paracalanus parvus, Centropages hamatus, Temora longicornis, Acartia clausi* and *Oithona similis* from the surface layers of Loch Striven to depths below 10 m in the months of July and August[56, 90, 91]; a similar downward movement occurs in Scoresby Sound, East Greenland[57]. A limited migration of *Calanus* and *Sagitta* occurs at Spitzbergen when light intensity changes, beneath the midnight sun, are sufficient; however, in July and August, the proportions of *Sagitta, Limacina, Beroë* and *Mertensia* in the surface waters are apparently related to date rather than light intensity at the time of sampling[92].

Environmental Tolerances of the Holoplankton

The relationship between temperature and the distribution of *Acartia* spp. on the eastern coast of North America is interesting. In Chesapeake Bay, *A. tonsa* dominates for much of the year, but the occurrence of *A. clausi* is limited to a short period in winter: *A. longiremis*, a continental shelf species, is normally absent from coastal waters south of Cape Cod, including Chesapeake Bay. In Long Island Sound *A. clausi* is more abundant than *A. tonsa*: both are less abundant in the more oceanic Block Island Sound, but *A. longiremis* is present. In Delaware Bay, which is roughly midway between the Chesapeake and Long Island Sound, *A. tonsa* is more important than *A. clausi*[93].

A few *holeuryhaline*, or *holeurysaline*, species can exist in the whole range of salinity from fresh to fully saline sea water; they occur in much smaller numbers than the euryhaline species[94]: the rotifers *Synchaeta vorax* and *Encentrum marinum* and the crustacean *Mysis oculata* have this ability[94, 95]. In the Laguna Madre, a number of holoplankton species which normally occur in a brackish habitat, have been found living in hypersaline conditions, and have therefore a wide range of salinity tolerance: ctenophores, *Beroë ovata* (10–ca. 80‰) and *Mnemiopsis mccradyi* (10–ca. 75‰), and copepods *Acartia tonsa* (10–80‰) and *Metis japonica* (ca. 25–80‰)[96].

It is of considerable interest that when freshwater animals are introduced to water of a low salinity and at a moderate temperature, the survival time is increased 2–3-fold with a 10°C decrease in temperature[74]; this effect closely parallels that due to toxic substances introduced into the aquatic environment by waste disposal and other processes (p. 440). However, the temperature at which an animal has been living is crucial to a consideration of salinity tolerance. Of the copepod genus *Acartia*, *A. tonsa* is most tolerant to changes of salinity by dilution, *A. discaudata* has a poor tolerance, and *A. bifilosa* is intermediate between these two. The tolerance of all three species declines rapidly with increasing departure from the environmental water temperature[97]. During the spring and summer of 1966, a decrease in species diversity and an abundance of rotifers, cladocerans, and copepods was associated with an abnormally low temperature and salinity in the coastal waters of the White Sea[98].

Laboratory studies of *Pseudocalanus minutus* show that, except at the highest temperatures (11·43–11·44°C) there is little difference in the mortality and development rate of eggs at a salinity of 25‰, compared to 35‰. An increase in temperature accelerated development: at a salinity of 28·6‰, a 50% hatch takes place in 11·25 and 2·92 days at 0 and 11·94°C

4

respectively[99]. Over a period of 6 yr, *Pseudocalanus elongatus*, maintained in a pH range of 7·6–7·9 (tolerance range 7·4–8·5) had an average life of 71 and 4 days at temperatures of 5·0–7·0 and 19·5–21·0°C, respectively[100]. The copepods have a variable tolerance to oxygen deficiency. In the River Tyne estuary, the euryhaline *Eurytemora hirundoides* may be abundant in completely anoxic waters[12]. The euryhaline species *Gladioferens pectinatus*, *Sulcanus conflictus* and *Temora turbinata* normally occur in Day's Lagoon, Queensland; in November 1951, deoxygenation occurred, consequent upon the collapse and decay of a dense bloom of the blue-green alga *Nostoc linckii*, and resulted in death of the copepod population. *Sulcanus conflictus* recovered after normal conditions were re-established, but neither of the other two species were present up to November 1952; although the means by which *S. conflictus* survived is unknown, it apparently has some more resistant stage than the other two species[74]. Most species of marine protozoa are more dense than normal sea water, in which *Noctiluca* (S.G. 1·014) floats[101]; variations in the salinity of estuarine waters may have an important influence on the distribution of these organisms (see also p. 84).

Mero-Plankton

The temporary, or meroplankton, is composed of species which do not spend the whole life cycle in the plankton; unlike the tychopelagic species, the meroplanktonic stage cannot live associated with the bottom sediments. It is composed of the larvae of benthic chordates and invertebrates, and nekton, particularly fish, on the one hand and of the sexual jellyfish stage of the hydrozoan and scyphozoan coelenterates, e.g. *Sarsia*, *Cyanea*, on the other.

The larval plankton is extraordinarily diverse[103, 104]. Opinions differ regarding the role of these larvae: some hold that it is a dispersive phase concerned only with ensuring a widespread distribution of the species, others that it is a means by which the young organism can take advantage of large supplies of "baby-food", e.g. nanoplankton which is so important to oyster larvae. Still others, hold that by first releasing and then feeding upon large numbers of these larvae the benthos is extending its food supply. There is no doubt that whichever of these postulates is true, larvae are very abundant at certain times of the year. In Southampton Water, cirripede nauplii dominate the zooplankton for most of the year: abundant in early spring, *Balanus balanoides* and *B. crenatus* are succeeded, in summer, by very large populations of *Elminius modestus*.

Studies of the behaviour and migrations of planktonic larvae have given an insight into the effect of estuarine circulations upon adult stocks. An important variety of fish, the croaker, *Micropogon undulatus*, spawns in the ocean off Chesapeake Bay where the reaction current may carry the newly hatched fish, over 100 miles upstream, to the upper limits of the salt water intrusion; here they gather and then gradually move downstream as they grow. The first movement is passive because the young larvae are poor swimmers; the movement downstream is apparently an active process; the extent to which this upstream transport benefits the croaker is uncertain[105, 106].

In a very fine oyster seed area of the James River, Virginia, the oysters are so small that an effective contribution to the annual set of spat is unlikely. The reaction current of this Type B, partially mixed estuary, could carry the larvae 20 miles upstream from the adult brood stock to the rich, seed area: the length of larval life corresponds with the transport time predicted from velocity measurements of the reaction current. The reaction current reaches its upper limit, on the bars, in the shallow waters of the north east side of the estuary; consistent with the larvae being derived from a downstream source, a greater set of larvae is found than on the south west side[107, 108]. In the Miramichi estuary, New Brunswick, the nauplii of *Balanus improvisus* occurred in the surface layers downstream, whereas the heavier cyprid larvae were concentrated in the lower layers towards the head of the estuary[109]. At Winchester, near the mouth of the Umpqua River, Oregon, crabbers are advised to work around high water, i.e. at the time of the greatest intrusion of salt water at the bed of the estuary; either the crabs migrate with, or are more active and easiest to catch in, the more saline water[110]. At Whitstable, Kent, the residual currents must carry many of the pelagic larvae away from the area, and are adverse to the maintenance of adult stocks.[111]. In the Crouch and Roach estuaries, Essex, oyster larvae are transported upstream, although the rivers appear to be vertically homogeneous[22]: this may be due either to an upstream residual movement, such as that found in the Mersey (p. 59), or, like sediments, if there is a sufficient asymmetry in the ebb and flood tidal current velocities, the larvae will be moved upstream (p. 51). In Milford Haven, herring move into the upper reaches to spawn. However, the greatest densities and youngest larval stages occur at a maximum distance downstream from the spawning ground: a distribution resulting only if the young larvae seek and remain close to the surface[112].

Environmental Tolerance of the Meroplankton

The scyphozoan medusa *Aurelia aurita* which is abundant in many British estuaries, e.g. Tamar[113], Thames[60], Blackwater[Y], Mersey[X], and Solway Firth[114, 115], during the summer months, still produces normal germ cells at a salinity of 6‰, but the scyphistoma stage does not develop[94].

Table 4.5. Order of tolerance of polychaete larvae compared with penetration of adults up an estuary[117]

Recovery experiments		Continuous experiments		Penetration of adults up an estuary
Hypotonic	Hypertonic	Hypotonic	Hypertonic	
Nereis	Nereis	Nereis	Nereis	Nereis
Pomatoceros	Pomatoceros	Scoloplos	Pomatoceros	Scoloplos and Pomatoceros
Notomastus	Notomastus	Pomatoceros	Scoloplos	
Phyllodoce	Phyllodoce	Notomastus	Phyllodoce	Phyllodoce
		Phyllodoce	Notomastus	Notomastus

It is convenient to consider polychaete larvae as a whole, though not all have a planktonic stage. The eggs and larvae of *Nereis* are particularly resistant to a lower salinity and exposure to anaerobic conditions[116]. Larvae of *Nereis diversicolor* are fully tolerant of sea water at a salinity of 10‰, whereas some individual *Scoloplos* and *Pomatoceros* larvae will survive for 2 days, but those of *Notomastus* and *Phyllodoce* do not survive for more than a day; larval survival at a reduced salinity and estuarine penetration by the adults are compared in Table 4.5. The ability of *Pomatoceros* to resist salinity change is greatest at 14°C; in a column of sea water these larvae tend to collect at a salinity of 12‰[117]. Adults of the three barnacle species, *Balanus balanoides*, *Elminius modestus* and *Chthamalus stellatus* occur widely in the coastal plain estuaries and sea lochs of the British Isles,

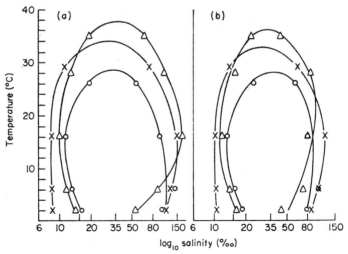

Fig. 4.4. Two-dimensional contours for 50% survival in relation to temperature and salinity for nauplii of the three species: *Balanus balanoides* (○), *Elminius modestus* (×), *Chthamalus stellatus* (△). Contour for (a) 12 hr survival; (b) 24 hr survival (after Bhatnagar and Crisp[118]).

the larvae have an ability to resist salinity changes over a wide, but not extreme, range of temperature (Fig. 4.4); all species are completely immobilized by salinities < 12 and > 50‰, but the larvae of *Elminius modestus*, Australasian in origin, seem to be particularly resistant[118], possibly one of the secrets of its success in European waters. All three species can withstand temperatures and salinities unlikely to be encountered in the normal environment (see also p. 249). The nauplii of the estuarine barnacle species *Chelonobia patula* is tolerant of salinities within the range 15–20‰ to 50‰, i.e. broadly similar to *B. balanoides*[119–121]. The nauplii of *B. eburneus* and *B. amphitrite*, can tolerate salinities between 10 and 50‰ indefinitely; at 5‰ both species will continue to swim for approximately ½ hr, but with a photonegative response[120]: an important adaptation to ensure survival in estuarine conditions where higher salinities occur in the deeper, darker water layers. The size of the cyprid larvae of *Balanus crenatus* and *Verruca stroemia* decreases depending upon how late in the year a brood is produced; in the spring *B. balanoides* and *B. crenatus* may produce cyprids of differing sizes depending upon the differing environments of the adults[121].

Because of the elegant work of Costlow and his collaborators the effect of salinity and temperature upon larval development of estuarine crabs is now comparatively well understood[122–128]. Temperature has a more pronounced effect upon the length of larval development than upon mortality: *Sesarma cinereum* develops to the first crab stage in 35–39 days at 20°C, but only takes 20–22 days at 30°C[123]. The larvae of the xanthid crab, *Rhithropanopeus harrisii*, which now occurs widely in brackish waters on both sides of the North Atlantic, take 26–41 and 11–18 days to reach the first crab stage at 20 and 30°C respectively; the time range is a function of salinity[126]. The larvae of *Ovalipes ocellatus* reach the first crab stage in 26·1–27 and 18 days at 20 and 25°C respectively, the rate of development is independent of salinity in the range 25–40‰: this larva could probably complete its development in estuaries, but not in the coastal waters inhabited by the adult[127]. Experimental studies upon the larvae of the calico crab *Hepatus epheliticus* which ranges from the mouth of Chesapeake Bay to Campeche, Mexico, suggest that it could not complete its development in estuarine salinities lower than 30‰, although it would be successful in coastal waters[125].

Application of response surface techniques to this work showed that the different larval stages of a crab may have very different requirements to ensure its survival. *Sesarma cinereum* depends essentially upon the fourth stage zoea for the successful completion of its life history: here, survival and the moult to the megalops can only take place in estuaries (Fig. 4.5). Fourth stage zoea washed out to sea or into tidal pools of high salinity

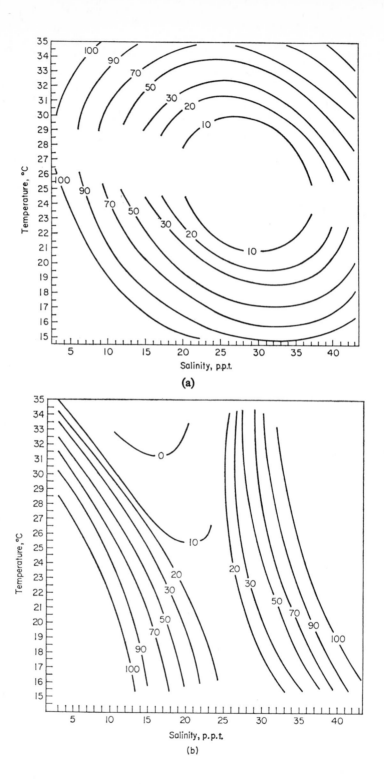

(a)

(b)

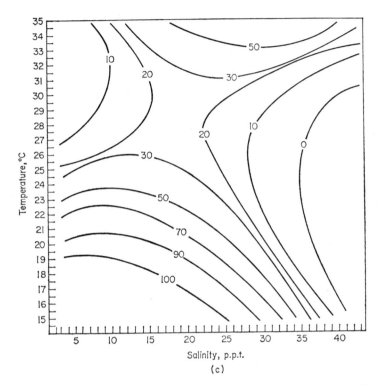

Salinity, p.p.t.

(c)

Fig. 4.5. Estimated percentage mortality of *Sesarma cinereum* larvae based on the fitted response surface to observed mortality under twelve different combinations of salinity and temperature (a) 1st zoeal stage, (b) 4th zoeal stage, and (c) megalops stage (after Costlow and Bookhout[124]).

would not survive, but once the megalopa stage is reached the larvae can withstand a wide range of salinity and temperature. Unlike *S. cinereum*, *Panopeus herbstii* can develop successfully in a wider range of salinities at all stages of development[126].

The larvae of the molluscs *Mytilus californianus* die in dilute sea water which the adult can tolerate indefinitely[94]. Normal straight-hinged larvae of the American oyster, *Crassostrea virginica*, develop in the temperature range 17·5–30°C; at 15°C the larvae survive for some days without growth, and at 33°C only 45% of the eggs reach the straight-hinged larval stage. The growth of larvae from Long Island Sound oysters is best at a salinity *ca.* 17·5‰, good at 15‰, but deteriorates at lower salinities and is accompanied by a high mortality at 10‰, to which older larvae are more tolerant. Development of eggs, from oysters accustomed to the brackish waters of

Chesapeake Bay, is best at salinities of 12–15‰, but ceases above 22·5‰. Like the barnacles, the adult environment influences the growth and survival of oyster larvae[23, 24]. Upon release, the larvae of *Ostrea edulis* can survive several days starvation, and then grow to a normal size once suitable food (p. 79) becomes available[128].

A salinity $\geqslant 26\cdot5$‰ is the optimum for the growth of straight-hinged larvae of *Mercenaria mercenaria*, hard-shelled clam. At $\leqslant 17$ and at 35‰ no eggs and 1% develop, respectively, but at 22·5‰, 80% of the eggs develop normally. However, once a larva has reached the plankton some will reach metamorphosis at a salinity of 17·5‰[23, 24].

The amount of silt present in sea water can influence the development of *Crassostrea* and *Mercenaria* larvae. The percentage of oyster eggs developing to the straight-hinged larvae falls from 95% to 0% at 0·125 and 2·000 g/litre respectively; similarly for clams the percentage falls from 95% to 0% at 0·125 and 3·000 g/litre respectively; silt is more harmful to oyster eggs than kaolin or Fuller's earth. Silt is also more harmful to oyster larvae than clam larvae: growth of oyster larvae is affected by 0·75 g/litre and at 1·5 g/litre silt is negligible, however clam larvae grow normally at 1·0 g/litre silt while the majority survived for 12 days and some show growth at 3·0 and 4·0 g/litre[129].

The cosmopolitan ascidian *Ciona intestinalis* occurs in the mouths of many estuaries, e.g. Firth of Clyde, River Blyth. The sessile adult has a planktonic larva which is reared easily in the laboratory; the eggs hatch in 24 hr at 16–20°C, normally the tadpole larvae have a free-swimming period of the order of 12 hr[130].

The developmental and hatching periods of fish are generally shorter at higher than at lower temperatures, but different species develop at differing rates: at an average temperature of 6·1°C, the developmental period of brown trout (*Salmo trutta*) is 88 days, of rainbow trout (*Salmo gairdneri*) is 61 days, of cod (*Gadus morhua*) is 14 days, and pollack (*Pollachius virens*) is 9 days[131]. Development of most pleuronectid eggs is retarded by a lowered salinity[94], and a high mortality of plaice *Pleuronectes platessa* eggs takes place at salinities below 17·5‰, most die in the blastula or early gastrula stage; after hatching, yolk-sac larvae tolerate salinities of 15–60‰ for 1 week; after metamorphosis, the tolerance to high salinities is reduced and the larvae can withstand a lower salinity, i.e. 2·5–45‰ for 1 week. Larvae which are migrating to the estuarine nursery ground experience salinities well within their tolerance limits[132].

Fertilization, development and hatching of herring eggs takes place in salinities ranging from 5·9 to 52·5‰. Both spring and autumn spawned larvae will tolerate salinities in the range 1·4–60·1‰ and 2·5–52·5‰ for 24 and 168 hr, respectively[133]. After metamorphosis, herring, of length

9–24 cm, will tolerate salinities in the range 6 to 40–45‰[134]. Herring larvae, of length 6–8 mm, acclimatized to temperatures between 7·5 and 15·5°C have lower and upper lethal temperature limits of $-0·75$ to $-1·8$°C and 22 to 24°C, respectively[135].

Studies of North American fish have shown that the euryhaline eggs of *Gadus macrocephalus* hatch best at a salinity of 19‰, and a temperature of 5°C: temperature changes of 1°C influence the hatching of the eggs to a degree equivalent to a salinity change of 1·2‰[136]. Investigations of the survival of eggs of the sole, *Parophrys vetulus*, in the salinity range 10–40‰, showed that over the geographical range of the species, the effect of salinity was likely to be of little importance, but near the limits the effects of temperature is critical[137].

Seasonal Changes in the Abundance of Zooplankton

In the open sea, the holoplankton, never entirely absent, generally experiences spring and autumn peaks of abundance: the spring peak is normally greater than that of the autumn. In estuaries, however, the pattern of events may differ[42]. Like the phytoplankton, species of zooplankton are abundant successively, but phenomena like the development and succession of *Acartia* spp. and *Eurytemora* spp. from small sources[66, 67], and the burgeoning of autochthonous species, influence the relative balance of the species. The meroplankton is very important and relationships with the microbenthos are exaggerated: some species can, upon excessive reproduction, "overflow" into the plankton, other species, e.g. the ostracods *Leptocythere pellucida* and *Cytherura sella*, may assume a planktonic life to avoid adverse conditions[138].

In Southampton Water, the zooplankton is dominated by meroplanktonic cirripede larvae, viz. *Balanus* spp. and *Elminius modestus*, but the major changes are due to the holoplankton[42]. At Whitstable, in the mouth of the Thames estuary, *Acartia* spp. are less and harpacticoid copepods more prominent than in Southampton Water[42, 61, 62]. Generally, the annual zooplankton production is very variable, and a series of observations is an essential prerequisite before making generalizations[42, 59, 63, 139]. The occurrence of the individual species of the genus *Acartia* spp. is interesting in this context. Generally, *A. clausi* is abundant earlier in the year than *A. tonsa*[42, 66, 140], however *A. clausi* may be most abundant at any one situation, and in any one year, in spring[42, 56, 59, 68, 140], summer[56, 59, 60, 61, 63, 68], and autumn[59]. In the Don estuary, Sea of Azov, it is most abundant from October to May, and is succeeded by *A. latisetosa*[141]. In Southampton Water the spring outburst of *A. clausi* is associated with *A. bifilosa*, and the late summer and autumn outburst of *A. tonsa* with *A. discaudata*[42]. *Eurytemora affinis* and *Eurytemora hirund-*

4*

oides (the latter is considered to be a variety of the former by some authors[71]) are normally most abundant in April–May in the Columbia River estuary, but may have a number of inferior maxima in other months of the year, viz. January, June, July, September, November, December; the population may be sharply depleted at times when floods occur[59, 131]. The harpacticoid copepod *Euterpina acutifrons* occurs commonly in the summer and early autumn plankton of Southampton Water and Whitstable[42, 60, 62]; but curiously, at Whitstable, its nauplii may be abundant in the shore soils. Food is plentiful here and the relatively large size of this nauplius may make a planktonic mode of life difficult[86, 138].

The small copepods commonly found in Loch Striven, viz. *Pseudocalanus minutus*, *Paracalanus parvus*, *Microcalanus pygmaeus*, *Centropages hamatus*, *Temora longicornis*, *Acartia clausi* and *Oithona similis* begin to reproduce about the time of the spring diatom outburst in March and April and a succession of broods is produced throughout the summer, in most cases the broods are not distinct except for those of *Microcalanus* which begins breeding before the spring diatom outburst and has distinct broods like *Calanus*. In general the production of eggs and nauplii depends upon the presence of diatoms which also help the development of later stages[56]. In other areas viz. Whitstable[60, 61], Plymouth[68], Narragansett Bay[63] and the Columbia River[59], *Pseudocalanus minutus* may be common in spring[60, 61], abundant from March to October, (with greatest numbers from April to August[68]), may occur throughout the year with a maximum in March and April[63], and throughout the year with maxima in spring and autumn[59]. *Paracalanus parvus* may be common in the spring plankton[60, 61], and abundant from April to December (maxima May–June and August–October)[68]. *Temora longicornis* may be common throughout the year[60, 61], and abundant from April to September (maxima May–June and August–September)[68]. *Oithona similis* may be abundant from April to October (maxima early May, late July and mid-September)[68] and abundant throughout the year with maxima in spring and autumn[59]. *Centropages hamatus* may be abundant throughout the year[60, 61].

The marine Cladocera belonging to the genera *Evadne* and *Podon* may be abundant in the plankton of some areas such as the Firth of Clyde[142, X], but may be poorly represented in others, e.g. Thames estuary[61, 62]. The mean fertility of parthenogenetic females is a characteristic of the individual species; there is, however, a positive correlation between size and the fertility of the parthenogenetic females of *Evadne nordmanni* and *Podon intermedius*. The fertility of *Evadne nordmanni* undergoes seasonal variations, which are not correlated with the abundance of diatoms[142].

The ctenophore *Pleurobrachia pileus* occurs throughout the year in some estuaries and has a peak of abundance in spring and early summer, it may be succeeded, in summer, by *Beroë cucumis*[42, 60, 115].

Meroplanktonic species show a similar succession to the holoplankton species, for example, in Southampton Water, the larvae of *Balanus* spp. are abundant in February and March just before the *Acartia bifilosa* and *A. clausi* outburst, whereas the larvae of *Elminius modestus* are dominant in the summer[42]. However the succession of the scyphozoan medusae is both interesting and has far reaching consequences upon the holoplankton and the fisheries. In European waters, *Aurelia aurita* is abundant from early June to the end of August, and is succeeded from July to early September by *Chrysaora isosceles* and *Cyanea capillata* (or on the northern North Sea coasts *C. lamarcki*) which in early September and October are succeeded by *Rhizostoma octopus*[42, 60, 115, Y]. In Narragansett Bay, *Aurelia* occurs in June, but curiously here, and in the adjacent Niantic River, *Cyanea capillata* is recorded earlier than in Britain, viz. from April to June, and earlier still in Chesapeake Bay, viz. from December to May[143]. The production of the medusae of hydroids appears to be correlated to the seasonal hydrographic regime[144].

The reduction of *Acartia* numbers, in Southampton Water, in summer, is due to the abundant *Pleurobrachia pileus* and *Aurelia aurita*[42]; by feeding upon young fish *Cyanea* may have an important influence upon fish recruitment[145]. *Aurelia aurita* and *Rhizostoma octopus* are both innocuous to humans, but *Cyanea capillata*, *C. lamarcki* and *Chrysaora isosceles*, known to sting, are a nuisance to fishermen, yatchsmen and swimmers alike; handling of tentacle contaminated gear invariably leads to transfer to the face, by insensitive hands. Nevertheless, despite the lethal nature ascribed to *Cyanea capillata* by Sir Arthur Conan Doyle in Sherlock Holmes adventure of "The Lion's Mane", only discomfort is caused. Apparently the only truly lethal jellyfish are the Cubomedusae, *Chironex fleckeri* and *Chiropsalmus quadrigatus* which live in Northern Australian waters; even the notorious Portuguese man-of-war has not, in fact, been known to kill anyone[146]. Many authors consider these large jellyfish to be allochthonous, being carried into the estuary by hydrographic forces: an explanation which is not entirely convincing for large estuaries like the Solway, true though it may be for smaller ones like the Blackwater. Clearly more work upon these large, easily observed animals, would be profitable; indeed they may make useful plankton indicators.

Plankton Indicators

Investigations by Meek of the distribution of chaetognaths off the Northumberland coast were the genesis of the concept of plankton species acting as indicators of particular water masses and of their movements: here fluctuations in the distribution of *Sagitta elegans* and *S. setosa* depend upon the magnitude of inflows of water from the Atlantic[147]. Later, Russell[148-151] and Fraser[152-155] were prominent in elaborating this concept. Plankton indicators may be used to detect upwellings of deep water as well as lateral movements[155], and areas where fishing may[156], or may not[157-159] be successful. Now, work with plankton indicators has been performed in many parts of the world, much of it with the species used by the earlier workers and including chaetognaths, pteropods, euphausiids and copepods[160-165]. However, a water mass may be characterized by its dinoflagellates[166], diatoms[12, 167], tintinnid protozoa[168], worms[169], or tunicates[170]. In the River Blackwater estuary, a marked increase of *Sagitta setosa* after a prolonged easterly wind may be taken to indicate the presence of increased amounts of water from The Wallet and the open sea[Y]. Alternatively, the presence of epiphytic species, e.g. *Rhabdonema* or *Licmophora*, in the offshore plankton indicate that there has been, or is, an estuarine influence. Other species, e.g. *Striatella interrupta*, indicate the presence of brackish water[167]. The influence of water from the Irish Sea northwards to the Minches can be deduced from the presence of *Sagitta setosa* in these waters (p. 62).

By means of diatoms, it has been demonstrated that Port Hacking, an Australian estuary, is influenced by remarkably distant sources: in 1940, the presence of *Bacteriastrum delicatulum, Climacodium frauenfeldianum* and *Stephanopyxis turris* suggested an incursion of mixed Coral Sea and East Australian water; in 1941 oceanic influences were suggested by *Rhizosolenia castracanei, R. alata, R. styliformis, Climacodium frauenfeldianum* and the dinoflagellate *Goniodoma polyedricum*; in 1942, oceanic influences were present through much of the year, but in July–September estuarine neritic influences were characterized by *Chaetoceros secundum* and *Coscinodiscus granii*, and in December this influence reached a maximum characterized by *Rhizosolenia robusta* with *Coscinodiscus granii*; in 1943, the indicators of the East Australian current and of the Coral Sea were present once more; and in 1944, indicators of the East Australian current were present[12].

During March 1957, a large number of drift bottles and indicators released in the south-east part of the Firth of Clyde suggested a clockwise upper water movement (Fig. 4.6)[171]. Zooplankton samples taken simultaneously conformed to the hydrographic system. In general, the water movements in this period were characterized by a north bound flow from the Irish Sea, through the North Channel to the Atlantic; with some inflow into the Firth of Clyde around Corsewall Point and Loch Ryan, and a loss from the Firth to the North Channel and Atlantic around the Mull of

Kintyre. At the beginning of April this orderly structure became chaotic with a thrust from the Atlantic becoming evident by the 8–11th, at the same time some Clyde water escaped south through the North Channel to the Irish Sea. By the 21–23 April, the thrust to the Firth of Clyde was less marked, but a strong flow to the Irish Sea continued. The influx of Atlantic water was characterized by the medusa *Aglantha* and the pteropod *Limacina retroversa*. The inflow from the Atlantic to the Irish Sea continued to the end of May with an increased thrust into the Firth of Clyde[X]. In this area, then, we have evidence of intermittent water movements, both north and south bound, from planktonic sources[76, 77, X].

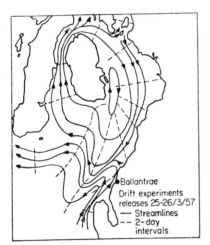 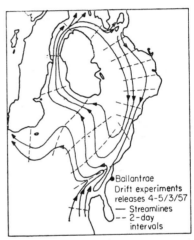

Fig. 4.6. Upper water movements in the Firth of Clyde deduced from drift-bottle and drift-indicator returns March–April 1957 (after Craig[171]).

Although plankton indicators are extremely useful there are some areas in which the plankton is very uniform, e.g. the Black Sea, where the definition of species which can be used may present some difficulty[70].

Although limited, present information indicates that an estuary cannot be considered in isolation. It must be considered in relation to the movements of water into it, often from very distant sources. For example, Port Hacking in New South Wales is about 1500 miles from the Coral Sea; on a lesser scale the Minches are about 280 miles from the Irish Sea. Therefore, distant, or very distant, sources may have significant effects upon individual estuaries. This appreciation of outside forces affecting estuaries will become more evident in Chapter 6. Before concluding however, it is worthwhile recalling the influence of different qualities of sea water upon the development of the larvae of the polychaete *Ophelia* and the sea urchin *Echinus* (Chapter 1). Water of a good quality, produced by the mixing of oceanic and neritic waters, is characterized by the chaetognath, *Sagitta elegans*; the eggs of the cod, haddock, whiting, plaice and long rough dab, and the larvae of the cod, haddock, whiting, plaice, lemon sole, long rough dab, *Gadus esmarkii*, other gadoids and other flatfish all survive better in *S. elegans* water, than when it is absent[172].

References

1. Harvey, H. W. (1945). "Recent Advances in the Chemistry and Physics of Sea Water", University Press, Cambridge.
2. Braarud, T. and Føyn, B. (1958). *Nytt Mag. Bot.* **6**, 47–73.
2a. Caspers, H. (1959). *Estratto Dall'Archivio Di Oceanografia e Limnologia.* **XI**. Suppl. 153–169.
3. Mommaerts, J. P. (1969). *J. mar. biol. Ass. U.K.* **49**, 749–765.
4. Kawamura, T. (1966). *Tech. Pap. Bur. sport Fish. Wildl. Wash.* **1**, 1–37.
5. Marshall, N. and Wheeler, B. M. (1965). *Ecology* **46**, 665–673.
6. Wood, E. J. F. (1964). *Nova Hedwigia.* **8**, 461–527.
7. Margalef, R. (1957). *Investigación pesq.* **6**, 39–52.
8. Davis, C. C. (1950). *Ecology* **31**, 519–531.
9. Braarud, T., Føyn, B., Hasle, G. R. (1958). *Hvalråd. Skr. (Norske Vidensk. Akad.)* **43**, 1–102.
10. Patrick, R. (1967). *In* "Estuaries" (G. H. Lauff, ed.), *Publs Am. Ass. Advmt. Sci.* No. **83**, 311–315.
11. Ketchum, B. H. (1954). *Ecology* **35**, 191–200.
12. Crosby, L. H. and Wood, E. J. F. (1958). *Trans. R. Soc. N.Z.* **85**, 483–530.
13. Lloyd, B. (1925). *J. Ecol.* **13**, 92–129.
14. Maghraby, A. M. el and Perkins, E. J. (1956). *Ann. Mag. nat. Hist.*, Ser. 12, **9**, 561–568.
15. Lloyd, B. (1926). *J. Ecol.* **14**, 92–110.
16. Hendey, N. I. (1954). *J. mar. biol. Ass. U.K.* **33**, 537–560.
17. Aleem, A. A. (1949). "Distribution and Ecology of British Littoral Diatoms", Ph.D. Thesis, University of London.
18. Margalef, R., Saíz, F., Rodríguez-Roda, J., Toll, R. and Vallés, J. M.ª (1952). *Publnes Inst. Biol. apl., Barcelona* **10**, 133–143.
19. Nordli, E. (1953). *Blyttia* **11**, 16–18.
20. Nordli, E. (1957). *Oikos* **8**, 200–265.
21. Braarud, T. (1951). *Physiologia Pl.* **4**, 28–34.
22. Waugh, G. D. (1960). *Proc. malac. Soc., Lond.* **34**, 113–122.
23. Loosanoff, V. L. (1963). *Comml. Fish. Rev.* **25**, 1–11.
24. Loosanoff, V. L. (1958). *In* "XVth International Congress of Zoology". Section 3, Paper 10, 1–3.
25. Walne, P. R. (1963). *J. mar. biol. Ass. U.K.* **43**, 767–784.
26. Walne, P. R. (1966). *Fishery Invest., Lond.* Ser 2. **25** (4), 1–53.
27. Cassie, V. (1959). *Tuatara* **7**, 107–118.
28. Atkins, W. R. G. (1923). *J. mar. biol. Ass. U.K.* **13**, 119–150.
29. Nielsen, E. Steeman (1964). *J. Ecol.* **52**, (*J. Anim. Ecol.* 33), 119–130.
30. Cassie, R. M. and Cassie, V. (1960). *N.Z. Jl. Sci.* **3**, 173–199.

31. Droop, M. R. (1969). *Proc. Challenger Soc.* **4** (1), 17–18.
32. Riley, G. A. (1946). *J. mar. Res.* **6**, 54–73.
33. Rochford, D. J. (1951). *Aust. J. mar. freshwat. Res.* **2**, 1–116.
34. Smidt, E. L. B. (1951). *Meddr. Kommn. Danm. Fisk.-og Havunders. Serie*: *Fiskeri.* **11**, 1–151.
35. Sanders, H. L., Goudsmidt, E. M., Mills, E. L. and Hampson, G. E. (1962). *Limnol. Oceanogr.* **7**, 63–79.
36. Wernstedt, C. (1943). *Vidensk. Meddr. dansk. naturh. Foren.* **106**, K. 241.
37. Smyth, J. C. (1955). *J. Ecol.* **43**, 149–171.
38. Williams, B. R. H., Perkins, E. J. and Hinde, A. (1965). H.M.S.O., U.K.A.E.A., P.G. Report 611 (CC).
39. Grontved, J. (1949). *Meddr. Kommn. Danm. Fisk.-og Havunders, Serie*: *Plankton.* **5** (2).
40. Barlow, J. P. (1958). *J. mar. Res.* **17**, 53–67.
41. Lorenzen, C. J. (1963). *Limnol. Oceanogr.* **8**, 56–62.
42. Raymont, J. E. G. and Carrie, B. G. A. (1964). *Int. Revue ges. Hydrobiol. Hydrogr.* **49**, 185–232.
43. Savage, P. D. V. (1965). *Br. phycol. Bull.* **2**, 551–516.
44. Savage, P. D. V. (1967). *Rep. Challenger Soc.* **3** (19), 41–42.
45. Perkins, E. J., Bailey, M. and Williams, B. R. H. (1964). London: H.M.S.O., U.K.A.E.A., P.G. Report 604 (CC).
46. Cassie, V. (1960). *N.Z. Jl. Sci.* **3**, 137–172.
47. Harvey, H. W. (1947). *J. mar. biol. Ass. U.K.* **26**, 562–579.
48. Harvey, H. W. (1949). *J. mar. biol. Ass. U.K.* **28**, 155–163.
49. Goldberg, E. D. (1952). *Biol. Bull. mar. biol. Lab. Woods Hole.* **102**, 243–248.
50. Ryther, J. H. and Kramer, D. D. (1961). *Ecology* **42**, 444–446.
51. Curl, H. (1962). *Limnol. Oceanogr.* **7**, 422–424.
52. Curl, H. and McLeod, G. C. (1961). *J. mar. Res.* **19**, 70–88.
53. Lewin, J. C. (1954). *J. gen. Physiol.* **37**, 589–599.
54. Bentley, J. A. (1960). *J. mar. biol. Ass. U.K.* **39**, 433–444.
55. Trevallion, A. (1967). *J. mar. biol. Ass. U.K.* **47**, 523–532.
56. Marshall, S. M. (1949). *J. mar. biol. Ass. U.K.* **28**, 45–95.
57. Digby, P. S. B. (1953). *J. Anim. Ecol.* **22**, 289–322.
58. Rogers, H. M. (1940). *J. Fish. Res. Bd. Can.* **5**, 164–171.
59. Haertel, L. and Osterberg, C. (1967). *Ecology* **48**, 459–472.
60. Newell, G. E. (1954). *Ann. Mag. nat. Hist.* Ser. 12, **7**, 321–350.
61. Maghraby, A. M. el (1956). "The Inshore Plankton of the Thames Estuary", Ph.D. Thesis, University of London.
62. Maghraby, A. M. el and Perkins, E. J. (1956). *Ann. Mag. nat. Hist.* Ser. 12, **9**, 481–496.
63. Frolander, H. F. (1964). *J. Wat. Pollut. Control Fed.* **36**, 1037–1048.
64. Tait, R. V. (1968). "Elements of Marine Ecology", Butterworths, London.
65. Whitehouse, J. W. (1968). *In* "Hydrobiological Studies of the River Blackwater in Relation to Bradwell Nuclear Power Station", Central Electricity Generating Board, London.
66. Jefferies, H. P. (1962). *Limnol. Oceanogr.* **7**, 354–364.
67. Jefferies, H. P. (1967). *In* "Estuaries" (G. H. Lauff, ed.), *Publs Am. Ass. Advmt. Sci.* **83**, 500–508.
68. Digby, P. S. B. (1950). *J. mar. biol. Ass. U.K.* **29**, 393–416.

69. Scott, T. (1910). *In* "27th Annual Report, Fishery Board for Scotland", 74–99.
70. Einarsson, H. and Gürtürk, N. (1959). *Et Ve Balik Kurumu Balikcilk Arastirma Merkezi Raporlari, Series Marine Research.* **1** (8), 1–28.
71. Green, J. (1968). "The Biology of Estuarine Animals", Sidgwick and Jackson, London.
72. Bayly, I. A. E. (1964). *Aust. J. mar. freshwat. Res.* **15**, 239–247.
73. Bayly, I. A. E. (1965). *Aust. J. mar. freshwat. Res.* **16**, 315–350.
74. Thomson, J. M. and Dunstan, D. J. (1968). *Crustaceana* Supplement 1, 82–86.
75. Bosch, H. F. and Taylor, W. R. (1968). *Crustaceana* **15**, 161–164.
76. Williamson, D. I. (1956). *J. mar. biol. Ass. U.K.* **35**, 461–466.
77. Vannucci, M. (1956). *Glasg. Nat.* **17**, 243–249.
78. Segerstråle, S. G. (1965). *Commentat. biol.* **28** (7), 1–28.
79. Kuhl, H. and Rheinheimer, G. (1968). *Kieler Meeresforsch.* **24**, 27–37.
80. Motoda, S. (1953). *Mem. Fisheries, Hokkaido University.* **1**, 1–56.
81. Russell, F. S. (1925–34). *J. mar. biol. Ass. U.K.* **13**, 769–809; **14**, 101–159, 387–414, 415–440, 557–608; **15**, 81–103, 429–454, 829–850; **16**, 639–676; **17**, 391–414, 767–784; **19**, 569–584.
82. Russell, F. S. (1927). *Biol. Rev. Biol. Proc., Camb. Phil. Soc.* **2**, 213–262.
83. Johnson, W. H. (1938). *Biol. Bull. mar. biol. Lab. Woods Hole* **75**, 106–118.
84. Hasle, G. R. (1950). *Oikos* **2**, 162–175.
85. Hasle, G. R. (1954). *Nytt Mag. Bot.* **2**, 139–146.
86. Perkins, E. J. (1965). *Trans. J. Proc. Dumfries. Galloway nat. Hist. Antiq. Soc.* Ser. 3, **42**, 6–13.
87. Segerstråle, S. G. (1964). *Oceanogr. Mar. Biol. A. Rev.* **2**, 373–392.
88. Lance, J. (1962). *J. mar. biol. Ass. U.K.* **42**, 131–154.
89. Grindley, J. R. (1964). *Nature, Lond.* **203**, 781–782.
90. Fish, C. J. (1936). *Biol. Bull. mar. biol. Lab. Woods Hole* **70**, 193–216.
91. Ussing, H. H. (1938). *Meddr. Grønland.* **100**, 1–108.
92. Digby, P. S. B. (1961). *J. Anim. Ecol.* **30**, 9–25.
93. Bowman, T. E. (1961). *Chesapeake Sci.* **2**, 206–207.
94. Kinne, O. (1964). *Oceanogr. Mar. Biol. A. Rev.* **2**, 281–339.
95. Hollowday, E. D. (1949). *J. mar. biol. Ass. U.K.* **28**, 239–253.
96. Hedgpeth, J. W. (1967). *In* "Estuaries" (G. H. Lauff, ed.), *Publs Am. Ass. Advmt. Sci.* **83**, 408–419.
97. Lance, J. (1963). *Limnol. Oceanogr.* **8**, 440–449.
98. Pertzova, N. M. and Sakharova, M. I. (1967). *Oceanology* **7**, 1068–1075.
99. McLaren, I. A., Walker, D. A. and Corkett, C. J. (1968). *Can. J. Zool.* **46**, 1267–1269.
100. Corkett, C. J. and Urry, D. L. (1968). *J. mar. biol. Ass. U.K.* **48**, 97–105.
101. Nicol, J. A. C. (1967). "The Biology of Marine Animals", Pitman, London.
102. Bull, H. O. (1931). *Nature, Lond.* **127**, 406.
103. Thorson, G. (1946). *Meddr. Kommn. Danm. Fisk-og. Havunders, Serie: Plankton,* **4**, 1–523.
104. Lebour, M. V. (1947). *J. mar. biol. Ass. U.K.* **26**, 527–547.
105. Wallace, D. (1940). *Trans. Am. Fish. Soc.* **70**, 475–482.
106. Haven, D. S. (1957). *Ecology* **38**, 88–97.
107. Pritchard, D. W. (1951). *Trans. 16th N. Am. Wildl. Conf.,* 368–376.
108. Pritchard, D. W. (1952). *Proc. Gulf Caribb. Fish. Inst.* 5th Annual Session, 1–10.

109. Bausfield, E. L. (1952). *In* "Proceedings of the American Society of Limnology and Oceanography", Meeting at Ithaca, New York, 8–10 September 1952, pp. 420–436.
110. Burt, W. V. and McAlister, W. B. (1959). *Res. Briefs Fish Commn. Ore.* **7**, 14–27.
111. Evans, F. and Newell, G. E. (1957). *Ann. Mag. nat. Hist.* Ser. 12, **10**, 161–173.
112. Nelson-Smith, A. (1964). *Rep. Challenger Soc.* **3** (16).
113. Hartley, P. H. T. (1940). *J. mar. biol. Ass. U.K.* **24**, 1–68.
114. Perkins, E. J. and Williams, B. R. H. (1962). *Trans. J. Proc. Dumfries. Galloway nat. Hist. Antiq. Soc.* Ser. 3, **40**, 77–78.
115. Perkins, E. J. (1968–71). *Trans. J. Proc. Dumfries Galloway nat. Hist. Antiq. Soc.* Ser. 3, **45**, 15–43; **46**, 1–26; **48**, 12–68.
116. Dales, R. P. (1950). *J. mar. biol. Ass. U.K.* **29**, 321–359.
117. Lyster, I. H. J. (1965). *J. Anim. Ecol.* **34**, 517–527.
118. Bhatnagar, K. M. and Crisp, D. J. (1965). *J. Anim. Ecol.* **34**, 419–428.
119. Crisp, D. J. and Costlow, J. D. (1963). *Oikos* **14**, 22–34.
120. Barnes, H. (1953). *J. Anim. Ecol.* **22**, 328–330.
121. Barnes, H. (1953). *J. mar. biol. Ass. U.K.* **32**, 297–304.
122. Costlow, J. D., Jr. and Bookhout, C. G. (1959). *Biol. Bull. mar. biol. Lab. Woods Hole* **116**, 373–396.
123. Costlow, J. D., Jr., Bookhout, C. G. and Monroe, R. (1960). *Biol. Bull. mar. biol. Lab. Woods Hole* **118**, 183–202.
124. Costlow, J. D., Jr. and Bookhout, C. G. (1962). *In* "Biological Problems in Water Pollution, Third Seminar". U.S. Dept. of Health, Education and Welfare, Public Health Service Publication No. 999-WP-25: 77–86.
125. Costlow, J. D., Jr. and Bookhout, C. G. (1962). *J. Elisha Mitchell scient. Soc.* **78**, 113–125.
126. Costlow, J. D., Jr., Bookhout, C. G. and Monroe, R. J. (1966). *Physiol. Zool.* **39**, 81–100.
127. Costlow, J. D., Jr. and Bookhout, C. G. (1966). *J. Elisha Mitchell scient. Soc.* **82**, 160–171.
128. Millar, R. H. and Scott, J. M. (1967). *J. mar. biol. Ass. U.K.* **47**, 475–484.
129. Loosanoff, V. L. (1961). *Gulf Caribb. Fish. Inst.* 14th Annual Session, pp. 80–95.
130. Berrill, N. J. (1947). *J. mar. biol. Ass. U.K.* **26**, 616–625.
131. Lagler, K. F., Bardach, J. E. and Miller, R. R. (1962). "Icthyology", Wiley, New York and London.
132. Holliday, F. G. T. and Jones, M. P. (1967). *J. mar. biol. Ass. U.K.* **47**, 39–48.
133. Holliday, F. G. T. and Blaxter, J. H. S. (1960). *J. mar. biol. Ass. U.K.* **39**, 591–603.
134. Holliday, F. G. T. and Blaxter, J. H. S. (1961). *J. mar. biol. Ass. U.K.* **41**, 37–48.
135. Blaxter, J. H. S. (1960). *J. mar. biol. Ass. U.K.* **39**, 605–608.
136. Forrester, C. R. and Alderdice, D. F. (1965). *J. Fish. Res. Bd. Can.* **23**, 319–340.
137. Alderdice, D. F. and Forrester, C. R. (1968). *J. Fish. Res. Bd. Can.* **25**, 495–521.
138. Perkins, E. J. (1958). "The Microfauna of the Shore", Ph.D. Thesis. University of London.

139. Haertel, L. (1969). *In* "Ecological Studies of Radio-activity in the Columbia River Estuary and Adjacent Pacific Ocean", Progress Report 69-9; RL01750-54, pp. 128–129. Dept. of Oceanography, Oregon State University, Corvallis.

140. Hedgpeth, J. W. (1966). *Am. Fish. Soc. Spec. Publ.* No. 3, 3–11.

141. Zenkevitch, L. A. (1963). "Biology of the Seas of the U.S.S.R.", Allen and Unwin, London.

142. Cheng, C. (1947). *J. mar. biol. Ass. U.K.* **26**, 551–561.

143. Bowman, T. E., Meyers, C. D. and Hicks, S. D. (1963). *Chesapeake Sci.* **4**, 141–146.

144. Edwards, C. (1965). *J. mar. biol. Ass. U.K.* **45**, 443–468.

145. Fraser, J. H. (1969). *Proc. Challenger Soc.* **4** (1): 34–35.

146. Idyll, C. P. (1971). "Abyss", 2nd edn. Crowell, New York.

147. Meek, A. (1928). *Proc. zool. Soc., Lond.* 1928, 743–776.

148. Russell, F. S. (1935). *J. mar. biol. Ass. U.K.* **20**, 309–332.

149. Russell, F. S. (1936). *Cons. perm. int. Explor. Mer, Rapp. et Proc.-Verb.* **100** (3), 7–10.

150. Russell, F. S. (1939). *J. Cons. perm. int. Explor. Mer* **14**, 171.

151. Russell, F. S. (1952). *Cons. perm. int. Explor. Mer, Rapp. et Proc.-Verb.* **131**, 28.

152. Fraser, J. H. (1937). *J. Cons. perm. int. Explor. Mer* **12** 311–320.

153. Fraser, J. H. (1952). *Mar. Res.* 1952. No. 2, 1–52.

154. Fraser, J. H. (1952). *Cons. perm. int. Explor. Mer, Rapp. et Proc-Verb.* **131**, 38–43.

155. Fraser, J. H. (1955). *Mar. Res.* 1955, No. 1, 12 pp.

156. Hardy, A. C., Henderson, G. T. D., Lucas, C. E. and Fraser, J. H. (1936). *J. mar. biol. Ass. U.K.* **21**, 147–291.

157. Stephen, G. A. (1949). *J. mar. biol. Ass. U.K.* **28**, 555–581.

158. Fraser, J. H. (1962). *Cons. perm. int. Explor. Mer, Rapp. et Proc.-Verb.* **153**, 121–123.

159. Fraser, J. H. (1963). *Cons. perm. int. Explor. Mer, Rapp. et Proc.-Verb.* **154**, 175–178.

160. Haida, T. S. (1957). *U.S. Fish and Wildl. Serv., Spec. Sci. Rept. Fish.* **215**, 1–13.

161. Furnestin, M-L. (1957). *Année Biol.* **33**, 345–366.

162. Biere, R. (1959). *Limnol. Oceanogr.* **4**, 1–28.

163. Lebrasseur, R. J. (1959). *J. Fish. Res. Bd. Can.* **16**, 795–806.

164. Abramova, V. D. (1959). *U.S. Fish and Wildl. Serv., Spec. Sci. Rept. Fish.* **327**, 77–103.

165. Bary, B. M. (1959). *Pacif. Sci.* **13**, 14–54.

166. Wood, E. J. F. (1954). *Aust. J. mar. Freshwat. Res.* **5**, 171–351.

167. Crosby, L. H. and Wood, E. J. F. (1959). *Trans. R. Soc. N.Z.* **86**, 1–58.

168. Hada, Y. (1957). *Inf. Bull. Planktol. Japan* **5**, 10–12. Hakodate, 1957.

169. McGowan, J. A. (1960). *Deep Sea Res.* **6**, 125–139.

170. Ganapati, P. N. and Bhavanarayana, P. V. (1958). *Curr. Sci.* **26**, 347–348.

171. Craig, R. E. (1957). *Annls biol., Copenh.* 1957, 62–63.

172. Fraser, J. H. (1962). *Proc. R. Soc.*, Series A, **265**, 335–341.

X.   Perkins, E. J. Previously unpublished work.

Y.   Perkins, E. J., Lewis, B. G., Williams, B. R. H., Howard, T. E., Davis, D. S., Burdett, J. R. and Carrie, B. G. A. Unpublished work on the C.E.G.B. survey of the River Blackwater estuary, Essex.

5

Shores and Sediments

Although rocky shores are found in fiordic and coastal plain estuaries, they are not very common in the latter where sedimentary shores predominate. Before considering the properties of these sediments, which influence the biota, it is essential to define what is meant by shoreline and beach, and thus by offshore sediments.

Coastlines may be classified according to their formative agencies[1-3]: terrestrial erosion structures, subsequently drowned by deglaciation or downwarping, or construction from the terrestrial deposits of rivers, glaciers and wind, lava flows and volcanic explosions, and diastrophic activity give rise to *primary* or *youthful coasts*. Conversely, *secondary* or *mature coasts* arise when the formative processes are marine.

The beach may be defined as a zone extending from the upper, landward limit of effective wave action to the low-water mark, but this definition is too narrow. More properly, it consists of the whole of that area affected by normal wave action; thus, it may be considered to extend from a depth of 30 ft below the water level at the lowest tide (E.L.W.M.S.T.) to the edge of the permanent coast[4]. It will be evident, that this practical limit includes the "seabed" of many coastal plain estuaries. Accordingly, it represents a real transition zone, between the land and the sea, and one which is particularly subject to the action of both waves and tides. Composed of unconsolidated, sedimentary materials, subject to erosion and deposition, beaches are characteristically in an unstable equilibrium[3]: an equilibrium which may be readily upset by the introduction of new factors, e.g. engineering works.

The extent of a beach depends upon the tidal range, i.e. small narrow beaches are concomitant with small tides and large, wide beaches with big tides, but within these limits, the nature of the individual beach depends upon the sources from which its sediments are derived, and the interplay of sediment erosion, transportation and deposition, which the tidal cycle modulates. That part of the beach which is exposed by the receding tide, i.e. the *intertidal zone*, is the *shore* as it is generally understood. Evidently, because the tidal range is variable, (Chapter 3), the manner in which the shore is covered or exposed is variable too; in some estuaries, particularly at the landward end, vast areas of sand and mud flats may be

exposed, e.g. Solway Firth, Morecambe Bay, Thames estuary. Consequently, several zones may be defined in this intertidal area, (Fig. 5.1)[5],
and are colonized by characteristic communities of plants and animals.
However, the position occupied by individual species may be controlled
by other factors, for example, the beautiful top-shell *Calliostoma zizyphinum*
or the velvet swimming crab, *Portunus puber* may be found intertidally in
South Devon, but only below the low-water mark in the Solway Firth[6]
(see Chapter 6). The gross consequences of tidal movements upon the
intertidal zone are summarized in Fig. 5.2[7].

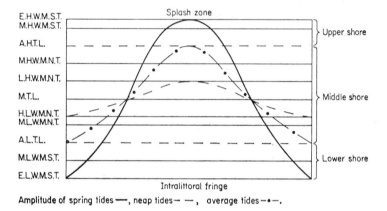

Fig. 5.1. The standard tidal levels of the shore used to describe the zonation of
littoral organisms (after Barrett and Yonge[163]).

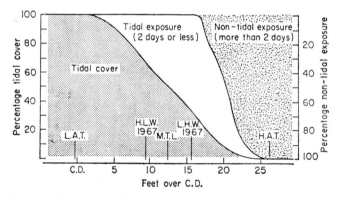

Fig. 5.2. Relationship between height and time covered by sea, or exposed to air.
Based on predictions for Milford Haven, 1967, given by the Admiralty Hydrographic Department (1966). H.L.W. = highest low water (neaps). L.A.T. = lowest
astronomical tide. L.H.W. = lowest high water (neaps) (after Dalby[7]).

Physical Properties of Estuarine Deposits

In the British Isles, early work on the particle structure of marine sediments was influenced by surveys of terrestrial soils and produced results which were not applicable to the marine environment. Indeed, it was not until 1956 that the advantages of describing the relative proportions of the different sizes of soil particles, i.e. *soil grades*, making up a soil, by mechanical analysis employing sieves based on the Udden-Wentworth scale, Table 5.1, were realized in Britain[8], much later indeed than the same conclusion had been reached by American workers. Each successive sieve

Table 5.1. Udden-Wentworth soil size classification

Name	Grade limits (diameter in mm)
Boulders	> 256
Cobbles	256–64
Pebbles	64–4
Granules	4–2
Very coarse sand	2–1
Coarse sand	1–0·5
Medium sand	0·5–0·25
Fine sand	0·25–0·125
Very fine sand	0·125–0·0625
Silt	0·0625–0·0039
Clay	≤ 0·0039

mesh is approximately one half the size of its predecessor, and the results of the mechanical analysis can be interpreted on the phi scale, where $\phi = -\log_2$ of the particle diameter in millimetres. Consequently, the symmetrical distribution curves which result can be analysed statistically[9–11]. In some circumstances, mechanical sieving may be replaced by use of settlement tubes[12], and Coulter counters[13]: the latter may be used also to investigate the size distribution of a wide variety of living organisms, e.g. algal populations, and invertebrate eggs and larvae. A special series of sieves, in which, in any one grade, the largest particle is 1·5 times the diameter of the smallest, may be used when it is necessary to define small differences in the particle size of fine deposits[14, 15]; this system is not generally used. French workers use a series of sieves in which the mesh of each grade is 1·25 times that of the next sieve in the series, results are analysed by means of histograms and cumulative curves: the number of sediment particles (n) retained by each sieve may be obtained thus:

$$\log n = 1\cdot60683 - 3 \log d_2 + \log p$$

where d_2 is the average mesh size, and p the weight of sediment retained by the sieve[16, 17].

Although mechanical analysis of the soil grade, including organic and calcareous matter, is important, analysis of organic content and other constituents may be necessary[18].

A first consequence of soil grade and the mixture of grades which constitute individual soils is changes in the property of *porosity*. If a soil consisted entirely of closely packed, uniform, impermeable spheres then the volume of the interstitial space, or porosity, is 26% of the total, whatever the size of the spheres[19]. However, estuarine sediments do not consist of uniform spheres, but of particles which are varied in both shape and size, and influence the porosity of the individual soil. The factors which affect the porosity are complex and depend upon grain size, absence of uniformity in grain size, proportions of the different grain sizes, grain shapes, method of deposition and the subsequent processes of compaction solidification[20-22]. Muds with a high silt and clay content have a low porosity, because the larger spaces are filled by these small particles; freshly deposited sand and clay may have a greater porosity initially, but later become compacted and of a low porosity[20, 23, 24].

Determinations of the porosity may depend upon measuring the relative amounts of water and air present in a soil[17, 25]. However, studies of surface areas and pore size distributions by gas adsorption methods have shown that the surface areas of sediments are greater than those calculated from particle size, which do not reveal irregularities such as micropores[26, 27].

The permeability, P, of a soil is related to the porosity, thus:[23, 28]

$$P\alpha\frac{n^3}{(1-n)^2}$$

where n is the porosity.

However, this relationship holds only when all other factors, especially grain size, are constant, and the permeability of soils is studied best by use of Darcy's Law, thus:

$$v=p\phi$$

where v is the velocity of the moving water, p is referred to variously as the permeability, the Darcy coefficient of permeability, the coefficient of permeability and the transmission constant and ϕ is the hydraulic gradient.

In normal use where a varying hydraulic head is used, the coefficient of permeability, p, is most readily obtained by use of the following equation:

$$p=\frac{al}{AT}\log_e\frac{h_1}{h_2}\ \text{cm/sec}$$

where a is the cross sectional area of the water reservoir, A is the cross-section and l the length of the porous material under consideration, h_1, h_2 the hydraulic heads at the start and after time T respectively[29].

In general, the permeability of a soil is proportional to its grade[14, 24, 30, 31], (at Whitstable, Kent, the coefficients of permeability of fine sand varied between 1.2×10^{-2} and 9.6×10^{-4} cm/sec, and of London Clay varied between 9.5×10^{-7} and 1.38×10^{-8} cm/sec), and the steepness of the beach[30]. A sand which has just settled is more permeable than one which has been settled for some time and has compacted; consequently the coefficient of permeability is inversely proportional to the time since settlement occurred[24] (the coefficient of permeability of fine sand from Whitstable ranged from 7.0×10^{-3} to 1.4×10^{-3} cm/sec when first settled and then compacted, respectively). Entrapped air may influence the permeability, also[32].

Because soil porosity and permeability influence the amount of water present in a soil, these factors are also related to *soil hardness*, which is dependent upon the amount, density and viscosity of the interstitial water[33]. Depending upon the amount of water present a soil may be *thixotropic* or *dilatant*. A thixotropic soil is one which exhibits decreased resistance with increased rate of shear, whereas a dilatant soil offers increased resistance with an increased rate of shear: the best known manifestation of the latter is the whitening of sand around the foot of a person walking on a shore. A *quicksand* is the most extreme case of thixotropy, and as civil engineers well know may be induced in any granular material subject to seepage forces[34, 35]. A mass of sand may be very stable and hard, or become quicksand depending upon the direction of flow. This may be evident in certain beaches, especially those in bays, where tidal conditions combined with the effect of underlying impermeable strata provide conditions for quicksand during rising tides[31, 35].

The ratio of water to solids under fully saturated conditions is defined by the *void ratio*[34], e, thus

$$e = \frac{\text{Volume of water}}{\text{Volume of solids}}$$

and is related to the porosity[34] by

$$\text{Porosity} = \frac{e}{1+e}.$$

The *velocity of percolation* or *seepage velocity* of water through a soil is defined by

$$v_s = \left(\frac{1+e}{e}\right) pi$$

where V_s=seepage velocity, e=void ratio, p=coefficient of permeability and i=hydraulic gradient[34].

When the pressure due to the rising flow of water in sand is sufficient to balance the downward force due to gravity, the condition of the sand is unstable, i.e. a quicksand. The hydraulic gradient at which instability commences, the *critical hydraulic gradient* (ci) is related to the void ratio, e, by the expression.

$$ci = \frac{G_s - 1}{1 + e}$$

where G_s is the specific gravity of the soil[34]. Clearly, a quicksand may develop in a granular material of any grain size, although they more usually occur in fine sand, under the appropriate conditions[35]. It is not a function of the presence of silt/clay particles[24, 33], as some early authors suggested[36].

The critical hydraulic gradient normally approximates to 1, and exactly equals 1 when $G_s = 2\cdot65$ and $e = 0\cdot65$, which are fairly average values[34].

On the shore exposed by the retreating tide the soil hardness may be measured by means of the Chapman Penetrometer[33], while on water covered beaches the Carruthers Penetrometer[30] may be used. The former gives consistent results although it may be criticized because of its small size which is an advantage where access is on foot; its greatest disadvantage is restriction of use to periods of tidal exposure. Nevertheless, its use has provided a valuable insight into conditions of life on the shore.

Perhaps the most evident feature of any sandy shore is the ripple system which orientates itself at right angles to the direction of the waves. On the shore at Whitstable, Kent, which has a low gradient (1/600) this system, aided by the low permeability of the sand and the even lower permeability of the underlying London Clay, retains a considerable amount of surface water during the exposure period. On this shore the ridges are always softer than the troughs[31]. This is also true at sites in the Solway Firth— Rockliffe and Auchencairn Bay; Bay of Luce—Philip and Mary; Firth of Clyde—Ardmore and Lochgoilhead; and Loch Ewe, Wester Ross— Firemore Bay, where a sufficient depth of water is retained by the system. At Firemore Bay, where ripples had a height of 3 cm, those which were well drained were harder than the troughs; at Findhorn, Moray Firth, there was no significant difference between the two[X]. At Whitstable, the shore became softer during the autumn, and hardened during the summer[31].

Many shores soften with the advancing tide[35]. The notorious quicksands of the Solway Firth and the migration of buoys on the Goodwin Sands are a result of this phenomenon[37]. At Whitstable, the sand became markedly softer as the incoming tide approached, signalling its presence up to 150 ft away, became very much softer as the tide edge reached the measuring station, and then hardened again[31]. This phenomena has been observed in the Solway Firth at Rockcliffe, the Firth of Clyde at Ardmore and Lochgoilhead, and at Firemore Bay, Loch Ewe[X]. Apart from the implications of this phenomenon with respect to the tidal rhythm of littoral diatoms and dinoflagellates (p. 203), there are some consequences with respect to interstitial salinity and water movements which would repay more detailed investigation (see p. 112).

The hardness of the soil may influence its residents. The burrowing of the lugworm, *Arenicola*, is facilitated by thixotropic soils and the time taken to burrow depends upon soil hardness[33, 38]: at Ardmore and Firemore Bay, *Arenicola* density and soil hardness are inversely related, although results elsewhere are inconclusive[X]. The converse is also true. The more stable soil associated with *Zostera* grass patches is always harder than the surrounding soil. Soil contiguous to the head shaft and faecal tumulus of *Arenicola* is always softer than the circumadjacent soil: that of the head shaft being the softer. The soil surrounding the siphon shaft of *Scrobicularia plana*, in Auchencairn Bay, is always harder than the circumadjacent soil. Measurements made upon patches of the alga *Merismopedia* gave inconclusive results[X].

The *capillarity* or *capillary lift* of a soil is related to porosity and permeability. It is essentially a surface tension phenomenon; consequently, water will rise higher in a column of fine dry soil than in a coarser soil, when the lower end of each is placed in water: in some sandy beaches, the capillarity will raise water 6–10 cm above the ground water level[19, 22, 39–41]. Evaporation of water from the soil surface will promote a continued upward movement of water in the soil[40, 41].

Fundamentally, the passage of water through a soil, i.e. the degree of drainage, is dependent upon porosity, permeability, soil compactness and capillarity. However, the amount of *interstitial* or *pore water* present is affected by many additional factors. The particle composition of a soil is very variable, both horizontally and vertically[41, 43–46]. In non-tidal beaches wave action makes a significant contribution to the pore water[39, 41]; ground water from the land is important also[41]. Although wave energy is dissipated by this means in both tidal and non-tidal beaches, recruitment of pore water by wave action is relatively unimportant in the former[39, 41]. Fluctuations in the water table follow a tidal pattern[30], and the circulation of water may be very intense[41, 47–49]. G. P. Wells noted in a Special Lecture read to the Challenger Society on 16 January 1969, that due to selection of fine sediments for food, the lugworm, *Arenicola*, produces a coarsening of soil at depth in the shore, and clearly has a pronounced effect upon water circulation within the soil. On shores with a gradient, such as that at Whitstable, Kent, drainage may be very slow and frequently surface water is held by the sand ripple system for the whole of the exposure period. In such shores, the black sulphide layer (p. 119) is well developed and indicates that the movement of water through the soil is limited. Nevertheless, the fluctuations of soil hardness (p. 110) indicate that some tidal movement of interstitial water does occur, but that it represents a movement of anaerobic water within the soil which undergoes little exchange with the well aerated water above.

Other physical factors, viz. radiation and temperature, which influence shore life, are not intrinsic properties of the soil. Like the sea, the soil surface reflects light and has a characteristic albedo. A dry sand has twice the reflectance of a wet one $(D:W::18\% :9\%)$[50], and shore soils, like the sea, have a high albedo at low solar altitude, declining to a lower, more constant value at high solar altitude[51].

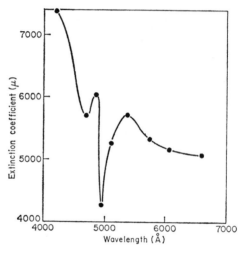

Fig. 5.3. The extinction coefficient of light of different wavelengths in 0·7 mm thick layer of soil from the River Eden estuary, Fife (after Perkins[51]).

Although first demonstrated in 1927[52], there is an unfortunate misconception that light does not penetrate soil; this phenomenon is described in a number of works[51, 53–57]. The concept of extinction coefficient is applicable to soils and sea water alike and in neither is all light equally penetrative: in soil from the River Eden estuary, Fife, a minimum of absorption occurs at 4950 Å (Fig. 5.3)[51]. Larvae of the lugworm, *Arenicola*, which live in shore sediments, are sensitive to light of approximately this wavelength[58]. For maximum photosynthetic activity, sediment dwelling-littoral diatoms require only 12 g/cal/cm²/hr; on a cloudless noon in midsummer, the compensation depth is probably greater than 3·0 m, while at 2·0 mm depth a photosynthetic efficiency of $\geqslant 90\%$ is attained[56].

Temperature

The temperature of a sandy shore is determined by insolation, evaporation, wind, rain, tidal inundation and the amount of pore water (the specific heat of sand ranges from 1/10 to 1/3 cal/g depending upon the amount of water present[19, 24, 39-41, 59]). There is a gradient of temperature across the shore with maximum and minimal fluctuations occurring near the H.W.M. and L.W.M., respectively[63]; the upper shore may be very hot in summer and very cold in winter[41, 45, 60, 61], and in sandy shores is a function of the decreasing amounts of interstitial water[62, 63]. At the time of the neap tides, organisms which dwell in the *Fucus spiralis* and *Pelvetia canaliculata* zones on rocky shores, and in the *Spartina townsendii* and *Salicornia* spp. zones, and above, on sandy shores, live in a near terrestrial temperature regime.

Marked vertical temperature gradients develop within the surface 10 cm[19, 24, 44, 45, 59, 63, 64], but at 12 cm depth the temperature is that of the sea[17]: during the day the temperature decreases with depth[19, 24, 44, 45, 59, 63, 64], but the gradient is reversed at night[45, 63]. The structure of the vertical temperature gradients depends upon the time at which the shore is exposed[19, 24, 59], and on the meteorological conditions prevailing[24, 59]. Small amounts of low and medium cloud, i.e. ≤ 4/10 have little effect, in amounts ≥ 4/10 low and medium cloud have a significant effect; high cloud, even in large amounts has little effect[24, 59]. During night exposures, the shore soil temperatures may be expected to fall, but can be maintained by advective processes[59]. On the land an increase in cloud at night is accompanied by a rise in temperature[65]: this has not been observed on shores, but may be expected. At any one station, the temperature gradient structure is modified by that of the inundating tide edge (p. 34), and in an early morning exposure by receding water cooled by the adjacent land at night[21, 43]; during the period of coverage the shore assumes the temperature of the main body of sea water[47]. Marked temperature gradients develop in the air above the soil: at Whitstable, Kent, soil and air temperatures at 1 ft above it were rarely equal, discrepancies ranged from − 1·7 to + 5·5°C; the former tended to occur in the autumn, and the latter in spring and early autumn[24, 59].

In temperate regions, the temperature is subject to substantial annual variations in both tidal[17, 24, 41, 47, 59, 66] and non-tidal shores[41, 63]. At 6 cm depth at Authie dans la Manche the temperature ranges from 4 to 23·5°C[17]. On the Swedish coast, sand may be frozen to 60–70 cm depth in the coldest months, although the insulating surface layers prevents the temperature of layers, deeper than 2 cm depth, from cooling much below 0°C[41, 63, 67]; here the greatest diurnal range occurs in May, when the warm, insolated surface layers overlay cold deeper layers, this disparity decreases towards the autumn, when with the onset of frost cooling the layers, it increases once more[41, 63].

The Chemical Properties of Estuarine Deposits

Salinity

Fundamentally, the interstitial salinity of shore deposits represents an equilibrium between that of sea water at the time of coverage, and fresh water seeping out from the land[39, 41, 63], Fig. 5.4, but this basic system is subject to modification by many factors. In general, there is a horizontal salinity gradient from L.W.M. to H.W.M., which is most marked in non-tidal shores[41, 44, 45, 63, 68-71]; in tidal shores with a steep gradient the circulation of interstitial water may be rapid, but evidence would suggest that this is not true in shores with a low gradient which retains a layer of surface water during the exposure period (p. 110). Like a small scale replica of the main estuarine water system, the progression of tides through the spring, neap, spring cycle must influence the distribution of the horizontal salinity gradient. Certainly, the very sandy soils of the *machair* of the Scottish west coast experience an intrusion of saline water at the time of the spring tides, which passes beneath dwellings and grass alike; on the other hand, the "internal tide" experienced in soils with over-lying surface water throughout the exposure period, and a well-developed black-sulphide layer (p. 119) experience a very different regime. It is in these soils too, that the infauna, e.g. *Arenicola*, which irrigate their burrows, must play a significant role in maintaining the interstitial salinity at approximately that of the inundating water, unlike the condition of soils from situations which are much affected by the ground water outflow[17, 39, 41, 60, 63, 72, 73]. Early studies showed that at a depth of 6 in. below a freshwater outflow, the interstitial salinities remain high[74-76]. Conversely, interstitial water of a low salinity is relatively unaffected by surface inundation with more saline water[77, 78]. In the Blyth estuary, salinity variations in tidal waters cause little change of interstitial salinity in a 12 hr cycle, although seasonal changes are reflected[73]. Similar seasonal changes have been observed in the Tamar[77], but on the whole the relationship between interstitial salinity and that of the open water is poorly understood. However, in the Tamar it is possible to define three major sections of the estuary thus[79]:

1. A lower region, "marine dominated", characterized by wide flats, relatively high salinities, and interstitial salinities little affected by floods or intertidal elevation.

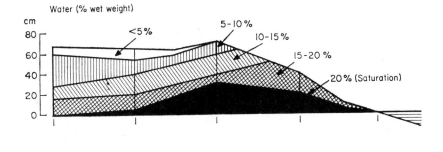

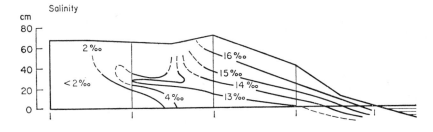

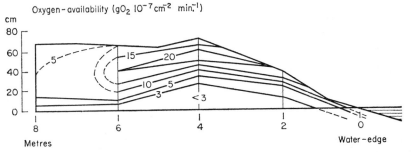

Fig. 5.4. The vertical and horizontal distribution of pore water, salinity and oxygen availability (measured as diffusion rate) on a tideless beach in the Øresund (after McIntyre[162]).

2. A middle region, with intermediate salinities, and interstitial salinities markedly affected by floods and tidal level.

3. Upper reaches, with low salinities at all seasons, absence of flats, interstitial salinity little affected by intertidal level or season.

These considerations apply primarily to the macrobenthos which lives deep within the soil. However, a very important constituent of shore life, particularly in estuaries, is the microbenthos which lives in the greatest numbers within 1 cm of the surface, although many nematodes extend below this. Clearly, the scale of interstitial salinity changes will be different and much greater, on the whole, than in the deeper layers. In this case too, changes in the salinity of the sheet of water remaining on the surface of the soil during the exposure period will be of great importance to those forms which emerge during the exposure period, or are to be found within the surface 2 or 3 mm, e.g. diatoms, dinoflagellates, ostracods, and harpacticoid copepods.

The salinity of these superficial layers is influenced by rainfall, evaporation, and wind[24], but changes are greatest around H.W.M.[39]. Where a shore is covered throughout the exposure period by a layer of water, the wind may move water across the shore, thus bringing about changes in salinity. In the overlying layer of water, rainfall may bring about rapid changes of salinity. At Whitstable in Kent, near continuous rain reduced the salinity of this water from 31‰ to 24·3‰ in 55 min. It is possible to give some definition of the likelihood of the occurrence of such events. A very simple analysis which considered the probability of rain falling in a given exposure period, p_1, and of the probability that when rain is falling a shore exposed at a particular point will be affected, p_2, is given in Table 5.2[80]; this simple analysis takes no account of intensity nor of the spring/neap, equinox/solstice tidal cycle, and its amplification would be very valuable. Rainfall probably has more influence upon tropical shores than temperature[60, 64].

Table 5.2. Probability of rain falling on given points of the shore at Whitstable, Kent (after Perkins[80])

Duration of exposure (hours)	P_1	P_2
1	0·083	0·23
6	0·21	0·57
11	0·34	0·97

Whenever the humidity falls below 100%, evaporation occurs and is enhanced by the wind. At Whitstable, in September 1955, evaporation fell from 0·51 ml/min./1000 cm² at 18·7–19·6°C, wind force 4, to 0·16 ml/min./1000 cm² at 18·7–19·6°C, wind force 1, and then rose to 0·79 ml/min./1000 cm² at 19·6–20·2°C, wind force 4[X].

Such changes in the salinity of the overlying water may be expected to affect the interstitial salinity, in particular of the superficial layers. However, it has been shown that under a maximum gradient in fine sand the rate of diffusion of sodium chloride is 0·5 cm/day[81]. During the exposure period and away from the burrows of the macrofauna, changes due to diffusion are probably slight, and much less than those due to soil movements in the period of coverage.

When a soil is covered by a thin layer of water only, evaporation takes place rapidly until the surface layer has gone, and then much more slowly from the interstices of the soil[X]. The rate of evaporation is proportional to the soil grade[14, 19, X], and a decrease occurs earlier in fine compared to coarse sediments: a coarse sand, mean diameter 0·9 mm retains little water, whereas a fine sand, mean diameter 0·15 mm retains about 14% of its water against evaporation[14]. In this case, as in the case of the layer of overlying water, the temperature and wind speed effect the rate of evaporation. In the Blyth estuary, changes in the salinity of the interstitial water of the surface layers of 1·15‰/hr and 4·5‰/hr occurred in December and March, and June and September respectively: in the last two months the interstitial salinity was raised much above that of normal sea water, to 49·1‰ during the short exposure periods studied[73]. In a two month study, September–October, of salinity in a *Salicornia-Spartina* marsh at Mission Bay, San Diego, California, the water retained on the marsh had a higher salinity than that of the bay (*ca.* 34‰) for 75% of the time, exceeding 40‰ for 37% of the time, exceeding 45‰ for 10% of the time and had a recorded maximum value of 50‰[82, 83]. The movement of such water off a marsh can produce "slugs" of high salinity water in the main body of an estuary[82, 83], and produced a salinity of 39‰ in the River Blackwater in 1960.[X] Because the specific heat of sea water is inversely proportion to the salinity (Chapter 1), changes in salinity which result from evaporation may be expected to influence the rate at which evaporation occurs.

Waves losing energy as they break on a shore induce a circulation through the soil, especially in storm conditions[48, 49], but under normal conditions their influence may be small[41, 63].

Oxygen

In general, marine beaches have a low oxygen content, although even in the blackened sulphide layer oxygen is rarely absent (Table 5.3)[45, 47, 60, 66, 84–87]. In contrast to the relatively well oxygenated interstitial water of the more steeply sloping beaches of coarser sand, the poorly drained beaches of low gradient, composed of a high percentage of fine sand, are poorly oxygenated. The oxygen content of a shore sediment may be directly correlated with the percentage content of the 250–125 μm grade present[88], indeed more than 30% of fine sand induces a marked reduction in the oxygen content of the interstitial water[22, 43, 45, 85, 88, 89], which can only rise above 20% of the air saturation level when the content of fines falls below 10%[88].

In part, the oxygenation of a beach depends upon the turbulence of

Table 5.3. The oxygen content of the interstitial water at four sites at
Whitstable, Kent (after Brafield[85])

Station	Average O_2 concentration (ml/litre)	Percentage fine sand	Presence of black layer	Presence of surface water	Flat or sloping
A	0·26	95·50	+	Persisted	Flat
D	0·31	50·75	+	Persisted	Flat
C	0·62	23·75	−	Persisted	Flat
B	3·93	4·00	−	None	Sloping

the water above the beach[39, 41, 43, 44, 85, 90, 91], therefore the *oxygen availability* is increased by the individual wave and by an increase in the wave frequency[90]. Rain also contributes to the oxygenation of shore soils[53, 60, 92]. *Oxygen availability* is the most appropriate expression defining the oxygenation of the substratum[41, 93–96]: oxygen availability= oxygen diffusion rate = diffusion + flow[41, 89, 90]. Clearly this concept is concerned with oxygen content, temperature, the circulation of water within a soil and influence of soil composition upon this circulation, i.e. by its influence upon porosity, permeability and percentage air volume[41, 45, 90], the soil grade structure influences the aeration and availability of oxygen in a soil[20, 23, 41, 45, 85, 88, 90]. Horizontal gradients may develop[39, 41, 88, 97], but are more important in non-tidal and freshwater beaches[40–42, 62, 92] (Fig. 5.4); vertical gradients develop and are dependent upon the development of the black, sulphide layer within the soil[85, 98]. During a daytime exposure period the oxygen levels of the surface layers of a soil rise, due to the photosynthetic activity of diatoms and other algae, whereas within the soil the oxygen content falls[85]. As one might expect from the temperature dependent migration of the black layer nearer to the surface in summer[85, 98], oxygen concentrations of the interstitial water change with season[85]. On those shores which become frozen in winter, oxygen levels remain appreciable presumably due to a flow of water through the soil[90]. In central Long Island Sound, the submarine sediments pass through a seasonal cycle of oxygen utilization which reaches a maximum in the late summer. Only a small part of this oxygen uptake can be ascribed to the dominant macrofauna, consequently most of the energy consumption in these sediments must be due to the microbenthos[99].

Sulphur

Sulphur plays a vital role in the chemistry of sediments, and thus in estuarine biology. Its chemistry, in the marine environment, is closely linked with, and is influential upon, its precursor in the periodic table, oxygen. In a reduced state it is important in the release of soluble phosphate[100–102], thus

$$FePO_4 + 2H_2S \rightarrow Fe(SH_2) + H_2PO_4^-.$$

The pellicle of *Enteromorpha*, which is rich in phosphate, will yield significant amounts of soluble phosphate when treated with H_2S. Conversely, the sulphydryl ion is vital to the formation of diatom frustules, and the growth and reproduction of many species[101].

Under conditions of oxygen deficiency the sulphate-reducing bacteria, e.g. *Desulphovibrio*, use sulphate as a hydrogen acceptor to produce H_2S[103–106]. These bacteria are widely and abundantly distributed in the marine environment, especially in sediments; they are obligatory anaerobes which, although not rapidly killed by, cannot reproduce in the presence of free oxygen[107]. Following large diatom blooms, when large amounts of decaying organic matter may consume all the oxygen present, and in stagnant water basins, sulphate-reducing bacteria flourish: in the Black Sea, there is no free oxygen below 200 m depth and the H_2S in solution forms a toxic layer 1800 m thick; similar conditions may arise below the euphotic zone in the Caspian Sea and the Norwegian fjords[3]. In the Sea of Azov, the seasonal production of H_2S may be sufficient to kill large numbers of fish and other aquatic animals[108]; periodically, production of H_2S by the bed of Walvis Bay, S. Africa, is sufficient to be lethal to the fauna and flora almost to the surface of the overlying water[109]. The production of H_2S in poorly aerated laboratory systems is probably a cause of difficulty in maintaining laboratory stocks of animals, both because of its intrinsic toxic properties[110], and a tendency to remove oxygen by its spontaneous oxidation[110, 111].

In marine soils, the H_2S combines with the oxides of iron to form sulphides which colour the anaerobic layer black, and there is good evidence to suggest that there are four important iron sulphides of which only two persist in aqueous suspension, viz. Fe(OH)SH—hydrotroilite which is black, but non-smelling, and Fe(SH)$_2$—disulphydryl iron which occurs in black foetid mud[102]. The blackened sulphide containing soils of the seashore and sea bottom are of a widespread distribution[98], but the sulphides formed in anaerobic conditions are oxidized rapidly in the presence of oxygen to yield the iron oxides and sulphur[84, 102] and the black layer normally occurs at some depth within the soil[84, 98]. The factors involved in the production of the black layer are complex and include the presence of sufficient moisture, nutrients and iron oxides in stable soils[84, 112, 113]. The depth of the black layer below the soil surface indicates the depth to which significant amounts of oxygen penetrate,

5

either by diffusion processes or by the circulation of aerated water[14], although total anaerobosis in the black layer is rare (p. 117). At Whitstable, Kent, the depth of the black layer from the soil surface varies seasonally and is inversely related to temperature[85, 98]: in summer, the oxygen concentrations of the interstitial water are lower than in winter when the black layer is further from the surface[85]. This seasonal variation is evidently due to the effect of temperature upon the velocities of the reactions involved, increased respiration of the fauna, increased bacterial activity, and a decreased solubility of oxygen[98]. In the River Eden estuary, Fife, the shore soils were so stable that the blackened sulphide containing layer occurred within 1–2 cm of the soil surface even during the winter[X]. In the Ouse estuary, Sussex, the occurrence of the black layer only 5 mm from the soil surface significantly reduced the abundance of the diatom *Pleurosigma aestuarii*, but not of *P. balticum*[114]; the somewhat deeper layer in the Eden estuary, Fife, apparently had no similar effect upon *P. aestuarii* there[115]. It is of interest that the sulphate reducing bacterium *Desulphovibrio*, in common with other hydrogenase producing bacteria, viz. *Chromatium*, *Chlorobacterium* and *Rhodospirillum*, can fix free N_2. Because such bacteria are of widespread occurrence in marine sediments, they may play an important part in the fixation of nitrogen in the sea[104]; furthermore the ability of sulphate reducing bacteria to consume hydrogen (H_2) may account for its general absence in natural gas[105]. The bacterium *Beggiatoa* can oxidize H_2S and retain the sulphur within its cells, moreover the free sulphur which results upon exposure of the iron sulphides and H_2S to the air is rarely found in marine soils. This sulphur may be removed by direct oxidation to SO_3^{2-} and SO_4^{2-}[84], or by bacterial action: the thiobacterium *Thiobacillus concretivorus*, purple bacteria, and Beggiotoaceae can oxidize sulphur to SO_4^{2-}[102]. Although the oxygen requirements of all these organisms are limiting, they are tolerant of the ranges of salinity, temperature, pH and Eh found in estuaries[106].

pH of Sediment

In general terms, the pH of the interstitial water of marine sediments is about that of sea water[24, 41, 47, 60, 65, 66, 84], but like freshwater beaches, the pH of the non-tidal marine beaches decreases towards the land[41, 45]. The pH of the shore soil is very variable, both from point to point in one area, and from area to area[24, 41, 45]. In general, the pH of the overlying water is more uniform, but may be lower than that of the surface layer of the soil, especially during the day when photosynthetic activity is high[24, 111]. The pH normally falls with increasing depth, in the soil, and especially in the presence of the black sulphide layer[24, 98, 111].

However, a pH 9 has been recorded at 10 cm depth, and here differences of 1 pH unit within 1 cm have been measured[41]. Under extreme drainage the pH of a soil may fall sharply, apparently due to a "double layer" effect; the pH changes rapidly when a soil is removed from the shore also[X]. The pH recorded on shores ranges from pH 7·9–9·2 at the sediment surface[24, 111, X], and from 6·1–9·0 at some depth within the soil[24, 41, 111, X]. At a depth of 4 cm within the bottom sediments of Long Island Sound a pH range of 6·0–8·5 has been recorded[116].

Other Inorganic Constituents

Marine sediments play an important role in maintaining the balance of the inorganic ions in sea water. For example, sand may preferentially remove sodium and magnesium ions, and in compensation release calcium ions to the circulating water[117], while many elements, viz. V, Pb, Cu, Ru, Co, Cr, Mn, Ni, Fe and Zn, occur in significant amounts in offshore sediments[118]. The nickel content of sediments increases seawards from inshore waters and estuaries; indeed, the nickel from freshwater run-off is scavenged by living organisms, organic matter, clay minerals and iron oxide and eventually becomes incorporated in the sediments. The nickel present in shallow water sediments is halmirolysed and carried to the deep-water deposits by the scavenging action of manganic hydroxide[119]. Cadmium is concentrated by living organisms, organic matter and sediments, also[120].

Studies of the balance of chloride, iodide, sulphate and borate suggest a greater abundance in sea water than is possible by the weathering of the primary rocks; consequently, it seems likely that chlorine, boron and probably sulphur were primary constituents of sea water, and may have been more abundant in the sea water of earlier geological ages than they are now. Conversely, the content of Cu, Pb, As, Se and Hg in primordial rock is such that if they were not precipitated from sea water, the oceans would have been poisoned. Certainly, Se, As, Sb, Bi and Pb may be removed from aqueous solution by the freshly precipitated hydroxides of iron, and sedimentary iron ores are heavily contaminated with Se, As, and Pb, while Mo is associated with the sedimentary compounds of Mn. The oceanic phosphates concentrate a number of elements, viz. Cd, In, Bi and especially Zn. Compared with the other heavy metals, Cu is surprisingly abundant in sea water.[121] Of course, I, Cu, V, Zn and many other elements may be removed from sea water by biological action[121], and are vital to life processes in the sea (p. 10). Curiously, a depletion of Fe from the waters of an area in the mouth of the English Channel has been attributed to removal of this element by a dense bed of the lamellibranch mollusc *Pinna*[122], and such an effect can influence other organisms with a requirement for iron. Indeed, it is of interest that the brackish-water diatom, *Phaeodactylum* has a lower requirement for Fe than the phytoplankton *Asterionella japonica*, and may represent an adaptation to life in estuaries where sedimentation and other removal processes may be extreme[123].

Considerable light has been thrown on the relationships between the composition of sea water and sediments by the many studies undertaken in order to understand the effects of radioactive waste disposal and of fall-out. Although radioactivity is an emotive subject, the radioactive nuclides have proved a valuable tool both in the present context and in studies of sediment transport (p. 135). Certain features of the chemistry of these nuclides when added to sea water are outstanding. Upon such addition, these nuclides either remain in the ionic-fully soluble-state, or enter solid-particulate or colloidal-phases. Elements of Groups I, II, V, VI or VII in the Periodic Table usually occur as ionic forms, e.g. Na, K, Ca, Sr, Cu, Zn, Ni, P, S, and I; other elements, the rare earths included, occur as solid phases, e.g. Fe, Y, Zr, Nb, Ru, Ce, U and Np; still others, e.g. Mo, Sb, Te and Cs, appear to show the properties of both ionic and particulate nuclides[124–130].

Although some adsorption can occur on the surface of soil up to sand size, the bulk of adsorption is in association with the silt and colloidal soil particles[131, 132]. Those radioactive nuclides which occur in the solid phase normally become adsorbed upon the surfaces of suspended and other sedimentary materials, viz. sand, silt and organic detritus[126, 133–135]. The specific activity of the silt/clay fraction may be 150 times greater than that of sand[133–135], but the uptake per unit area is constant for all the soil grades[134]; the specific activities of the silt/clay fraction and detrital organic matter are similar[135]. The adsorption of trace elements in sediments occurs in proportion to the concentration of iron and manganese with which they are coprecipitated[136]; moreover, once nitrosyl ruthenium, a radioactive waste product, is adsorbed by a sediment it forms a complex with ferric hydroxide and is then difficult to remove[134].

Studies of the uptake of ferric iron, strontium, cupric copper, phosphate, iodide and sulphate upon samples of sediment from Chesapeake Bay showed, that with the exception of iodide, the solids removed appreciable amounts of these substances from solution under controlled, but varied conditions of pH, temperature, salinity, concentration of solids and specific activity; where strontium carbonate did not precipitate it was adsorbed[137]. In the Columbia River estuary, ^{65}Zn, ^{51}Cr, ^{46}Sc and ^{54}Mn are adsorbed upon the sediments. Here the ^{65}Zn, held by specific sorption, is not removed easily by washing with sea-water. A similar treatment has little effect upon adsorbed ^{51}Cr and ^{46}Sc, although about one-third of the ^{54}Mn is removed by this means[138]. However, the quantity of material which may be adsorbed by a suspended solid will depend upon the chemical composition and past history of the solids[139].

It is worth considering the migration of radionuclides in terrestrial soils, where all known radionuclides except Zr, Nb and Cs are more completely adsorbed from alkaline rather than acid solution; Ce is affected by the pH, particularly, for in alkaline solution adsorption is complete, but in acidic solutions, and especially in the presence of Al, Ce is effectively not retained by the soil. Similarly, the sorption of Sr is several times lower in an acidic than in an alkaline medium. The sorption of radiostrontium by montmorillonite is affected by certain cations which in order of effectiveness are $Al^{3+} > Fe^{3+} > Ba^{2+} > Ca^{2+} > Mg^{2+} \geqslant H^+ \geqslant NH_4^+ > K^+ > Na^+$, i.e. the greater the cation valency the greater its effect in reducing sorption; Zr and Nb are adsorbed equally well from acidic and alkaline solutions, Cs sorption is not dependent upon the pH of the system[140]. Sorption of Sr and Cs takes place to a high degree upon clays[141], and while other radioisotopes are firmly retained by soils, ruthenium and strontium wash off readily[140]. The behaviour of radiocaesium is interesting in that it is absorbed by sediments[142, 143] and in river sediments, in particular, it is strongly and irreversibly adsorbed such that further movement results only from the physical transport of the sediment[143]. In estuarine sediments, however, Cs tends to "migrate" from the surface layers, and occurs only at depth within the sediment; this is not a function of soil movement, since it occurs in the stable soils of salt marshes, as well as in more mobile sediments[135].

Nutrients

Even in 1934, a knowledge of the relatively large amounts of plant nutrients, viz. phosphates, nitrates, sulphates and iron, present in soils, led to the conclusion that the psammolittoral, i.e. sedimentary, habitat could be classified as extremely eutrophic[144], and recent investigations have confirmed this view. Estuarine sediments are rich in phosphate which may be liberated to overlying waters either under storm or reducing conditions (p. 118), and thus have an important effect upon the productivity of these waters. In these sediments, the phosphate is considered to exist in three states: (1) interstitial phosphate, which is present in pore water and is liberated upon agitation; (2) adsorbed phosphate, which is released by chemical processes; and (3) insoluble phosphate bound by such ions as Fe^{3+} and Ca^{2+}[131, 145]. In Langston Harbour, England, the sand banks are an important reservoir of nutrients, notably in the form of finely divided $FePO_4$, which is released by storms[146]. Silt adsorbs significant amounts of phosphate, and the amount of phosphate present in a soil is proportional to the silt content[132]. In the freshwater environment, in general, plant production is limited by the amount of available phosphate, but in the sea, and in estuaries in particular, nitrate and nitrite concentrations are generally low[131, 147–151] and although influenced by run-off from the land[152], may limit phytoplankton production[23, 153]. In sediments, however, this position is modified by the activities of bacteria[104] and blue-green alga, e.g. *Calothrix*[154], which can fix free nitrogen.

Table 5.4. The nutrient content of newly accreted silt, 0–0·5 cm depth, from the *Spartina* marsh, Bridgewater Bay (after Ranwell[156])

Sampling site	Distance seaward (m)	% oven dry weight of silt				
		K	Ca	P	N	Organic carbon
Level *Spartina*	30	2·2	5·11	0·11	0·30	5·66
Level *Spartina*	150	2·1	5·17	0·10	0·29	6·39
Spartina on ridge	230	1·3	5·62	0·08	0·14	4·68
Bare mud on ridge	230	1·1	6·11	0·07	0·11	3·80

Nutrient levels are high in the accreting marsh at Bridgewater Bay, Somerset, which receives a supply of 800 kg/ha/yr of N, 250 kg/ha/yr of P and 6500 kg/ha/yr of K. The nutrient levels decrease to the seaward (Table 5.4), each zone apparently receives mud from the zone immediately to the seawards, and the source of silt for accretion is probably of a local origin[155, 156]: here plant growth is probably not limited by these nutrients[156]. In the Solway Firth more distant silt sources influence the plant life of the marshes, and the indications here are that plant production in estuarine sediments and salt marshes may be dependent upon hydrographical processes taking place at some distance in the open sea (p. 149). Indeed attempts at land reclamation in Morecambe Bay failed apparently owing to a lack of nutrient bearing silt[157].

Although adsorbed material can occur on sedimentary particles up to sand size, the bulk is associated with silt and colloidal soil particles (p. 122), which act as an important carrier of vitamin B_{12} in addition to phosphate and metallic ions[158]. Although clays adsorb organic nutrients such as thiamine or glucose, they are still available to bacteria, *Escherichia coli* and yeasts, *Cryptococcus albidus* and *Rhodotorula* sp.[159]. The vital role of a steady supply of silt and clay in successful land reclamation has long been recognized in Germany, where these sediments have been used to enhance the fertility of arable soils: here too they are considered to possess therapeutic properties[160]. Curiously folk lore ascribed similar healing properties to the muddy sediments of the Severn estuary and Bristol Channel. To summarize, the silt and clay grades (i.e. < 0·0625 mm diameter) of marine sediments are known to be important ecologically [161] and to have an important role related to production in estuarine areas, but with the exceptions quoted above their role is poorly understood.

References

1. Johnson, D. W. (1919). "Shore Processes and Shoreline Development", Wiley, New York.
2. Shepard, F. P. (1937). *J. Geol.* **45**, 602–604.
3. Sverdrup, H. U., Johnson, M. W. and Fleming, R. H. (1960). "The Oceans: Their Physics, Chemistry and General Biology", Prentice-Hall, New York.
4. Bascom, W. N. (1960). *Scient. Am.* **203** (2), 81–94.
5. Tait, R. V. (1968). "Elements of Marine Ecology", Butterworths, London.
6. Perkins, E. J. (1968). *Trans. J. Proc. Dumfries. Galloway nat. Hist. Antiq. Soc.* Ser. 3, **46**, 1–26.
7. Dalby, D. H. (1968). *Fld Stud.* **2**. (Suppl.), 31–37.
8. Morgans, J. F. C. (1956). *J. Anim. Ecol.* **25**, 367–387.
9. Krumbein, W. C. (1932). *J. sedim. Petrol.* **2**, 140–149.
10. Krumbein, W. C. (1936). *J. sedim. Petrol.* **6**, 35–47.
11. Inman, D. L. (1952). *J. sedim. Petrol.* **22**, 125–145.
12. Emery, K. O. (1938). *J. sedim. Petrol.* **8**, 105–111.
13. Sheldon, R. W. and Parsons, T. R. *Fisheries Research Board of Canada, Manuscript Report Series (Oceanographic and Limnological)* No. 214, 36 pp.
14. Webb, J. E. (1958). *Phil. Trans. R. Soc.* (B), **241**, 307–419.
15. Newell, R. (1965). *Proc. zool. Soc., Lond.* **144**, 25–45.
16. Prenant, M. (1960). *Cah. Biol. mar.* **1**, 295–340.
17. Renaud-Debyser, J. and Salvat, B. (1963). *Vie Milieu* **14**, 463–550.
18. Piper, C. S. (1947). "Soil and Plant Analysis", a monograph from the Waite Agricultural Research Institute, 368 pp. University of Australia, Adelaide.
19. Bruce, J. R. (1928). *J. mar. biol. Ass. U.K.* **15**, 535–552.
20. Frazer, H. J. (1935). *J. Geol.* **43**, 910–1010.
21. King, F. H. (1898). *U.S. geol. Surv. A. Rep.* **19**, 208–218.
22. Ruttner-Kolisko, A. (1962). *Schweiz. Z. Hydrol.* **24**, 444–458.
23. Green, J. (1968). "The Biology of Estuarine Animals", Sidgwick and Jackson, London.
24. Perkins, E. J. (1958). "The Microfauna of the Shore", Ph.D. thesis, University of London.
25. Davant, P. and Salvat, B. (1961). *Vie Milieu* **12**, 405–471.
26. Slabaugh, W. H. and Stump, A. D. (1964). *J. geophys. Res.* **69**, 4773–4778.
27. Slabaugh, W. H. and Stump, A. D. (1964). *J. phys. Chem.* **68**, 1251–1253.
28. Franzini, J. B. (1951). *Trans. Am. geophys. Un.* **32**, 443–446.
29. Fireman, M. (1944). *Soil Sci.* **58**, 337–353.
30. Emery, K. O. and Forster, J. F. (1948). *J. mar. Res.* **7**, 644–654.
31. Perkins, E. J. (1958). *J. Ecol.* **46**, 71–81.
32. Christiansen, J. E. (1944). *Soil Sci.* **58**, 355–365.
33. Chapman, G. (1949). *J. mar. biol. Ass. U.K.* **28**, 123–140.
34. Capper, P. L. and Cassie, W. F. (1963). "The Mechanics of Engineering Soils", 4th edn. McKay, Chatham.

35. Kolbuszewski, J. (1963). *New Scient.* 19 December 1963, 685–686.
36. Freundlich, H. and Juliusberger, F. (1935). *Trans. Faraday Soc.* **31**, 769–774.
37. Carruthers, J. N. (1954). *J. mar. biol. Ass. U.K.* **33**, 637–643.
38. Chapman, G. and Newell, G. E. (1947). *Proc. R. Soc. B.* **134**, 431–455.
39. Pennak, R. W. (1951). *Annls Biol.* **27**, 217–248.
40. Pennak, R. W. (1940). *Ecol. Monogr.* **10**, 537–615.
41. Jansson, B.-O. (1968). *Akademisk Avhanling.* Stockholm 1968. 1–16.
42. Ruttner-Kolisko, A. (1956). *Verh. dt. zool. Ges.* Hamburg, 1956: 421–427.
43. Jansson, B.-O. (1967). *Oikos* **18**, 311–322.
44. Jansson, B.-O. (1967). *Helgoländer wiss Meeresunters.* **15**, 41–58.
45. Jansson, B.-O. (1966). *Veröff Inst. Meeresforsch. Bremerh.* **2**, 77–86.
46. Rullier, F. (1957). *C.r. hebd. Séanc. Acad. Sci., Paris* **245**, 936–938.
47. Salvat, B. (1967). *Mem. Mus. natn. Hist. nat., Paris, N.S., Zool.* **45**, 1–275.
48. Debouteville, C. D. (1955). *C.r. hebd. Séanc. Acad. Sci., Paris* **240**, 460–462.
49. Debouteville, C. D. (1955). *C.r. hebd. Séanc. Acad. Sci., Paris* **240**, 555–557.
50. Ångstrom, A. (1925). *Geogr. Ann.* **7**, 323–342.
51. Perkins, E. J. (1963). *J. Ecol.* **51**, 687–692.
52. Sassuchin, D. N., Kabanov, N. M. and Neiswestnova, K. S. (1927). *Russk. gidrobiol. Zh.* **6**, 59–83.
53. Neel, J. K. (1948). *Trans. Am. microsc. Soc.* **67**, 1–53.
54. Aleem, A. A. (1950). *New Phytol.* **49**, 174–188.
55. Taylor, W. R. (with Palmer, J. D.) (1963). *Biol. Bull. mar. biol. Lab. Woods Hole* **125**, 395.
56. Taylor, W. R. (1964). *Helgoländer wiss. Meeresunters.* **10**, 29–37.
57. Hopkins, J. T. (1963). *J. mar. biol. Ass. U.K.* **43**, 653–663.
58. Russell, F. S. (1936). *J. Cons. perm. int. Explor. Mer, Rapp. et Proc. Verb.* Cl (2): 3–8.
59. Perkins, E. J. (1964). H.M.S.O., U.K.A.E.A., P.G. Report 456 (CC).
60. Ganapati, P. N. and Chandrsekhara Rao, G. (1962). *J. mar. biol. Ass. India* **4**, 44–57.
61. Macintyre, R. J. (1963). *Trans. R. Soc. N.Z., General.* **1**, 89–103.
62. Wiszniewski, J. (1934). *Archwm. Hydrobiol. Ryb.* **8**, 149–272.
63. Jansson, B.-O. (1967). *Ophelia* **4**, 173–201.
64. Linke, O. (1939). *Helgoländer wiss. Meeresunters.* **1**, 301–348.
65. Drimmel, J. (1953). *Arch. Met. Geophys. Bioklim. Ser. B.* **5**, 18–40.
66. Renaud-Debyser, J. (1963). *Vie Milieu* (suppl.) **15**, 1–57.
67. Jansson, B.-O. (1968). *Ophelia* **5**, 1–71.
68. Remane, A. and Schulz, E. (1934). *Schr. naturw. Ver. Schlesw.-Holst.* **20**, 399–408.
69. Brinck, P., Dahl, E. and Wieser, W. (1955). *K. fysiogr. Sällsk. Lund Förh.* **25**, 1–21.
70. Ax, P. (1966). *Veröff Inst. Meeresforsch. Bremerh.* **2**, 15–65.
71. Munch, H. D. and Petzold, H. G. (1956). *Wiss. Z. Ernst Moritz Arndt-Univ. Griefwald, Math. Nat. R.* **5**, 413–429.
72. Alexander, W. B., Southgate, B. A. and Bassindale, R. (1932). *J. mar. biol. Ass. U.K.* **18**, 297–298.
73. Capstick, C. K. (1957). *J. Anim. Ecol.* **26**, 295–315.
74. Reid, D. M. (1930). *J. mar. biol. Ass. U.K.* **16**, 609–614.
75. Reid, D. M. (1932). *J. mar. biol. Ass. U.K.* **18**, 299–306.
76. Nicol, E. A. T. (1935). *J. mar. biol. Ass. U.K.* **20**, 203–261.

77. Smith, R. I. (1955). *J. mar. biol. Ass. U.K.* **34**, 33–46.
78. Callame, B. (1960). *Bull. Inst. océanogr., Monaco* **1181**, 1–19.
79. Smith, R. I. (1956). *J. mar. biol. Ass. U.K.* **35**, 81–104.
80. Perkins, E. J. (1957). *Proc. R. phys. Soc. Edinb.* **25**, 53–55.
81. Kitagawa, K. (1934). *Mem. Coll. Sci. Kyoto.* Ser. A. **17**, 37–42.
82. Phleger, F. B. and Bradshaw, J. S. (1966). *Science, N.Y.* **154**, 1551–1553.
83. Bradshaw, J. S. (1968). *Limnol. Oceanogr.* **13**, 26–38.
84. Bruce, J. R. (1928). *J. mar. biol. Ass. U.K.* **15**, 553–565.
85. Brafield, A. E. (1964). *J. Anim. Ecol.* **33**, 97–115.
86. Emery, K. O., Tracey, J. I. and Ladd, H. S. (1954). *Prof. Pap. U.S. geol. Surv.* 260A: 265 pp.
87. Amoreux, L. (1963). *Cah. Biol. mar.* **4**, 23–32.
88. Brafield, A. E. (1965). *Vie Milieu (Sér. Oceanogr.)* **19**, 889–897.
89. Jansson, B.-O. (1966). *Vie Milieu (Sér. Biol. mar.)* **17**, 143–186.
90. Jansson, B.-O. (1967). *J. exp. mar. Biol. Ecol.* **1**, 123–143.
91. Gordon, M. S. (1960). *Science, N.Y.* **132**, 616–617.
92. Ruttner-Kolisko, A. (1955). *Wett. Leben* **7**, 16–22.
93. Ruttner-Kolisko, A. (1961). *Verh. int. Verein theor. angew. Limnol.* **14**, 362–368.
94. Lemon, E. R. and Erickson, A. E. (1952). *Proc. Soil Sci. Soc. Am.* **16**, 160–163.
95. Lemon, E. R. and Erickson, A. E. (1954). *Soil Sci.* **79**, 383–392.
96. Odén, S. (1962). *Grundförbättring* **3**, 150–178.
97. Pennak, R. W. (1942). *Ecology* **23**, 446–456.
98. Perkins, E. J. (1957). *Ann. mag. nat. Hist.* Ser. 12. **10**, 25–35.
99. Carey, A. G. (1967). *Bull. Bingham oceanogr. Coll.* **19**, 136–144.
100. Baas-Becking, L. G. M. and Mackay, M. (1956). *Proc. K. ned. Akad. Wet.* **B59**, 109–123.
101. Wood, E. J. F. and Davis, P. S. (1959). *N.Z., D.S.I.R., Information Series* No. 22, *N.Z. Oceanogr. Inst. Mem.* **3**, 31–35.
102. Baas-Becking, L. G. M. (1959). *N.Z., D.S.I.R., Information Series* No. 22, *N.Z. Oceanogr. Inst. Mem.* **3**, 48–64.
103. Sisler, F. D. and ZoBell, C. E. (1950). *J. Bact.* **60**, 747–756.
104. Sisler, F. D. and ZoBell, C. E. (1951). *J. Bact.* **62**, 117–127.
105. Sisler, F. D. and ZoBell, C. E. (1951). *Science, N.Y.* **113**, 511–512.
106. Baas-Becking, L. G. M. and Wood, E. J. F. (1955). *Proc. K. ned. Akad. Wet.* **B58**, 160–181.
107. ZoBell, C. E. and Rittenberg, S. C. (1948). *J. mar. Res.* **7**, 602–617.
108. Bunker, H. J. (1936). "A review of the Physiology and Biochemistry of the Sulphur Bacteria", H.M.S.O., London.
109. Copenhagen, W. J. (1934). *Fish. Mar. biol. Surv. Un. S. Afr.*, Rep. No. 11, 3–18.
110. Perkins, E. J., Williams, B. R. H. and Hinde, A. (1964). H.M.S.O., U.K.A.E.A., P.G. Report 578 (CC), 26 pp.
111. Bradshaw, J. S. (1968). *Limnol. Oceanogr.* **13**, 26–38.
112. Ellis, D. (1925). *Jl. R. tech. Coll. Glasg.* (1925), 144–156.
113. Kimata, M., Kadota, H., Hata, Y. and Miyoshi, H. (1957). *Bull. Jap. Soc. Scient. Fish.* **22**, 701–707.
114. Hopkins, J. T. (1964). *J. mar. biol. Ass. U.K.* **44**, 333–341.
115. Perkins, E. J. (1960). *J. Ecol.* **48**, 725–728.

5*

116. Schafer, C. T. (1968). *Atlantic Oceanographic Laboratory, Bedford Institute.* Report A.O.L. 688, 48 pp.
117. Stowell, F. P. (1927). *J. mar. biol. Ass. U.K.* **14**, 955–966.
118. Forster, W. O., Naidu, J. and Wagner, J. (1969). *In* "Ecological Studies of Radioactivity in the Columbia River Estuary and the Adjacent Pacific Ocean", Progress Report 69–9, 1969: 88–91, Dept. of Oceanography, Oregon State University, Corvallis.
119. Laevastu, T. and Thompson, T. G. (1956). *J. Cons. perm. int. Explor. Mer* **25**, 125–143.
120. Mullin, J. B. and Riley, J. P. (1956). *J. mar. Res.* **15**, 103–122.
121. Goldschmidt, V. M. (1937). *J. chem. Soc.* 1937, 655–673.
122. Cooper, L. H. N. (1948). *J. mar. biol. Ass. U.K.* **27**, 279–313.
123. Hayward, J. (1968). *J. mar. biol. Ass. U.K.* **48**, 295–302.
124. Revelle, R. and Schaefer, M. B. (1957). *Academy of Sciences. Nat. Res. Counc. Publ. No. 551*, 1–25.
125. Greendale, A. E. and Ballou, N. E. (1954). *U.S.N.R.D.L.*—436. 16 February 1954.
126 Bowen, V. T. and Sugihara, T. T. (1958). *In* "Peaceful Uses of Atomic Energy. Proc. 2nd International Conference held at Geneva, September 1958", U.N. Publication, **18** 434–438.
127. Fontaine, Y. (1960). *Centre d'Etudes Nucleaires de Saclay* 1960. C.E.A.—1588. Unclassified.
128. Sugihara, T. T. and Bowen, V. T. (1960). "Radio-active Rare Earths from Fall-out for Study of Sea Particle Movement. Radioisotopes in the Physical Sciences and Industry 1", Proc. Conf. held by I.A.E.A. and U.N.E.S.C.O. at Copenhagen, 6–17 September 1960: 57–65.
129. Freiling, E. C. and Ballou, N. E. (1962). *Nature, Lond.* **195**, 1283.
130. Perkins, E. J. and Williams, B. R. H. (1964). *U.K.A.E.A.* Chapelcross Health Physics Note No. 2.
131. Rochford, D. J. (1951). *Aust. J. mar. Freshwat. Res* **2**, 1–116.
132. Rochford, D. J. (1958). *Archo. Oceangr. Limnol.* **11** suppl. 1959, 171–177.
133. Dunster, H. J. (1958). *In* "Peaceful Uses of Atomic Energy. Proc. 2nd International Conference held at Geneva, September 1958", U.N. Publication, **18**, 390–399.
134. Jones, R. F. (1960). *Limnol. Oceanogr.* **5**, 312–325.
135. Perkins, E. J. and Williams, B. R. H. (1966). H.M.S.O., U.K.A.E.A., P.G. Report 587 (CC): 63 pp.
136. Goldberg, E. D. (1954). *J. Geol.* **62**, 249.
137. Carritt, D. E. and Goodgall, S. (1954). *Deep Sea Res.* **1**, 224–243.
138. Johnson, V., Cutshall, N. and Osterberg, C. (1967). *Wat. Resour. Res.* **3**, 99–102.
139. Carritt, D. E. *et al.* (1959). *National Acad. of Sciences. Nat. Res. Council* Washington, D.C. Publ. No. 655.
140. Spitsyn, V. I. *et al.* (1958). *In* "Peaceful Uses of Atomic Energy. Proc. 2nd International Conference held at Geneva, September 1958", U.N. Publication, **18**, 439–448.
141. Ioanid, G. *et al.* (1958). *In* "Peaceful Uses of Atomic Energy. Proc. 2nd International Conference held at Geneva, September 1958", U.N. Publication, **18**, 598–604.

142. Bogatyrev, I. O. (1962). *Bull. Acad. Sci. U.R.S.S.* Ser. Biol. 1962. No. 1, 122–126 *Nucl. Sci. Abst.* 1962, **16**, 1847.
143. Morgan, A. and Arkell, G. M. (1961). U.K.A.E.A., A.E.R.E.—R 3555. Unclassified.
144. Stagenberg, M. (1934). *Archwm Hydrobiol. Ryb.* **8**, 273–284.
145. Jitts, H. R. (1959). *Aust. J. mar. Freshwat. Res.* **10**, 7–21.
146. Oliver, J. H. (Reported by L. H. N. Cooper) (1961). *In* "Bacteriological Proceedings: Symposium on Marine Microbiology", Society of American Bacteriologists, D 20, p. 48.
147. MacGinitie, G. E. (1935). *Am. Midl. Nat.* **16**, 629–765.
148. Howes, N. H. (1939). *J. Linn. Soc. (Zool.)* **40**, 383–445.
149. Nash, C. B. (1947). *J. mar. Res.* **6**, 147–174.
150. Harvey, H. W. (1960). "The Chemistry and Fertility of Sea Waters", The University Press, Cambridge.
151. Perkins, E. J., Bailey, M. and Williams, B. R. H. (1964). H.M.S.O., U.K.A.E.A., P.G. Report 604 (CC).
152. Spencer, R. (1956). *Aust. J. mar. Freshwat. Res.* **7**, 193–253.
153. Stewart, K. M. and Rohlich, G. A. (1967). "Eutrophication—A Review", State of California Water Quality Control Board, Publication No. 34.
154. Jones, K. and Stewart, W. D. P. (1969). *J. mar. biol. Ass. U.K.* **49**, 475–488.
155. Ranwell, D. S. (1964). *J. Ecol.* **52**, 79–94.
156. Ranwell, D. S. (1964). *J. Ecol.* **52**, 95–105.
157. Steers, J. A. (1962). "The Sea Coast", Collins, London; New Naturalist Series, No. 25.
158. Burkholder, P. R. and Burkholder, L. M. (1956). *Limnol. Oceanogr.* **1**, 202–208.
159. Button, D. K. (1969). *Limnol. Oceanogr.* **14**, 95–100.
160. Hantzschel, W. (1939). *In* "Recent Marine Sediments: A Symposium", (Trask, P. D., ed.), Am. Assoc. of Petroleum Geologists, Tulsa, Oklahoma; Thomas Murby, London.
161. Sanders, H. L. (1958). *Limnol. Oceanogr.* **3**, 245–258.
162. McIntyre, A. D. (1969). *Biol. Rev.* **44**, 245–290.
163. Barrett, J. and Yonge, C. M. (1970). "Pocket Guide to the Sea Shore", Collins, London.
X.   Perkins, E. J. Previously unpublished work.

6

Sediment Transport

Listen! you hear the grating roar
Of pebbles which the waves draw back, and fling,
At their return, up the high strand,

Matthew Arnold "Dover Beach"

Sediment may be transported by biological agents viz. pink shrimp, *Pandalus montagui*; shore crab, *Carcinus maenas*; and fish—sole, *Solea solea*; plaice, *Pleuronectes platessa*; whiting, *Merlangius merlangus*; pouting, *Gadus luscus*; and skate, *Raja batis*, but this process is unimportant except in areas where sedimentation, by other processes, is very slow[1]. Although the contribution of living organisms to sediment transport may be insignificant, filter feeders, e.g. mussel, *Mytilus edulis*, may be influential in the formation of estuarine muds, by removing sedimentary materials from the overlying water and depositing them as deep, extremely fine-grained, slimy, argillaceous pseudo-faecal mud, upon which walking may be difficult[2]. Many other benthonic, e.g. polychaete worms, *Nereis*, *Heteromastus*; lamellibranch molluscs, *Cardium*, *Nucula*, *Abra*; gastropod molluscs, *Littorina*, *Hydrobia*; and planktonic, e.g. *Calanus*; Euphausiids; Mysids and many larvae, organisms contribute to the formation of sediments in this way[2–6a]. The faecal pellets of planktonic organisms break down relatively quickly, but those of the bottom dwellers are highly resistant and may retain their form for upwards of 100 years[6a]. Such biological agencies are considered to be responsible for the accelerated rate of removal of radioactive nuclides, such as the rare earths, which were liberated to the sea in nuclear bomb tests[7, 8].

Although, living organisms can influence the composition of sediments significantly, the dominant factor in sediment transport is the motion of the water itself: for example, running water not only erodes the land, but transports the eroded materials to the sea.

In a water current, of low velocity, where the particles move in a parallel way, without cross currents, the flow is *laminar*. In rivers and estuaries, however, the critical velocity above which laminar motion cannot be maintained, is always exceeded and all the currents have a *turbulent motion*. Such motion is characterized by eddies, and, in some cases, a movement against the net direction of flow of the current.

Because of the turbulent motion of the currents, the erosion and transportation of deposited sediment and detritus may be very variable, and the processes of erosion, transportation and deposition are closely related.

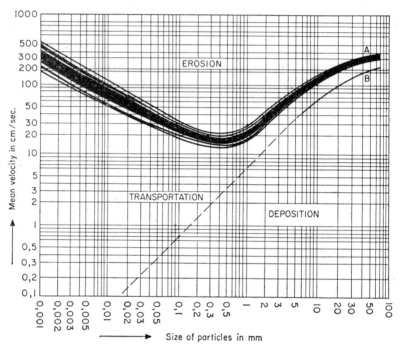

Fig. 6.1. Approximate curves for erosion and deposition of uniform material (after Hjulström[9]).

An approximate relationship between the average velocity across a transverse profile of a river and these phenomena is given by Fig. 6.1[9], but because of its approximate nature is represented by a series of bands rather than lines. Nevertheless it does indicate that "fine-sand", 0·3–0·6 mm in diameter, is the easiest grade to erode, with increasing difficulty at other grade sizes. Stiff clays are eroded with the least facility, whereas large, immovable stones can promote erosion by the swirling motions which their intrusion into the current generates[9]. More recent work has shown that an eroding current moves the fine sand grade (i.e., 0·25–0·125 mm) most easily[10]. The eroding power of water is reduced by suspended matter, especially colloids[9].

Once particles of sediment are torn from the bed they remain suspended at much lower velocities than that at which erosion occurs, but below the *lowest transportation* velocity, they are deposited: indeed, the water velocity which induces erosion may be decreased about 30% before deposition begins. Sedimentary materials may be transported in a number of ways depending upon the water velocity. With increasing velocity, these particles may be moved by *sliding, rolling, saltation* and *suspension*. Sliding is rarely seen, whereas rolling is the normal mode of transportation in rivers; here, however, the particle velocity is much less than that of the water. With a further increase in velocity some of the particles undergo a hopping motion, i.e. saltation, and some of these particles tend to lift into suspension, but only in a part of the path of motion does the particle have the velocity of water. However, when transportation by suspension occurs these particles are considered to have the same velocity as the transporting water mass[9]. Clearly, all sediment particles in suspension will be influenced by their size, and therefore their settlement velocity. Consequently, coarse sand will be concentrated near the stream bed whereas the finer grained constituents are more uniformly distributed in the water column: at a height of 30 cm above the bottom only 10% as much "coarse sand" is present as at a height of 2 cm, and at 5 m only 0·3% as much as at 2 cm. At this height of 5 m approximately 50% of "fine sand" and 97% of "silty loam" is in suspension as at a height of 2 cm; no differences throughout the 5 m column would be noted for clay. The settlement velocity is influenced by the size, shape and specific gravity of the particles, and the specific gravity, temperature and viscosity of the medium, through which they settle. Because an increased temperature increases the settlement velocity, sediments are transported more readily in winter than in summer, and conversely, because of their influence upon the viscosity of the medium, the presence of significant amounts of fine, suspended material leads to a marked decrease in the settlement velocity of the larger particles. Finally, the height to which a particle of sediment is carried into suspension is dependent upon the turbulence of the water[9].

In estuaries, there is a broad relationship between current velocity and the amount of sediment carried in suspension[11, 12]. As in rivers, however, the relationship is complex and confused in detail, with a marked variability related to large local time changes, seasonal, tidal and horizontal gradients[12–14], and large differences in sediment content may be recorded due to the passage of different water cells across the station[12, 14]. Variability in the load of suspended organic matter is dependent upon wind mixing, tidal scouring and river discharge, but not upon diatom populations[13]. Furthermore, the pattern of sediment suspension and transport within an estuary may be modified dramatically by events taking place, in the sea, outside the estuary (p. 152).

In general, as one might expect, the sediment load decreases with distance from the sea bed[11, 12, 14] but this is not always true[12, 14], presumably for the reasons discussed above. It may be greater at the time of the equinoxes[11, 12] and during winter gales, but a decline in the amount of suspended material due to increased temperature is not easy to observe, and may be most evident in outer estuarine areas[15]. At low and high water slack tides, the suspended solids may fall out of suspension[11, 12, Y], but this is not always true[14].

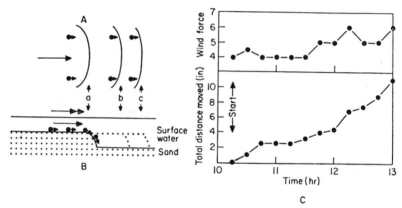

Fig. 6.2. Movements of soil on the shore, at Whitstable, Kent, due to the action of wind on standing surface water layers (after Perkins[16]).

Before considering the influence of estuarine water movements upon the sediments of estuaries, it should be noted that there are two kinds of sediment movement in which these large water bodies play no part. First, wind blown sand is responsible for the formation of dunes in outer estuarine areas. Secondly, many estuarine shores retain a layer of surface water during the exposure period. The wind acting upon these thin layers can disrupt the ripple system and affect the microbenthos, locally; under these conditions a rolling of fine sand in the direction of the wind may persist throughout the exposure period[16] (Fig. 6.2).

The transport of sediments by water movements in estuarine areas may be due to reaction currents, wave action—either directly or by longshore drift—and to tidal currents. Clearly each of these forces may act either independently or in unison to a varying degree.

The sediment transported in estuarine and coastal waters is derived from three principal sources: (1) the bed of continental shelf areas; (2) that carried by the inflowing rivers; and (3) an addition which results from dumping by man. In countries like Britain, where the rivers are short, the direct contribution from these rivers may be relatively small[106], and where an ice sheet has first ground sediment from the land and then dumped in the adjacent sea the amount of material reworked into estuaries from this source may be very considerable[104].

In contrast, the large rivers of the world, e.g. Amazon, Mississippi and Nile, carry a considerable burden of silt which is derived from the uplands and is deposited on the margin of the river in the lowlands and as a delta at the mouth. Such an alluvium exerts a considerable influence upon the productivity of the delta, the lower river margins and the fisheries of the adjacent sea. The annual average of such material carried by the Mississippi is 10 593 200 000 ft^3 and by the Yellow River 1.89×10^9 tons[16a]; small wonder then that the Aswan High Dam with its lack of sluices should be a cause for concern[16b]. Those river-borne particles of sediment which are negatively charged become flocculated upon contact with the positively charged metal ions present in sea water. However, organic materials which carry a positive charge remain in suspension for prolonged periods[16c]. Customarily, man has dumped unwanted wastes in the sea and in 1968, for example, the amount of waste dumped in New York Bight and Long Island Sound reached some 9.6×10^6 tons, which represented the largest single source of sediments contributed to the North Atlantic Ocean from the North American continent[16d].

Transport by Reaction Current

The transport of sedimentary materials by reaction currents, in the Göta-Älf, led to their recognition by Ekman (p. 55). Where the development of reaction currents is pronounced, the mud and sand, transported upstream, are deposited at the head of slack water[17]. The need to understand more thoroughly the process of siltation in the Thames estuary led to a revolution in the methods of investigating this phenomenon[18]. Here the essential problem was that to keep the Port of London open 3×10^6 tons of mud were dredged annually; an amount estimated to be several times the amount of silt brought down by the upland waters and effluents[19]. Of these dredgings, about 250 000 tons were dumped on mud flats in the estuary and remainder taken out to sea, where it was known to disperse. A glass containing the radioactive ^{46}Sc was ground to a particle size, comparable with that of the sediment, and released at Tilbury at high water. Within a day it had travelled 4 miles upstream, and in a further 6 days it had travelled some 14 miles upstream to the dredged cut at Barking where it was covered by mud and lost. While being transported by the reaction current, the mud was soft and fluid, but at the point of sedimentation it became compacted and solid. Previous observations with current meters, both in the Thames estuary itself and its model, had shown that under low river flow conditions, the Barking "Mud Reaches" represented the upstream limit of the net landward drift near the sea bed[19, 20]. The use of ^{46}Sc in this project led to its wide use in studies of sediment movement on shores, in estuaries and in nearshore waters[21-44]. Subsequent studies employed ^{110}Ag, ^{51}Cr, ^{198}Au, a sand labelled by glue containing ^{32}P, and radioactive diatoms[45-51].

Pioneered in 1939[52], a parallel development which used fluorescent dyes attached to sand grains with glue took place, especially in studies of beach movement[53-56]. More recently, "Tenite" pellets, a vari-coloured plastic developed by the Eastman Kodak Company, have been used in beach studies[57]. In addition to these "additive methods" the components of a sediment can be used as natural indicators of sediment movement[58] and on the Bermuda Platform, the foraminiferan *Homotrema rubrum* (Lamark) has been used for the same purpose.[59] It is of interest that water may also be characterized by its particle content[60].

Work in estuaries, other than the Thames, has shown that fine grained sediments tend to be trapped in the salt wedge and that this leads to shoaling, particularly in side basins[17, 61]. In the River Tyne estuary, Northumberland, liquid effluents which are carried in the upper layers are transported seawards, but solids sink into the lower layer which has a net movement upstream. In other words, the river acts as a natural sedimentation tank, from which removal of deposited material may only occur when the River Tyne is in flood[62].

Estuarine circulations depend upon the amount of run-out from the land, and where this undergoes marked seasonal changes, the circulation type may change also (p. 59) and have marked effects upon the sedimentation regime of the estuary. Such seasonal changes in sedimentation have been observed in Yaquina Bay, Oregon. Here maximum deposition occurs in the winter and spring when river run-off is greatest. At these times littoral drift, coastal winds and the partially-mixed, type B, estuarine circulation system promote the transport of beach, nearshore marine and dune sands into the estuary. During the summer and autumn, the deposition of sediment is slight because littoral drift brings little to the mouth of the estuary, and with the discharge of fresh water to the estuary reduced, the now well-mixed estuarine system has a slow net drift seawards, at all depths. In these conditions, the river carries little sediment into the estuary, but that introduced is carried slowly towards the mouth[59, 63].

Failure to recognize the importance of such changes in estuarine circulation patterns, and their influence upon sedimentation can have costly results. In the years prior to 1956, a hydroelectric development was completed on the upper Cooper River, in which water from the Santee watershed was diverted into the Cooper. The flow of the Cooper, before this diversion, was relatively low, and the tidal range was great enough to maintain vertical homogeneity and a net flow seawards at all depths in the estuary. This slow net movement of water seawards was sufficient to remove the fresh water and silt from the estuary. With the development of the Santee-Cooper project, the increased river flow led to a partial stratification of the estuary and the development of a reaction current at the bed. In consequence, the river brought down more sediments which were held in the estuary, while the reaction current introduced large amounts of sand from seawards of the estuary, with the result that the cost of dredging Charleston Harbour rose from $10 000 p.a. to *ca.* $1 000 000 p.a.[64, 65]. Viewed in this light, the economic advantages of this power generation scheme is not all that it might appear superficially.

The density currents, which arise from differences in salinity and density between two interconnected or periodically separated bodies of water, are a major factor in the siltation of docks and tidal basins. This is not a consequence of a fresh water inflow as in estuaries, and the absence of such an inflow is characteristic of this phenomenon. In the Thames, Tilbury Tidal Basin adjoins and is permanently open to the estuary. Here the fluctuating density differentials between the two bodies of water were an important factor in the inevitable siltation problem: to maintain a required depth of 24 ft below L.W.S.T., *ca.* 400 000 yd³ of silt are dredged from the basin annually[66]. In the reach in which Tilbury is situated, a heavy burden of suspended silt, especially in the lower layers, is carried by these estuarine waters[67]. At the entrance to the basin, the silt and mud concentrations often reach 50 000 p.p.m., but even outside such "fluid mud layers" the sediment load suspended near the bed was four or five times that at the surface during the early stages of the flood tide. During the flood tide, the main inflow, to the basin, occurs at the bed when water, heavily burdened with silt, passes into the relatively still waters of the basin, and deposits its sediment load: a phenomenon which is most marked towards the end of the flood tide period. After high water, loss of water occurs at all depths and is greatest near the bottom towards the end of the ebb tide period. During the outflow, a slow eddy develops in the basin, and is accompanied by an inward flow at the bed. Such a circulation implied that more silt laden water entered at the bed of the basin than was necessary to the tidal prism, and that sedimentation took place at this time also[66]. It can be shown that the velocity (v) of the wave front of a fluid intruding into a fluid of lower density is given thus:

$$v = 0 \cdot 707 \sqrt{\frac{\sigma_1 - \sigma_2}{\sigma_1 + \sigma_2} g d}$$

where σ_1 and σ_2 are the specific gravity of the denser and less dense fluids respectively, d = water depth, and g = acceleration due to gravity[66]. Coriolis' Force, which has a direct effect upon estuarine circulations (p. 57) may influence estuarine sediments by other means also. The meandering of rivers, and therefore of channels in estuaries like the Solway Firth, is probably induced by its action. Moreover, in the Mississippi Delta silt deposits more readily on the right side of a pass than on the left, a pattern consistent with the effect of Coriolis' Force: similar effects have been noted in the mouths of the Nile, Po, Danube and Volga, while rivers entering the Swiss lakes undergo a marked deflection to the right[68].

Transport by Wind-Induced Mechanisms

Direct Wave Action

The depth to which wave action affects the sea is of fundamental importance to the study of shore processes and coastal biology. At a depth equal to one wave length, oscillatory waves move in an orbit diameter equal to 1/535 of that of the sea's surface, and this, i.e. the wave height, is in turn dependent upon the fetch (p. 42). Evidently, the effect of wave action declines rapidly with depth. Nevertheless, waves which have abraded the bottom may change colour on passage into water of 40–50 ft depth. Coarse material on Chesil Bank, S. England, is subject to movement at 50 ft depth in heavy storms. Divers have been tossed back and forth at 100 ft depth. Sand from 75 to 80 ft depth of water may be left on the decks of ships. At the Bell Rock Lighthouse, North Sea, stones weighing two tons or more are often thrown up on to the rocks from "deep water"[69]. In general, waves may be considered to affect the seabed as far as the edge of the Continental Shelf[70]. A comparison of mud accumulation and the intensity of wave action in the North Sea, indicates that the depth at which mud begins to accumulate increases in proportion to the increasing intensity of wave action[71]. By inference, the upper limit of mud accumulation is a measure of the maximum depth of wave disturbance. In the North Sea, wave action reaches to a depth of 20, 50 or 100 ft in the protected areas of the Moray Firth, the Firth of Forth, and along parts of the Netherlands coast, whereas in exposed areas appreciable disturbance due to wave action may reach depths greater than 300–500 ft[69]. In the southern North Sea, especially, the continual movement of the seabed sorts out the finer sedimentary materials. However, the relatively weak tidal currents and greater depth of the middle and northern sections, combined with the effects of, all but the greatest, storms failing to reach depths in excess of 50–60 m result in relatively unworked sediments[17].

Beaches consist of unstable aggregations of sedimentary materials which undergo changes of orientation depending upon waves and currents. Nevertheless two major beach forms can be distinguished: the *berm* which is above water, and the *bar* which is below water[72]. (Fig. 6.3). The height and energy of a wave influences the depth to which sand is disturbed: this increases by approximately 1 cm depth of soil for a 1 ft increase in wave height at the break point. The depth of disturbance is greatest in the shallow waters of the swash zone and surf zone, and becomes greater as the size of the beach material increases[73, 74].

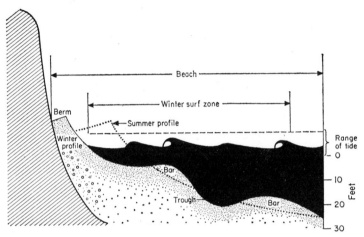

Fig. 6.3. Profile of a beach is characterized by a berm (the deposit of material at the top of the beach) and bars. In winter heavy surf removes sand from the berm and deposits it on the bars; in summer, light surf builds the berm. Vertical scale is exaggerated 25 times (after Bascom[80]).

In 1847, Stokes[75] showed that due to a slight progressive movement of water particles in the direction of wave propagation there is a slight advantage in favour of forward movement in the particles of an oscillatory wave. Field and laboratory investigation confirmed this movement which provides the means whereby loose sediment, present in the shallow offshore zone, may be transported shorewards[72, 76]. The differential between the forward and backward velocities of the water particles provides the mechanism which affects sediment particles of individual sizes and settling rates selectively[76]. Experiments in a wave channel revealed that waves with a steepness > 0·03 produce storm profiles of the beach, erode the foreshore and build up longshore bars, whereas those waves with a steepness < 0·02 add materials to the foreshore to produce ordinary or summer beach profiles which lack longshore bars. The rate of sand movement, and thus of profile change, is greatest immediately after a change in wave steepness[77]. Other laboratory studies of the equilibrium profiles of beaches demonstrate that if effects such as obliquity of wave approach, changes in water level and the complexity of natural wave trains are excluded, the conditions which govern beach formation and bottom profiles, in nature, may be summarized, thus:

1. wave steepness is an important factor governing onshore and offshore movement of beach material;
2. for a specific class of beach material, the tendency for the material to move shorewards increases as the wave steepness decreases; and
3. for any specific condition of wave size and steepness, the tendency for beach material to move shorewards increases if the grain size of the material is increased[78].

On the U.S. Pacific Coast beaches, the slope of the sandy beach-face ranges from 1 : 4 to 1 : 100, changing in response to changes in wave steepness: an eroding beach will flatten and a building beach will steepen.

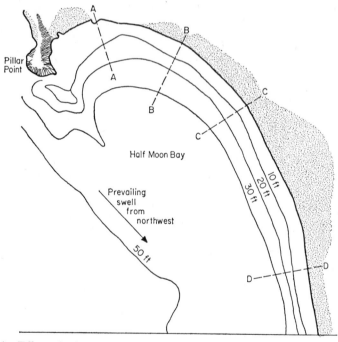

Fig. 6.4a. Effect of a headland on a beach eroded by a littoral current. The headland is Pillar Point and the sheltered area is Half Moon Bay. The profiles of the beach as at locations marked above are traced in the graph at Fig. 6.4b (after Bascom[80]).

Reference points	Sand size (Millimeters)	Slope
①	0.17	1.41
②	0.20	1.30
③	0.39	1.13.5
④	0.65	1.8

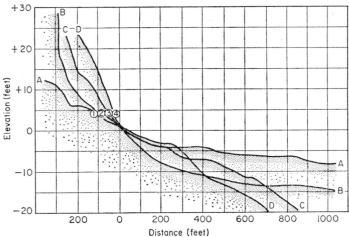

Fig. 6.4b. Profiles of beach at locations marked on Fig. 6.4a. The average size of the sand grains at each location and the slope of the bottom are listed in table above (after Bascom[80]).

The distribution of sand size along a profile is predictable, with reasonable accuracy, if the median diameter at the reference point is known. Large grains are present where turbulence is at a maximum, and a decrease in turbulence is accompanied by a decrease in the size of the sand grains. Finally the slope of a beach face is related to the median diameter of the sand and the wave energy at that point[79, 80] (Fig. 6.4).

A wave constructed beach undergoes seasonal changes in profile. In winter, the heavy surf removes sand from the berm and deposits it on the bars, whereas in summer the light surf builds the berm and bar formation is not so evident[80] (Figs. 6.3, 6.5). Evidence can always be found to contradict any law which attempts to relate beach structure to wave conditions, but that least open to dispute states that low waves enlarge sand beaches, and steep waves accompanying on-shore winds erode them[81, 82]. In general, the changes in sand distribution at the H.W.M. are more easily observed than those below the L.W.M., a situation which arises because the material formerly at the H.W.M. can be dispersed over so wide an area, e.g. between the 20 ft and 30 fm contours, once it passes below the L.W.M.[82]. In one instance, however, on the coast of

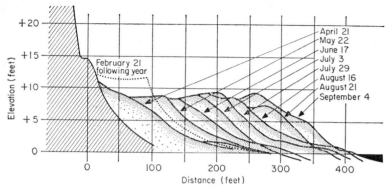

Fig. 6.5. Growth of the berm at Carmel, Calif., during the spring and summer is indicated by this series of dated slopes, based on actual measurements. Vertical dimension is exaggerated 10 times. The dotted line shows how the berm was cut back during following winter (after Bascom[80]).

California, it was shown that steep, winter waves removed sand from shallow water and moved it seawards within an area bounded by the 20 ft contour. In summer, on the other hand, sand returned under the action of long, low waves[83].

Shingle like sand, is influenced by these processes, and although long-shore drift was formerly considered to be of an overwhelming importance with respect to shingle, there is now reason to believe that direct onshore movements may be significant also[84]. It is of interest that, under the appropriate conditions, waves of oscillation and of translation may transport debris towards the land: shingle and ballast deposited at 10–20 fm depth, in the North Sea, and some 7–10 miles off Sunderland have been thrown upon the shore by storm waves; elsewhere, pig lead, from a wreck 1 mile offshore, has been thrown up on the shore during a violent storm[69]. Similarly, remarkable banks of shingle have been thrown up near the mouths of the Nith and Urr and in Auchencairn Bay, all of which are tributaries of the Solway Firth[85].

Just inshore of the breaker, where waves plunge, a sand *bar* and *trough* is formed. If the breakers reform and plunge again, then at this second plunge line, a second line of bars and troughs form. Bars and troughs found to the seawards of the breakers are not a part of the existent equilibrium slope of the beach, but were formed by waves of a greater height in an antecedent wave train which broke in deeper water. Such a system may be slow to disappear. The genesis of bars and troughs may be explained in part, by the abrupt increase in turbulence as the wave plunges: the trough is created by erosion, but the bar is an erosion remnant. An alternative, but less convincing explanation, proposes that sand is deposited at the bar because it is brought up in the breaker from deep water faster than it is carried on to the upper beach[82]. In models, bars may make an orderly advance upshore to the upper beach[81], array up the shore on to the upper beach, or make a migration towards or away from the shore depending upon wave steepness[82]. A small tidal range is an essential prerequisite for the growth of these *plunge line bars* which are therefore better developed at neap tides than at springs. It follows that they are most completely developed in "tideless" bodies of water, e.g. the Mediterranean and the Great Lakes of North America. In normal weather conditions, the growth of bars may occupy several weeks, whereas in storm conditions it make take as many days[82]. Another type of bar, the *swash bar*, is not only not reduced by the tides, but also depends upon them for its existence. A swash bar is created by the swash from breakers which cause a local build up on an otherwise gently sloping shore. Near the upper limit of the swash, the slope constructed is steeper than elsewhere. Moreover, at those situations where the tidal level tends to remain constant, i.e. at H.W. and L.W., the swash will have a greater opportunity of creating a local steep slope. Indeed nearly all sandy shores are characterized by a bar at low water mark neap tides[82].

At the seabed, water may move heavy materials, such as shingle, by rolling them along the bed, unlike sand which may remain suspended indefinitely under the influence of water movement at all depths[82].

It has been shown by experimental studies that for a given beach, i.e. slope and roughness, a given incident wave and a given sediment, the offshore movement of sand is slower and weaker than the onshore movement. On those beaches which have a constant slope the net transport of sediments due to wave action must be shoreward; significant movement offshore was due to rip currents and suspended load subject to density currents. Because of the increased turbulence, the median particle size increases upshore as occurs on natural beaches. Furthermore, there is on an experimental beach a tendency for many sizes to accumulate in the breaker region as occurs in the field[86, 87]. Some interesting experiments have been performed in an attempt to replenish beaches which have lost sand by erosion. These attempts have been characterized by dumping the replenishing materials offshore, and do not appear to have been very successful: some material did move onshore[88], but, it was considered that elsewhere in the absence of offshore bars, replenishment would not occur, and the dumped material would be lost permanently to the deeper water[89].

Ripples are a feature of most shores and show an orientation normal to the incidence of the waves. Normally, they have a wave length of a few inches, but some giants with a wave length of 10 ft have been observed[80].

On sand bars, they exist on the seaward slope and progress steadily shorewards. An equilibrium condition is never reached, especially in tidal seas where both bar and ripples make a response to the changes in level related to the tidal cycle[82]. Ripples are often used as a source of evidence of sediment movements during the period when a shore is covered. The observation and interpretation of such data requires great care, for the ripples created by waves and wavelets during the ebbing tide can give a completely erroneous picture[90]. Moreover, the relationship between current velocity, sand grain size and the dimensions of ripple marks is still unexplained[80]. Not only ripples, but beaches of sand and shingle in wide bays and on open coasts are orientated at right angles to the direction of the dominant waves[91-93].

Although not a result of wave action, a strong offshore wind will induce an onshore drift or current at the seabed, while an onshore wind will induce a drift or current seawards. In Torquay Harbour, strong onshore winds drive surface water shorewards and induce a strong bottom current flowing out of the habour; the reverse happens when a strong wind blows offshore. This phenomenon is familiar to European bathers who have found warmer water offshore than inshore. Such bottom currents may be feeble, but under favourable conditions, as in channels between shallows or in shallow bays, the velocities may be sufficient to move fairly coarse debris, especially where waves disturb the bottom. Sand and debris migrate shorewards and seawards with offshore and onshore winds respectively[69, 74, 82].

Longshore or Beach Drifting

When a wave train is obliquely incident upon a shore, a lateral transfer of water results and gives rise to a longshore current (p. 47). Such a longshore current may reach a velocity of several knots and may be capable of carrying large amounts of sediment. If the flow of sediments is impeded either by direct obstructions, in the path of the longshore current, or by interference with the waves that maintain the current, the equilibrium of the beaches is disturbed and sedimentation or erosion will follow. A knowledge of the relationship between the character of ocean waves and the direction and magnitude of the longshore current has proved helpful in the design of jetties. During the 1949-45 war considerable difficulty was experienced in landing amphibious craft in the presence of longshore currents. At times these currents were of such a magnitude that the landing craft broached even though the breakers had been negotiated successfully. Often it was a greater hazard than the breaking waves. On a straight beach with parallel contours, the velocity (v) of the longshore current may be derived, thus:

$$v = k \left[(mH^2/T) \sin 2\alpha \right]^{1/3}$$

where m is the average beach slope, H, L and T are the height, length and period of the breaking wave, k is a friction parameter, and α the angle formed by the wave crest with the shoreline[94].

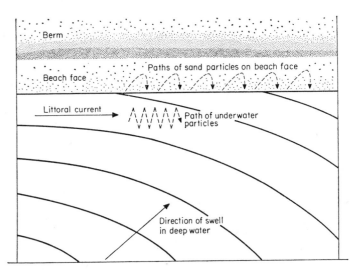

Fig. 6.6. Littoral current, a current running parallel to the beach, is set up when waves move toward the beach at an oblique angle. Under such conditions sand grains lifted by the surf, normally moved at right angles to the beach, are transported with the current (after Bascom[80]).

Evidently, the individual wave breaking obliquely upon a shore will translocate sediment, either sand or shingle, in a path conforming to that of the swash across the shore and gravity pulling the backwash directly down the shore. Therefore, the sedimentary materials will be moved in a parabola, with a resultant movement away from the direction of wave advance (Fig. 6.6) and thus undergo beach drifting[69, 80]. This phenomenon was first recognized in 1834 by Palmer who also appreciated that smaller particles describe larger parabolas and are carried further down the slope, in the backwash, than the larger particles[69].

The direction of beach drifting depends on a number of factors, such as:

1. The direction of groundswell approach, when little refraction occurs, and waves generally are less influential in movements of debris.

2. The direction of prevailing or dominant storm winds, when these develop powerful waves and ground swells are weak, much refracted or approach normal to the shore.

3. The direction of the greatest fetch is also important, weak winds acting over a long fetch may develop larger waves than strong winds blowing over a short fetch[69].

Because the fetch of the prevailing westerly winds is only 40 miles, compared with the 180 miles for the easterly winds, the direction of beach drifting on the sand spit which encloses Toronto harbour, Lake Ontario, is westwards, i.e. against the prevailing winds[69]. On the south coast of England, the prevailing W.S.W. wind, blowing over the longest fetch, induces a drift up the English Channel, i.e. from west to east. On the east coast, however, the prevailing winds are offshore and the drift depends upon fetch at the individual situation. At Great Yarmouth, which is partially sheltered from northerly winds with the longest fetch, the beach drifting is variable, because the fetch from all other directions is approximately equal. Conversely, Spurn Head, Yorkshire, is not so protected from the north, and the direction of beach drifting is from north to south[82].

Beach drifting may occur either on a *prograding beach* to which materials are added, or on a *retrograding beach* from which materials are removed, or on a *beach at grade* in which there is no resultant change[69]. It is considered by some authorities to be the most important force in the construction of shingle and sand beaches. On any given part of the coast the direction of littoral drift, but not the quantity of material transported, is usually known. A widespread belief that the littoral drift occurs in the direction of the flood current, is true for certain places only. The influence of the tide results from a disparity between the flood and ebb currents in shallow water (p. 149). In the absence of wave action and in the presence of this asymmetrical tidal curve, sedimentary materials, if moved, tend to move more with the flood than the ebb, with a resultant drift in the direction of the flood current. With the effect of turbulence due to waves more material would be carried into suspension than due to the influence of the tides alone. Where winds blow along an estuary equally in both directions, the effects of wave action would be nullified and the movement of sedimentary materials would predominantly remain in the direction of the flood tide[82].

Although it may not always be true, movements due to littoral drifting are considered to be greatest in the upper beach, but they are difficult to measure. In general, the amount of material in motion is measured by the rate of accretion of the seabed, of sand or shingle spits, or at natural or artificial obstacles which prevent the further movement of sedimentary material. The accumulation of materials at a newly erected groyne may lead to the construction of a beach at the expense of ones further down the coast where erosion may supervene. Indeed, the characteristic accumulation of sedimentary materials on one side of a groyne, and the removal on the other indicate the magnitude and direction of littoral drifting[80, 82]. The rate of arrival of shingle to replace that which has been removed for construction purposes will give a measure of this process, e.g. at Rye 48 000 yd³ arrives per year. The inherent difficulty of forecasting the scale of sedimentary movement due to littoral drifting is illustrated by the events at Madras, where prior to the construction of the harbour, the drift was estimated to be 243 000 tons/yr. In 1904, on partial completion of the harbour this was revised to 550 000 tons/yr, but in fact it amounted to 1 000 000 tons/yr[82].

Of course, littoral drifting is not solely a shore phenomenon. At Santa Barbara where a littoral current flows constantly, a breakwater was built to create a quiet area for the shelter of boats. Unfortunately, the cessation of turbulence within the breakwater caused all the material carried in suspension by the littoral current to be deposited there. On average, about 800 yd³ of sand are dumped in the harbour daily, and four times that amount under storm conditions. To prevent filling of the harbour and erosion to the south, a dredge pumps the sand across the quiet water to a downstream beach. Here wave action raises it into suspension once more, and the littoral current carries it along until it reaches Santa Monica which has similar problems. On the southern shore of Long Island, littoral drifting, due to the prevailing winds and unrefracted waves from the North Atlantic, sweeps sand from Montauk Point, at the island's eastern tip, to the Rockaway spit, at the entrance to New York harbour. There is rapid erosion at Montauk Point, and, before the present shore-line structures were built, Rockaway Spit was increasing at a rate of 200 ft/yr[80].

Unlike the amount of sediment movement, the velocity of the drift is measured more easily. At Scolt Head, Norfolk, it varies from 5 to 25 m.p.h. depending upon the wind velocity[95]. At Anapa, on the Black Sea, the velocity of sand displacement was, on average, about three-quarters of the current velocity at the surface of the water, and could reach 3 km/hr; here, despite the high velocities observed, some of the marked sand remained at the site of injection for protracted periods. The paths followed, by the sedimentary particles, were linked to grain size. A slower movement in zig-zags of large amplitude was also observed. In places where erosion of the seabed occurred, the sand travelled distances in excess of 3 km in 3 months of calm, summer weather. In experiments on the Polish coast, a free transport of sand along the coast occurred in periods of light swell, but it was held by the groynes in rough weather[56]. On the Delaware coast, littoral drift was studied by means of "Tenite" pellets, these studies showed that the velocity of the drift depended upon the situation and time of release[57] (Table 6.1).

Table 6.1.   Littoral transport of "Tenite" pellets on the Delaware coast (Schuster[57])

(Total time elapsed in minutes : seconds for a pellet transport of 150 m)			
		Hours of release	
Location	1200	1300	1400
Cape Henlopen	4 : 35	5 : 50	6 : 45
Indian River	7 : 00	8 : 25	12 : 10
Fenwick Island	7 : 03	9 : 10	12 : 09

Tidal Transport Mechanisms

Tidal currents are powerful agents of sediment erosion and transport. For example, a current of 0·6 kt will erode all grades of soil between 0·05 and 2 mm. While soils of the 2 mm grade are transported by currents exceeding 0·3 kt, those of the 0·05 mm grade are transported by all currents which exceed 0·008 kt[9]. Similarly, fine gravel, shingle and angular stones, of 1–1·5 in. diameter, are moved at 1 kt, 2·5 kt and 3·5 kt respectively[69]. Moreover, unlike wave action, tidal currents do not diminish with increasing depth: the sands and gravels at a depth of 120–130 fm in "Beaufort's Dyke", in the North Channel between the Galloway coast of Scotland and Northern Ireland, are moved backwards and forwards by currents generated under the action of waves and tides[96]. Similar effects occur in Long Island Sound[69, 97].

In general, it has long been known that tidal currents can bring about significant movements of marine soils[98]. On the coast of Ireland, a number of beaches and submarine banks are formed mainly by tidal currents[99]. A study of Caister Shoal, off California, Norfolk, using the radioactive ⁴⁶Sc glass showed that while waves could assist in bringing materials into suspension, they had no significant effect upon the actual direction in which the particles moved. A comparison of the envelope of movement of the radioactive material with the tidal currents revealed that the ebb current had a maximum velocity of 1·4–1·5 ft/sec (0·8–0·9 kt), but the maximum flood current velocity ranged from 3·0–3·5 ft/sec (1·8–2·0 kt), and that the movement of the radioactive material was in the direction of the flood tide[100]. Elsewhere in the southern North Sea, sand streams, deduced from a study of sand waves, move in preferred directions for up to 100 miles, and complex patterns associated geographically with large sand banks may be inferred from the relative strengths of flood and ebb streams. The movement of such bottom materials, induced by dominant tidal streams, is filling the Wash with sediment[101, 102].

Such movements of sediment in a preferred direction, in areas outside an estuary, result from an asymmetrical tidal curve. In estuaries where the flood stream has a greater maximum velocity than the ebb, then there is a steady net transference of sedimentary materials upstream[69, 82, 103, 104]. Here deposition will exceed erosion, not merely until the regions adjacent to the main channels are silted up, but until the channels themselves may be blocked[69]. Indeed, a difference in velocity not exceeding $\frac{1}{2}-\frac{3}{4}$ kt, even at springs, is sufficient to ensure a steady net transference of sediment towards the head of the estuary[82, 104], while above M.S.L. the effects become more pronounced[103]. Where the ebb tide is considerably augmented by the outflow of a river, as in a Type A salt-wedge estuary, a net loss of sediment from the estuary results[69].

The influence of dominant flood tidal streams in transporting sediments into estuaries, bays and gulfs has been recorded in many countries of the European and American continents[105–116]. However, it must be emphasized that this transportation of sediment is a result of asymmetry in the tidal curve and *is not due to a residual current.*

It is not yet fully appreciated that much sedimentary material is introduced into estuaries from the offshore region, and that in much of N.W. Europe and Britain, at least, relatively little sediment is introduced by the rivers flowing into the estuary. As long ago as 1862, it was appreciated that the natural deposits of sediment in the Humber estuary probably resulted from the erosion of the sea coast of Holderness and subsequent transportation by the tides into the Humber estuary, and that these sediments were being deposited at a distance of 13–49 miles away from the site of erosion[98]. However, it was Skertchley, in 1877[106], who showed clearly that the amount of sediment deposited in the Wash could not be accounted for by the negligible amounts contributed by the inflowing rivers. Such deposition of silt transported into an estuary from the sea may have pronounced effects upon the estuary: the capacity of the Mersey, above Liverpool, decreased in 20 yr by some 5% (68×10^6 tons), the source of which could only have been the sea[111]. In the Wash, 64 000 acres were accreted in 1700 yr, although the rate of accretion was variable[106] (Table 6.2).

In the Solway Firth, the rivers contribute little sedimentary material, except some clay at times of flood[117]. Nevertheless, the Solway has been filling with sediment for centuries and continues to do so[116], and from 1856–1946 some 1780 acres were added to the Inner Solway[118]. Here much of the sediment is reworked material derived from the Irish Sea where it was dumped at the time of the Pleistocene Ice Age[104, 116]. Here, ^{106}Ru arising from the Windscale effluent, becomes attached to

Table 6.2.   Accretion at the base of the Wash (according
to Skertchley[106])

Date	Rate of accretion (ft/annum)		
	Maximum	Minimum	Mean
Between second and seventeenth centuries	15·84	1·76	7·29
Eighteenth century	89·76	21·12	48·65
Nineteenth century	70·41	13·20	31·68
Mean rates for 1700 years	59·00	3·88	10·73

silt on the bed of the Irish Sea, and is then intermittently transported upstream[116]. In many estuaries, large amounts of sediment remain in suspension throughout the tidal cycle: in the Dutch Waddensea a total quantity of 8×10^6 kg dry weight of silt is transported through the Marsdiep in every tidal phase. Here silt is accumulating in the upper estuary, but the rate of accumulation is difficult to define[11].

It is a characteristic of all estuaries and protected bays that the inflowing tide sweeps the finer materials far upstream so that the silt and clay are deposited in mud flats and salt marshes; at the same time, coarser sand is moved landwards a much shorter distance, often forming bars along the channels. Clearly, the outflowing current of the ebb tide will tend to remove sedimentary materials, particularly in closer proximity to channels, and in the exposed outer areas wave action prevents the settlement of finer grade materials. This tendency of estuaries to accumulate finer materials towards the head of the shore and towards the head of the estuary contrasts markedly with a shore influenced primarily by wave action. The dominant effect of tidal action upon the formation and soil grade structure of estuarine beaches has been observed in many parts of the world[69]: in the British Isles—The Wash[106, 118], Tamar and Lynher[110], Morecambe Bay and Liverpool Bay[119], Aberlady Bay, Firth of Forth[120], Tees and Tay[121], the Crouch[122], and the Solway Firth[116]; in N.W. Europe[123]—Jade and Dollart Bays[2], the Dutch Waddensea[11] and the Danish Waddensea[124]; in North America—Barataria Bay, Mississippi estuary[125], Yaquina Bay, Oregon[63, 126], and Barnstable Harbour, Massachusetts[115]; in South America—Gulf of Paria[114]; in South Africa—Zwartkops estuary[127]; and in Australian estuaries[128].

In general, the deposition of sediment occurs around the time of slack water, and in the upper area of the beach at the high tide slack water. However, in the River Avon, near Bristol, deposition occurs during two-thirds of the ebb tide instead of at high water[129]. Indeed, it is important to realize that because of the asymmetry of the tidal curve, the slower currents of the ebb tide provide a greater opportunity for sedimentation than during the flood. During the nineteenth century, land reclamation schemes took advantage of this effect: banks and sluices to regulate the water flow enclosed marshlands which were built up rapidly by means of these "warping" techniques[98, 105, 130].

Although all estuarine sediments are liable to reworking, two principal sites of deposition can be distinguished. At a site of *primary deposition*, which is normally protected from the prevailing wind or the longest fetch, sediments are deposited for a relatively brief period before being reworked. At a site of *secondary deposition*, which is normally an accreting edge of a marsh or the salt marsh itself, sediments may be deposited either after reworking from a primary site of deposition or directly upon being transported into the estuary. Unlike the primary site of deposition, sediments deposited at a secondary site of deposition represent either a permanent deposit or one which will not be reworked for some years[116]. The most stable estuarine sediments are those in which the higher salt marsh plants and their precursors achieve a rapid colonization. In an area like the Solway Firth, the channel migrations and consequent reworking of sediments, may be very pronounced and appear to be cyclical in nature[116, 117, 131–133]. In addition to the direct reworking of the sediments during the channel shift, the sedimentation/erosion regime of the marsh edge is much influenced by the proximity of the channel[X]

Both the rate of reworking[133], and the erosion of a marsh edge may be accelerated by storms, although in the latter case much of the erosion is induced by wave action on the ebb tide, and especially around the level of high water neaps[69]. At Cross Canonby, Outer Solway Firth, a gale in February 1967 induced a marked erosion of the "dune" area which continued to 1971, simultaneously the whole of the shore was set into a continuous motion which makes day to day recognition difficult[X]. Under very extreme conditions, the mouth of a river may migrate many miles; that of the Hwang-ho is perhaps the best known, but others have taken place over centuries[134].

Not all reworking and erosion is so spectacular, although the effects may be equally far reaching. Just as living organisms may be responsible for the deposition of sedimentary materials (p. 131) so they may rework considerable amounts of sediment: assuming, as a conservative estimate, that 15 lugworms, *Arenicola marina*, per m^2 each ingest $20\,cm^3$ of sediment per tidal cycle during the 500 tidal cycles per year, then these worms will rework $150\,000\,cm^3$ per annum; by these means, the *Arenicola* alone rework the entire thickness of sandy sediment, above the level of the feeding depth, once in every 20 months. Because the worm rejects coarse particles, a coarse layer is formed at depth within the soil[135, 136].

Not all the influences which induce erosion and reworking are marine in origin. Man, since early times, has affected estuarine areas, but his

6

influence has increased since the early nineteenth century, especially by the construction of engineering works. These works can induce deposition or erosion, either by providing shelter which permits increased deposition or by causing shifts in a channel which induce erosion[85, 117]. Erosion may be induced by overgrazing of marshes[117], and increasing use or misuse during leisure activities is responsible for the breakdown of vegetation and subsequent erosion of the supra-littoral and salt marsh fringes. This in turn may lead to a loss of sediment from the shore[85, X]. In the Solway Firth, this problem has become especially serious since 1964, but it is not confined to Britain or N.W. Europe. In North America, this problem has become more acute following the introduction of the "dune-buggy". In an age which has become conservation conscious it cannot be emphasized too strongly that lack of forethought concerning the use of these "fragile" areas may be much more damaging than the effects of pollution: a statement as true of the marine environment[85, X], as it is of the land[137].

At any one situation, in an estuary, the structure of the sediments is influenced by the processes of sedimentation and erosion, and all the reworking that this implies. Consequently many soils are laminated due to varying sources of material and the variations in current direction and velocity during the tidal cycle[118, 136] (Fig. 6.7).

The infauna of a sediment can influence the sedimentation, erosion and reworking of that sediment. However, sediment movements can influence not only the infauna, but the whole ecology of an estuary. The Solway Firth, as we have already seen, has been filling with sediment for centuries. This has resulted in a decline of the fisheries in the Inner Solway, and the fishing grounds for shrimp and plaice have undergone a pronounced movement downstream; other fish species which formerly penetrated the Inner Solway no longer do so. Navigation has become more difficult and of the few fishing families which remain, many work as far afield as the Oban area[116, 138, X]. Few animals live in the southern shore of the Solway, around Cross Canonby, because of the rapid sediment movements. At Beckfoot less exaggerated movements, from 1963 onwards, led to a replacement of *Macoma balthica* by *Tellina tenuis*; subsequently both declined, and this once rich feeding area became impoverished (p. 442).

Although the transport of sediments by tidal currents is of considerable importance in estuaries which border on the North Sea and the Irish Sea, wave action offshore influences the amount of silt available for transport to these estuaries. The relative lack of wave action reaching to the bed of the Irish Sea has resulted in shore and marsh soils which have a significantly lower silt content than those which border the North Sea. Although the role of silt is not yet fully understood, it is appreciated that

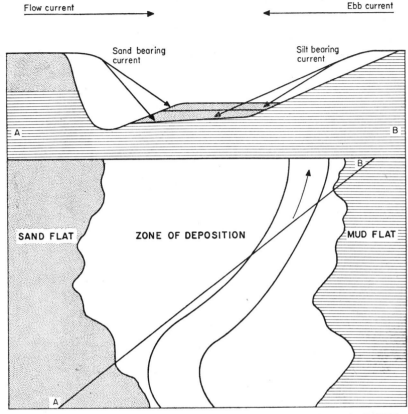

Flow current ⟶   Ebb current ⟵

Sand bearing current   Silt bearing current

A   B

B

SAND FLAT   ZONE OF DEPOSITION   MUD FLAT

A

Fig. 6.7. Diagram indicating origin of type of lamination (after Kindle[118]).

it carries a vital nutrient load. In consequence, the salt marshes bordering the Irish Sea are floristically impoverished compared with those which border the North Sea[116, 139]. Furthermore, this lack of silt caused the failure of land reclamation schemes in the Morecambe Bay area which is now considered to be worthless for such purposes[92]. Such an example confirms what has already been noted on p. 62, viz. that an estuary is a complex system which cannot be treated as a self-contained unit. Indeed, it may be influenced by events which occur at a considerable distance from the site at which events are observed.

The degree to which either wave or tidal action, or both combined, influence the sedimentary regime of an estuary may be difficult to assess. However, much can be learned from a simple rule, viz. on a wave built shore the median particle size decreases from H.W.M. to L.W.M., whereas on a shore constructed predominantly by tidal mechanism, the median particle size increases from H.W.M. to L.W.M. Further difficulty is removed when it is appreciated that resulting from the accretion of centuries, the sediments at the head of an estuary are deposited in a gentle gradient. Downstream where accumulation is not so great, there is a steep interface at which erosion frequently occurs. Such an interface must be subject to wave action and some longshore drift is usually evident, as the fetches here at the time of high water are usually considerable. At a slightly lower tidal level, shelter conferred by the marginal and mid-stream banks will reduce wave action and therefore longshore drift. It is advisable in such cases to use caution in making an extrapolation from effects observed at the interface to the whole beach area. Similarly, erosion of marshes may occur largely because a supply of fertilizing silt is insufficient, but sedimentation in the estuary as a whole may be widespread and rapid[116].

Ebb and Flood Channel Systems

Those bays and estuaries in which ebb and flood channel systems exist show a considerable diversity of morphometry. Common to all, the seabed is fashioned into a series of inter-digitating channels of two main types. Those running into the estuary from the seaward, gradually narrowing and shoaling as they approach the head of the estuary, are termed *flood channels*. On the other hand, the opposing *ebb channels* are often deepest at their inner end, shoaling to the seawards and terminating in a submarine bar. Channels of both types may run side by side for a considerable distance to form an interlocking system, in which the individual units are separated by elongated sand banks[104]. This avoidance, typical of ebb and flood channels, results because the ebb stream may continue to run for some time after the flood has begun to make.

During the Pleistocene Ice Age sediment was dumped in the North and Irish Seas. Since then, much of this material has reworked into the adjacent estuaries and now chokes them. Here, with an adequate supply and thickness of unconsolidated sediment and rectilinear tidal streams, the distinctive configuration has developed. The English Channel was not affected by this Ice Age, and the estuaries bordering it do not have these systems. In estuaries, tributary to the North and Irish Seas, where, despite run-out from the land, the flood current is stronger than the ebb, the flood channels are more deeply cut and are more stable than the ebb channels[104].

Flood channels and flood channel barbs are an important influence in the transport and deposition of sediments and areas of heavy silt deposition are often associated with flood channel barbs[116, 140]. Flood channel elements may be traced a considerable distance offshore and influence the supply of sediment to an estuary[141]. For these reasons, flood channel systems should not be used as spoil grounds.

These systems are present in the Thames, Wash, Humber, Tay, Solway Firth and Liverpool Bay, where they are of considerable concern to the shipping companies which use these estuaries[104, 140, 142]. Where harbours are to be constructed or enlarged these systems can be considered to advantage[143]. Nevertheless, it should be remembered that these systems are unstable, and small changes in the relative strengths of currents or of sediment supply can have repercussions throughout the whole.

References

1. Sheldon, R. W. and Warren, P. J. (1966). *Nature, Lond.* **210**, 1171–1172.
2. Hantzschel, W. (1939). *In* "Recent Marine Sediments: A Symposium" (P. D. Trask, ed.), Am. Assoc. of Petroleum Geologists, Tulsa, Oklahoma; Thomas Murby, London.
3. Wetzel, W. (1937). *N. Jahrb. f. Minerol. Beil.* Bd **78**, B.S. 109–122.
4. Schultz, E. (1932). *Kieler Meeresforsch.* 1936–37, **1**, 359–378.
5. Damas, D. (1935). *Annls Soc. géol. Belg.* **58**, S.B. 143–B151.
6. Moore, H. B. (1930). *J. mar. biol. Ass. U.K.* **16**, 595–607.
6a. Moore, H. B. (1931). *J. mar. biol. Ass. U.K.* **17**, 325–358.
7. Carritt, D. E. and Harley, J. H. (1957). *Nat. Acad. Sci. Nat. Res. Counc.* Washington, D.C. Publ. No. 551, 60–68.
8. Osterberg, C., Carey, A. G. and Curl, H. (1963). *Nature, Lond.* **200**, 1276–1277.
9. Hjulström, F. (1939). *In* "Recent Marine Sediments: A Symposium" (P. D. Trask, ed.), Am. Assoc. of Petroleum Geologists, Tulsa, Oklahoma; Thomas Murby, London.
10. Inman, D. L. (1949). *J. sedim. Petrol.* **19**, 51–70.
11. Postma, H. (1954). *Archs néerl. Zool.* **10**, 1–106.
12. Perkins, E. J., Bailey, M. and Williams, B. R. H. (1964). H.M.S.O., U.K.A.E.A., P.G. Report 604 (CC).
13. Burt, W. V. (1955). *J. mar. Res.* **14**, 47.
14. Spencer, J. F. Personal Communication.
15. Perkins, E. J. and Williams, B. R. H. (1964). U.K.A.E.A., Chapelcross Health Physics Department Note No. 2.
16. Perkins, E. J. (1965). *Trans. J. Proc. Dumfries. Galloway nat. Hist. Antiq. Soc.* Ser. 3, **42**, 6–13.
16a. Dorst, J. (1960). "Before Nature Dies", Collins, London.
16b. Nicholson, M. (1970). "The Environmental Revolution", Hodder and Stoughton, London.
16c. Darnell, R. M. (1967). *In* "Estuaries" (G. H. Lauff, ed.), *Publs Am. Ass. Advmt. Sci.* **83**, 376–381.
16d. Pearce, J. B. (1970). *In* "F.A.O. Technical Conference on Marine Pollution and its Effects on Living Resources and Fishing", F.I.R.: MP/70/E-99. Rome 9–18 December 1970.
17. Luders, K. (1939). *In* "Recent Marine Sediments: A Symposium" (P. D. Trask, ed.), Am. Assoc. of Petroleum Geologists, Tulsa, Oklahoma; Thomas Murby, London.

18. Putman, J. L., Smith, D. B., Wells, H. M., Allen, F. and Rowan, G. (1954). U.K.A.E.A., A.E.R.E. 1/R. 1576. Harwell, December 1954.
19. Putman, J. L. and Smith, D. B. (1956). *Int. J. appl. Radiat. Isotopes* 1, 24–32.
20. Hydraulics Research Station (1956). H.R.S./P.L.A. Paper, 20, 24 pp.
21. Goldberg, E. D. and Inman, D. L. (1955). *Bull. geol. Soc. Am.* 66, 611–613.
22. Hours, R., Nesteroff, W. D., Romanovsky, V. (1955). *C.r. hebd. Séanc. Acad. Sci., Paris* 240, 1798–1799.
23. Hours, R., Nesteroff, W. D. and Romanovsky, V. (1955). *Trav. Cent. Rech. Étud. océanogr.* 1 (11).
24. Inose, S. and Shiriashi, H. (1955). *In* "Premiere Conference de Geneve sur les applications pacifiques de l'energie atomique, 1955," U.N. Publication, 8P/1053.
25. Boldyrev, V. L. (1956). *Proc. Assoc. for Maritime Schemes, Ministry of the Mercantile Marine of the U.S.S.R.* 3, 1956.
26. Hydraulics Research Board (1957). *Hydrauls Res.* 1956. H.M.S.O., London.
27. Inose, S. and Shiriashi, H. (1956). *Dock Harb. Auth.* 36 (434) : 284–288.
28. Kidson, D., Smith, D. B. and Steers, J. A. (1956). *Nature, Lond.* 178, 257.
29. Medviediev, V. S. and Aiboulatov, A. (1956). *News Issues of the Academy of Sciences of the U.S.S.R., Geographic Series* No. 4, 1956.
30. Allen, F. H. and Grindley, J. (1957). *Dock Harb. Auth.* 38, 302–306.
31. Amano, R. and Hamada, T. (1957). *In* "Proc. 19th International Congress of Navigation", Section II Communication, Part 3, London 1957.
32. Arlman, J. J., Santema, P. and Svasek, J. N. (1957). "Movement of Bottom Sediments in Coastal Waters by Currents and Waves; Measurement with the Aid of Radioactive Tracers in the Netherlands", Progress Report, June 1957. Issued by Deltadienst Rykswaterstaat, Ministry of Transport and Waterstaat, The Netherlands.
33. Davidson, J. (1957). *Svensk geogr. Arsb.* 33, Lund 1957.
34. Forest, G. (1957). *J. Mar. march.* Special Number, "Nouveautes Techniques Maratimes", 223–229.
35. Forest, G. and Jaffry, P. (1957). *In* "Congress of the Assn. Int. Res. Hyd.", Lisbon 1957.
36. German, J., Forest, G. and Jaffry, P. (1957). *In* "Proc. 6th Conf. on Coastal Engineering", Gainesville (U.S.A.) December 1957.
37. Hydraulics Research Station (1957). "Radioactive Tracers for the Detection of Offshore Beach Movements", H.R.S. Paper. Wallingford, 1957.
38. Smith, D. B. and Eakins, J. D. (1957). *In* "International Conference on Radio-isotopes in Scientific Research", UNESCO/NS/RIC/63, Paris, September 1957.
39. Hours, R. (1958). "Application des Radioelements au Genic Civil Journees d'Information sur les Applications Industrielles des Radioelements et Rayonnements", Conservatoire Nationaledes Arts et Metieres, Paris, February 1958.
40. Hours, R. and Jaffry, P. (1958). Publ. C.E.A.-S.A.R./58-312, RH/GC. Saclay. March 1958.
41. Inman, D. L. and Chamberlain, T. K. (1958). *In* "Peaceful Uses of Atomic Energy. Proc. 2nd International Conference held at Geneva, September 1958", U.N. Publication No. P/2357.

42. Zhadin, V. I., Kuznetsov, S. I. and Timofeev-Resovsky, N. V. (1958). *In* Peaceful Uses of Atomic Energy. Proc. 2nd International Conference held at Geneva, September 1958", U.N. Publication 27: 200–207.
43. Hours, R. and Jaffry, P. (1959). Centre d'etudes nucleaires de Saclay. Rapport C.E.A. No. 1269: 30 pp.
44. Jaffry, P. and Hours, R. (1959). Centre d'etudes nucleaires de Saclay. Rapport C.E.A. No. 1271: 475–498.
45. Gibert, A. (1955). "Essai sur la Possibilite d'Employer ^{110}Ag dans l'Etude du Transport du Sable par la Mer", Ministerio dan Obras Publicas, Laboratorio Nacional de Engenharia Civil, Publ. No. 63, Coimbra, 1955.
46. Gibert, A. (1957). *In* "Proc. 7th Congress of the Association of Hydraulics Research", Communication No. D35, Lisbon 1957.
47. Gibert, A. and Cordeiro, S. (1958). *In* "Peaceful Uses of Atomic Energy. Proc. 2nd International Conference held at Geneva, September 1958", U.N. Publication No. P/1820.
48. Schulz, H. and Strohl, G. (1960). *In* "Proc. 12th General Assembly of the International Union of Geod. and Geophys.", August 1960, Helsinki, Finland.
49. Gibert, A. and Cordeiro, S. (1962). *Int. J. appl. Radiat. Isotopes.* **13**, 41–45.
50. Klein, A. (1960). *In* "Proc. 12th General Assembly of the International Union of Geod. and Geophys.", August 1960, Helsinki, Finland.
51. Claske, R. (1966). *Sci. J.* **2** (2): 20–21.
52. Wasmund, E. (1939). *Geologie Meere Binnengewäss.* **3**.
53. Hermanovitch, D. E. and Zabelina, E. K. (1956). *Proc. Inst. Oceanogr.* **40**.
54. Vendron, S. L. *et al.* (1957). "Use of Fluorescent Substances for the Investigation of Material Along the Banks of Impounding Reservoirs", *River Transport*, No. 40.
55. Zenkovitch, V. P. and Yegoron, E. N. (1957). *Proc. Inst. of Oceanology of the Academy of Sciences of the U.S.S.R.* **21**.
56. Zenkovitch, V. P. (1960). *Dock Harb. Auth.* **40**, 280–283.
57. Schuster, C. N. (1963). "Our Ever-changing Coastline", *Univ. of Delaware, Estuarine Bulletin* **7** (1), 3–12.
58. Byrne, J. V. and Kulm, L. D. (1967). *Proc. Am. Soc. civ. Engrs, J. Waterways and Harbour Division* **93** (WW2, May 1967), 181–194.
59. Mackenzie, F. T., Kulm, L. D., Cooley, R. L. and Barnhart, J. T. (1965). *J. sedim. Petrol.* **35**, 265–272.
60. Jerlov, N. G. (1960). *Assoc. d'Oceanogr. Physique U.G.G.I., Proc.-Verb.* No. 7: 178.
61. Linder, C. P. (1960). *In* "Proc. 12th General Assembly International Union Geod. and Geophys.", 25–26 August 1960, Helsinki, Finland. *Trans. Am. geophys. Un.* **41** (4), 1960.
62. Editor (1961). *Surveyor* **120** (3579), 5–6.
63. Kulm, L. D. and Byrne, J. V. (1967). *In* "Estuaries" (G. H. Lauff, ed.), *Publs. Am. Ass. Advmt. Sci.* **83** 226–238.
64. Carritt, D. E. (1956). *Trans. 21st N.Am. Wildl. Conf.*, March 5–7, 1956, 420–436.
65. Burt, W. V. and McAlister, W. B. (1959). *Res. Briefs Fish Commn. Ore.* **7**, 14–27.
66. Allen, F. H. and Price, W. A. (1959). *Dock Harb. Auth.* **40**, 72–76.
67. Inglis, C. C. and Allen, F. H. (1957). *Proc. Instn civ. Engrs.* **7**, August 1957.

68. Bates, C. C. and Freeman, F. (1952). *In* "Proc. 3rd Conference Coastal Engineering", Camb., Mass. 1952 (12), 165–175.
69. Johnson, D. W. (1919). "Shore Processes and Shoreline Development", Wiley, New York.
70. Cornish, V. (1910). "Waves of the Sea and Other Water Waves", T. Fisher Unwin, London and Leipzig.
71. Stevenson, T. (1886). "The Design and Construction of Harbours", 3rd edn. Edinburgh.
72. Bascom, W. N. (1954). *In* "Proc. 4th Conference Coastal Engineering", Council on Wave Research, Engineering Foundation, Chicago, 1954.
73. King, C. A. M. (1951). *J. sedim. Petrol.* **21**, 131–140.
74. King, C. A. M. (1959). "Beaches and Coasts", Edward Arnold, London.
75. Stokes, G. G. (1847). *Trans. Camb. phil. Soc.* **8**, 441–445.
76. Grant, U. S. (1943). *Am. J. Sci.* **241**, 117–123.
77. Scott, T. (1954). *Dept. of the Army, U.S. Corps of Engineers, Beach Erosion Board, Tech. Memo.* No. 48. 1954.
78. Rector, R. L. (1954). *Dept. of the Army, U.S. Corps of Engineers, Beach Erosion, Board, Tech. Memo.* No. 41, 1954.
79. Bascom, W. N. (1951). *Trans. Am. geophys. Un.* **32**, 866–874.
80. Bascom, W. N. (1960). *Scient. Am.* **203** (2): 81–94, August 1960.
81. Bagnold, R. A. (1947). *J. Instn civ. Engrs*, 1946–47, Paper No. 5554, No. 4.
82. Russell, R. C. and MacMillan, D. H. (1952). "Waves and Tides", Hutchinson, London.
83. Grant, U. S. and Shepard, F. P. (1940). *In* "Proc. 6th Pacific Science Congress", **2**, 801–805.
84. Steers, J. A. (1957). *In* "Proc. 6th Conference on Coastal Engineering", Miami, Florida, 1957. Review *Dock Harb. Auth.* **40** (463), May 1959.
85. Perkins, E. J. (1968). *Trans. J. Proc. Dumfries. Galloway nat. Hist. and Antiq. Soc.* Ser. 3, **45**, 15–43.
86. Eagleson, P. S., Dean, R. G. and Peralta, L. A. (1957). *Mass. Inst. Technol. Hydrodynamics Laboratory, Dept. of Civil and Sanitary Engineering, Tech. Rept.* No. 26.
87. Ippen, A. T. and Eagleson, P. S. (1955). *Dept. of the Army, U.S. Corps of Engineers. Beach Erosion Board, Tech. Memo.* No. 63.
88. Hall, J. V. (1953). *In* "Proc. 3rd Conference Coastal Engineering, Cambridge, Mass. 1952", Chap. 10, 119–136.
89. Currier, L. W. (1953). *In* "Proc. 3rd Conference Coastal Engineering, Cambridge, Mass. 1952", Chap. 9, 109–118.
90. Dixon, E. E. L. *et al.* (1926). "The Geology of the Carlisle, Longtown and Silloth District", Memoirs of the Geological Survey of England and Wales, 1926, 113 pp.
91. Lewis, W. V. (1938). *Proc. geol. Assoc.* **49**, 107–127.
92. Steers, J. A. (1946). "The Coastline of England and Wales", University Press, Cambridge.
93. Steers, J. A. (1962). "The Sea Coast", Collins, London, New Naturalist Series No. 25.
94. Putnam, J. A., Munk, W. H. and Traylor, M. A. (1949). *Trans. Am. geophys. Un.* **30**, 337–335.
95. Steers, J. A. (1959). "Scolt Head Island", Heffer, Cambridge.
96. Kinahan, H. C. (1900). *Proc. R. Ir. Acad.* Ser. 3, **6**, 27.

97. Dana, J. D. (1895). "Manual of Geology", New York.
98. Oldham, J. (1962). *Minut. Proc. Instn civ. Engrs* **21**, 454–465.
99. Kinahan, G. H. (1883). *Proc. R. Ir. Acad.*, Ser. 2, **3**, 191–195.
100. Reid, W. J. (1958). *Dock Harb. Auth.* **39**, 84–88.
101. Stride, A. H. (1959). *Dock Harb. Auth.* **40**, 145–147.
102. Stride, A. H. and Cartwright, D. E. (1958). *Dock Harb. Auth.* **38**, 323–324.
103. Cornish, V. (1922). *Jl R. Soc. Arts.* **60**, 1123.
104. Robinson, A. H. W. (1956). *J. Inst. Navig.* **9**, 20–46.
105. Mitchell, H. (1872). *Rept. U.S. Coast Survey for* 1899, Appendix 5, p. 82.
106. Skertchley, S. B. J. (1877). "The Geology of the Fenland", Memoirs of the Geological Survey of England and Wales, London.
107. Dawson, J. W. (1878). "Acadian Geology", London.
108. Sollas, W. J. (1883). *Q. Jl geol. Soc. Lond.* **39**
109. Crosby, W. V. (1903). *Technology Quart.* **16**, 89.
110. Percival, E. (1929). *J. mar. biol. Ass. U.K.* **21**, 721–742.
111. Southgate, B. A. *et al.* (1938 [1961]). "Effect of Discharge of Crude Sewage into the Estuary of the River Mersey on the Amount and Hardness of the Deposit in the Estuary", D.S.I.R. Water Pollution Research Tech. Paper No. 7.
112. Louderback, G. D. (1939). *In* "Proc. 6th Pacific Science Congress. California 1939", **2**, 783–793.
113. Lucht, F. (1953). *Dt. hydrogr. Z.* **6**, 186–207.
114. Van Andel, T. J. and Postma, H. (1954). "Recent Sediments of the Gulf of Paria", Reports of the Orinoco Shelf Expedition, **1**, 1–242, North-Holland Publishing Co., Amsterdam.
115. Ayers, J. C. (1959). *Limnol. Oceanogr.* **4**, 448–462.
116. Perkins, E. J. and Williams, B. R. H. (1966). H.M.S.O., U.K.A.E.A., P.G. Report 587 (CC).
117. Marshall, J. R. (1962). *Scott. geogr. Mag.* **78**, 81–99.
118. Kindle, E. M. (1930). Report of the Committee on Sedimentation 1928–29, Reprint and Circular. Series of the National Research Council Number 92, National Academy of Sciences—National Research Council, Washington, D.C. 1930.
119. Fraser, J. H. (1935). *Proc. Trans. Lpool biol. Soc.* 1935, 48–65.
120. Nicol, E. A. T. (1935). *J. mar. biol. Ass. U.K.* **20**, 203–261.
121. Alexander, W. B., Southgate, B. A. and Bassindale, R. (1935). "Survey of the River Tees. Pt. II: The Estuary—Chemical and Biological", D.S.I.R. Water Pollution Research Tech. Paper No. 5.
122. Sheldon, R. W. (1968). *Limnol. Oceanogr.* **13**, 72–83.
123. Spärck, R. (1935). *J. Cons. perm. int. Explor. Mer.* **10**, 3–19.
124. Smidt, E. L. B. (1951). *Meddr Kommn. Danm. Fisk.-og Havunders.* **11**, 1–151.
125. Krumbein, W. C. and Aberdeen, E. (1937). *J. sedim. Petrol.* **7**, 3–17.
126. Kulm, L. D. and Byrne, J. V. (1966). *Mar. Geol.* **4**, 85–118.
127. Macnae, W. (1957). *J. Ecol.* **45**, 113–131; 361–387.
128. Rochford, D. J. (1951). *Aust. J. mar. Freshwat. Res.* **2**, 1–116.
129. Browne, W. R. (1881). *Minut. Proc. Instn civ. Engrs.* **46** (20), 1881.
130. Beazeley, A. (1900). "The Reclamation of Land from Tidal Waters", London.
131. Abernethy, J. (1862). *Proc. Instn civ. Engrs.* **21**, 309–344.

6*

132. Marshall, J. R. (1962). *Trans. J. Proc. Dumfries. Galloway nat. Hist. Antiq. Soc.* Ser. 3, **39**, 000.
133. Wilson, J. B. (1967). *Scott. J. Geol.* **3**, 329–371.
134. D'Arrigo, A. (1949). *Dock Harb. Auth.* **30**, 182–185.
135. van Straaten, L. M. J. (1952). *Kon. Ned. Akad. Wetensch. Proc.* Ser. B, **5**, 500–516.
136. Klein, G. de V. and Sanders, J. E. (1964). *J. sedim. Petrol.* **34**, 18–24.
137. Pontin, J. (1970). "Wildlife Importance of Chobham Common", Daily Telegraph, Letters, 3 March 1970.
138. Williams, B. R. H., Perkins, E. J. and Hinde, A. (1965). H.M.S.O., U.K.A.E.A., P.G. Report 611 (CC).
139. Chapman, V. J. (1960). "Salt Marshes and Salt Deserts of the World", Hill, London, Plant Science Monographs.
140. Robinson, A. H. W. (1964). *The East Midland Geographer* **3**, 307–321.
141. Perkins, E. J., Bailey, M. and Williams, B. R. H. (1964). H.M.S.O., U.K.A.E.A., P.G. Report 605 (CC).
142. Robinson, A. H. W. (1964). *Dock Harb. Auth.* **44**, 325–327.
143. Elliot, F. E., Tressler, W. L. and Myers, W. H. (1953). *In* "Proc. 3rd Conference on Coastal Engineering, Cambridge, Mass., October 1952", Chap. 4, 48–53.
X.   Perkins, E. J. Previously unpublished work.
Y.   Perkins, E. J., Lewis, B. G., Williams, B. R. H., Howard, T. E., Davis, D. S., Burdett, J. R. and Carrie, B. G. A. Unpublished work on the C.E.G.B. Survey of the River Blackwater Estuary, Essex, 1959–61.

7

Benthos

All organisms, plant and animal, which live on, and are closely associated with, the seabed belong to the *benthic division* or *benthos*. In neritic waters, i.e. to the edge of the continental shelf at 200 m depth, the inhabitants of the shore live in the *littoral zone*, but below E.L.W.M.S.T. the inhabitants live in the *sub-littoral zone*. The benthos which lives at depths greater than 200 m may live in the *bathyal zone*, i.e. on the continental slope from 200 to 3000 m depth, in the *abyssal zone*, i.e. from 3000–6000 m, or in the *hadal zone*, i.e. the great trenches at depths ranging from 6000 to 12 000 m[1].

The benthos may be divided conveniently into a number of subgroupings which may refer to both plants and animals or to one kingdom only. All the larger organisms, i.e. those retained by sieves of 2·0–0·5 mm mesh, are referred to as the *macrobenthos*; those organisms which pass through a sieve of 1·0–0·5 mm, but are retained by one of 0·04–0·1 mm are referred to as the *meiobenthos*; those which pass through these small size sieves are referred to as the *microbenthos* which is composed essentially of diatoms, bacteria and protozoa[2]. In some cases, it is convenient to distinguish only between those forms which are retained by, or pass through a 1 mm sieve, viz. *macrobenthos* and *microbenthos* respectively[3, 4]. The meiobenthos (or microbenthos by the second definition), falls into two distinct groups, viz. the *temporary meiofauna*, or young macrofauna from any group which has benthic juvenile stages, and the *permanent meiofauna* which when adult is small enough to be considered meiofauna. The permanent meiofauna, does not include large protozoans, but includes nearly all rotifers, gastrotrichs, kinorhynchs, nematodes, archiannelids, tardigrades, harpacticoid copepods, ostracods, mystacocarids, and halacarines, together with representatives from the turbellarians, oligochaetes, polychaetes, hydrozoans, nemertines, bryozoans, gastropods, holothurians and tunicates. The juveniles which compose the temporary meiofauna may be very abundant: in the Danish Waddensea as many as $5·5 \times 10^5$, $1·7 \times 10^4$ and $7·2 \times 10^4$ O-group individuals of *Hydrobia ulvae* (Pennant), various species of bivalve mollusc and *Cardium edule* L., respectively, were recorded at the time of maximum abundance of each[2, 5]. At the time of maximum abundance, in April 1955, $1·22 \times 10^5$ juvenile polychaetes/m² including

Phyllodoce maculata, Eteone longa, Scoloplos armiger and *Capitella capitata* were recorded at Whitstable, Kent[4].

All those animals and plants which live upon the surface of a substratum are referred to as the *epifauna* and *epiphyta* respectively. Those animals which live within a substratum are referred to as the *infauna*. Animals which live permanently attached to a substratum, but which do not have a peduncle, or stalk, are referred to as *sessile*, while those animals which are capable of movement, but do so infrequently, are referred to as *sedentary*. By these definitions, the acorn barnacles which are capable of a limited rotational movement can be considered sessile, whereas the mussel, *Mytilus*, which can undertake much greater movements can be considered sedentary. Those animals which move actively and undertake considerable migrations during the life span are referred to as the *vagile epifauna*.

Clearly, that part of the flora which performs photosynthesis is confined to the photic zone, but that which does not, e.g. bacteria and fungi, is not. During the Galathea expedition viable bacteria were found living at concentrations of 10^5–10^6/ml of mud at depths of 10 200 to 10 400 m in the Philippine Trench and Kermadec–Tonga Deep[6]. Fungi have been found to 6000 m depth[7].

In inshore waters, however, the flora is diverse and includes bacteria, fungi, yeasts, lichens, algae and flowering plants. Most species are confined

Table 7.1. Estuarine zones in which some common intertidal species reach their upper limits in Milford Haven (after Nelson Smith[8])

Zone I	Zone II
Oceanic–marine, sandy or rocky substrata	Marine–polyhaline, mixed substrata
Salinity always 32–34‰	Salinity 20–30‰
Lichina pygmaea	*"Verrucaria mucosa"*
Fucus vesiculosus evesiculosus	*Porphyra umbilicalis*
Himanthalia elongata	*Laurencia pinnatifida*
Laminaria hyperborea	*Gigartina stellata*
Saccorhiza polyschides	*Rhodymenia palmata*
Alaria esculenta	
Corallina officinalis	

Zone III	Zone IV
Marine–mesohaline, rocky substrata	Polyhaline–mesohaline, muddy substrata
Salinity 10–28‰	Salinity <10‰
Lichina confinis	*Pelvetia canaliculata*
Fucus serratus	*Fucus spiralis*
Laminaria digitata	*F. vesiculosus*
L. saccharina	*Ascophyllum nodosum*
Catenella repens	
Chondrus crispus	
"Lithothamnia"	

to the shore and the upper levels of the sub-littoral zone: vertical zonation is readily observed, especially on rocky shores.

The factors which influence the distribution of plants and animals in an estuary are essentially the same: (1) tidal changes; (2) physical and chemical conditions of the water; (3) nature of the substratum; (4) degree of exposure to wave and current action; (5) the effects of grazing and predation; and (6) fishing. However, of all these factors, the consequences of a gradual change in salinity from the mouth to the head of the estuary is the primary factor to which all forms of estuarine life have to respond. In general all living organisms respond to decreasing salinity, by a gradual decline first of the less and then of the more tolerant species in the increasingly extreme conditions.

It seems appropriate, therefore, to consider first the general response of plants and animals to these changing conditions of salinity and then, in subsequent chapters, to review the general biology of the primary benthic habitats, i.e. rocky and sandy/muddy substrata, or where it seems more appropriate as the primary divisions described above.

Distribution of Benthic Organisms in Relation to Salinity

Attached Algae

The diversity of attached algal species decreases with distance moved upstream, but the abundance depends upon the amount of solid substratum present. Clearly, then, these algae are more abundant in the upper reaches of a fiordic estuary than in a coastal plain estuary, because, unlike the latter, sedimentary areas and salt marshes are poorly developed in the former. This decline in diversity may be illustrated by reference to Milford Haven[8] (Table 7.1): a similar decline has been observed in the Solway Firth[X] and the Bay of Fundy[9, 10]. The degree of penetration of brackish inlets is summarized in Table 7.2[11]; the species *Fucus ceranoides* is normally found where freshwater outlets run across the shore, and occurs upstream of the limit reached by *F. spiralis* and *F. vesiculosus* wherever the substratum in the bed of the estuary is suitable[11–14].

The ability of attached algae to resist a wide range of salinities is related to the position which they normally occupy on the shore. Algae of the intertidal zone generally, e.g. *Cladophora spinulosa* and *Porphyra leucosticta*, are tolerant to salinities ranging from 10–300% of normal sea water. Among species which live at the low-water mark, some, e.g. *Chaetomorpha linum* and *Callithamnion granulatum*, can tolerate from 0–260% of the salinity of normal sea water, but others, e.g. *Ceramium berneri*, are less resistant and have a tolerance ranging from 30–190% of normal sea water. In the sub-littoral zone, tolerances range from 10–200% of normal sea

Table 7.2.   Order of penetration of some Fucaceae into various estuaries
(after Nelson Smith[11])

Ria de Vigo	Ria de Camariñas	Ria del Barquero	La Rance	Milford Haven	Severn Estuary	River Exe	River Tees	River Tay	Solway Firth[X]	River Dee, Aberdeenshire	Hardangerfjord	
4	2	1	2	3=	2=	2	N	2	3=	2=	4	*Pelvetia canaliculata*
1	1	3	3	2	2=	4	N	N	1	1=	5	*Fucus spiralis*
3	3	2	4	3=	2=	3	2	3	3=	2=	2	*Ascophyllum nodosum*
2	4	4	1	1	1	1	1	1	2	1=	1	*Fucus vesiculosus*
N	N	N	5	4	3	5	N	N	4	A	3	*Fucus serratus*

1 indicates furthest and larger numbers proportionally less penetration upstream. A = absent, N = insufficient data.

water with *Cladophora prolifera* to 80–150% of normal sea water with *Taonia atomaria*[15]. A similar response is displayed in relation to resistance to desiccation. When exposed to the atmosphere for 13 hr typical shore dwellers such as *Porphyra umbilicalis* and *Cladophora rupestris* tolerated a minimum relative humidity of 83–86%. *Polysiphonia nigrescens* and *Plumaria elegans*, resident at the low-water mark, withstood 86–94% relative humidity, but *Halarachnion ligulatum* and *Plocamium coccineum*, from the sub-littoral, could only withstand a minimum relative humidity of 96·8–98·8%[15]. The red algae which have a low resistance to seawater temperature change and a ready tolerance to exposure, e.g. *Plumaria elegans*, have a high lipid content in the cell wall, unlike those species which have high resistance to seawater temperature change and a poor tolerance of exposure, e.g. *Antithamnion plumula*[16].

Animals

Clearly the diurnal and seasonal fluctuations of salinity, in addition to the longitudinal and lateral salinity variations, of an estuary present severe problems in adaptation to would-be colonizers. These problems are overcome in a number of ways. The vagile epifauna migrates to escape adverse conditions. Some species may be sheltered by their mode of existence, e.g. the lug-worm, *Arenicola*, while others can withstand periods of adverse low salinity by firmly closing their shell, e.g. the mussel *Mytilus*. Nevertheless, reduced to their simplest condition, all marine organisms are a mixture of salts, proteins, fats and carbohydrates invested by a semi-permeable membrane, and are, therefore, potentially subject to osmosis should the salinity change.

In the open sea, with a relatively stable salinity, at or about 35‰, life presents no especial difficulty. However, in estuaries the problem assumes quite different proportions, for the salinity may, indeed does, change rapidly, and an organism may find itself in an environment containing a significantly lower concentration of salts than are present in its body. Osmosis results, and if uncontrolled, this phenomena leads to swelling of the animal body which may eventually burst. The ability to control osmotic flooding is possessed by marine animals to varying degrees. This ability, known as *osmo-regulation*, is linked with an ability to control the ionic composition of the body liquids, i.e. *ionic regulation*. By means of the latter, an animal may maintain its body fluids at a composition which differs significantly from that of sea water: the importance of ionic regulation may be judged against the extent to which the ionic composition of estuarine waters may differ from that of the open sea (Chapter 2).

Animals which are unable to osmo-regulate to any significant degree and whose bodies are always isosmotic (i.e. having the same osmotic concentration) as the external environment are known as *poikilosmotic forms*. On the other hand, those organisms which can maintain an internal osmotic environment significantly different from the external environment are known as *homoiosmotic forms*. Maintenance of an internal concentration greater than that of the external environment is known as *hyperosmotic regulation*. Other animals can perform *hyposomotic regulation*, i.e. maintain the body fluids at a lower concentration than that of the environment, still others can regulate hyperosmotically at low salinities and hyposmotically at high salinities, e.g. the prawn *Palaemonetes varians*, which is often abundant in the brackish ponds of salt marshes, has this remarkable ability. Finally, organisms may be able to perform the function of osmo-regulation within narrow limits only when they are said to be *stenohaline*; those animals which are said to be *euryhaline* can osmo-regulate within wide limits. Estuarine faunas are derived from five principal components in a manner summarized in Table 7.3.

Most marine animals, adapted to the relatively uniform conditions of the open sea, are poikilosmotic forms, which belong to the *stenohaline marine component* and are able to penetrate estuaries to a limited degree only. For example, the spider crab, *Hyas araneus*, is common and widespread in the Outer Solway, it is however, a poikilosmotic form and does not penetrate the Inner Solway to any marked degree[20] The starfish, *Asterias rubens*, is a stenohaline form, with little power of ionic regulation. Those which inhabit the North Sea have a salinity tolerance of 23‰; in contrast,

Table 7.3. Estuarine fauna components (after Day[17], Remane[18], Green[19])

---

1. *The freshwater component* restricted to waters of low salinity.
   (i) The stenohaline freshwater component, not penetrating salinities greater than $0 \cdot 5\%_0$.
   (ii) The euryhaline freshwater component
      (a) first grade, penetrate to a salinity of $3\%_0$
      (b) second grade, penetrate to a salinity of $8\%_0$
      (c) third grade, penetrate to a salinity $> 8\%_0$.
2. *The marine component.*
   (i) The stenohaline component, not penetrating a salinity below 30%.
   (ii) The euryhaline component, extending from the sea over the middle reaches of the estuary wherever conditions are suitable.
      (a) first grade, penetrate to a salinity of 15%
      (b) second grade, penetrate to a salinity of 8%
      (c) third grade, penetrate to a salinity of 3%
      (d) fourth grade, penetrate to a salinity of $< 3\%_0$.
3. *The estuarine component* comprising the few species which are restricted to estuaries, and penetrate neither fresh or fully saline waters.
4. *The migratory component* which spends only a part of its life in the estuary.
   (i) Marine species which enter the estuary for a part of their life cycle either to feed, e.g. plaice and flounder, or to spawn, e.g. herring and roker.
   (ii) Species which migrate through estuaries from the sea to freshwater, or vice versa.
      (a) anadromous species, ascending the rivers to spawn, e.g. salmon
      (b) catadromous species, descending to the sea to spawn, e.g. eels.
5. *The terrestrial component.*
   (i) Tolerant of submersion.
   (ii) Intolerant of submersion.

---

those which inhabit the Baltic Sea have a salinity tolerance of $8\%_0$, and are apparently of a distinct "physiological race"[21, 22]. The crab, *Carcinus maenas* is poikilosmotic at high salinities and homoiosmotic at low salinities; it occurs commonly and widely in many estuaries, and has a salinity tolerance of $2\%_0$[23].

The response of an animal to salinity and temperature changes may vary in its life time, or at different stages in its life history, both with respect to planktonic (p. 90) and benthonic animals. The polychaete, *Nereis diversicolor*, has an homoiosmotic adult which can live in salinities as low as $4-5\%_0$. In the Tamar estuary, *N. diversicolor* reaches a maximum population density in that part of the estuary where the greatest salinity variations, both with season and with intertidal height, occur. At its upstream limit, it withstands, regularly, a salinity of less than $0 \cdot 5\%_0$. Indeed, in this estuary it lives at a salinity less than half of that which limits its distribution in the Baltic Sea[24]. Of a group of animals, which includes *Macoma balthica*, *Mytilus edulis*, *Cardium edule*, *Balanus improvisus* and *Corophium volutator*, it penetrates furthest upstream in many British and Danish estuaries. In the Gulf of Finland, however, it is the first species of this assemblage to drop out, apparently because the hydrographical conditions in spring when it breeds, i.e. low temperature, low salinity, and a net seaward water movement, are adverse to the survival of its naked larvae. In contrast, its associates are summer breeders, or their larvae have a long life in the meroplankton, and are thus able to colonize the low salinity areas in the summer months[25]. Similarly, the larvae of the mussel, *Mytilus californianus*, die in dilute sea water in which the adults live indefinitely in the laboratory[26]. The polychaete *Nereis limicola*, an inhabitant of Californian estuaries, has a response to reduced salinity similar to *N. diversicolor*; however *N. limicola* can control its body volume in fresh water and 1/20 sea water (i.e. Cl = 1‰) at a temperature of 13°C, but it is unable to do so at 1–2°C[25]. In the Severn estuary and Bristol Channel, the Solway Firth and in Dutch waters, the shrimp *Crangon crangon* is found in the shallow, brackish waters, inshore, during the summer, but with the onset of cold weather in autumn, it leaves these regions for the deeper and more saline waters of the adjacent Irish and North Seas. The adult of this shrimp exposed to a temperature of 4°C survives best at a salinity of 33‰, but at a temperature of 20°C survives best at a salinity of 28–29‰. In summer, at least, its salinity range is greater than the laboratory optima, nevertheless its migrations are a response to maintain a constant difference between the osmotic pressures of its blood and that of sea water[25, 27–29].

In the Thames, Wash and other broad, shallow estuaries, but not on the Dutch coast apparently[30], *Pandalus montagui* undergoes a massive inshore migration towards the end of April[31]. On the coast of Northumberland, no such massive migration is undertaken. Here, within the depth of 20–30 fm, the population is composed predominantly of 0-group males and females, and a smaller number of 1-group females; on the other hand, at 50 fm depth, the population is composed predominantly of 1- and 2-group males and females: a difference which is explicable in terms of sex change and migration. After breeding, many of the 0-group males, in shallow water, change sex, those that do not migrate to deep water and mature a second time; no mature 1-group males are found in shallow water. After breeding, many of the males in shallow water migrate offshore also, but whereas the males migrate in March, the females do not move into deep water until October[32]. Over the whole of its range, then, the migrations of *P. montagui* are complex, and it seems likely that the temperature-salinity responses of this species vary with age and sex. Clearly, this is an interesting animal worthy of much greater attention than has been paid it hitherto.

Apparently some estuarine animals regulate better at lower temperatures than at higher temperatures, cf. *N. limicola*, for in addition to *Pandalus montagui*, the shrimp *Crangon allmani* and the spider crab, *Hyas araneus*, migrate offshore from the Dutch coast in spring and return in the lower temperatures of the autumn[30], but further investigation may reveal complexities similar to that exhibited by *P. montagui* off the coast of Northumberland. Nevertheless, *Heteropanope tridentata*[33, 34], and *Gammarus duebeni*[35] tolerate lower salinities better at low than at high temperatures. *Hydrobia ulvae*, which in the River Crouch estuary is confined between H.W.M.S.T. and the 2·8‰ isohaline contour, also tolerates low salinities better at lower temperatures, and like *H. tridentata* and *G. duebeni*, does not have to undertake a seasonal vertical migration to a level of higher salinity[36].

Brackish water populations readily adapt to different salinity regimes, and many transfers to new localities have occurred naturally or as a result of the direct intervention of mankind. The Chinese mitten crab, *Eriocheir*, has become widespread in European waters[25]. In 1939–41, some 65 000 of the polychaete worm *Nereis diversicolor* were introduced into the Caspian Sea from the Black Sea: by 1959, it was so abundant that the number consumed annually was considered sufficient to load a freight train 72 miles long[25, 37]. The bivalve mollusc, *Dreissensia polymorpha*, is considered by recent authors to have spread to the rivers of Europe from the Caspian Sea, since about 1810[25]; this despite evidence that it was already present in the River Volga and the Black Sea in 1754, and was common in central Germany before 1780[38]. Nevertheless, the retention of a planktonic larva may be evidence that it has invaded fresh water in relatively recent times.

The gastropod mollusc, *Potamopyrgus jenkinsi* was considered to have invaded the fresh waters of Europe, from an estuarine source, sometime in the last century. This now seems to be uncertain, although, without question, it underwent a rapid expansion of distribution[40, 41]. The adults, of this species, live in estuaries as well as the fresh waters of lakes and streams. They are tolerant of all salinities, from fresh water to full sea water, in all of which they move and feed actively and in which new levels of internal concentration are re-established very rapidly. However, the young are more sensitive: those of the freshwater population are born only in salinities from 0 to 12‰, whereas the brackish populations are born in salinities from 0 to 18‰[42–44]. Few larvae, from freshwater or brackish origins, can withstand a salinity > 18‰, although a previous acclimation improves the survival rate[43].

Some animals, like *Nereis diversicolor*, are transferred from the estuaries of one country to those of another for an economic return: the Portuguese oyster, *Crassostrea angulata*, imported to British estuaries from the Tagus, in Portugal, for fattening and sale, at a time when the native is not available, is an example of such a traffic. The Portuguese oyster, does not reproduce readily in British waters, unlike some inadvertent introductions which do. Both the tingle, *Urosalpinx cinerea*, and the slipper limpet, *Crepidula fornicata*, were imported upon oysters from N. America, and by predation and competition respectively have become important pests of oyster fisheries, especially in S.E. England[45]. Perhaps the best documented invasion by an exotic species, is that of the Australasian barnacle, *Elminius modestus* Darwin. In 1946, this species was discovered in the Portsmouth area, although it is known to have been present prior to 1945. The exceptional shipping conditions at the beginning of the 1939–45 war, when many ships assembled to allow the formation of convoys could have allowed barnacles on ships in the Australasian trade adequate time to liberate sufficient larvae to colonize Southampton water; a process assisted by the succeeding warm summer of 1940. In 1943–44, a secondary centre was established in the Thames estuary and the species subsequently spread to much of the coastline of England, Wales and N.W. Europe[46] (Fig. 7.1). Since 1956, it has become established in Ireland, as colonies around Belfast, Dublin and Cork[47], and in Scotland settlements have occurred at Millport[48] and Finnart, Loch Long[X], in the Firth of Clyde, on the west coast, and at Rosyth[49] and Kirkcaldy[X] in the Firth of Forth on the east coast. Curiously, despite the more adverse climate, *Elminius* is apparently better established in the Firth of Forth than in the Firth of Clyde.

The introduction of *Elminius* markedly affected the balance of other species endemic to N.W. European estuaries. On the shore the main ecological effects resulted in competition for space with *Balanus balanoides* which underwent a gradual decline in competition with *Elminius*. It disappeared from some estuaries, but made a recovery to achieve some sort of balance in others. Unlike *Balanus balanoides*, *Elminius* does not accumulate food reserves for winter breeding, but rapidly converts the food into nauplii, which enter the plankton during the summer. The success of *Elminius* nauplii in obtaining food in competition with other larvae, such as *Balanus improvisus*, *Polydora ciliata*, *Littorina littorea*, *Crepidula fornicata* and *Ostrea edulis*, may be judged by the number of its cyprids which develop and settle; indeed the intensity of *Elminius* settlement adversely affects the growth of early settlements of *Ostrea* spat[46]. The attributes which permit *Elminius* to compete successfully with endemic species are summarized in Table 7.4.

An interesting case which parallels the effect of competition between the native *B. balanoides* and the immigrant *Elminius modestus*, concerns the

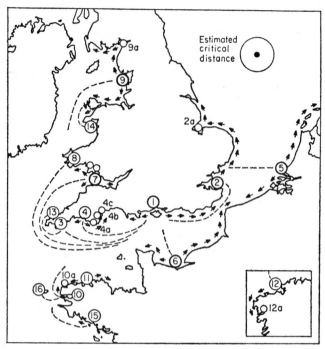

Fig. 7.1. Illustration of the probable order in which centres of dissemination of *Elminius* were established. Major centres responsible for the colonization of large areas of coast are shown by circles 1, 2, 5, 6, 7 and 9. More localized centres also set up by remote dispersal are shown by circles 3, 4, 8, 10, 11, 12, 13, 14, 15 and 16. Subsidiary centres probably reached by marginal dispersal are shown by small circles. Arrows show direction of marginal dispersal, dotted lines the probable routes of remote dispersal.

Inset: north-west Spain.

1, Southampton Water, ?1940–43; 2, Thames estuary, ?1943–44 [2a, the Wash, 1948]; 3, Helford River, ?1944–46; 4, Plymouth Sound, 1946 [4a, River Dart, 1948; 4b, Salcombe, 1948; 4c, River Exe, 1951]; 5, South Holland, 1946; 6, Seine estuary, ?1944–49; 7, Bristol Channel, 1946–47; [7a, Llanelly Bay, 1949; 7b, Swansea Bay, 1949]; 8, Milford Haven, 1947; 9, Morecambe Bay, 1948 [9a, Solway Firth, 1953]; 10, Bade de Brest, ?1944–52 [10a, l'Aber Wrach, l'Aber Benoit, ?year]; 11, Roscoff area, ?1944–52; 12, north-west Spain, ?1950–53 [12a, Noya, ?year]; 13, St. Ives Bay, 1951–54; 14, Tremadoc Bay, 1953–54; 15, south Brittany, 1954–57; 16, Ile d'Ouessant, 1956 (after Crisp[46].)

oysters *Ostrea edulis*, the native European, and *Crassostrea angulata*, the Portuguese oyster. Until the middle of the nineteenth century, *C. angulata* was confined to the Mediterranean and was absent from the Atlantic coast of France. However, once it gained a foothold in this new environment, it spread rapidly and came to occupy large areas occupied by *Ostrea*

Table 7.4. Ecological requirements of *Elminius*, compared with native species (prepared by D. J. Crisp)

Species	*Elminius modestus*	*Balanus balanoides*	*Balanus improvisus*	*Balanus crenatus*	*Balanus perforatus*	*Chthamalus stellatus*
Settlement season in S.W. Britain	May–Nov.	March–May	May–Sept.	March–May	Aug.–Sept.	July–Sept.
Levels occupied in S.W. Britain	H.W.–L.W.S.	H.W.N.–L.W.S.	L.W.N.–Sublittoral	L.W.–Sublittoral	M.T.L.–L.W.S.	H.W.S.–L.W.N.
Tolerance of:						
Low salinity[1]	10–14‰	14‰	1–2‰	14‰	?	17–20‰
Siltation	++	+	+++	++	+	+
Low temperature[2]	–6°C	–18°C	?	–8°C	?	–9°C
High temperature[3]	45°C	43°C	46°C	37°C	45°C	50°C
Desiccation[4]	46 hr	54 hr	?	17 hr	?	165 hr
Mechanical change	+	++	++	++	+++	+
Cirral beat activity[5] 20°C	17–18	5–6	ca. 9	ca. 10	ca. 7	ca. 6

Key: 1. Salinity at which acclimated individuals cease to beat. Foster, B. A. (1969a). *Phil. Trans. R. Soc.* B **256**, 377–400.
2. Winter values of median lethal temperature 18 hr exposure in air. Crisp, D. J. and Ritz, D. A., private communication.
3. Median lethal temperatures at constant rate of increase of 0·2 deg/min. Foster (1969b). *Mar. Biol.* **4**, 326–332. Also Southward, see 5.
4. Median lethal times at 0% humidity for 6 mm individuals. Foster (1971). *Mar. Biol.* **8**, 12–29.
5. Beats per 10 sec. Southward, A. J. (1955). *J. mar. biol. Ass. U.K.* **34**, 403–422; *J. mar. biol. Ass. U.K.* **36**, 323–334.

alone formerly: like *B. balanoides*, *Ostrea* has disappeared completely from some areas. Apparently, this success is a consequence of several factors, viz. a greater fecundity, a faster growth rate and an ability to feed more rapidly than *O. edulis* which was deprived of living space and food therefore[50].

Of all the transfer experiments performed perhaps the one of greatest interest was that which involved the cross-transfer of *Mytilus edulis* from the North Sea at a salinity of 30‰ and from the Bay of Kiel, Baltic Sea, at a salinity of 15‰. Initially, the oxygen consumption of gills isolated from these transferred animals was less than that of the native mussel gills. However, the respiratory rate shifted progressively until after 4–7 weeks the native and transferred mussels had the same respiratory characteristics. Clearly there is no evidence for physiological races, here, but the persistence of a depressed rate of respiration, long after osmotic equilibrium is accomplished, suggested that physiological factors, other than osmoregulation alone, are implicated in the mechanism of euryhalinity in marine invertebrates[25, 51, 52]. Clearly, this result has important consequences in relation to the acclimation of field stocks for laboratory toxicity studies (p. 487).

Although it may be convenient to use simple two parameter systems in studies of environmental tolerances, it is important to appreciate that the environment is a multivariate system. For example, when subjected to varying combinations of salinity, temperature and oxygen tension sufficient to cause a 50% mortality, salinity is limiting to the lobster, *Homarus americanus*, only within a certain temperature range, Fig. 7.2. However, low oxygen concentrations have an adverse effect especially at high temperatures[25, 53].

The introduction of freshwater animals into brackish conditions is informative also. The freshwater hydroid, *Cordylophora*, inhabits a wide range of salinities $\geq 0‰$, and thrives at a salinity of 17‰ in the Baltic Sea. At this salinity, growth is uniform over the temperature range 10–20°C, but a change of 3°C, at either end of the range, i.e. to 7° and 23°C, reduces growth significantly. On the other hand, growth optima occur at 20°C in a salinity of 17‰ and at 10°C in a salinity of 10‰: in fresh water, growth at 10°C is superior to that in any salinity at 23°C[25, 26, 54]. The duration of the period between fertilization and hatching of the eggs of the killifish, *Cyprinodon macularius* increases with increasing salinity to 75‰. The retardation in hatching is inversely proportional to the amount of available oxygen, so that increased aeration will reduce the retardation due to salinity increases. At hatching, the length of the young fish is maximal in fresh water, and decreases with increased salinity. However, at 25°C, the final sizes in fresh water and at a salinity of 35‰ do not differ. At a

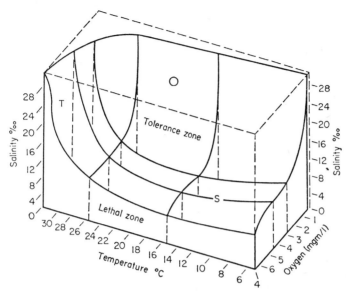

Fig. 7.2. Diagram of the boundaries of lethal conditions set by combination of high temperature, low salinity, and low oxygen for the lobster, *Homarus* (after McLeese[53]).

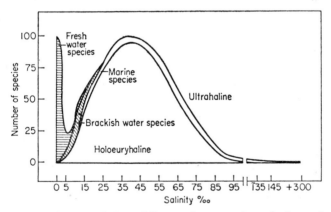

Fig. 7.3. Suggested extrapolation of Remane's curve through the entire salinity spectrum (after Hedgpeth[58]).

low salinity and at moderate temperatures, the survival time of freshwater animals is increased 2–3 fold by a 10°C decrease in temperature[26].

Without discussing in detail the mechanisms of osmoregulation adopted by animals, it should be noted that animals introduced into a lowered salinity can passively oppose reduced salinities by means of membranes having a reduced permeability to water and salts. They may also react functionally by actively absorbing salts from the surrounding medium so as to keep up the internal salt level in the face of various conditions tending to cause internal dilution. In the former case the element calcium is of particular importance[55, 56]. Although calcium may be important to the osmoregulation of *Potamopyrgus jenkinsi*, the haemolymph osmotic concentration of these snails from calcium rich fresh water is lower than those from brackish waters, and factors other than the calcium level are involved[43].

Characteristically, the diversity of species decreases with distance moved upstream in an estuary. There are characteristic breaks in the fauna at salinites of 5 and 18‰, i.e. at the oligohaline–mesohaline and the mesohaline–polyhaline boundaries respectively[25, 57]. From a minimum, at a salinity of 5‰, animal species numbers increase seawards to a maximum at *ca.* 35‰, i.e. the normal salinity of sea water, and decline once more with the increasing salinity of hypersaline or negative estuaries[58] (Fig. 7.3). In the Solway Firth where 15 species of lamellibranch mollusc characteristic of the N.E. Irish Sea, decline to 8, 5, 3 and 2 at 10, 15–21, 23–25 and 26–28 miles from the open sea, respectively the discontinuity zone, between inner and outer areas, i.e. between 10 and 15 miles upstream, is not apparently due to salinity alone, but to the relative lack of water in

Table 7.5. Estuarine zones in which some common intertidal species reach their upstream limits in Milford Haven (after Nelson Smith[8])

ZONE I	ZONE II
Oceanic–marine, sandy or rocky substrata Salinity always 32–34‰	Marine–polyhaline, mixed substrata Salinity 20–30‰
Littorina saxatilis tenebrosa	*Actinia equina*
Patella depressa	*Spirorbis borealis*
P. aspera	*S. rupestris*
Chthamalus stellatus	*Monodonta lineata*
Balanus perforatus	*Gibbula umbilicalis*
	Littorina neritoides
	L. saxatilis saxatilis

ZONE III	ZONE IV
Marine–mesohaline, rocky substrata Salinity 10–28‰	Polyhaline–mesohaline, muddy substrata Salinity < 10‰
Spirorbis tridentatus	*Littorina saxatilis rudis*
S. pagenstecheri	*L. s. jugosa*
Pomatoceros triqueter	*Mytilus edulis*
Gibbula cineraria	*Elminius modestus*
Littorina littorea	
L. littoralis	
Nucella lapillus	
Balanus balanoides	
B. crenatus	

the inner area at the time of low water[20]. Nevertheless, salinity boundaries which are characteristic of the individual system exist in many estuaries[8]. The influence of salinity in controlling the upstream distribution of animal species in Milford Haven is given in Table 7.5, and the ability of some common littoral animals to penetrate estuarine waters is summarized in Table 7.6.

Table 7.6. Order of penetration of some common animals into various estuaries (after Nelson-Smith[11])

	Ria de Vigo	Ria de Camariñas	Ria del Barquero	La Rance	Milford Haven	River Tamar	River Tees	River Tay	Solway Firth[X]	River Dee, Aberdeenshire	Hardangerfjord
Mytilus edulis	1 =	3	3	2 =	1	2 =	2	2 =	2 =	2 =	1
Littorina littorea	1 =	1	1	4 =	2	3 =	3 =	2 =	2 =	2 =	2
Patella vulgata	3	4	2	3 =	4	3 =	3 =	3 =	5	3	4
Balanus balanoides	4	2	A	1	3 =	2 =	1	1	1	1	3 =
Pomatoceros triqueter	2	5	4	4 =	4	1	3 =	4	4	5	3 =
Actinia equina	5	6	5	2 =	5	3 =	4	3 =	3	4	5

1 indicates the furthest and larger numbers proportionately less penetration upstream. A = absent.

In the Baltic Sea, the rock pools of Bornholm are populated by a curious mixture of essentially freshwater and marine species[59]: at Randkløve, with a salinity of 3·1‰, the following animals occurred in association:

Nais elinguis
Gammarus duebeni
Bidessus geminus
Helophorus guttulus
Oecetis ochracea (Imagines)
Hydrometra stagnorum
Callicorixa producta
Culex pipiens

Chironomid larvae
Ephydra riparia
Theodoxus fluviatilis
Hydrobia ulvae
Radix ovata
Mytilus edulis
Anguilla anguilla

Although the number of animal species which penetrate an estuary declines with distance moved upstream, the individual species within a genus or a series of related genera, may show a different response to lowered salinity. A classic example concerns the amphipod genera *Gammarus* and *Marinogammarus*[60-62] (Fig. 7.4). A similar relationship is found in the amphipod genus *Bathyporeia*: in the Elbe, *B. guilliamsoniana* is represented only in the more saline waters of the outer estuary, whereas *B. sarsi* and *B. pelagica* occur in the deeper, brackish water of Cuxhaven, but *B. pilosa* is abundant even in the oligohaline zone[63]; a broadly similar relationship is found in the Irish sea area[64-66]. The shrimp genus *Palaemonetes* is confined to coasts and countries bordering upon the Atlantic.

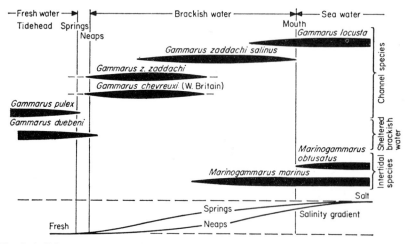

Fig. 7.4. Diagram representing the relative distributions of some related gammarids in estuarine and contiguous regions of rivers (after Nicol[55]).

P. kadiakensis and *P. paludosus* inhabit fresh waters to the west and the east of the Alleghanies, respectively: the former species can tolerate slightly saline water, but *P. paludosus* cannot withstand the strictly freshwater habitat of *P. kadiakensis* once its internal balance has been upset. The species *P. pugio*, *P. intermedius* and *P. vulgaris* occupy a range of habitat from brackish to fresh water, brackish, and marine to brackish, respectively[67]. The European species, *P. varians* inhabits the extreme environment of salt marsh pools. Although a species may penetrate a considerable distance upstream in an estuary, this ability may not be related to its ability to colonize the more exposed parts of a shore[11], (Table 7.7).

   That the genera and the species within a genus of a phylum should occupy different ecological niches and differ in their tolerance to reduced salinity, allows these organisms to make fullest use of the wide variety of opportunities provided by the habitat. In this connection the isopod

Table 7.7. Horizontal and vertical "order of resistance" of some common estuarine species (after Nelson-Smith[11])

River Dee, Aberdeenshire Horizontal and vertical order	Milford Haven Horizontal order (miles above Thorn Is.)	Milford Haven Vertical order (ft above C.D. at Lawrenny)
B. balanoides	Mytilus 18	Patella 21
L. littorea	L. littorea 15	L. littorea 19
Mytilus	Patella 14	Mytilus 17
Patella	B. balanoides 14	B. balanoides 15

The organism at the head of each list is either found highest on the shore or penetrates furthest up the estuary.

Jaera albifrons is of some interest. This species is composed of a group of subspecies, viz. *Jaera albifrons albifrons*, *J. a. ischiosetosa*, *J. a. forsmani*, *J. a. praehirsuta* and *J. a. syei*. In Milford Haven, these subspecies achieve a separation and freedom from direct competition, with each other, by occupying a variety of niches. *J. a. albifrons* is occasionally found on algae, but is most common beneath stones, in wet areas, around H.W.M.N.T. and below, on sheltered shores; *J. a. ischiosetosa* is the most euryhaline of the subspecies, it occurs normally beneath stones, in strong streams flowing over sheltered shores, above L.W.M.N.T.; *J. a. forsmani* is probably the least euryhaline of the subspecies, it occurs beneath stones, in well-drained areas, from H.W.M.N.T. to L.W.M.S.T., near the mouth of the estuary; *J. a. praehirsuta* is a markedly euryhaline subspecies which tolerates almost fully marine shores and freshwater streams on the shore: it ranges from H.W.M.S.T. to L.W.M.S.T., but is most common at H.W.M.N.T. and below where it lives among algae, but occurs under stones, particularly on the upper shore. The distribution of these species upstream in Milford Haven is consistent with these habits[68] (Fig. 7.5).

Where a genus is near the limit of its geographical range, the extent to which the species occupy an estuary depends only partly upon a response to salinity; moreover, the ability to penetrate an estuary is not related to the geographical extreme which the individual species can occupy. The isopod genus *Sphaeroma* has a sub-tropical origin, but the individual species penetrate northern latitudes to a variable degree. *S. serratum* ranges from West Africa, the Canaries, Madeira, the Azores and Morocco to the Black Sea; in N.W. Europe its limits are Donegal, on the Atlantic coast of Ireland, Anglesey, in the Irish Sea, and in the English Channel it has not reached the Straits of Dover. *S. monodi* is more euryhaline than *S. serratum*, it has not penetrated the Mediterranean, but in the north it has reached the Irish coast of the North Channel and the Belgian coast.

S. rugicauda ranges from Morocco to Shetland, eastern Norway, and the Swedish and Polish coasts of the Baltic, but does not penetrate the Mediterranean. *S. hookeri* extends into the Baltic, but unlike *S. rugicauda* penetrates the Mediterranean. However, *S. hookeri* may be divided into two subspecies viz. *S. h. hookeri* and *S. h. mediterranea*, of which the latter occurs typically in brackish lagoons in the Mediterranean, whereas the former occurs on the coasts of N.W. Europe. In estuaries, *S. serratum* predominates at the

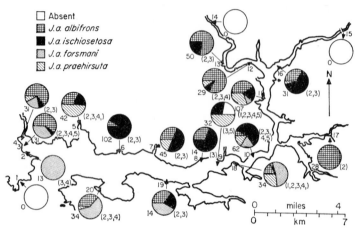

Fig. 7.5. Relative abundance of male *Jaera albifrons albifrons*, *J. a. forsmani*, *J. a. ischiosetosa* and *J. a. praehirsuta* at various localities in Milford Haven, Pembrokeshire. Figures on left below each circle give total males collected; figures in parentheses on right indicate approximate levels where samples were taken, thus: 1, HWS; 2, HWN; 3, MTL; 4, LWN; 5, LWS. Localities are numbered thus: 1, Mill Bay; 2, Dale Cross; 3, Black Rock; 4, Gann Flat; 5, Sandy Haven; b, Gelliswick Bay; 7, Milford Haven Pill; 8, Hazelbeach; 9, Neyland; 10, Burton Hawn; 11, Black Tar Beach; 12, West Hook; 13, Little Milford; 14, Haverfordwest; 15, Blackpool; 16, Landshipping; 17, Cresswell; 18, Pembroke Ferry South; 19, Martins Haven; 20, Angle Bay (after Naylor and Haahtela[68]).

mouth, especially in rock crevices, *S. monodi* occurs in the middle reaches where the salinity is reduced, especially among shingle and stones on sand, and *S. rugicauda* and *S. hookeri* occur far up estuaries near to the tidal limit[69].

Unlike the vagile epifauna[65, 70], the ability of a sessile or a relatively sedentary species to penetrate unfavourable parts of an estuary depends upon the environmental extremes at the situation which it is attempting to colonize, since the response to these conditions is relatively inflexible.

Although a sessile animal may be unable to move away when conditions in an estuary become unfavourable, they may show some degree of acclimation to conditions in which the salinity is lowered. In British waters, the acorn barnacles *Elminius modestus, Balanus balanoides* and *B. improvisus* taken from habitats with a low salinity, and *B. crenatus* which has been acclimated to a low salinity, are able to tolerate lower salinities than individuals of the same species which have been taken from or are acclimated to high salinities. *Elminius* and *B. balanoides* from the shore and *B. crenatus* from around L.W.M. or the infra-littoral, can tolerate salinities of 14–17‰, if they have been acclimated first. Estuarine *B. improvisus* can tolerate a salinity *ca.* 2‰, if it has been allowed a period of acclimation. The salinity acclimation shown by these species is typical of osmoconformers which lack specific organs for effective regulation. In the case of *B. improvisus* its ability to live in low salinity situations is dependent upon a wide tissue resistance rather than an ability to osmoregulate[7].

Three other species of acorn barnacle occupy characteristic positions in estuarine areas. *Balanus amphitrite* lives on banks in the more saline outer levels and occurs at higher levels in the intertidal zone than *B. eburneus*. *Chelonobia patula* lives on crabs in conditions which range from fully saline to much reduced salinity. The crabs are immersed for most of the time and therefore *B. amphitrite* and *C. patula* are likely to be exposed to the most severe and the least desiccation respectively. The salinity range for the 50% successful development of the embryos of these barnacles of $32 \cdot 5 \pm 16 \cdot 5‰$, $34 \pm 21 \cdot 5‰$ and $34 \cdot 2 \pm 18 \cdot 7‰$ for *C. patula*, *B. amphitrite* and *B. eburneus* respectively, is consistent with these habits.

The response of estuarine animals to changes in the salinity regime of the individual estuary may have striking results. The Solway Firth has been filling with sedimentary materials since the Pleistocene Ice Age, with dramatic consequences in the time during which scientific records have been kept, or for which information is available from fishermen. Since the early years of this estuary, sedimentation, and the consequent fall in salinity, particularly around the time of low water when fishing is carried on, is such that the limit of plaice penetration has moved *ca.* 10 miles downstream. Other fishery species have moved downstream also, although less spectacularly than the plaice. The once flourishing clipper port of Annan, locus of an important shrimp fishery formerly, is now largely abandoned[70, 73].

In the Baltic Sea, the salinity has increased markedly since the 1930s. In consequence, the benthic species *Priapulus caudatus, Scoloplos armiger, Terebellides stroemi, Diastylis rathkei* and others underwent a marked increase in abundance in the northern and eastern parts of the open Baltic, by 1949–50[74].

Lagos Harbour and the adjacent creeks are typical of sheltered estuarine waters and are colonized by species which differ from those inhabiting the exposed rocky coasts of the Lagos Moles, Ghana and the Cameroons.

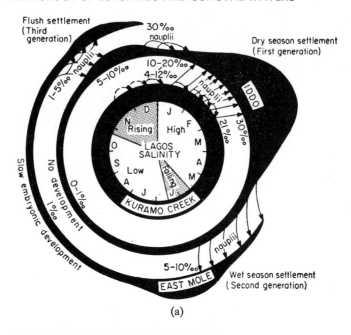

(a)

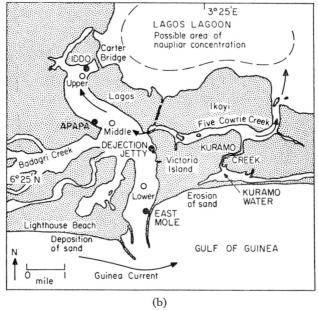

(b)

Fig. 7.6. The life history of *Balanus pallidus stutsburi* at Lagos. (a) The annual cycle of events in the life of *Balanus pallidus stutsburi* in Lagos. The widths of the circles

Lagos harbour, itself, forms the tidal reach of a marginal lagoon system which extends for 160 miles along the western coast of Nigeria and into Dahomey. In the harbour, the annual range of water temperature is only 4°C, viz. from 25–29°C, but the tidal influence and rainfall, which is seasonal, give rise to diurnal and annual salinity fluctuations of 20‰ and 34‰ respectively. The barnacle, *Balanus pallidus stutsburi*, the oyster, *Gryphaea gasar* and the serpulid polychaetes *Mercierella enigmatica* and *Hydroides uncinata* are abundant, in the harbour, during the early part of the high salinity season. However, during April and May, i.e. at the end of the high salinity season, they suffer a high mortality, and by the end of the low salinity season, in October, very few of the original settlers remain alive. In the Kuramo area, nearby, where conditions are less extreme and the salinity ranges from 6·9 to 26·5‰, these species survive throughout the year, and it seems likely that the brood stock, in this area, provide the larvae which recolonize Lagos harbour. Spatially, the distribution of these animals has vertical and horizontal components. It is considered that *B. pallidus stusburi* settles in regions of lower salinity than *Mercierella enigmatica* or *Hydroides uncinata*; *Gryphaea gasar* occupies a situation between the barnacles and serpulids[75]. The optimal range of salinity for *B. pallidus stutsburi* varies seasonally between 4 and 10‰. The high mortality rate, viz. 85–100%, in the high salinity season, may be due to the fact that for normal life the highest salinity which this barnacle can tolerate is 20‰: at this time the environmental salinity ranges from 20–30‰, moreover it is at this time that maximum settlement occurs. In consequence, a complex cycle of recruitment has evolved: there is an alternation of the main breeding population between the upper and lower regions of the harbour, and three or four generations are produced annually[76] (Fig. 7.6).

In the high salinity season, the oyster *Gryphaea gasar* is apparently an important sub-littoral inhabitant of Lagos Harbour. However, it suffers 100% mortality during the year, and it is maintained, in the harbour, only by an influx of larvae from the creeks. Its high mortality in the high salinity season is probably due to the salinity at this time, for it lives longest in the salinity range 0–15‰[77].

---

at Iddo and East Mole are an indication of the relative sizes of the populations. In Kuramo Creek variations in population numbers are not known. Seasons of high and low salinity in Lagos Harbour are shown in the centre of the figure. (b) Map of Lagos Harbour (1956) illustrating possible routes of *Balanus pallidus stutsburi* nauplii between Kuramo Creek and Lagos Lagoon. ●Marks the positions of the sampling stations at Iddo, Apapa, Dejection Jetty, East Mole and Kuramo Creek; and ○ indicates the positions of the plankton sampling stations in the upper, middle and lower regions of the harbour (after Sandison[76]).

In itself, salinity may not be the only controlling factor in the water: the paucity of species present in the Black Sea, compared with the Mediterranean, has been attributed to the peculiar composition of Black Sea water. This situation arises from the transport of a large amount of mineral ions, especially K^+, Ca^{2+} and Mg^{2+}, into the basin, by rivers which flow over sedimentary rocks bearing these minerals[74].

The annual restocking of the former area by a breeding population living elsewhere is necessitated by the extremes of salinity which it experiences. However, the phenomenon is more widespread than this interesting case would imply, and is indeed linked to another, viz. the intermittence of successful larval survival and settlement. The populations of those estuaries, which contain little water at the time of low tide and off the mouth of which pronounced residual currents are present, must of necessity be restocked from external sources. However, in those estuaries which drain only partially the non-tidal circulation may ensure that the larvae of a particular species may not settle in close proximity to the parent stock. In the James River estuary, Chesapeake Bay, heavy spatfalls of seed oysters, i.e. those which require to be transferred to fattening grounds elsewhere, takes place some 20 miles upstream from the parent stock. Although the distances involved are less, a similar settlement of seed oysters takes places in the upper reaches of the River Blackwater estuary, Essex. The classic work by Buchanan showed that in coastal waters a breeding stock of the heart urchin, *Echinocardium cordatum*, is confined to clean sand at the low-water mark, but that non-breeding descendants are distributed over a wide area of fine, organic rich sediments further off-shore[79].

The bivalve *Tellina tenuis* may show marked geographical variations in its ability to spawn and settle. Given that spawning, larval survival and settlement are generally successful, however, the conditions at the time of settlement in exposed areas may inhibit the settlement or the survival of spat[80].

In the Solway Firth, *Macoma balthica* achieves a successful annual settlement in the inner areas, but downstream on the Cumberland coast of the outer Solway only an intermittent recruitment occurs: in the years since 1966, such success has been achieved in 1967–68 and 1971–72. Such a failure may in part be due to adverse sedimentary movements, but is apparently due also a failure in recruitment from populations upstream. Settlement from the spawning of 1971 persisted until the spring of 1972, but no further settlement occurred in that year: 1971 seems to have been a good year generally for the bivalve molluscs of the Solway area since heavy spatfalls of *Tellina tenuis*, *T. fabula*, *Abra* spp. and *Nucula* spp. were also recorded[X].

References

1. Deacon, G. E. R. (1962). "Oceans", Paul Hamlyn, London.
2. McIntyre, A. D. (1969). *Biol. Rev.* **44**, 245–290.
3. Weis-Fogh, T. (1948). *Natura jutl.* **1**, 139–270.
4. Perkins, E. J. (1958). "The Microfauna of the Shore", Ph.D. thesis, University of London.
5. Smidt, E. L. B. (1951). *Meddr kommn Danm. Fisk.-og Havunders* **11**, 1–151.
6. Zobell, C. E. (1953). *In* "XIV Intern. Zool. Congr. Copenhagen 1953", *I.U.B.S. Deep Sea Colloquium.*
7. Morita, R. Y. (1965). *In* "The Fungi" (G. C. Ainsworth and A. S. Sussman, eds), Vol. 1, pp. 551–557, Academic Press, New York.
8. Nelson-Smith, A. (1965). *Fld Stud.* **2**, 155–188.
9. Stephenson, T. A. and Anne (1954). *J. Ecol.* **42**, 14–45.
10. Stephenson, T. A. and Anne (1954). *J. Ecol.* **42**, 46–70.
11. Nelson-Smith, A. (1967). *Fld Stud.* **2**, 407–434.
12. Knight, M. and Parke, M. W. (1931). "Manx Algae", *L.M.B.C. typ. Br. mar. Mem. Pl. Anim.* **30**.
13. Perkins, E. J. (1960). *J. Ecol.* **48**, 725–728.
14. Perkins, E. J. (1971). *Trans. J. Proc. Dumfries. Galloway nat. Hist. Antiq. Soc.* Ser. 3, **48**, 12–68.
15. Biebl, R. (1952). *J. mar. biol. Ass. U.K.* **31**, 307–315.
16. Boney, A. D. and Corner, E. D. S. (1959). *J. mar. biol. Ass. U.K.* **38**, 267–275.
17. Day, J. H. (1950). *Trans. R. Soc. S. Afr.* **33**, 53–91.
18. Remane, A. (1958). *In* "Die Biologie des Brackwassers" (A. Thienemann, ed.), pp. 1–213, Schweizerbartsche Verlag, Stuttgart.
19. Green, J. (1968). "The Biology of Estuarine Animals", Sidgwick and Jackson, London.
20. Perkins, E. J. (1968). *Trans. J. Proc. Dumfries. Galloway nat. Hist. Antiq. Soc.* Ser. 3, **45**, 15–43.
21. Binyon, J. (1961). *J. mar. biol. Ass. U.K.* **41**, 161–174.
22. Binyon, J. (1962). *J. mar. biol. Ass. U.K.* **42**, 49–64.
23. Perkins, E. J., Gribbon, E. and Murray, R. B. (1969). *Trans. J. Proc. Dumfries. Galloway nat. Hist. Antiq. Soc.* Ser. 3, **46**, 27–28.
24. Smith, R. I. (1956). *J. mar. biol. Ass. U.K.* **35**, 81–104.
25. Smith, R. I. (1959). *In* "Mar. Biol., Biol. Colloquium held at O.S.U. Corvallis, 1959", 59–69.
26. Kinne, O. (1964). *Oceanogr. Mar. Biol. A. Rev.* **2**, 291–339.
27. Broekema, M. M. M. (1941). *Archs néerl. Zool.* **6**, 1–100.
28. Lloyd, A. J. and Yonge, C. M. (1947). *J. mar. biol. Ass. U.K.* **26**, 626–661.
29. Verwey, J. (1957). *Année biol.* **33**, 129–149.
30. Verwey, J. (1956). *In* "Colloque international de biologie marine; Station Biologique de Roscoff".
31. Mistakidis, M. N. (1957). *Fishery Invest., Lond.* Ser. 2, **21**, 52 pp.
32. Allen, J. A. *J. mar. biol. Ass. U.K.* **43**, 665–682.
33. Otto, J. P. (1934). *Zool. Anz.* **108**, 130–135.
34. Kinne, O. and Rotthauwe, H. W. (1952). *Kieler Meeresforsch.* **8**, 212–217.
35. Kinne, O. (1952). *Kieler Meeresforsch.* **9**, 134–150.
36. Newell, R. (1964). *Proc. zool. Soc. Lond.* **142**, 85–106.
37. Zenkevitch, L. A. (1957). *Mem. geol. Soc. Am.* 67, **1**, 891–896.
38. Jeffreys, J. G. (1862). "British Conchology", Vol. 1, van Voorst, London.

39. Macan, T. T. and Worthington, E. B. (1951). "Life in Lakes and Rivers", Collins, London. New Naturalist Series, No. 15.
40. Duncan, A. (1966). *Verh. int. Verein. theor. angew. Limnol.* **16**, 1739–1751.
41. Klekowski, R. Z. and Duncan, A. (1966). *Verh. int. Verein. theor. angew. Limnol.* **16**, 1753–1760.
42. Duncan, A. and Klekowski, R. Z. (1967). *Comp. Biochem. Physiol.* **22**, 495–505.
43. Duncan, A. (1967). *Polskie Archwm Hydrobiol.* **14**, 1–10.
44. Duncan, A. and Klekowski, R. Z. (1967). *Comp. Biochem. Physiol.* **22**, 495–505.
45. Chipperfield, P. N. J. (1951). *J. mar. biol. Ass. U.K.* **30**, 49–71.
46. Crisp, D. J. (1958). *J. mar. biol. Ass. U.K.* **37**, 483–520.
47. Crisp, D. J. and Southward, A. J. (1959). *J. mar. biol. Ass. U.K.* **38**, 429–437.
48. Barnes, H. (1960). *Nature, Lond.* **186**, 989–990.
49. Meadows, P. S. (1969). *Hydrobiologia* **33**, 65–92.
50. Loosanoff, V. L. (1955). *Science, N.Y.* **121** (3135), 119–121.
51. Schleiper, V. (1955). *Kieler Meeresforsch.* **11**, 22–33.
52. Lange, R. (1968). *Nytt. Mag. Zool.* **16**, 1–13.
53. McLeese, D. W. (1956). *J. Fish. Res. Bd. Can.* **13**, 247–272.
54. Kinne, O. (1957). *Verh. dt. zool. Ges., Zool. Anz., Suppl.* **20**, 445–449.
55. Nicol, J. A. C. (1968). "The Biology of Marine Animals", Pitman, London.
56. Albert, A. (1968). "Selective Toxicity", Methuen, London.
57. Remane, A. (1934). *Verh. dt. zool. Ges. Zool. Anz., Suppl.* **7**, 34–74.
58. Hedgpeth, J. W. (1967). *In* "Estuaries" (G. H. Lauff, ed.), *Publs Am. Ass. Advmt Sci.* **83**, 408–419.
59. Johnsen, P. (1946). *Vidensk. Meddr dansk naturh. Foren.* **109**, 1–53.
60. Bassindale, R. (1942). *J. Anim. Ecol.* **11**, 131–144.
61. Bassindale, R. (1943). *J. Ecol.* **31**, 1–28.
62. Spooner, G. M. (1947). *J. mar. biol. Ass. U.K.* **27**, 1–52.
63. Caspers, H. (1959). *Estrato Dall'Archivio di Oceanografia e Limnologia, XI* (*Suppl.*) 153–169.
64. Perkins, E. J. (1956). *Ann. Mag. nat. Hist.* Ser. 12, **9**, 112–128.
65. Perkins, E. J. and Williams, B. R. H. (1963). *Trans. J. Proc. Dumfries Galloway nat. Hist. Antiq. Soc.* Ser. 3, **40**, 75–87.
66. Fincham, A. A. (1969). *J. mar. biol. Ass. U.K.* **49**, 1003–1024.
67. Hedgpeth, J. W. (1966). *Am. Fish. Soc. Spec. Publ.* No. 3, 3–11.
68. Naylor, E. and Haahtela, I. (1966). *J. Anim. Ecol.* **35**, 209–216.
69. Harvey, C. E. (1969). *J. Anim. Ecol.* **38**, 399–406.
70. Williams, B. R. H., Perkins, E. J. and Hinde, A. (1965). H.M.S.O., U.K.A.E.A., P.G. Report 611 (CC).
71. Forster, B. A. (1970). *Phil. Trans. R. Soc.* Series B, 256 (810).
72. Crisp, D. J. and Costlow, J. D. jr. (1963). *Oikos* **14**, 22–34.
73. Perkins, E. J. and Williams, B. R. H. (1966). H.M.S.O., U.K.A.E.A., P.G. Report 587 (CC).
74. Segerstråle, S. G. (1964). *Oceonogr. Mar. Biol. A. Rev.* **2**, 373–392.
75. Sandison, E. E. and Hill, M. B. (1966). *J. Anim. Ecol.* **35**, 235–250.
76. Sandison, E. E. (1966). *J. Anim. Ecol.* **35**, 379–389.
77. Sandison, E. E. (1966). *J. Anim. Ecol.* **35**, 363–378.
78. Pritchard, D. W. (1952). *Proc. Gulf Caribb. Fish. Inst.* 5th Annual Session, November 1952, pp. 1–10.
79. Buchanan, J. B. (1966). *J. mar. biol. Ass. U.K.* **46**, 97–111.
80. McIntyre, A. D. (1970). *J. mar. biol. Ass. U.K.* **50**, 561–575.
X. Perkins, E. J. Previously unpublished work.

8

Microbenthos

The microbenthos are taken here to be that group of living organisms which in general pass through sieves of 0·04 to 0·1 mm mesh size[1]. This grouping is not exclusive since the larger protozoa and lichens are discussed most conveniently under this heading, even though they are exceptions to the general classification.

Bacteria

On the basis of their tolerance or requirement for sodium chloride, all bacteria can be divided into three main groups[2], viz:

1. The *halophobic*, or salt-sensitive, group which either grow poorly or are killed by a salt content of up to 8%.
2. The *halotolerant* group which are able to grow in water with a range of salt content up to saturation, i.e. from 0 to 32% NaCl. This group, which is comprised mainly of micrococci and aerobic spore bearers, may be stimulated by low concentrations of NaCl.
3. The *halophilic* group which have a salt requirement ranging from 2 to 32%. Within this group three subdivisions may be recognized, viz. slight, moderate and extreme halophiles which require 2–5%, 5–20% and 20–30% of NaCl, respectively.

Most marine bacteria are either halotolerant or slightly to moderately halophilic. In Kamogawa Bay, Japan, the heterotroph flora was composed of *Vibrio* spp. 37·3%, *Pseudomonas* spp. 29·8%, *Achromobacter* spp. 21·3%, *Aeromonas* spp. 0·4%, *Photobacterium* spp. 0·4%, *Flavobacterium* spp. 2·1%, *Bacillus* spp. 5·5%, *Micrococcus* spp. 0·4%, "Coryneforms" 0·4%, and miscellaneous 2·3%. On the other hand, off the coast of Aberdeen, the populations were composed differently, with *Micrococcus* spp. 31%, *Achromobacter* spp. 22%, "Coryneforms" 12%, *Pseudomonas* spp. 10%, *Flavobacterium-Cytophaga* spp. 7·5%, *Vibrio* spp. 5·5% and miscellaneous 12%. In both areas, individual samples are highly variable. Further, the vertical distribution of these organisms in sea water is characterized by an irregularity which seems to depend upon the clarity of the water, the convergence of water masses, boundary currents, thermoclines and areas of estuarine influence[2].

The bacteria may be classified by other criteria. Mud and water taken from depths greater than 1000 m are inhabited by *barophiles*, i.e. pressure loving forms, and belonging principally to the *Pseudomonas* group, e.g. *Pseudomonas xanthrocus*[2]. Species which have a growth optimum at approximately 20°C, i.e. ≤ 20°C, but grow well at 0°C, are termed *psychrophiles*, whereas those species which have a growth optimum at temperatures ≥ 20°C are termed *psychrotolerant*[2, 3]. These psychrotolerant species may be divided into *mesophiles* and *thermophiles* which require for growth the temperature ranges 20–55–65°C, and > 55–65°C respectively: most bacteria are mesophilic.

The obligate psychrophile *Vibrio marinus* has an optimum growth temperature of 15°C and a maximum growth temperature of 20°C. Above this temperature the cells appear abnormal; even at 20·5°C, i.e. < 1°C above the maximum growth temperature, the number of viable cells decreases. Apparently this loss of viability is due to failure to control cell wall permeability, and the consequent leakage of the contents of the cell[4–6]. Indeed, it has been postulated that, to a large degree, the general temperature classes of bacteria are limited in their growth range by the stability of their membranes to heat[7].

Luminescent bacteria may be very beautiful; some are symbiotic and responsible for the luminescence of deep sea fish, viz. the species *Photoblepharon*, *Anomalops*, *Monocentris*, *Malacocephalus* and *Ceratias*. About ten species are responsible for this phenomenon: principally species of *Vibrio* and *Aeromonas*[2, 8]. Other luminous bacteria may be associated with beach and sand fleas, gammarid amphipods, shrimps, and midges[2, 8, 9]. The amphipod *Corophium volutator* bearing luminescent bacteria is one of the many interesting things to be seen in a drained intake culvert of a power station[X].

In the sea, bacteria with the heterotrophic habit outnumber those which are autotrophic[10, 11]. Bacteria, in general, are more abundant in association with sediments than the water above, and are more abundant in neritic than they are in oceanic water[3, 11, 12]. The limits of hydrogen ion concentration for growth of micro-organisms is pH 4·0 to 9·0; bacteria, in general, prefer media of pH near neutrality, and cannot usually tolerate pH values much below 4–5 (exceptions are the acetic acid bacteria and those responsible for acid mine drainage problems)[13]. Bacteria isolated from the surface layers of the sea have a requirement for amino acids, whereas those isolated from the sediment are dependent, under experimental conditions, upon nutrients derived from marine sediment extracts. A significant proportion of the strains of bacteria isolated from sediments and oceanic waters, both surface and deep, require vitamins for growth, viz. thiamin, biotin and cyanocobalamin; some strains require folic acid, nicotinic acid, riboflavin, pantothenic and p-aminobenzoic acid[11]. On the other hand, bacteria synthesize vitamins required in the nutrition of marine animals from deep and shallow sea alike[8, 14]. Bacterial synthesis of the B vitamins occurs in the gut of many animals; vitamin B_{12} is synthesized thus in chum salmon and Pacific herring. Many marine bacteria synthesize various carotenes, the precursor of vitamin A, and ergosterol, provitamin D[8]. In culture, a persistent association was recorded between the dinoflagellate, *Gymnodinium breve*, a vitamin B_{12} auxotroph, and a dominant type of bacterium found to be a vitamin B_{12} producer[11]. In estuaries, the production of vitamin B_{12} is related to the marsh muds rather than to the water[15].

Although bacteria are important producers of the external metabolites which other organisms require, they play other important roles in the biology of the sea. The choice of substratum, at settlement, by the larvae of the polychaetes *Ophelia bicornis*[16], *Spirorbis borealis*[17], the archiannelid *Protodrilus symbioticus*[18], the oyster *Ostrea edulis*[19], and polyzoa[20] is influenced by the presence of favourable or unfavourable bacterial films. These bacterial films are most complete on surfaces which are either not liable to abrasion or are present upon the particles in stable sediments. When exposure to wave or current action is prolonged, then these microorganisms are swept off the sediment surfaces and carried into suspension in increasing amounts: indeed soil particles subjected to prolonged movement may be worn down by self-abrasion[21].

Aggregates of bacteria in sea water and attached to surfaces may be broken up and studied by the use of surfactants[21, 22]; however, the number and size of colonies found in bacterial counts of sea water depend upon technique and the peptone content of the culture medium[23]. Studies in the Firth of Clyde and of the sands of Morar, Inverness-shire, showed sand grains to be colonized by bacteria, blue-green algae (Cyanophyceae), *Merismopedia*, *Anabaena*, *Microcystis*, and diatoms (Bacillariophyceae). The microbial flora may alter slightly to depths of 15 cm below the sediment surface, or change within a few millimetres. It is sparse towards the high-water mark, but there is little difference between that of sands of the lower shore and the sub-littoral. Colonies of these microorganisms may be composed of single or mixed species and may range from 5 to 150 cells per colony[24]. Counts of individual micro-organisms range from 25×10^3 to $259 \times 10^3/mm^2$ of sand grain surface, i.e. from 140×10^6 to $1183 \times 10^6/g$ of dry sand: the viable counts were 0·2 to $40/mm^2$ and $2·6 \times 10^3$ to $241 \times 10^3/g$ respectively. No relationship between bacterial numbers and particle size was observed[21].

Although the total numbers of bacteria per gram of sand quoted by earlier workers are probably inaccurate[21], it has long been recognized that bacteria contribute an appreciable amount of nourishment to sediment-dwelling animals[25–31]. A large proportion of the inhabitants of mud flats are equipped with the specialized mechanisms necessary to ingest microscopic particles of detritus and bacteria[32], and an appreciable proportion of mud-eating animals probably feed upon bacteria[32, 33]. A number of benthic animals have been maintained exclusively upon a bacterial diet: *Dendrostoma zostericola*[27], *Urechis caupo*[32], *Corophium volutator*[31], *Emerita analoga*[27], and *Mytilus californianus*[27]. It was subsequently shown that the deposit feeding molluscs *Hydrobia ulvae* and *Macoma balthica*, formerly recognized as detritus feeders, do not feed upon detritus, but upon the bacteria which break it down[34]. The large numbers of bacteria available in sediments indicate the importance of this food source[21, 28, 31]; according to one estimate 11 g dry weight of bacteria is produced per ft^3 of sediment per day and is available for nutrition[28].

Of course, bacteria have many other functions, and are active in the breakdown of dead material particularly[28]; fungi and actinomycetes may assist or be more effective in the breakdown of plant material[35]. Bacteria are responsible for the production of petroleum[10, 12], oxidize hydrocarbons[36, 37], oxidize ammonia in sea water to nitrite[38] and the sulphate reducing bacteria present in sediments are capable of nitrogen fixation (p. 120). When sediments are deposited, modification by bacteria occurs and may have a marked effect upon the organic and inorganic constituents[39, 40]. In estuaries, the most obvious bacterial influence upon sediments is that due to the sulphate reducing bacteria which produce the black anaerobic layer and thus influence the biota and the phosphate cycle (p. 118).

Effects may be far reaching, or of considerable economic importance to man, when bacteria act as an agent either of disease or of spoilation[2, 12]. Fish spoilage is caused primarily by bacteria which invade the tissues from the digestive tract or the external slime: species of *Achromobacter*, *Pseudomonas*, *Flavobacterium* and *Micrococcus* predominate in this process[2, 12].

The human disease, botulism, caused by *Clostridium botulinum* Type E, has been responsible for a number of deaths consequent upon the ingestion of uncooked smoked or salted fish; in North America, unchilled vacuum packed cisco trout caused 19 deaths from this cause. This problem seems most likely to arise in waters which receive drainage from large land masses: the organism responsible has been found in sediments off north-west Canada, in the Baltic, in the Caspian Sea and in the Great Lakes area[2].

In some circumstances chitonoclastic bacteria, which perform a vital function in breaking down chitin from dead molluscs, coelenterates, protozoa, crustacea and fungi (Crustacea alone produce several million tons annually)[12], can be detrimental to man's interest. Fishermen, from the Isle of Man, know edible crabs (*Cancer*) which are typified by dirty, discoloured shells, broken claws and a characteristic pitting of the carapace, i.e. black necrosis, as "Grannies" which when eaten have a bitter taste and a powerful purgative effect. A black necrosis may occur widely in the shore crab *Carcinus*[41], and has been observed in the shrimp *Crangon*[X].

Cellulose-decomposing, or cellulolytic, bacteria are generally present in sea water and sediments, especially in the latter: most are aerobes; anaerobes are rarely detected in marine materials. The majority belong to the *Vibrio* and *Cytophaga* groups, though some belong to *Pseudomonas*, *Micrococcus*, *Flavobacterium*, *Nocardia* and *Streptomyces* spp.[2, 12]. Cellulose decomposers, accompanied by other physiological groups of bacteria, participate in the rapid rotting of wooden piles, board boat bottoms, cordage, viz. fish nets, lines and ropes in the marine environment[12]. The problem of loss of tensile strength in ropes due to this cause has long been recognized and, although nylon and polypropylene have been introduced with great success, ropes of natural fibres are still very widely used. In Japan, *ca.* 1956, about 75% of nets made from natural fibres were replaced annually for this reason. The intensity of the attack upon these materials depends upon water quality, viz. pH, salinity, dissolved organic matter content and the amount of suspended solids, and varies with season and geographical location. In the Baltic and North Seas, fishing nets deteriorate more rapidly in flowing water than in stagnant water and, disregarding the degree of pollution or of eutrophication, at a rate which is strongly correlated with temperature; below 5°C the deterioration decreases markedly. In stagnant water, nets rot more rapidly in summer and if the water is eutrophic. In the North Sea, a decrease in deterioration is related to increased distance from the land. Deterioration may be reduced by "cutching" of nets in the presence of mercuric and copper salts, phenol, pentachlorophenol, and 5 nitro-2 furfural-semi-carbazide, which are all effective at concentrations < 1%. The properties of modified cellulose molecules have been examined for this purpose: cellulose triacetate has been found to be particularly resistant to the attack of cellulolytic bacteria[2].

Fungi
Marine fungi, like others, depend on organic substrates for their growth. The principal groups represented in the sea include members of the lower aquatic Phycomycetes, the Myxomycetes, the Ascomycetes and the Fungi

Imperfecti. In all, only a few hundred species of fungi are marine, and are therefore only a small proportion of fungal species as a whole. However, they have an ecological role which is comparable to that of the marine bacteria. The origins of the marine species seem to be obscure, but the marine Pyrenomycetes and Fungi Imperfecti are predominantly attributable to well known terrestrial genera. The marine Phycomycetes include the Saprolegniaceae which appear to be primarily freshwater forms; the Pythiaceae seem to have the same origin, but they are more tolerant of salt water and have achieved a greater success in colonizing the sea. The Basidiomycetes are virtually absent from the sea[42].

Chytrid Phycomycetes are well represented in the sea, and can tolerate a wide range of salinity. Fewer species live in sediments of the sub-littoral than those of the shore. The species of saprophytic lower Phycomycetes are essentially the same in the sea, the sea strand, in freshwater and on land: the general chytrid flora is essentially the same from the polar circle to the equator. All chytrids isolated from saline soils grow equally well in fresh and saline media, although most species develop more readily in fresh water. An exception, *Rhizophidium halophilium*, attacks baits more readily in sea water than in fresh water; *Rh. carpophilum*, the most common species found, attacks baits in fresh and salt water with equal vigour. *Olpidium maritimum* will grow at all salinities from 0–28‰, but sporulation is most prolific between 7 and 13‰.

Although most anisochytrid Phycomycetes are found as parasites on freshwater hosts or in soil, *Anisolpidium ectocarpii* parasitises the brown alga *Ectocarpus*: both have been found in water at a salinity of 22·9‰. *A. sphacellarum* and *A. rosenvingii* are parasitic on Phaeophyceae (brown algae) also. The bulk of the lagenid Phycomycetes are parasitic and occur both in the sea and in fresh water. The species *Sirolpidium bryopsidis* and *Pontisma lagenoides* occur widely: *Sirolpidium zoophthorum* is an obligate estuarine species[42].

In the sea, the parasitic species of Saprolegniales are more successful than the saprophytes. *Eurychasma Dickinsonii* parasitises the algae *Ectocarpus* and *Striaria*. A curious association includes *Ectrogella eurychasmoides* which is parasitic upon the diatom *Licmophora Lyngbyei*, an epiphyte of *Pylaiella littoralis*. The species *Haliphthorus milfordensis* parasitizes the eggs of *Urosalpinx*, the tingle, and *Pinnotheres*, the pea crab: since it will reproduce only in sea water, or an approximate synthetic equivalent, it is considered to be an obligate marine species. *Alkinsiella dubia* infects the eggs of marine crustacea. In the Baltic, the parasitic species *Leptolegnia baltica* caused an epidemic infection of the copepod *Eurytemora hirundoides*, while in the estuaries of southern England, a related species *L. marina* is a destructive parasite of *Pinnotheres*, itself a parasite of mussels. Saprophytic

Saprolegniaceae penetrate brackish waters to a limited degree only. In the River Rheidol, Wales, about a dozen species lived above the tidal limit, but towards the mouth the number decreased: *Aphanomyces laevis* penetrated furthest downstream, but developed only upon baits exposed in waters of which the salinity was $\leqslant 4\%_0$. *Saprolegnia ferox* occurred in the tidal portion of the river also, but lost its powers of sexual and asexual reproduction at 3 and $7\%_0$ respectively; it cannot form gemmae at $13\%_0$ and mycelia formation is feeble at $25\%_0$.

Phycomycetes of the order Peronosporales include one family, the Pythiaceae, which has only a few marine species. In the River Weser, at Bremerhaven, the Pythiaceae are represented by seven sexual species, but these decrease in number towards the sea. The most tolerant species, judging by oospore formation, are *Pythium marinum*, *P. maritimum* and *P. salinum*: all tend to settle on nets, baskets and ropes. *P. maritimum* is euryhaline, with growth and reproductive optima in mesohaline conditions (i.e. $> 7\%_0$ to $\pm 18\%_0$), and is the best adapted to brackish conditions. Parasitic Pythiaceae have been found in *Calanus finmarchicus* in some Norwegian fiords. *Pythium thalassium* is parasitic and saprophytic in eggs of the pea crab, *Pinnotheres*.

Phycomycetes of the orders Mucorales and Eccrinales occur in marine soils and as parasites of marine animals respectively. *Ichthyosporidium hoferi*, which belongs to the order Entomophthorales, parasitizes many species of marine fish; it is possible that *Calanus finmarchicus* is an intermediate host of this parasite[42].

The Myxomycetes are a curious group, regarded as plant by the botanist and animal (i.e. Mycetozoa, members of the protozoan class Sarcodina) by the zoologist. This group contains the species *Labyrinthula macrocystis*, which together with the Pyrenomycete *Lulworthia halima* was associated with the dramatic disappearance of the eel grass, *Zostera marina*, from the coasts of America and Europe in the 1930s: the actual cause of the decline of the *Zostera* has never been settled satisfactorily[42]. What is more certain, although not all authors are in agreement[43], is that the cataclysmic decline of *Zostera* was paralleled by the destruction of its inhabitants, e.g. *Rissoa membranacea*, *Cantharidus striatus* and *Haliclystus auricula*[44]. At Salcombe, also, changes in the sediment distribution resulted and many infauna species declined, viz. *Acrocnida brachiata* and its commensal *Mysella bidentata*; *Phascolosoma elongatum*; *Amphitrite edwardsi* and its commensal *Lepidasthenia argus*; the synaptids, *Leptosynapta inhaerens* and *Labidoplax digitata*; *Upogebia deltaura* and *U. stellata*; *Callianassa subterranea*; *Owenia fusiformis* and *Notomastus latericeus*, and one disappeared, viz. *Lepton squamosum*. On the other hand, *Echinocardium cordatum* and its commensal *Montacuta ferruginosa* remained common, and *Caesicir-*

7*

rus neglectus and *Magelona* spp. were abundant[44]. In both North America and Europe, the *Zostera* has recovered only very slowly from the epidemic[44-48]. The genus *Labyrinthula* is generally stenohaline, but *L. macrocystis* has a wide range of salinity tolerance, viz. 0–32‰. Other genera of myxomycete, viz. *Plasmodiophora* and *Tetramyxa*, occur on marine angiosperms[42].

Actinomycetes, including 17 strains which decompose laminarin and alginate, have been isolated from decomposing brown seaweeds, lagoon muds, salt marsh soils and sands[35, 42]. The genera *Nocardia*, *Micromonospora* and *Streptomyces* were recorded from sub-littoral sediments, wood and other substrates. Apparently these were terrestrial rather than autochthonous forms: about 50% of the isolates showed some degree of antibiosis against selected gram positive and gram negative bacteria[42].

Fungi Imperfecti occur commonly in the sea, growing saprophytically on seaweeds, on decaying salt marsh plants and in salt marsh soil; of these a terrestrial *Aspergillus*, which grows in the sea, is as much a marine as a terrestrial fungus[42]. In estuaries, vegetated areas have a richer fungal flora than bare mud, e.g. 17 species grow on bare mud compared with 27 species in soil colonized by *Salicornia stricta*. Typical salt marsh species occur more frequently in the upper regions of the intertidal zone, whereas the more widespread and transient species occur more frequently on the lower shore. This difference may be explained by the transient species being washed out from the land and finding favourable conditions for

Table 8.1. Yeasts most frequently isolated from various marine environments (after Morris[56])

Species	Sea water	Animals (warm and cold blooded)	Plants
Candida guilleriermondii	9	3	0
C. parapsilosis	8	6	1
C. tropicalis	7	4	1
Black yeasts	7	3	0
Debaryomyces Kloeckeri	6	6	1
Rhodotorula glutinis	6	4	1
Rh. mucilaginosa	5	7	1
Rh. minuta	3	4	0
Rh. rubra	2	5	0
Cryptococcus albidus	5	1	1
Cr. laurentii	5	2	1
Cr. diffluens	0	4	1
Torulopsis candida	5	4	1
T. famata	3	4	1
Trichosporon cutaneum	3	4	1

growth among the non-specific detritus low on the shore, while the upper shore is colonized by fungi adapted to live on salt marsh plants[49-53]. In bare muds, *Mucor* and *Penicillium* occur most frequently, together with *Aspergillus fumigatus* and *Trichoderma viride*. Some species characteristically occur in salt marshes, although other species may be present: *Dendryphiella salina* is found associated with roots of *Salicornia*[49, 52] which are penetrated only upon the death of the *Salicornia*. The species *Dendryphiella salina* and *Ascochytula obiones* occur in association with *Halimione portulacoides*[54]. The leaves of this plant may be colonized by 20 species of fungi, of which only *Dendryphiella salina* and *Ascochytula obiones* are characteristic of salt marshes; these species enter a reproductive phase only when the leaves of *Halimione* become moribund[55]. The occurrence of fewer species of fungi in the *Spartina* zone, compared with the *Salicornia* zone, may be a consequence of the differing soil environments[50]; *Nectria inventa* and *Scopulariopsis* spp. are more abundant in this zone than elsewhere[52].

Hemiascomycetes include the Saccharomycetales or yeasts and the Endomycetales: the former are well represented in the sea, but the latter have only one or two marine members[42].

Up to 1968, some 89 species of yeast from 15 genera and some which have not yet been given a specific rank had been recovered from sea water. Of these, the most widely distributed species are *Candida guilleriermondii*, *C. parapsilosis*, *C. tropicalis*, Black yeasts, *Debaryomyces Kloeckeri*, *Rhodotorula glutinis*, *Torulopsis candida*, *Rhodotorula mucilaginosa*, *Cryptococcus albidus*, and *Cr. laurentii*. High counts are recorded from inshore waters, especially estuaries, but smaller counts are recorded from offshore waters: the high counts may be associated with the presence of detritus which provide a substrate which the yeasts can colonize. Indeed, the abundance of yeasts at depth compared with the surface layers in some areas, a homogeneous distribution in others, and in the Pacific Ocean, a maximum abundance near the surface and a development in layers like plankton at depth seem to imply such a relationship.

In comparison with sea water, rather less attention has been paid to the yeasts upon marine animals, but the distribution in sea water, plants and animals are compared in Table 8.1. The species *Rhodotorula texensis*, *Rh. marina*, *Rh. penaeus*, *Trichosporon cutaneum* var. *penaeus*, *Tr. diddensii* and *Tr. lodderi* were isolated from the shrimp *Penaeus setiferus*. Species of *Rhodotorula*, *Candida* and *Torulopsis* were isolated from Japanese clams. In general, the yeasts most commonly found upon fish are *Debaryomyces Kloeckeri*, *Candida parapsilosis*, *C. tropicalis* and *Rhodotorula* spp. The species *Candida guilleriermondii* has been isolated less frequently from fish than from sea water, whereas for *Rhodotorula* spp. the reverse is probably true; the *Cryptococcus* spp. occur with reasonable frequency in both situations, but the species composition is different.

Just as the number of yeasts recorded from marine animals is much smaller than the number recorded from sea water, so the number isolated from marine plants is smaller still. The following species have been isolated from marine plants: *Candida brumptii*, *C. parapsilosis* var. *intermedia*, *C. tenuis*, *C. tropicalis*, *Candida* sp., *Cryptococcus albidus*, *Cr. diffluens*, *Cr. laurentii*; *Cr. microfugosus*; *Cr. neoformans*, *Rhodotorula glutinis*, *Rh. infirmo-miniata*, *Rh. mucilaginosa*, *Trichosporon cutaneum*, *Tr. infestans*, *Torulopsis famata* and *Metschnikowella zobellii*. Some fifteen varieties of "white Torula" and two "red Torula" and yeast-like fungi have been isolated from the surfaces of the algae *Laminaria saccharina*, *Alaria esculenta*, *Fucus vesiculosis* and other seaweeds: of these, some species of "Torula" and *Endomyces vernalis* may be autochthonous.

Studies of marine yeasts are beset with culture difficulties which make comparison of the individual worker's investigations difficult at the present time. Moreover, there are many outstanding questions concerning the role of yeasts in the sea; not least of these questions is the problem of establishing whether or not there are any true marine yeasts[42, 56].

Euascomycetes are primarily represented, in the sea, by the Pyrenomycetes, which are essentially parasites, saprophytes and endophytes of seaweeds. The range of these species colonizing the commonest species of brown algae is given in Table 8.2. Marine Pyrenomycetes occur on

Table 8.2. Marine Pyrenomycetes on some littoral Phaeophyceae on British Coasts (after Wilson[42])

Marine Pyrenomycete	Phaeophyceae
Mycosphaerella Ascophylli Cotton	*Pelvetia canaliculata* Dcne and Thur.
Stigmatea Pelvetiae Suth.	*Pelvetia canaliculata* Dcne and Thur.
Pharcidia Pelvetiae Suth.	*Pelvetia canaliculata* Dcne and Thur.
Pleospora Pelvetiae Suth.	*Pelvetia canaliculata* Dcne and Thur.
Orcadia pelvetiana Suth.	*Pelvetia canaliculata* Dcne and Thur.
Didymosphaeria pelvetiana Suth.	*Pelvetia canaliculata* Dcne and Thur.
Dothidella Pelvetiae Suth.	*Pelvetia canaliculata* Dcne and Thur.
Mycosphaerella Ascophylli Cotton	*Ascophyllum nodosum* Le Jol
Orcadia Ascophylli Suth.	*Ascophyllum nodosum* Le Jol
Trailia Ascophylli Suth.	*Ascophyllum nodosum* Le Jol
Didymosphaeria fucicola Suth.	*Fucus vesiculosus* L.
Orcadia sp.	*Fucus vesiculosus* L.
Lulworthia fucicola Suth.	*Fucus vesiculosus* L.
Lulworthia salina (Linder) Cribb and Cribb	*Fucus vesiculosus* L.
Hypoderma Laminariae Suth.	*Laminaria saccharina* Lamour
Ophiobolus Laminariae Suth.	*Laminaria digitata* Lamour
Pleospora laminariana Suth.	*Laminaria* sp.
Rosellinia laminariana Suth.	*Laminaria* sp.

other brown, green and red algae, upon marine angiosperms, viz. *Zostera marina*, and on salt marsh plants. Records from marine animals are rare, but *Laboulbenia marina* has been recorded on marine Carabidae. A few species are systemic endophytes of seaweeds. In associations like those between *Guignardia ulvae* and *G. alaskana*, and *Ulva californica* and *Prasiola borealis*, respectively, the fruiting bodies of the fungus develop all over the algal thallus. However, in associations like that between *Mycosphaerella ascophylli* and the algae *Ascophyllum nodosum* and *Pelvetia canaliculata*, the fungal mycellium permeates the whole alga and the fungal fruiting bodies develop only upon the receptacles of the alga; furthermore the ascospores and oospheres are discharged nearly simultaneously. These dual organisms, lichens and the composite organism *Cladophora-Blodgettia* bear some similarities[42]. Some marine Pyrenomycetes have ascospores which by their form and the possession of oil drops, as food reserves, are adapted to float in the sea; by their form also, entanglement with the objects which they encounter is facilitated and the presence of mucilage sheaths and mucilaginous appendages may promote attachment. Many of these ascospores germinate after a short period in the sea[42].

Lignicolous fungi belong principally to the Fungi Imperfecti and the Pyrenomycetes; species of the former belong to the genera *Diplodia*, *Helicoma*, *Speira*, *Botryophialophora*, *Phialophorophoma* and *Orbimyces* among others, while species of the latter belong to the genera *Samarosporella*, *Ceriosporopsis*, *Remispora*, *Lentecospora*, *Halosphaeria*, *Peritrichospora*, and *Halophiobolus*. Floras composed of these genera develop upon wood blocks immersed in the sea for varying periods. Cellulose, pectin and starch are the sources of carbon preferred by 14 species of lignicolous marine fungi, but sugars can be utilized also. Growth of the mycelia was optimal at a high temperature, only three species had optima below 25°C. All of these fungi could grow in hypersaline water. In a study of growth of seven lignicolous marine fungi, only one, *Halophiobolus opacus*, required sea water, but five species required thiamine; asparagine was generally preferred as a nitrogen source to ammonium compounds or nitrates. In culture, a medium containing glucose, thiamine, biotin and asparagine yields good growth of these organisms; wood flour and powdered cellulose may be substituted for glucose and an increased reproductive capacity may result. Almost all the lignicolous marine fungi can tolerate a range of salinity from brackish to that of full salinity, but many Pyrenomycetes can tolerate hypersaline conditions also.

Most marine fungi soften only the surface layers of the wood which is then gradually eroded by the sea. Species of *Ceriosporopsis* and *Lulworthia* give rise to a "soft rot" in the surface layers; *Lindra inflata* attacks the medullary rays, and some Indian species, notably *Halosphaeria quadricornuta*, rapidly destroy the light, porous wood of fishing craft[42]. Attacks by the wood boring crustacean *Limnoria lignorum*, the gribble, frequently follow attacks by marine fungi, and young gribble have been seen in the empty perithecial cavities of *Ceriosporopsis cambrensis*. *Limnoria* may tunnel into fresh, fungus-free wood, but this alone is an insufficient source of food. In nature, fungal mycelium is invariably found in wood round the tunnels of *Limnoria* whose excrement may contain the conidia of *Helicoma*, which has been eaten by the crustacean. It has been shown experimentally that the fungus is a source of food, but, like wood, it is insufficient when taken alone. Furthermore, the fungal mycelium makes some fraction of the wood utilizable, possibly by decomposing parts of the cellulose. For a balanced diet, therefore, a gribble requires both wood and the marine fungi growing in it. This, apparently, is a case of "ectosymbiosis" whereby the gribble gains nutritional benefit, and the fungus has its spores dispersed within a rich food supply. Possibly this association could permit a new and different approach to the control of *Limnoria*[42].

The fungus *Dermocystidium marinum* is found in oysters along the U.S. coast from Delaware Bay to Texas. The fungus is pathogenic and flourishes at high temperatures and moderate to high salinities. It is usually absent from areas which have a summer salinity < 15‰. Established infections are inhibited by temperatures below 15°C and salinities below 12–15‰, but the spores in oyster tissue are resistant to freezing and drying. Warm winters followed by hot, dry summers produce a high mortality in the oysters which is contributed to by the gregarine, *Nematopsis*, the boring polychaete, *Polydora*, and the boring sponge, *Cliona*. The occurrence of *Dermocystidium*-like bodies in numerous other bivalves raised the problem of host specificity, but cross infection experiments have failed to produce evidence of conspecificity: on the evidence available, bivalve molluscs are apparently subject to attack by a group of related fungal parasite species[57]. Curiously, successive observations have shown numerous fungal cells in the parapodia of the polychaete *Neanthes succinea* which have fed upon tissues of oysters dying of *Dermocystidium* infection[57].

Lichens, which are a symbiotic association between an Ascomycete fungus and an alga, form an important element of some maritime communities, particularly in fixed dunes, shingle beaches and rocks and cliffs[42, 59, 60]. These organisms show a zonation from the grey-green, stiff, brittle *Ramalina* at high levels in the splash zone, to the orange, leafy *Xanthoria parietina* in the lower splash zone, to the black encrusting *Verrucaria* which may extend to or below the E.L.W.M.S.T. Such genera as *Verrucaria* constitute the truly marine lichens: one of these *V. serpu-*

loides forms a sub-littoral association with a *Lithophyllum* sp. in the Antarctic[42]. *Lichina pygmaea*, which grows on the shore from E.H.W.S.T. to M.T.L.[58, 59], is apparently very slow growing; at Church Reef, Wembury, Devon, areas cleared within lichen mats showed no significant recolonization in a four year study[61]. Possibly growth is periodic and related to climatic cycles. Clearly such organisms are very sensitive to adverse events such as an oil spill[62]. Although *Lichina* lives on shores exposed to fairly heavy wave action on the southern coast of Britain, it does occur on very sheltered and unshaded shingle beaches in the Scottish west coast sea lochs. Here it is found among dense growths of the fucoids *Pelvetia canaliculata* and *Fucus spiralis*: in some places it may replace the latter species. In the still waters of Loch nan Cilltean it lives mingled with the uppermost plants of *Ascophyllum nodosum* f. *mackai*[63]. In estuaries the upstream distribution of lichens is related to salinity also (p. 162).

Algae

The algae of estuaries exhibit a wide range of habit from the single-celled, motile tychopelagic species to the large, attached, multicellular species, so conspicuous on rocky shores. The minute, single celled littoral algae are important contributors to primary production, especially in inner estuarine areas. Many of these minute forms reveal themselves by a discoloration of the littoral soil: these include the blue-greens, Myxophyceae, diatoms, Bacillariophyceae, and dinoflagellates, Dinophyceae. The blue-green *Merismopedia* sp. may give rise to green patches of the soil[X], and commonly occurs attached to sand grains in company with another species *Anabaena* sp.[24].

Blue-Green Algae

In the freshwater environment the development of intense blooms of blue-green algae, predominantly *Anabaena* spp. and *Anacystis cyanea* (= *Microcystis aeruginosa*), but including many other species of algae, takes place in eutrophic, nutrient enriched waters. At best, such blooms are a nuisance; at worst, the toxins produced may be harmful to man and animals[64], and, even if they are not lethal, allergic symptoms may develop.

In estuaries, the blue-green algae are widely distributed in developing salt marshes, and, like bacteria[65] and fungi[50], the number of isolates decreases from upper to lower levels. At Gibraltar Point, Lincolnshire, 67% of all the species present occurred at all levels in the salt marsh and there was no evident correlation between the majority of the blue-green algae and the dominant angiosperms. It therefore seems unlikely that the distribution of blue-green algal species has been affected significantly by the recent rapid spread of *Spartina townsendii*.

Of the algae ubiquitous throughout the marsh, the species *Microcoleus chthonoplastes, Nostoc commune, Nodularia harveyana, Oscillatoria brevis, O. formosa, Phormidium autumnale* and *P. tenue* are apparently the most robust and least sensitive to environmental change. Although the species *Lyngbya aestuarii* and *Oscillatoria amphibia* are associated with the Chlorophyceae zone, these commonly occur in mature regions of salt marsh, but *Calothrix scopulorum* is restricted to the lower *Spartina* and *Salicornia* zones, a concurrence which is due largely to the influence of common environmental factors, e.g. salinity and desiccation[66]. In the shallow, lagoonal flats of Texas, *Lyngbya confervoides* form extensive mats[67].

In pure culture, the species *Nostoc commune, N. entophytum* and *Calothrix scopulorum* all fix nitrogen[66, 68], an ability shared by *Nodularia* spp. and probably most other species of *Nostoc*[66]. At Gibraltar Point, between 17 and 33% of all blue-green algal species can assimilate elemental nitrogen; in fresh water, some 25% of all blue-green species have this facility. In salt marshes generally, and in areas uncolonized by higher plants in particular, nitrogen fixing blue-green algae may play an important role in the improvement of the nitrogen status of littoral soils[66]. *Calothrix scopulorum* liberates its extracellular nitrogen as peptides and free amino acids[68], which are removed from solution by marine algae, fungi and bacteria. *Chlorella* sp. utilizes several amino acids[69]; indeed amino acids and peptides may act as the sole source of nitrogen for the growth of *Chlorella marina*[70]. A number of workers have shown that littoral and brackish algae, in particular, utilize or even require organic nitrogen for growth. For example, *Nannochloris* spp. and *Stichococcus* spp. from polluted waters at Long Island grow equally well when provided with urea, uric acid, amino acids, or ammonium-nitrogen as nitrogen sources[71]; chrysomonads, such as species of *Hymenomonas, Syracosphaera* and *Ochromonas*, utilize a variety of amino acids[72]; and some flagellates from brackish pools, e.g. *Hemiselmis virescens*, require amino nitrogen for growth[73]. Such nitrogen sources may not be the only extracellular products formed, for the blue-green algae are powerful producers of vitamin B_{12}[70] which, together with other vitamins, is required by a variety of marine algae[74, 75]. It is of interest too, that organic aggregates formed in sea water can serve as the sole source of nutrients for the brine shrimp, *Artemia salina*[70, 76]. Furthermore, by using ^{15}N as a tracer, a nitrogen transfer between blue-green algae and grasses has been demonstrated[77].

Diatoms

Diatoms inhabit a wide range of estuarine situations including the surfaces of algae and higher plants, of animals, e.g. hydroids, bryozoa and tunicates, on the surface of muds in the photic zone and on the surface of rocks[78-80]. They may exist as individual cells, or as filamentous colonies; some, viz. the naviculoids, some nitzschids and pleurosigmas, are motile although the mechanism of this movement is uncertain[81]. One species, *Navicula endophytica*, is unusual in that it inhabits the receptacles of *Ascophyllum nodosum*, but is not an epiphyte on this seaweed and lacks a planktonic form[82]. Benthic diatoms recovered from a depth of 110 m, in Loch Sween, may be evidence that such diatoms have a lower compensation point than do planktonic forms[29, 83]. The salinity tolerances of a wide variety of species are known[79, 80, 84, 85].

In the River Ouse estuary, Sussex, most solitary epiphytic diatoms grew well on chalk and large algae, but most filamentous diatoms grew well on large algae, except for filamentous *Navicula* which flourished on chalk. An *Achnanthes*—blue-green community frequently occurred on wood. In general however, the survival of the diatom depends upon the water-holding capacity of the substratum, which helps it to resist desiccation. Concrete provides a firm substratum which permits rapid growth in winter; in summer, however, most diatoms are destroyed because it dries out so readily[86]. Similarly, in Tasmania, rocks exposed to the sun are rapidly cleared of diatoms, whereas shaded rocks are not[87].

The diatom flora of the substratum and of the large algae which it supports may be very different[88]: in the Stour, the concrete wall was colonized by a *Navicula grevillei–N. ramosissima* community, whereas, in January and February, *Fucus spiralis*, both stipe and frond, bore an *Achnanthes*–blue-green community together with free-solitary *Cocconeis scutellum* and *Licmophora juergensii*. With the exception of *Fucus spiralis*, which has more silt and more abundant diatoms than *F. vesiculosus*, diatom abundance is inversely related to the amount of silt present on the plant surface. The abundance of the epiphytic diatoms undergoes seasonal change[86], Fig. 8.1, but diatom communities dwelling upon the same species of large attached alga in different areas show considerable differences in the specific composition[80, 86]. Four diatom genera, viz. *Fragilaria*, *Grammatophora*, *Biddulphia* and *Melosira* are able to exist in the epiphyte flora and in the shore plankton: to these forms the term *facultative epiphyte* may be applied[86].

On a sand or mud flat colonized by higher plants, epiphytic diatoms are continually falling on to the soil surface. However, the soil has its own population of epontic diatoms, but the two communities may be differentiated by means of the differing ratios of species on the plants and on the mud[78]. Littoral diatoms which live in the surface layers of a sedimentary substratum may be observed by means of the characteristic brown patches of coloration which they produce. These patches may be abundant every-

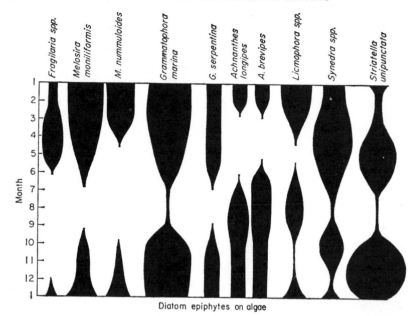

Diatom epiphytes on algae

Fig. 8.1. The seasonal distribution of ten diatoms that are common epiphytes upon large algae. The greatest width shows that *Striatella unipunctata* was common on five of the eight species of algae examined in November (after Hopkins[86]).

where in estuaries except at levels approaching the high-water mark and where conditions of drainage are such that a soil dries or tends to dry out during the period of exposure; these patches are not in evidence at the lower shore levels, i.e. near the channels, where, as in the Solway Firth, the currents and scour are such that little can live in these soils[X]. In the River Ouse estuary, Sussex, the three principal factors which control diatom distribution on a mud flat transect are: (1) the ability to tolerate desiccation at higher levels near the M.H.W.L.; (2) the ability to survive on short periods of illumination at levels near to the M.L.W.L.; and (3) the ability to tolerate the products of the black-sulphide layer whenever this is near the surface [89] (Table 8.3).

Although fine sediments support more diatoms than coarse sediments[90], differences in soil grade apparently do not produce significant qualitative effects in epontic diatom populations: at Whitstable, Kent, most species were of widespread occurrence except for *Surirella gemma* and *Nitzschia sigma* which were confined to soils with a higher silt content[91]. The former species was also abundant in the more silty soils of the River Eden estuary, Fife[92]. Salinity is a decisive factor in the distribution of many diatoms in brackish water areas[84]; populations at individual sites

Table 8.3. Apparent tolerance of littoral diatoms to desiccation, low illumination and sulphureous materials (after Hopkins[89])

Diatoms	Desiccation resistance	Tolerance of low illumination	Resistance to $SO_4^=$ reduction products
Navicula ammophila var. *flanatica*	Not in July	Yes	Yes
N. cyprinus	Not in July	Yes	No
N. cancellata	Yes	Yes	No
Stauroneis salina	Not in July	Yes	No
Nitzschia closterium	No	Yes	Unknown
Tropidoneis vitrea	Yes	Not good	No
Pleurosigma angulatum	Yes	No	Unknown
P. aestuarii	Yes	No	No
P. balticum	Not good	Not in February	Yes
Triceratium favus	No	Yes	Unknown
Campylodiscus sp.	Yes	No	Unknown

Table 8.4. The recorded distribution with respect to salinity of some benthic diatoms which occurred in small numbers with the dominant populations of *Pleurosigma aestuarii*, *P. angulatum*, *P. fasciola* and *Surirella gemma*, in the River Eden estuary November 1956–March 1957[(X)]

Diatom	Fresh	Brackish	Marine
Cyclotella kutzingiana	+	+	+
Biddulphia aurita			+
Biddulphia pulchella			+
Isthmia enervis			+
Rhaphoneis amphiceros			+
Diatoma elongatum	+	+	
D. vulgare	+		
Grammatophora serpentina			+
Tabellaria flocculosa	+		+
Achnanthes longipes			+
Cocconeis scutellum			+
Navicula forcipata			+
N. lyra			+
Scoliopleura tumida	+	+	+
Stauroneis salina			+
Gomphonema geminatum	+		
Bacillaria paxillifer		+	+
Surirella fastuosa			+
S. ovalis	+	+	

may be composed of a relatively few abundant species, together with many other species of rare occurrence which have their source in other areas (Table 8.4).[(X)] Of 14 species inhabiting a salt marsh in Georgia, all grew well at salinities between 10 and 30‰, and many continued to divide at salinities of 1–2‰ and 68‰.[(93)]

Soil-dwelling diatoms undergo seasonal changes of abundance (Fig. 8.2), and of species composition of the populations[79, 86, 92, 94]. Perhaps the most interesting feature of the soil-dwelling diatoms, and one which has aroused great scientific interest, is the ability of many to move. This ability is dependent upon the degree of drainage of the soil: at Whitstable, Kent, the diatoms were immobilized when the water content fell below 30% (generally to 20–25%) of the dry weight of the soil[94]. These communities reside mainly in the surface 2 mm of a sediment, and in bright light move less than 1 mm towards the surface where marked changes in colour are produced[90, 94]: communities in fine soils dwell nearer to the

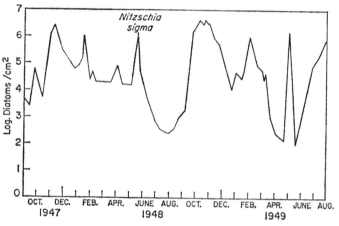

Fig. 8.2. Seasonal distribution of epontic diatoms in Botany Bay between 1947 and 1949 (after Crosby and Wood[79]).

surface than those in coarser grades of soil.[90] Diatoms move most rapidly between 10 and 17·5°C, but velocities are sufficient to enable the community to reach the soil surface between 7·5 and 20°C. Species which live near to the L.W.M., i.e. they are adapted to survive with relatively short periods of illumination daily, are able to move at lower temperatures than those, e.g. *Pleurosigma angulatum* and *Tropidoneis vitrea*, which are abundant at higher shore levels: *Nitzschia closterium* and *Stauroneis salina* are motile at 0°C, and two further species, *Navicula ammophila* var. *flanatica* and *N. cyprinus*, are motile at 5°C. At upper shore levels, resistance to desiccation is of more value to survival than rapid movement, and here the diatoms, e.g. *Nitzschia* and *Pleurosigma*, secrete a mucilaginous envelope for protection[89]. In the dark, diatom movement is reduced, species becoming immobilized in times which range from 4–24 hr; after a period of darkness movement is stimulated by physical shock. Fluctuations in pH, dissolved oxygen and soluble organic matter do not affect the appearance of the diatom community on the surface[90].

Diatoms produce coloured patches in response to light, but similar patches are produced by blue-green algae[95], dinoflagellates, euglenoids, and flatworms. Each of these organisms requires exposure to sunlight, either for itself or a symbiont, and, because many estuarine waters with a sediment load are turbid, early observers saw this response as a *tidal rhythm*[96-121], which resulted, eventually, in a belief that this response was of necessity due to an innate rhythm, itself independent of the environment.

This misunderstanding arose when meticulous early work upon *Convoluta roscoffensis*, a marine flatworm, which has an algal symbiont *Platymonas convolutae*[122, 123], was replaced by somewhat speculative conclusions based essentially on the consequences of exposing this organism to a radioactive radiation source of unknown composition and intensity[97] and which ignored valid work on the effects of "shock" in producing migrations[96, 103, 104]. At this time too, the effects of oxygen saturation, on invertebrate tissues[124], were unknown. Therefore the original "physiological rhythm" theory was born by misconception. Over the years, further fallacies were generated based on a failure to understand the transmission of wave shock in sedimentary and rock structures, the essentials of estuarine hydrography, and the lack of knowledge that many littoral diatoms are tychopelagic (pp. 71, 73).

The first rational analysis of this phenomenon, since 1903, was due to Aleem[94] who appreciated that the essential response of these shore-dwelling organism was to light, and who resurrected Bohn's original concept of "tidal shock" as an explanation for diatoms submerging before being covered by the incoming tide[96]. Further work, particularly with reference to euglenoids and diatoms[90, 125-127], has resulted in the conclusion that the tidal rhythm of littoral diatoms is a curtailed diurnal migration, and that the rhythm is inherently based upon a "biological clock".

It seems unreasonable to suggest that a "biological clock" does not exist, if only because living organisms undergo a cyclical exposure to day and night, and in a photo-synthesizing organism, at least the velocity of its chemical reactions must be geared to this cycle of change. However, this explanation seems too inflexible for an organism living in as complex an environment as that inhabited by these interesting organisms, and to which, for survival, they must respond. Furthermore, laboratory studies on the "maintenance" of rhythms in light intensities as low as 112 fc are somewhat irrelevant in the context of solar illumination, which can reach an intensity of 10 000 fc, and intensities of $\geqslant 1000$ fc occur frequently in Britain.

Again, dinoflagellates appear on the soil surface only in the hours after dawn and before dusk, never in the intense light of the middle day[111, 117]. Besides, diatom patches in the Solway Firth, and other areas, develop a greater intensity of colour in the darker months of the year when the solar altitude is low, and in summer on cloudy days rather than on the very bright days[128, X].

Other workers have doubted the "biological clock" hypothesis[92, 129] and have seen these migrations as a response to changing environmental conditions. The phototaxy of marine green algae and dinoflagellates is determined by the Ca–Mg–K ratio[130]. In Sierra Leone, the movements of *Euglena deses* point to a relationship with the evaporating power of the air and of the behaviour of surface films, i.e. response to drainage: here it was considered that protoplast contraction could be affected by changes in the water content of the soil[129]. Like *Euglena deses*, the diatom *Surirella gemma* responds to a change in the water content of the soil by a contraction of the chloroplast which in the diatom is succeeded, within 5 min., by a downwards migration[127]. It will be recalled that the water content of a soil can undergo marked changes ahead of an advancing tide edge (p. 110), and although the response of these soil dwelling organisms is essentially one to light, the ever-changing environment is transmitting signals continuously so that a "biological-clock" seems to be unnecessary to the performance of these movements: the delay in surfacing after dawn appears to be a support for the "biological-clock" theory[126], but this is a weak support as some of the delay, at least, can be explained in terms of the high albedo of soils when the solar altitude is low (pp. 18, 112) and consequently the penetration of the soil by light is poor, and is probably insufficient to induce an upward movement. Furthermore there is a considerable body of work performed in the early years of this century on the possibilities of "learning" in simple organisms and, although largely ignored, it might be well to recall in the context of persistent rhythms that a simple learning process could be involved as the tidal system is intrinsically a perfect conditioning mechanism, which could account for the short term persistence of field rhythms in the laboratory. Clearly much more work is required before these interesting rhythms can be explained adequately.

Protozoa

Ciliates

Ciliates may be classified according to the type of soil which they inhabit. Those species which inhabit sands with a mean particle size between 0·4 and 1·0 mm are termed *mesoporal*, and are generally somewhat "square"

Table 8.5.   Micro-habitats adopted by microbenthic ciliates. Species characteristic of sand only
(after Faure-Fremiet[131–136])

Mesoporal species	Microporal species	Euryporal species	Facultative sand-dwelling species
Mesodinium pulex f. pupula	Helicoprorodon (Chaenea) gigas	Loxocephalus intermedius	Trachelocerca phoenicopterus
Placus (Thoracophrya) buddenbrocki	Centrophorella fasciolata	Pleuronema coronatum	Lacrymaria olor v. marina
Remanella margaritifera	C. lanceolata	Histiobalantium marinum	Coleps tesselatus
Strombidium sauerbreyae	Geleia simplex	Holosticha fasciola	C. pulcher
Discocephalus rotatorius	G. decolor		Frontonia sp.
	G. fossata		Condylostoma patens
	G. orbis		Epiclintes ambiguus
	Condylostoma remanei		Diophrys scutum
	Blepharisma clarissimum		Uronychia transfuga
	Urostrongylum caudatum		Sonderia labiata
	Amphisiella lithophora		

in form, while those living in sand with a mean particle size of 0·1–0·4 mm are termed *microporal*, and are generally long and slender in form. *Euryporal* species, which are very varied in shape and size, inhabit both kinds of soil. Sand-dwelling ciliates have a cosmopolitan distribution, and although some species are so adapted to the interstitial environment that they cannot live elsewhere, others enter and leave the sand freely and have a widespread distribution on the seashore[131–133] (Table 8.5).

The ciliate protozoa can adapt to a wide range of habitat[132]; indeed five of the nine species of facultative sand dwellers listed in Table 8.5, viz. *Trachelocerca phoenicopterus*, *Frontonia* sp., *Condylostoma patens*, *Diophrys scutum* and *Uronychia transfuga*, live equally well in rock pools or the brackish pools of salt marshes. In the former, the temperature can rise from 18 to 27°C, the oxygen content vary from 0 to 16·7 p.p.m. and the pH range from 8·25 to 9·1[134]. In the latter, the salinities normally range from 11 to 23‰, but may rise to 31·1‰ in summer and fall to 8·6–2·2‰ in winter; these pools may be covered in ice in winter, and the temperature may rise to 28°C in summer; the pH may range from 7·3 to 8·3[30].

Table 8.6. Species of ciliate protozoa occurring commonly throughout the year in brackish salt marsh pools at Parkgate, Cheshire (after Webb[30])

Aspidisca crenata	*Lionotus folium*
Chlamydodon mnemosyne	*Loxophyllum helus*
C. triquetrus	*L. setigerum*
Diophrys appendiculatus	*Mesodinium acarus*
Euplotes charon	*Opisthotricha parallelis*
Frontonia marina	*Pleuronema marinum*
Holosticha kessleri	*Strombidium elegans*
Lacrymaria coronata	*Trachelocerca phoenicopterus*
L. olor v. *marina*	*Uronychia transfuga*

In general, the ciliate protozoa are more abundant at high than at low temperatures, but there are no major quantitative fluctuations associated with seasonal change. Many species occur commonly throughout the year (Table 8.6), but *Strombidium elegans* may be very abundant at temperatures between 4·5 and 7·0°C, and *Placus socialis* and *Euplotes harpa* have been found between April and October only. They show no numerical relationship to tidal level in the littoral zone, however they are always found in association with the algal and microbial films upon which the trophic structure of the protozoan communities depends: these films develop at phase boundaries and at the mud–water interface in particular[30].

Where pools of water stand in sheltered situations, a surface film composed of protein or lipoprotein may develop[137] and carry on its underside an association of diatoms and bacteria which may support as large and varied a population of protozoa as the mud itself. At Parkgate, Cheshire, this population included the flagellates *Euglena limosa* and *Oxyrrhis marina* and the ciliates *Chlamydodon triquetrus, Euplotes charon, Lacrymaria coronata, Holosticha kessleri, Loxophyllum helus* v. *multiverrucosum, Placus socialis,* and *Uronychia transfuga*[30].

The ciliate protozoa are generally unable to live in sediments with a grade diameter < 0.1 mm, or in coarse sediments the interstitial spaces of which are filled with finer materials[133]. The inter- and intra-littoral variations in number are often considerable, and may be the result either of a variable supply of food or of soil structure[133]. However, they may be very abundant, ranging from 10^6 to 4×10^7 per m^2[138, 139]. The microphagous feeders are usually more abundant than macrophagous forms[132], but the abundance of the individual species depends upon the amount of chosen food available, and when a given food species is depleted, the ciliate which fed upon it declines, encysts and is replaced by another in an active phase. Encystment does not provide security from predation for *Lacrymaria coronata* takes cysts of *Nassula citraea* and *Glenodinium* sp., and *Loxophyllum setigerum* ingests cysts of *Diophrys appendiculatus*[30].

Of 252 genera and species of algae and protozoa recorded in the Tamar and Plym estuaries, 87 were ciliates associated primarily with the bottom sediments, although some were recorded from the plankton alone, and others were taken from both habitats. Superficially, the Plym flats, with clay deposits and large numbers of sulphide bacteria, presented an unpromising environment for ciliate protozoa, but the populations were very rich. In contrast, the Tamar with a considerable development of large attached seaweed, a different development of sulphide bacteria and saprophytic flagellates was characterized by a small ciliate group. Like the Plym flats, however, the harbour sand was a haven for a large ciliate group associated with saprophytic flagellates and sand-dwelling dinoflagellates[140].

In the sand flats of Barnstaple Harbour, Cape Cod, the microporal ciliates are confined to a superficial layer of grey sand[136]. In the Baltic Sea, at Askö Island, with a salinity range of 6.1–$6.5\%_0$, the ciliate population, in soils with a low availability of oxygen, was largely confined to the surface 2 cm of the shore sand. Nevertheless, *Mesodinium pupula* was most abundant at depths greater than 3 cm within the soil, and reached depths of 6–7 cm on occasion. *Sonderia vorax* and *S. schizostoma* penetrated to some depth also. *S. vorax* feeds on bacteria, flagellates and diatoms, whereas *S. schizostoma* feeds almost exclusively on the sulphur bacteria *Beggiatoa*, consequently the range of depth distribution of *S. vorax* is wider than that of *S. schizostoma* which is confined to a narrow band beneath the surface. The distribution of *Mesodinium pupula* showed some relationship to Eh, viz. when Eh fell to -100 m.v. in the surface 2 cm this ciliate was most abundant at 3–4 cm depth. On the other hand, when the Eh differed little from 0 m.v. in the surface 8 cm its maximum abundance was reached at 5–6 cm depth[141].

Foraminifera

This group of rhizopod protozoans is predominantly marine in distribution and has only a few aberrant representatives which live in fresh water. These organisms have tests, or shells, which are usually calcareous and arranged in chambers, but may be composed of chitin or of sand grains. The calcareous tests are of importance since they are the primary constituent of chalk, and the various assemblages found in other geological formations are vital to the geologists concerned with drilling for oil.

Foraminifera can penetrate estuarine areas to some degree and some species are euryhaline. In the Gulf of Mexico a number of foramineral faunas have been defined[142, 143], thus:

Marine Marsh

1. All benthonic species.
2. 3–8 genera; 5–8 species; populations range from very low to very high.
3. Arenaceous forms constitute most or all of the population except in marshes on lagoon barriers.
4. Characteristic genera: *Ammoastuta, Arenoparrella, Discorinopsis, Jadammina, Miliammina,* and *Trochammina.*

Coastal Lagoon

1. No planktonic forms or occasional specimens only, depending upon the run-off regime.
2. 5–10 genera; 10–20 species; population size variable depending upon the rate of sedimentation.
3. Arenaceous specimens may constitute 5–75% of the population, with higher percentages in the inner lagoon.
4. Characteristic genera: *Ammotium, Elphidium, Quinqueloculina, Streblus, Triloculina.*

Nearshore Turbulent Zone and Beach

1. Population size smaller than in most other environments, with large populations locally.
2. Occasional planktonic specimens, but these are very rare.
3. Species are generally large and thick shelled, but they may be waterworn.
4. Characteristic genera: *Elphidium, Quinqueloculina, Streblus, Textularia.*

Inner Continental Shelf

1. Planktonic to benthonic ratio may be less than 0·1.
2. 5–15 genera; 10–25 benthonic species.
3. Arenaceous forms may be 10–25 ± % of the population.
4. Genera having highest frequencies: *Buliminella*, *Discorbis* (*Rosalina*), *Elphidium*, *Nouria*, *Streblus*.

In San Antonio Bay, the marsh and lagoon facies are broadly similar to that above, however the foraminiferal fauna of upper San Antonio Bay is less rich in species than the lower Bay and Mesquite, Aransas and Copano Bays[144]. Off the mouth of the Guadalupe River, in upper San Antonio Bay, the fauna is characteristically that of a coastal lagoon, but in the immediate mouth, *Palmerinella gardenislandensis* is an important element of the fauna. This species is typical of the fluvial-marine assemblage where the salt wedge invades the bottoms of the river channels in the Mississippi. Downstream, at the mouths of the passes of the Mississippi River, the fauna contains *Bolivina lowmani*, *Buliminella* cf. *B. bassendorfensis*, *Epistominella vitrea* and *Nonionella opima* in addition to *P. gardenislandensis*. To the seawards of the passes of the Mississippi delta, and where the river water and Gulf water mix, 90% of the deltaic marine fauna consists of the four species *B. lowmani*, *B.* cf. *B. bassendorfensis*, *E. vitrea* and *N. opima*[142].

Although the number of species is reduced in upper San Antonio Bay, the abundance of the species present is much increased. At the mouth of Guadalupe River, populations ranging from 302–3000/10 ml of wet sediment were recorded, whereas in the bay as a whole they range from 50–200/10 ml of wet sediment; similar populations are observed of the mouths of the Mississippi, Colorado and Brazos Rivers[145–147], the rivers of New Brunswick and Maine[148–150], and the Castries River, St. Lucia[151]

In the Mississippi Sound area, i.e. *ca.* 80 miles north east of the Delta, the foraminiferal fauna can be divided into four geographical and ecological facies[152], thus:

1. Open gulf facies which occur outside the barrier islands and are characterized chiefly by calcareous species—*Nonionella atlantica*, *Cibicidina strattoni*, *Elphidium*, various Miliolidae, *Discorbis* and "*Rotalia*" *beccarii*.
2. The sound facies which consists principally of *Ammobaculites*.
3. The estuary facies which are characterized by an abundance of *Ammobaculites exilis* and *Ammosclaria fluvialis*. *Miliammina fusca* is abundant.
4. The marsh facies which are characterized by *Ammoastuta inepta*, *Arenoparrella mexicana*, *Leptodermella*, *Trochammina comprimata*, *T. macrescens*, *Urnulina*, *Haplophragmoides subinvolutum* and *Recurvoides* sp. *Miliammina fusca* is abundant.

In each area, the distribution of the various species depends upon the balance between run-off and evaporation. Where run-off exceeds evaporation, open-ocean water masses and their microfaunas cannot invade the sound or bay; conversely, where rainfall is low and evaporation high, penetration by open ocean water masses and associated microfaunas occurs[143, 152].

The factors present in river borne materials which induce a high productivity of foraminiferans are believed to be numerous, but are to all intents unknown. Possible influences may be:

1. An abundance of particulate food may be introduced by river water into the marine environment to which the foraminiferan populations are adapted.
2. Abundant nutrients may be introduced by river water into the marine environment and cause high production of plant materials which provide food for Foraminifera.
3. In the Mississippi, the largest benthonic populations occur where there is a combination of high organic production and fast detrital sedimentation. Here the fine grained material may bring detrital food upon which the Foraminifera feed.
4. The river water may introduce trace materials, organic and inorganic, which are conducive to growth of populations[147].

Further north at Mason Inlet, North Carolina, where the salinity ranged from that of normal sea water at the bar, to almost that of fresh water in the upper reaches, there are no sharply defined faunal zones: the greatest upstream penetration was made by the species *Elphidium incertum, Quinqueloculina seminulum, Elphidium* aff. *incertum* v. *mexicanum, Quinqueloculina lamarckina* which were common; *Q. poeyana* occurred abundantly here; other species were present but were much less abundant[153].

In the Choptank River, a tributary of Chesapeake Bay, the density of *Elphidium clavatum* decreases progressively upstream, whereas *Ammobaculites exiguus* and *Ammonia beccarii* underwent little change[154].

British estuaries are penetrated by a wide variety of species, but five species are distributed more widely and penetrate further upstream than the others[155]. They are *Miliammina fusca, Trochammina inflata, Ammonia beccarii, Nonion depressus*, and *Elphidium striato-punctata*; the Foraminifera of British estuaries and of the Gulf of Mexico are broadly similar, and *M. fusca* is common to both areas.

In the Tamar and Plymouth Sound, *Protelphidium* is most abundant up-river and decreases in abundance downstream, whereas *Ammonia beccarii batavus* behaves in the opposite manner, to the seawards of Drakes

Island, but similarly upstream of it. Downstream from Drakes Island, more species are present with peaks of abundance at different parts of the Sound, e.g. *Quinqueloculina* spp. are most abundant at the Breakwater, but *Reophax fusiformis* and *Textularia sagittula* are conspicuous outside it. *Elphidium* spp. form a significant proportion of many populations. However the populations as a whole in this area are rather poor.

Just as there are geographical variations in the populations of Foraminifera in the Gulf of Mexico area so there are between individual estuaries in Britain. The estuary at Christchurch Harbour, where it was comparable with the Tamar, had a winter population dominated by *Protelphidium* sp., together with *Elphidium excavatum* and *Miliammina fusca*. In the Tamar, at this time, *Protelphidium* was less important and was almost equalled in abundance by *Ammonia beccarii batavus*: *Elphidium excavatum* was a regular member of the population. At this time the salinity at Christchurch was < 1‰, and in the Tamar was 25–31‰.

In spring, *Protelphidium* sp. still dominated the Christchurch populations together with *Miliammina fusca, Elphidium excavatum* and *E.* aff. *poeyanum*. The population in the Tamar was as before, but now included *Eggerella scabra*. At this time, the salinity at Christchurch was 2–29‰, and in the Tamar was 26–32‰.

In summer, the population in Christchurch Harbour had the same three dominant species, but *E.* aff. *poeyanum* was replaced by *A. beccarii batavus*, while in the Tamar, the *A. beccarii batavus–Protelphidium* sp. partnership prevailed. At this time, the salinity in Christchurch Harbour was 14–34‰, and in the Tamar was 33–35‰[156].

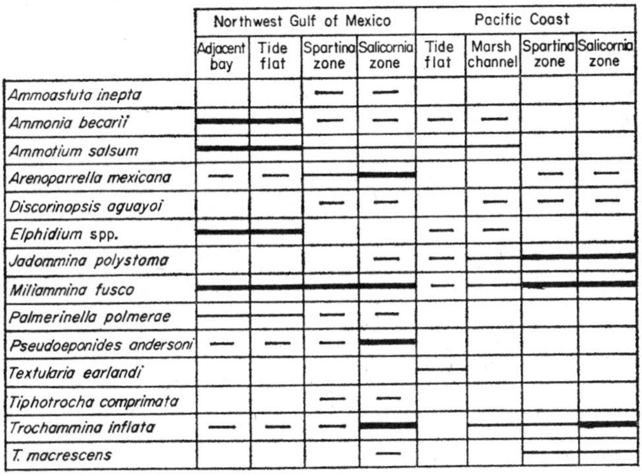

	Northwest Gulf of Mexico				Pacific Coast			
	Adjacent bay	Tide flat	Spartina zone	Salicornia zone	Tide flat	Marsh channel	Spartina zone	Salicornia zone
Ammoastuta inepta			—	—				
Ammonia becarii	▬	▬	—	—	—	—		
Ammotium salsum	▬	▬						
Arenoparrella mexicana	—	—		▬			—	—
Discorinopsis aguayoi			—	▬	—	—		
Elphidium spp.	▬	▬			—	—		
Jadommina polystoma				—	▬			
Miliammina fusca					—			
Palmerinella polmerae			—	—				
Pseudoeponides andersoni	—	—	—	▬				
Textularia earlandi								
Tiphotrocha comprimata				—	—			
Trochammina inflata	—	—	—	▬				▬
T. macrescens				—	▬			

Fig. 8.3. Generalized distributions of marsh foraminifera in the northwest Gulf of Mexico and the Pacific coast of North America (after Phleger and Bradshaw[161]).

Not only do species distributions differ from estuary to estuary, and within the individual estuary[142–159], but the species composition changes with tidal level, so that the population of a *Salicornia* zone is different from that of a tidal flat, and moreover, these differences are not the same on the Gulf of Mexico and Pacific coasts of the U.S.A.[160–162] (Fig. 8.3).

Although the species composition may undergo changes with season as in Christchurch Harbour and the Tamar, seasonal variations in density elsewhere are not so clear cut. In Upper San Antonio Bay such changes are marked, but none occur in lower San Antonio Bay[147]. No seasonal periodicity was observed in Narragansett Bay, but it was in the Choptank River[154]. Some species breed all the year round, e.g. *Elphidium clavatum* in Long Island Sound[154]. In Patos Lagoon, southern Brazil, the period of reproduction of each species is individual and not necessarily shared by any other of the many species present: here *Miliammina fusca*, *Trilocularena patensis* and *Elphidium excavatum* were most abundant in the autumn period, and other species reproduced all the year round, viz. *Rotalia beccarii parkinsoniana*, and *Buliminella elegantissima*. However, the peak of reproduction of the individual species may be different at different stations. *Miliammina fusca* reproduced at the BA station in autumn, but all the year round at the DE station. In this area as a whole, reproduction reached a maximum in autumn and winter[163, 164].

In the Argentine, *Elphidium macellum* reproduced most actively in November, but its largest total and living populations occurred in January. *Quinqueloculina seminulum* reproduced most actively in December, and its largest living population occurred in January also. However, as in the Patos Lagoon, *Rotalia beccarii* and *Buliminella elegantissima*, together with *Cornuspira involvens* do not show a seasonal periodicity[165]. More complex still the populations from Timbalier Bay, Louisiana, have three or four reproductive periods, with a minimum in winter[164]. Not all seasonal change is numerical, for *Elphidium crispum* exhibits differences in size and chamber shape which are related to the food supply, and the paucity of the Plymouth foraminiferan fauna is ascribed to this cause[156]. In general though, forms living at a lower temperature tend to grow large, and all forms living close to their optimum are smaller than those which are not. The generally smaller size of estuarine forms compared with their marine counterparts is probably a reflection of the availability of calcium since they are distinct from marine forms and are adapted to estuarine conditions[166].

From Long Island Sound to Prince Edward Island, on the eastern coast of North America, the foraminiferan populations show considerable temporal and lateral variation. In Smithtown Bay, Long Island, the living fauna is composed primarily of *Elphidium subarcticum*; *E. clavatum* and

Ammonia beccarii are second in the order of abundance. Living specimens of *E. margaritaceum* and *Trochammina inflata* may be transported here by algal rafting from the marshes at the mouth of the Nissequoque River[148]. In the Gulf of St. Lawrence, the dominant species are *Elphidium clavatum, Quinqueloculina seminulum, Eggerella advena* and *Buccella frigida.* The species *Trochammina lobata, Elphidium margaritaceum, E. subarcticum, Cibicides lobatulus, Ammotium cassis* and *Lagena gracillima* are present also[149]. The foraminiferan faunas of these areas bear a general resemblance to the others described, but, as elsewhere, the specific composition is characteristic of the individual area.

In the estuaries of New Brunswick and Maine, *Miliammina fusca* is generally the most abundant species; *Elphidium margaritaceum* is important here also. As in other estuaries and marshes, *E. margaritaceum* is an important species where the salinity exceeds 20‰. In these New England estuaries no intertidal foraminiferan species penetrated upstream of the 5‰ isohaline[150].

In the Gulf of California, which has some of the attributes of a negative estuary, the standing crops of the Foraminifera are greatest in the shallow waters at the head of the Gulf. This may be a result of river influence, hypersalinity or the effect of a silty substratum. Here, there are a number of successive foraminiferal facies whose distribution is related to depth[146].

Some interesting laboratory studies have been performed upon neritic foraminiferans, although it is not easy to relate these results to conditions experienced by these animals in the environment[167]. *Rotaliella heterocaryotica,* a species from the Adriatic, can utilize both dead *Chlamydomonas* cells and living *Nitzschia* as food, but a *Rotaliella* sp. required living *Nitzschia* to ensure survival. *R. heterocaryotica* and *Rotaliella* sp. grew more rapidly at 22·2 and 24·5°C than at 14·5°C, while the former failed to grow at salinities < 25‰ and > 37‰, and had an optimum for growth between 26 and 30‰[167]. *Streblus beccarii* v. *tepida* does not grow at temperatures < 10°C and > 30–35°C, and reproduces in the temperature range 20–30°C; growth and reproduction are normal within the salinity range 20–40‰, but outside these limits both activities decline, and cease outside the range 7 and 67‰ and 13 and 40‰ respectively. Near the tolerance limits the time taken to reach reproductive maturity increases, and at 13‰ it takes twice as long as within the normal salinity range[168]. *Elphidium crispum,* a common marine species, can survive at a salinity of 15‰ and a temperature of 8°C for at least 15 days, but dies rapidly at 16°C; at a salinity of 20‰, cultures survived for 38 days at both temperatures[166].

Ammonia beccarii, an estuarine species of cosmopolitan distribution, may extend up into the *Salicornietum* in some places. In the laboratory, it can reproduce between 20 and 32°C, with an optimum in the range 25–30°C. Growth ceases at 10°C, and is very slow even at 15°C, but when the temperature fluctuates between 15 and 20°C, the growth is no different from those animals maintained constantly at 20°C. In the field at Mission Bay, California, *A. beccarii* grows little in the period January to April/May, because the temperatures either do not reach 20°C, or only reach this value for *ca.* 5% of the time; from April/May to the end of October the temperatures are warm enough for growth to occur, but from November onwards growth ceases because the temperatures are too low once more[162].

The benthonic Foraminifera may be used as tracers of sediment movement and indicators of sediment sources, e.g. where marsh sediments have been washed out into estuarine channels or where shallow water sediments have been moved to deeper waters[143]. In those areas where fast rates of accumulation, based on accurate surveys, are known to occur these high rates are reflected in high living: total foraminiferal ratios, i.e. the relationship

$$\frac{\text{Number of living Foraminifera}}{\text{Total number of Foraminifera}} \times 100$$

is high[143]. Indeed an approximate index of the relative amounts of sediment deposition may be obtained by use of the formula:

$$\text{Sedimentation Index} = \frac{1}{\dfrac{\text{Total population of Foraminifera}}{\text{Gram of wet sediment}}} \times 100.$$

This index tends to mask any short-term fluctuations in the abundance of the living populations. Where values of the sedimentation index range between 0·01 and 0·1 then the area receives relatively little sediment and as the rate of sedimentation increases higher values of the index result. In western Long Island Sound, there is a rapid accumulation of sediment and the sedimentation index ranges from 10 to 100; south of mainland New York, however, the values range from 0·1 to 1·0. The very low values, i.e. 0·01 to 0·1 obtained along the Connecticut shore may not indicate a slow rate of accumulation so much as a high rate of foraminiferal production resulting from the nutrients introduced by the Mianus and Bryam Rivers[148].

References

1. McIntyre, A. D. (1969). *Biol. Rev.* **44**, 245–290.
2. Scholes, R. B. and Shewan, J. M. (1964). *Adv. mar. Biol.* **2**, 133–169.
3. Morita, R. Y. (1966). *Oceanogr. Mar. Biol. A. Rev.* **4**, 105–121.
4. Morita, R. Y. (1965). *J. gen. Microbiol.* **41**, 26–27.
5. Robison, S. M. and Morita, R. Y. (1966). *Z. allg. Mikrobiol.* **6**, 181–187.

6. Haight, R. D. and Morita, R. Y. (1966). *J. Bact.* **92**, 1388–1393.
7. Haight, R. D., Langridge, P., Morita, R. Y. and Becker, R. (1965). *Bact. Proc.* 1965, p. 20.
8. ZoBell, C. E. (1953). *In* "XIV Intern. Zool. Congr. Copenhagen 1953", *I.U.B.S. Deep Sea Colloquium.*
9. Harvey, E. N. (1952). "Bioluminescence", Academic Press, New York.
10. ZoBell, C. E. (1952). *J. sedim. Petrol.* **22**, 42–49.
11. Skerman, T. M. (1961). "Symposium on Marine Microbiology", Chap. 64, 685–698, (C. H. Oppenheimer, ed.), Thomas Springfield, Illinois.
12. ZoBell, C. E. (1959). *In* "Contributions to Marine Microbiology" (T. K. Skerman, ed.), *N.Z. Oceanogr. Inst. Memoir No. 3 (Information Series* No. 22), 7–20.
13. Rose, A. H. (1968). "Chemical Microbiology", Butterworths, London.
14. Starr, T. J., Jones, M. E. and Martinez, D. (1957). *Limnol. Oceanogr.* **2**, 114–119.
15. Burkholder, P. R. and Burkholder, L. M. (1956). *Limnol. Oceanogr.* **1**, 202–208.
16. Wilson, D. P. (1955). *J. mar. biol. Ass. U.K.* **34**, 531–543.
17. Meadows, P. S. and Williams, G. B. (1963). *Nature, Lond.* **198**, 610–611.
18. Gray, J. S. (1966). *J. mar. biol. Ass. U.K.* **46**, 627–645.
19. Cole, H. A. and Knight-Jones, E. W. (1949). *Fishery Invest., London* Ser. 2. **17**, 1–39.
20. Crisp, J. and Ryland, J. S. (1960). *Nature, Lond.* **185**, 119.
21. Anderson, J. G. and Meadows, P. S. (1969). *Hydrobiologia* **33**, 33–46.
22. Jones, G. E. and Jannasch, H. W. (1959). *Limnol. Oceanogr.* **4**, 269–276.
23. Floodgate, G. D. (1964). *J. mar. biol. Ass. U.K.* **44**, 365–372.
24. Meadows, P. S. and Anderson, J. G. (1968). *J. mar. biol. Ass. U.K.* **48**, 161–175.
25. Luck, J. M., Sheets, G. and Thomas, J. O. (1931). *Q. Rev. Biol.* **6**, 40–58.
26. ZoBell, C. E. and Landon, W. A. (1937). *Proc. exp. biol. Med.* **36**, 607–609.
27. ZoBell, C. E. and Feltham, C. B. (1938). *J. mar. Res.* **1**, 312.
28. ZoBell, C. E. and Feltham, C. B. (1942). *Ecology* **23**, 69–77.
29. Mare, M. F. (1942). *J. mar. biol. Ass. U.K.* **25**, 517–574.
30. Webb, M. G. (1956). *J. Anim. Ecol.* **25**, 149–175.
31. Perkins, E. J. (1958). *Ann. Mag. nat. Hist.* Ser. 13, **1**, 64–77.
32. MacGinitie, G. E. (1935). *Am. Midl. Nat.* **16**, 629–765.
33. Baier, C. R. (1935). *Arch. Hydrobiol.* **29**, 183–264.
34. Newell, R. (1965). *Proc. zool. Soc. Lond.* **144**, 25–45.
35. Chesters, C. G. C., Apinis A. and Turner, M. (1956). *Proc. Linn. Soc., Lond.* **166**, 87–97.
36. Grant, C. W. and ZoBell, C. E. (1942). *Proc. exp. Biol. Med.* **51**, 266–267.
37. Davies, J. A. and Hughes, D. E. (1968). *Fld Stud.* **2** (Suppl.), 139–144.
38. Spencer, C. P. (1956). *J. mar. biol. Ass. U.K.* **35**, 621–639.
39. ZoBell, C. E. (1938). *J. sedim. Petrol.* **8**, 10–18.
40. ZoBell, C. E. (1942). *J. sedim. Petrol.* **12**, 127–136.
41. Perkins, E. J. (1967). *Trans. J. Proc. Dumfries. Galloway nat. Hist. Antiq. Soc.* Ser. 3, **44**, 48–56.
42. Wilson, I. M. (1960). *Proc. Linn. Soc., Lond.* **171**, 53–70.
43. Stauffer, R. C. (1937). *Ecology* **18**, 427–431.
44. Wilson, D. P. (1949). *J. mar. biol. Ass. U.K.* **28**, 395–412.

8

45. Dexter, R. W. (1947). *Pl. Dis. Reptr.* **31**, 448–449.
46. Dexter, R. W. (1950). *Ecology* **31**, 286–288.
47. Dexter, R. W. (1951). *Pl. Dis. Reptr.* **35**, 507–508.
48. Dexter, R. W. (1953). *Ecology* **34**, 229–231.
49. Pugh, G. J. F. (1960). *In* "Ecology of Soil Fungi", 202–208, University Press, Liverpool.
50. Pugh, G. J. F. (1961). *Nature, Lond.* **190**, 1032–1033.
51. Pugh, G. J. F. (1962). *Trans. Br. mycol. Soc.* **45**, 255–260.
52. Pugh, G. J. F. (1962). *Trans. Br. mycol. Soc.* **45**, 560–566.
53. Green, J. (1968). "The Biology of Estuarine Animals", Sidgwick and Jackson, London.
54. Dickinson, C. H. and Pugh, G. J. F. (1965). *Trans. Br. mycol. Soc.* **48**, 381–390.
55. Dickinson, C. H. (1965). *Trans. Br. mycol. Soc.* **48**, 603–610.
56. Morris, E. O. (1968). *Oceanogr. Mar. Biol. A. Rev.* **6**, 201–230.
57. Andrews, J. D. and Hewatt, W. G. (1957). *Ecol. Monogr.* **27**, 1–25.
58. Southward, A. J. and Orton, J. H. (1954). *J. mar. biol. Ass. U.K.* **33**, 1–19.
59. Nelson-Smith, A. (1967). *Fld Stud.* **2**, 407–434.
60. Moyse, J. and Nelson-Smith, A. (1963). *Fld Stud.* **1**, 1–31.
61. Boney, A. D. (1961). *J. mar. biol. Ass. U.K.* **41**, 123–126.
62. Ranwell, D. S. (1968). *The Lichenologist* **4**, 55–56.
63. Lewis, J. R. (1957). *Trans. R. Soc. Edinb.* **63**, 185–220.
64. Mackenthum, K. M., Keup, L. E. and Stewart, R. J. (1968). *J. Wat. Pollut. Control Fed.* **40** (2), Pt. 2: R73–R81.
65. Turner, M. and Gray, T. R. G. (1962). *Nature, Lond.* **194**, 559–560.
66. Stewart, W. D. P. and Pugh, G. J. F. (1963). *J. mar. biol. Ass. U.K.* **43**, 309–317.
67. Sorensen, L. O. and Conover, J. T. (1962). *Publs Inst. mar. Sci. Univ. Tex.* **8**, 61–74.
68. Jones, K. and Stewart, W. D. P. (1969). *J. mar. biol. Ass. U.K.* **49**, 475–488.
69. Arnow, P., Oleson, J. J. and Williams, J. H. (1953). *Am. J. Bot.* **40**, 100–104.
70. Jones, K. and Stewart, W. D. P. (1969). *J. mar. biol. Ass. U.K.* **49**, 701–716.
71. Ryther, J. H. (1954). *Biol. Bull. mar. biol. Lab., Woods Hole.* **106**, 198–209.
72. Pinter, I. J. and Provasoli, L. (1963). *In* "Marine Microbiology", pp. 114–121, Thomas, Springfield, Illinois.
73. Droop, M. R. (1957). *J. gen. Microbiol.* **16**, 286–293.
74. Provasoli, L. (1958). *A. Rev. Microbiol.* **12**, 279–308.
75. Droop, M. R. (1959). *J. mar. biol. Ass. U.K.* **38**, 605–620.
76. Baylor, E. R. and Sutcliffe, W. H. (1963). *Limnol. Oceanogr.* **8**, 369–371.
77. Stewart, W. D. P. (1967). *Nature, Lond.* **214**, 603–604.
78. Hustedt, F. and Aleem, A. A. (1951). *J. mar. biol. Ass. U.K.* **30**, 177–196.
79. Crosby, L. H. and Wood, E. J. F. (1959). *Trans. R. Soc. N.Z.* **86**, 1–58.
80. Edsbagge, H. (1966). *Botanica Gothoburgensia* **VI**, 1–139.
81. Crosby, L. H. and Wood, E. J. F. (1958). *Trans. R. Soc. N.Z.* **85**, 483–530.
82. Hasle, G. R. (1968). *Br. phycol. Bull.* **3**, 475–480.
83. Smyth, J. C. (1955). *J. Ecol.* **43**, 149–171.
84. Kolbe, R. W. (1927). *Pflanzenforschung.* **7**, 1–46.
85. Hustedt, F. (1939). *Abh. naturw. Ver. Bremen.* **31**, 572–677.
86. Hopkins, J. T. (1964). *J. mar. biol. Ass. U.K.* **44**, 613–644.
87. Guiler, E. R. (1950). *Pap. and Proc. R. Soc. Tasm.* 1949, pp. 135–201.

88. Aleem, A. A. (1950). *J. Ecol.* **38**, 75–106.
89. Hopkins, J. T. (1964). *J. mar. biol. Ass. U.K.* **44**, 333–341.
90. Hopkins, J. T. (1963). *J. mar. biol. Ass. U.K.* **43**, 653–663.
91. Perkins, E. J. (1958). "The Microfauna of the Shore", Ph.D. Thesis, University of London.
92. Perkins, E. J. (1960). *J. Ecol.* **48**, 725–728.
93. Williams, R. B. (1964). *Ecology* **45**, 877–880.
94. Aleem, A. A. (1950). *New Phytol.* **49**, 174–188.
95. Noble, A. (1961). *J. mar. biol. Ass. India* **3** (1 and 2).
96. Bohn, G. (1903). *C.r. hebd. Séanc. Acad. Sci., Paris* **137**, 576–578.
97. Bohn, G. (1903). *C.r. hebd. Séanc. Acad. Sci., Paris* **137**, 883–885.
98. Bohn, G. (1903). *C.r. hebd. Séanc. Acad. Sci., Paris* **137**, 1292–1294.
99. Bohn, G. (1903). *Bull. Mus. Hist. nat., Paris* **9**, 352–364.
100. Bohn, G. (1903). *Bull. Mus. Hist. nat., Paris* **9**, 397–399.
101. Bohn, G. (1904). *C.r. hebd. Séanc. Acad. Sci., Paris* **139**, 610–611.
102. Bohn, G. (1904). *C.r. hebd. Séanc. Acad. Sci., Paris* **139**, 646–648.
103. Gamble, F. W. and Keeble, F. (1903). *Proc. R. Soc.* **72**, 93–98.
104. Gamble, F. W. and Keeble, F. (1903). *Q. Jl microsc. Sci.* **47**, 363–431.
105. Bohn, G. (1907). *C.r. Séanc. Soc. Biol.* **62**, 51–52.
106. Bohn, G. (1907). *C.r. Séanc. Soc. Biol.* **62**, 211–213.
107. Bohn, G. (1907). *C.r. Séanc. Soc. Biol.* **62**, 564–567.
108. Bohn, G. and Drzewina, A. (1928). *Annls Sci. nat. Zool.* **10**, 299–398.
109. Fauvel, P. (1907). *C.r. Séanc. Soc. Biol.* **62**, 242.
110. Fauvel, P. and Bohn, G. (1907). *C.r. Séanc. Soc. Biol.* **62**, 121–123.
111. Laurie, R. D. (1914). *83rd A. Rept. Br. Ass. Birmingham* **1913**, 509–510.
112. Whitehead, T. (1914). *Rep. Dove mar. Lab. for 1913* **2**, 107.
113. Jorgensen, O. (1918). *Rep. Dove mar. lab. for 1918* **7**, 57–59.
114. Bracher, R. (1919). *Ann. Bot.* **33**, 93–108.
115. Gard, M. (1919). *C.r. hebd. Séanc. Acad. Sci., Paris* **169**, 1423.
116. Gard, M. (1920). *Bull. Soc. bot. Fr.* **69**, 184–250; 241–250; 306–312.
117. Herdman, E. C. (1921). *Proc. Trans. Lpool biol. Soc.* **35**, 59–63.
118. Lebour, M. V. (1925). "The Dinoflagellates of Northern Seas", Marine Biological Association, Plymouth.
119. Faure-Fremiet, E. (1948). *Bull. biol. Fr. Belg.* **82**, 3–23.
120. Faure-Fremiet, E. (1950). *Bull. biol. Fr. Belg.* **84**, 207–214.
121. Faure-Fremiet, E. (1951). *Biol. Bull. mar. biol. Lab., Woods Hole* **100**, 173–177.
122. Parke, M. W. and Manton, I. (1967). *J. mar. biol. Ass. U.K.* **47**, 445–464.
123. Provasoli, L., Yamasu, T. and Manton, I. (1968). *J. mar. biol. Ass. U.K.* **48**, 465–479.
124. Fox, H. Munro and Taylor, A. E. R. (1954). *Nature, Lond.* **174**, 312.
125. Palmer, J. D. and Round, F. E. (1965). *J. mar. biol. Ass. U.K.* **45**, 567–582.
126. Round, F. E. and Palmer, J. D. (1966). *J. mar. biol. Ass. U.K.* **46**, 191–214.
127. Hopkins, J. T. (1966). *J. mar. biol. Ass. U.K.* **46**, 617–626.
128. Hopkins, J. T. (1965). *In* "Light as an Ecological Factor" (R. Bainbridge, ed.), pp. 205–209, Blackwell, Oxford.
129. Taylor, F. J. (1967). *J. Ecol.* **55**, 345–359.
130. Halldal, P. (1959). *Physiologia. Pl.* **12**, 742–752.
131. Faure-Fremiet, E. (1950). *Bull. biol.* **84**, 35–75.
132. Faure-Fremiet, E. (1950). *Endeavour* **9** (36), October 1960, 5 pp.

133. Faurę-Fremiet, E. (1951). *Biol. Bull. mar. biol. Lab. Woods Hole* **100**, 59–70.
134. Faure-Fremiet, E. (1948). *J. Anim. Ecol.* **17**, 127–130.
135. Faure-Fremiet, E. (1954). *Bull. Soc. zool. Fr.* **79**, 473–479.
136. Faure-Fremiet, E. and Tuffrau, M. (1955). *Hydrobiologia* **7**, 210–218.
137. Goldacre, R. J. (1949). *J. Anim. Ecol.* **18**, 36–39.
138. Fenchel, T. (1967). *Ophelia* **4**, 121–137.
139. Fenchel, T. (1968). *Ophelia* **5**, 73–121.
140. Lackey, J. B. and Lackey, E. W. (1963). *J. mar. biol. Ass. U.K.* **43**, 797–805.
141. Fenchel, T. and Jansson, B.-O. (1966). *Opehlia* **3**, 161–177.
142. Phleger, F. B. (1960). *In* "Recent Sediment, Northwest Gulf of Mexico, 1951–58", pp. 267–381, *Am. Ass. Petrol. Geol.*
143. Phleger, F. B. (1964). *Marine Geol.* **1**, 16–43.
144. Parker, F. L., Phleger, F. B. and Peirson, J. F. (1953). *Cushman Fdn foramin. Res. Spec. Pub.* No. 2, 1–75.
145. Phleger, F. B. (1956). *Contr. Cushman Fdn foramin. Res.* **7**, 106.
146. Phleger, F. B. (1964). *In* "Marine Geology of the Gulf of California—A Symposium", *Am. Ass. Petrol. Geol.* Memoir No. 3, 377–394.
147. Phleger, F. B. and Lankford, R. R. (1957). *Contr. Cushman Fdn foramin. Res.* **8**, 93–105.
148. Schafer, C. T. (1968). "Ecology of Benthonic Foraminifera in Western Long Island Sound and Adjacent Near-Shore Waters", Report A.O.L. 68-8: 48 pp. Unpublished manuscript.
149. Schafer, C. T. (1968). "Lateral and Temporal Variation of Foraminifera Popluations Living in Near-Shore Areas", Report A.O.L. 68-4: 28 pp. Unpublished manuscript.
150. Schafer, C. T. and Sen Gupta, B. K. (1969). "Foraminiferal Ecology in Polluted Estuaries of New Brunswick and Maine", Report A.O.L. 69-1: 24 pp. Unpublished manuscript.
151. Schafer, C. T. and Sen Gupta, B. K. (1968). *Maritime Sediments* **4**, 57–63.
152. Phleger, F. B. (1954). *Bull. Am. Ass. Petrol. Geol.* **38**, 584–647.
153. Miller, D. N. (1953). *Contr. Cushman Fdn Foramin. Res.* **4**, 41–63.
154. Buzas, M. A. (1969). *Limnol. Oceanogr.* **14**, 411–422.
155. Brady, G. S. and Robertson, D. (1870). *Ann. Mag. nat. Hist.* Ser. 4, **6**, 1–33.
156. Murray, J. W. (1965). *J. mar. biol. Ass. U.K.* **45**, 481–505.
157. Closs, D. and Madeira, M. L. (1966). *Bol. da Universidade Federal do Paraná, Zoologia* **II** (10), 139–162.
158. Closs, D. and Madeira, M. L. (1967). *Iheringia, Pôrto Alegre, Zool.* **35**, 7–31.
159. Closs, D. and De Medeiros, V. M. F. (1967). *Iheringia, Pôrto Alegre, Zool.* **35**, 75–88.
160. Phleger, F. B. and Ewing, G. C. (1962). *Bull. geol. Soc. Am.* **73**, 145–182.
161. Phleger, F. B. and Bradshaw, J. S. (1966). *Science, N.Y.* **154**, 1551–1553.
162. Bradshaw, J. S. (1968). *Limnol. Oceanogr.* **13**, 26–38.
163. Forti, I. R. S. and Roettger, E. (1966). *Arch. Oceanogr. Limnol.* **15**, 55–61.
164. Closs, D. and Madeira, M. L. (1968). *Esc. Geol. P. Alegre, Publ. Esp.* No. 15, 1–51.
165. Boltovskoy, E. (1964). *Int. Revue ges. Hydrobiol. Hydrogr.* **50**, 293–296.
166. Murray, J. W. (1963). *J. mar. biol. Ass. U.K.* **43**, 621–642.
167. Bradshaw, J. S. (1955). *Micropalaeontology* **1**, 351–358.
168. Bradshaw, J. S. (1957). *J. Palaent.* **31**, 1138–1147.
X.   Perkins, E. J. Previously unpublished work.

9

Meiobenthos

The meiobenthos is cosmopolitan and is found in fresh water as well as marine substrata. This group of living organisms has been found living at depths of 5030 m offshore, although the over-all density decreases with increasing depth[1–4]. In the intertidal zone, total meiobenthos populations range from 11×10^3 to $> 16 \times 10^6$ per m^2; the higher densities occur in the softer deposits of sheltered areas. Below the L.W.M., i.e. on the continental shelf, the populations range from 4×10^3 to $3 \cdot 2 \times 10^6$ per m^2. As on the shore the softer deposits are richest, with a decrease in abundance as the depth increases, however even in the abyssal zone $1 \cdot 6$–17×10^4 per m^2 have been recorded. On both the shore and the continental shelf, nematodes are normally the most numerous group with the harpacticoid copepods second in order of abundance[3]. No doubt, like the macrobenthos, these organisms will eventually be taken from the greatest depth of the sea. Not all groups of meiobenthos are equally well represented in both fresh and marine waters[5]; for example, mystacocarids, ostracods and acarines occur in marine soils, but are absent from or rare in freshwater sediments, whereas rotifers which are extremely abundant in fresh water, occur rarely in the sea and are confined to a few species.

The meiobenthos may be divided into the true interstitial and the benthic forms. The *true interstitial* meiobenthos inhabits the pore spaces of a soil and moves through the interstitial medium without disturbing the particles of the substratum; the *benthic meiobenthos*, on the other hand, is composed of more bulky forms than the interstitial group, and disrupts the soil structure by their passage through it. The distinction is more evident in reference to the harpacticoid copepods (Fig. 9.1): the true interstitial forms are characterized by the *Arenosetella* type, whereas the benthic forms are typified by the *Asellopsis* type. A comparison may be made here between the body forms of the harpacticoid copepods and those of the mesoporal and microporal ciliate protozoans (Fig. 9.2).

It is of interest that a meiofaunal organism, viz. the ostracod *Mesocypris terrestris*, was found in the terrestrial environment in 1953[6, 7]. Since then, rotifers, nematodes, tardigrades, gastrotrichs, turbellarians, aquatic oligochaetes (including enchytraeids), crustaceans, insect larvae and small molluscs have been found in the terrestrial meiobenthos: these populations

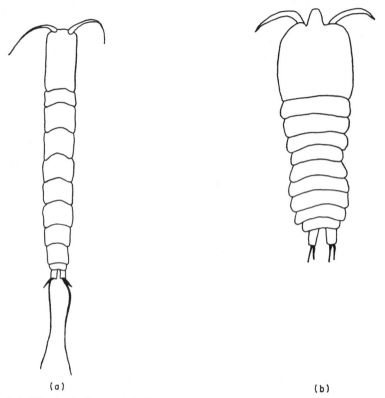

(a)   (b)

Fig. 9.1. The body form of the harpacticoid copepods. (a) *Arenosetella* interstitial type. (b) *Asellopsis* benthic type (after Perkins[29]).

are especially well developed in damp situations such as that provided by beech litter. All are much smaller than their counterparts in the freshwater and marine environments. The crustacea are represented by copepods, cladocerans and ostracods, and the parallel with the intertidal, marine environment is particularly strong when it is recalled that the harpacticoids occur most commonly while cyclopoid copepods occur only rarely, in both environments[8, 9], and the whole assemblage has more in common with that of the marine environment than of fresh water, and with the meiobenthos rather than the plankton.

A meiobenthos may live in the most difficult and exposed conditions. The island of Madeira is rocky and mountainous; its shores are formed largely of steep or even sheer cliffs, often hundreds of feet in height. Some of the cliffs fall straight into the sea, but others have small, stony beaches at their foot. These beaches are not easily accessible and the more accessible occur only at the mouths of rivers and along the shore at Funchal. They

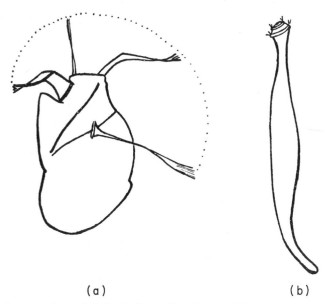

(a)   (b)

Fig. 9.2. A comparison of the body form of marine sand dwelling ciliates. (a) *Strombidium sauerbreyae* (length, (l), excluding cirri, 72 μm, length : width, 1·6 : 1), a mesoporal type in which most of the thigmotactic cirri are not included here, but the cirral corona is indicated by the dotted line, (b) *Helicoprorodon gigas* (length (l)=450 μm, length : width, 11·4:1), a microporal type.

are steep and composed either of coarse shingle or of tumbled masses of rock fallen from the cliffs above. The smoothest beaches are composed of gravel rather than true sand, and a similar material is found in some of the pools or beneath large boulders: apparently the finer materials are swept away by the strong tides. Here the rivers contribute little fine material to the beaches, moreover there is a pronounced dry season and much of the spring water is drawn off for irrigation and never reaches the sea. Nevertheless, in a 10 ml sample of gravel from a small pool near to H.W.M. at the mouth of the Ribeiro Seco, near Funchal, and where the fresh water influence was minimal, two nematodes were found in association with diatoms, listed in order of abundance, thus: *Achnanthes brevipes, Cocconeis scutellum, Actinocyclus fasciculatus, Navicula* sp., *Grammatophora marina, Striatella unipunctata, Stauroneis* sp., *Coscinodiscus perforatus* and *Licmophora lyngbyei*. While in a similar sample of grit from a pool at the M.T.L., 5 nematodes, 14 ostracods of the genus *Loxoconcha* and 3 harpacticoid copepods of the family Diosaccidae occurred in association with diatoms, listed in order of abundance thus: *Cocconeis scutellum, Grammatophora marina, G. serpentina, Navicula* sp., *Actinocyclus fasciculatus, Achnanthes brevipes, Melosira* sp., *Licmophora lyngbyei, Coscinodiscus perforatus, Striatella unipunctata, Surirella dives, Pyscidicula* (*Dictyopyxsis*) *brevis, Nitzschia closterium* and *Stauroneis* sp.[X]. Of these genera, *Grammatophora, Striatella, Achnanthes, Navicula, Nitzschia, Cocconeis, Licmophora* and *Surirella* are important primary colonizers, indeed diatoms can attach readily to moving surfaces[10]. Apparently the diatoms provided a surface which gave a foothold to the meiofauna. Indeed the greater abundance of clinging, crawling harpacticoid copepods and ostracods than nematodes is a marked contrast to the meiofaunal composition of sandy and muddy shores where the nematodes are dominant. Clearly in this Madeiran environment the nematode is at a disadvantage[X].

In estuaries the total population of meiobenthos increases with increasing shelter[3] (Table 9.1): the richest population of nematodes recorded, viz. 163 000 below 10 cm² of the surface, was found in a salt marsh in Georgia, U.S.A.[11].

As with the macrobenthos, the distribution of meiobenthos in an estuary is influenced by salinity. Even those most resistant of all animals, the nematodes, have relatively few species which are common to both the freshwater and the marine environments[12, 13]. In the Elbe estuary, where the change from the marine conditions of the North Sea to the freshwater zone of the river covers a distance of 292 km, 35% of the 258 meiofauna species identified were limited to a salinity $> 18\%_0$[14].

In the North River, Massachusetts, the densest populations occurred at Little Bridge where the average salinity was $28 \cdot 1\%_0$ ($\pm 6\%_0$). At Union Bridge where the average salinity was $17 \cdot 1\%_0$ ($\pm 15 \cdot 5\%_0$) the populations were only 22% smaller than at Little Bridge, but at Wanton Bridge, with an average salinity of $3 \cdot 9\%_0$ ($\pm 3 \cdot 0\%_0$), the populations were $< 10\%$ as abundant as those at Little Bridge[15].

In the River Eden estuary, Fife, where the soils had a peculiarly adhesive quality, and the fetch was such that little disturbance resulted even from gale force winds[16], a rich population of meiobenthos ($> 3193/10$ cm²) was recorded in February 1957. This estuary drains completely, except for a small flow in the main channel at the time of low water; the brackish conditions at the station were indicated by the presence of the alga *F. ceranoides* and reflected by the fresh water, brackish and marine components of the meiobenthos[X] (Table 9.2).

In the middle reaches of the River Blyth estuary, England, the nematode populations, derived from 19 species, reached a mean density of $2590/10$ cm² in mud with a comparatively stable and high salinity. Upstream, where the interstitial salinity fell to *ca.* $20\%_0$, the population derived from 37 species, fell from a transect mean of about 1300 to $508/10$ cm² with the increasing distance of the individual transect upstream[17, 18]. In the sandy shores of the Gulf of Finland, the nematode population is composed of

Table 9.1. Density of intertidal meiofauna—numbers below 10 cm² of surface (after McIntyre[3])

Deposit	Location	Nematoda	Copepoda	Ostracoda	Oligochaeta	Total Meiofauna
Sand	S. of Stockholm, Locality "A", Sweden	38–169	31–833	0	214–498	391–1529
Sand	S. of Stockholm, Locality "B", Sweden	48–88	24–129	0	92–109	173–402
Sand	Gothland, Sweden	8–14	0	0	17–112	225–1048
Sand	Scanian east coast, Sweden	1	2–45	0	16–160	49–318
Sand	Kattegat, Sweden	6–106	1–52	0	1–243	11–845
Sand	Øresund, Denmark	115–497	1–10	0	1–375	249–764
Sand	Hard Wadden, W. Denmark	10–1050	<1–840	0–61	+	13–1914
Sand	Soft Wadden, W. Denmark	31–367	13–389	4–12	+	50–768
Sand	Boulogne, France	3–367	3–192	0–<1	0	19–389
Sand	Acachon, France	67–204	5–331	<1–18	0–12	83–591
Sand	Eden estuary, St. Andrews, Scotland[x]	3163	17	13	+	3193
Sand	Blyth estuary, England	300–1320	*	*	*	—
Sand	Whitstable, England	1136–5220	13–486	21–742	0–7	1264–5817
Sand	Miami, U.S.A.	0–110	0–260	0	0	14–872
Sand	Porto Novo, S. India	594–1150	186–448	0	0–2	968–1960
Sand	E. coast of Malaya	346–8068	86–2416	0	0	700–10 212
Coral sand	E. coast of Malaya	32	144	0	12	244
Mud	Bristol Channel (Wales)	70–10 440	0–500	0–790	0–780	90–11 820
Mud	Southampton Water, England	*	81–1021	*	*	—
Mud	Blyth estuary, England	228–2830	*	*	*	—
Mud	Salt marsh, Massachusetts, U.S.A.	1440–2130	*	*	*	—
Mud	Salt marsh, Georgia, U.S.A.	460–16 300	*	*	*	—
Mud	Vellar estuary, S. India	307–3240	5–490	0–63	0	420–3815

* Not studied.

8*

Table 9.2. The components of the meio- and microbenthos of the sand flats of the outer Eden estuary, Fife, in relation to distribution in other areas[X]

	Abundance	Freshwater	Brackish	Marine
Protozoa				
Mesodinium pulex	Common		+	
Lacrymaria olor	Rare	+		+
Lionotus fasciola	Rare	+		+
Loxophyllum helus	Abundant		+	+
Spirostomum teres	Rare	+		+
Blepharisma clarissimum				
var. longissimum	Rare		+	+
Aspidisca turrita	Rare			+
Oligochaetes				
Paranais litoralis	Rare	+	+	+
Nais elinguis	Rare	+	+	
Limnodrilus heterochaetus	Rare		+	
Harpacticoid Copepods				
Ectinosoma tenerum	Rare			+
E. curticorne	Rare		+	+
E. gothiceps	Abundant			+
Tachidius discipes	Rare	+	+	+
T. incisipes	Rare		+	+
Stenhelia palustris	Rare		+	+
Nitocra typica	Rare		+	+
Mesochra lilljeborgi	Common	+	+	
Laophonte setosa	Common		+	
Heterolaophonte minuta	Rare		+	
Platychelipus littoralis	Abundant		+	

60% euryhaline marine species, 30% brackish water and ground-water species and 10% terrestrial species; in the ground water of the soils above the head of the shore 10% of the population are euryhaline marine species, 50% are brackish water and ground-water species (of the total 10% are characteristic of this habitat alone) and 40% are terrestrial species[19]. The salinity tolerances of the nematodes are poorly known. However, *Panagrolaimus salinus* is associated typically with the dead and decaying leaves of *Puccinellia maritima*, and other rotting vegetation in the Essex marshes. In experimental conditions this species can live and actively reproduce in any salinity from 0 to 35‰. In contrast, *P. rigidus*, an inhabitant of peat bog, can live and reproduce only in salinities of up to 10‰.[20].

The interstitial fauna of Swedish beaches shows a wide range of tolerance to salinity and temperature (Fig. 9.3) and the oxygen content of the interstitial waters controls the distribution of many species; for example, the harpacticoid copepod *Parastenocaris vicesima*, has a preference for soil of 125–250 μ grain diameter and has a minimum tolerance of oxygen supply of 1×10^{-7} g $O_2/cm^2/min$.

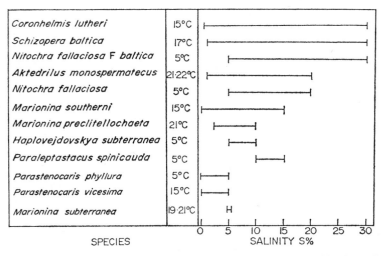

SPECIES   SALINITY S%

Fig. 9.3. Salinity optima, based on the tolerance experiments with the different species (after Jansson[21]).

In the Tamar estuary, the ostracod species *Carinocythereis* sp., *Pterygocythereis jonesi, Loxoconcha guttata* and *Costa* sp. inhabited localities with a marine salinity; *Leptocythere castanea* and *Loxoconcha impressa* lived in ranges of salinity from marine to oligohaline, while *Aurila convexa* was found in marine to mesohaline conditions only. *Candona candida* was the only ostracod representative from the freshwater environment. Both *Leptocythere castanea* and *Loxoconcha impressa* underwent a marked decrease in size which correlated with the sharp decrease in salinity between Cargreen and Weir. Green, about a mile further upstream. Although temperature may influence the growth of ostracods, e.g. *Philomedes globosus* is larger in the cold waters of Greenland than in the North Sea and Skagerrak, it is salinity and not temperature which controls the limit of penetration of many estuarine species[22].

The archiannelid *Trilobodrilus heideri* is photophobic, fairly tolerant of low salinities and responsive to water current, but it loses these senses of orientation when suddenly transferred to low salinities[23].

At Tvarminne, on the south-west coast of Finland, the oligochaete populations were more abundant in shallow water; sandy bottoms supported the largest and clay bottoms the smallest numbers. However, the abundance of the individual species was related to the substratum preference of that species (Table 9.3). Nevertheless, the species found there could be divided into three groups, viz. marine, brackish and freshwater species[24] (Fig. 9.4). Two other Baltic species, *Aktedrilus monospermatecus* and

Table 9.3. Distribution of oligochaetes at Tvarminne, in relation to depth
and quality of substratum (after Laakso[24])

Species	No. per m²	Depth, m	Substratum
Amphichaeta sannio	10 800	0–4	Fine sand + Chara
Paranais litoralis	6 213	0·1–0·4	Sand + gravel
Nais elinguis	13 600	0·8	Sand + Chara
Slavina appendiculata	1 000	1·0	Sand
Tubifex costatus	5 083	4·0	Sand + gravel
Psammoryctes barbatus	2 240	1·5	Fine sand
P. albicola	1 600	0·5	Plant remains
Limnodrilus hoffmeisteri	5 200	1·0	Sand with ooze
L. udekemianus	600	0·3	Ooze, fine sand + plants
Peloscolex ferox	3 200	0·3	Sand, plants
P. heterochaetus	538	2·0	Sand + ooze
Euilodrilus heuscheri	18 800	0·5	Plant remains
E. hammoniensis	2 267	1·5	Ooze
Rhyacodrilus coccineus	1 400	0	Sand + gravel
Aulodrilus pigueti	800	0·6	Fine sand
Clitellio arenarius	5 400	2·0	Sand, gravel + Chara
Lumbriculus variegatus	1 600	0·5	Plant remains

Marionina preclitellochaeta are euryhaline with a range of tolerance of
1·25–20‰ and 2·5–10‰ respectively[25].

Although the meiofauna is richer in sheltered areas where sediments are
fine and its numbers decline where the salinity is significantly reduced[3],
the sediment composition has an important influence upon distribution
within the individual transect, station or shore. In the upper reaches of
the Blyth estuary, a muddy sand was inhabited by 8 species of nematodes
with a total abundance of $1320/10$ cm², whereas the population of fine
sand composed of the 13 species reached a total abundance of $300/10$ cm²
only[18]. In the salt marshes of Georgia, 15 species were distributed across
the shore from the low water to the marsh grass area near the land. The
marine nematodes reached the highest density, viz. $12\,400/10$ cm², in the
Spartina zone (i.e. the zone in which the deposition of the fine sediment
would occur most rapidly) whereas lower down the shore where currents
prevented the sedimentation of the finer materials, and in the more sandy
sediment at the highest site, the nematode populations were much smaller,
viz. 2200 and $720/10$ cm² respectively[3, 11].

The classification of nematodes, by the structure of their mouth
parts[26, 27], has been of particular value in defining the status of nematode
populations in relation to soil grade. By means of this classification the
marine species may be assigned to one of four groups (Fig. 9.5), thus[26, (27)]:

1. *Group 1-A:* Without any mouth cavity (though sometimes with traces
   of it). Food obtained merely by means of the sucking power of the
   oesophagus. Consistency of material available as food most probably

soft or floating. Large and hard particles never found in the intestine. Probably selective deposit feeders. 97 genera.

2. *Group 1-B:* With cup-shaped, conical or cylindrical mouth cavity, without any armature. Food obtained as in group 1-A with help from active movement of lips and the anterior part of the buccal cavity. The food taken is similar to that of group 1-A, but includes diatoms in addition.

Non-selective deposit feeders. 73 genera.

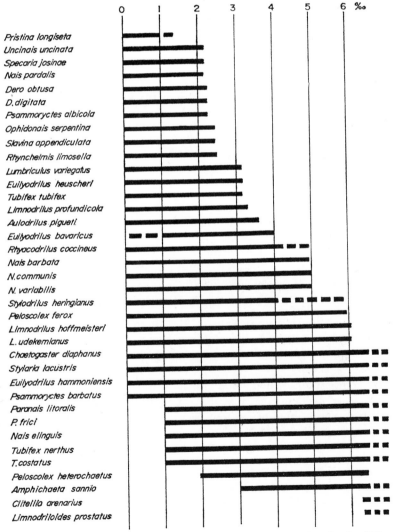

Fig. 9.4. The salinity tolerance of Finnish oligochaetes (after Laakso[24]).

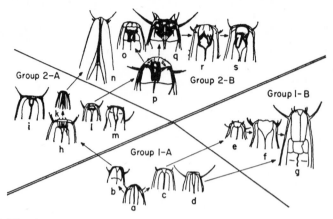

Fig. 9.5. The four principal types of buccal morphology of the free-living marine nematodes. a = *Oxystomatina*, b = *Anticoma*, c = *Terschellingia*, d = *Parachromagasteriella*, e = *Sabatiera*, f = *Paramonhystera*, g = *Bathylaimus*, h = *Paracanthonchus*, i = *Linhomoeus*, k = *Onchium*, l = *Chromadora*, m = *Microlaimus*, n = *Siphonolaimus*, o = *Halichoanolaimus*, p = *Enoplus*, q = *Oxyonchus*, r = *Oncholaimus*, s = *Eurystomatina* (after Wieser[26]).

3. *Group 2-A:* Mouth cavity provided with small armature. Food is scrapped off the bigger surfaces, or else the food object is pierced and the cell-liquid sucked through the hole in the wall.
   Epigrowth feeders. 104 genera.

4. *Group 2-B:* With big, powerful buccal armature. Mostly, but not all predators. Prey swallowed whole or pierced by means of buccal armature and liquid food sucked and swallowed.
   87 genera.

This classification is true in general, but there are some exceptions, e.g. *Dorylaimopsis metatypicus, Odontophora pugilator* and *O. papusi* all have a strong buccal armature, but are not predators, and feed by scraping food off sand grains.

The composition of nematode faunas in different habitats is related to the mode of feeding and may be classified simply by expressing the percentage composition of the four groups. In general, these compositions may be expressed thus[26, 27]:

1. Algae, exposed: Dominating group 2-A and 2-B (average of 57·99 and 26·33% respectively). The most important families: Chromadoridae and the Enoplidae respectively. Deposit feeders only very scarcely represented.

2. Algae, sheltered: Increase of deposit feeders (average of group 1-B = 37·23% against 5·64% in the former biotope), due to the increase of

deposits in the interstitial spaces of the algae. However, epigrowth-feeders still dominate in 5 out of 7 samples.

3. Littoral sand, fine and rich in deposits: Dominance of group 1-B (53·4%) and group 2-A (32·9%) due to rich content of deposits and to flourishing epigrowth.
4. Littoral sand, fine and poor in deposits: Decrease of deposit feeders and of epigrowth-feeders (33·6% and 8·2% respectively), increase of predators (48% against 14·3% in the former biotope).
5. Littoral sand, coarse and poor in deposits: Predators definitely dominating (69·8%).
6. Littoral, muddy sand and mud: Biotope not well defined. In muddy sands the composition of the fauna is similar to that of sands rich in deposits. In stiffer mud, however, epigrowth feeders of the families Cyatholaimidae and Desmodoridae predominate by far (up to 83%).
7. Sub-littoral, shells and coarse sand: Dominated by group 2-A and 2-B. Few deposit feeders (average 10·71%).
8. Sub-littoral, soft bottom (fine sand and various sorts of mud): Deposit feeders again dominate (62·27%) due to the extremely strong accumulation of detritus and other kinds of organic materials. Most important family: Comesomidae.
9. Special muds: The nematode-fauna appears to change completely in muddy habitats which are solid and not aerated. Lack of oxygen produces poverty in the population; this seems to be dominated rather by predators (above all *Sphaerolaimus* spp.) than by deposit feeders as in the former muddy habitat[26, 27].

At Whitstable, Kent, *Monohystera filicaudata* predominated with *Araeolaimus villosus* and *Euchromadora vulgaris* in the blackened, sulphide containing layer[28].

In a study at Whitstable, Kent, three principal sediment types were considered: (1) the sandy flats at half tide level where the soil contained *ca.* 80% of fine sand with 1–3% of the silt clay fraction; (2) the sandy flats near the low water level, which were coarser than (1) and contained shell fragments; and (3) the harbour mud which was much finer than (1).

The sandy flats at half tide level were inhabited by the harpacticoid copepods *Tetanopsis medius, T. smithi, Stenhelia palustris, Eudactylopus latipes, Asellopsis intermedia* and *Normanella tenuifurca*; the composition of the nematode population was Group 1-A 10%, Group 1-B 25%, Group 2-A 10% and Group 2-B 50%. In the sandy flats at the low water level, only harpacticoid copepods were analysed and these belonged to a wide variety of species not generally found at the higher shore level, viz. *Harpacticus obscurus, H. littoralis, Tisbe furcata, T. longicornis, Alteutha interrupta, A. oblonga, Parathalestris harpacticoides, Paradactylopus latipes, Dactylopodia dilatata, Enhydrosoma buchholtzi* and *Paronychocamptus curticaudatus*. In the harbour mud, the meiofauna was characterized by the harpacticoid copepods *Microarthridion fallax, Ameira parvula, Mesochra pygmaea, Enhydrosoma curticauda, Eurycletodes similis, Paramphiascella vararensis*; the composition of the nematode population was Group 1-A 0, Group 1-B 30%, Group 2-A 40% and Group 2-B 30%[29].

Similarly, in the Danish Waddensea, the distribution of the harpacticoid copepods was related to the nature of the substratum. The species *Tachidius discipes, Harpacticus flexus, Asellopsis intermedia* and *Laophonte* spp. occurred exlusively on sandy bottoms, whereas *Ectinosoma curticorne, Microarthridion littorale, Amphiascus* spp. and *Platychelipus littoralis* occurred exclusively on soft bottoms. *Canuella furcigera* was evenly distributed on soft and sandy bottoms[30].

In general, it would seem that the finer estuarine sediments do not have the diversity of meiobenthos species characteristic of somewhat coarser sediments. In these estuarine areas the benthic *Asellopsis*-type of harpacticoid copepod appear to be more suited to the environment than the slender interstitial *Arenosetella* type[29]. At St. Andrews, Fife, the gastrotrich *Turbanella hyalina* occurred commonly in the coarser soil grade of the East Sands, but was absent from the Eden estuary[X]. At Whitstable, Kent, no gastrotrichs, mystacocarids, coelenterates, viz. *Psammohydra* or *Halammohydra* or the bryozoan *Monobryozoon* were recorded in a two year period[29], and contrasted markedly with the more varied fauna of some beaches of a coarser soil grade in North Wales[31]; France, Italy, Spain, Tunisia and Algeria[32-37]; and in the Baltic[38, 39]. At Gothland, where the beach sand had a median diameter of 400 μm, harpacticoid copepods were absent, nematodes scarce, and the gastrotrich *Xentrichula velox* was the dominant animal; however, in a coarser, less homogeneous sand, at Stockholm, the fauna was more varied[39]. In the Øresund, where the median soil diameter of 0·25 mm more closely approached that at Whitstable, gastrotrichs, mainly *Turbanella hyalina*, became important 2 m from the water line[40].

In Buzzards Bay, Massachusetts, two distinct meiofaunal communities were recorded. The *Odontophora–Leptonemella* (nematode) community characterized sandy habitats, whereas silty habitats were characterized by *Terschellingia longicaudata* (nematode)–*Trachydemus mainensis* (kinorhynch) community[41]. Off the coast of Northumberland, the macrofaunal communities are correlated with depth rather than soil composition; in contrast, the nematode facies of the meiofauna are more sensitive to sediment composition. The species *Sabatiera ornata, Terschellingia longicaudata, Axonolaimus spinosus, Campylaimus inequalis* and *Halalaimus isaitshikovi* are eurytopic, but, when these species are excluded, a definite mud community can be recognized and is characterized by *Dorylaimopsis punctatus, Leptolaimus elegans, Sabatiera cupida* and *Oncholaimus skawensis*; the sand population was characterized by *Odontophora longisetosa, Sabatiera hilarula, Theristus setosus, Microlaimus honestus, Viscosia abyssorum* and *Desmodora norvegica*[42].

In Puget Sound, Washington, at least, the 0·2 mm median diameter of soil grade is an important distribution barrier, the archiannelids *Protodrilus flabelliger, Nerilla antennata,* the nematodes *Enoploides harpax, Onyx rugata, Nudora armillata* and *Theristus trecuspidatus* are limited to soils above this size, whereas the nematodes *Sabatiera* spp., *Theristus wimmeri, Pareurystomina pugetensis, Eurystomina repanda, Odontophora peritricha, Trileptium jacobinum,* the harpacticoid copepods *Dactylopodia glacialis, Heterolaophonte stromi,* and *H. discophora,* the hydrozoan *Protohydra leuckarti,* the polychaetes *Rhynchospio* cf. *arenicola, Boccardia* sp., the cheliferan *Leptochelia dubia,* and the cumaceans *Cumella vulgaris* (♀ only), *Diastylopsis tenuis* and *Lamprops krasheninikova* are limited to finer soils than this. This barrier may arise because at those soil grain sizes which exceed 0·2 mm, the interstitial spaces begin to fill up with finer materials, and the 0·2 mm median size may separate species which live in the interstitial spaces from those which live by burrowing[43]. Although the observed differences may be an effect of the "lebensraum" provided at a given pore size, it should be recalled that sediments smaller than 0·25 mm tend to be poorly oxygenated (p. 116); soils around this size are also the most readily moved of any sediment (p. 132), and all tend to be thixotropic (p. 109). Nevertheless, bearing in mind the forces which sort sediments on beaches (p. 138), it will be seen that the response of individual species to particular soil grades, whatever the direct cause, will result in a zonation such as that described below. Moreover, the differing soil distributions upon shores largely formed by wave processes, compared with those formed essentially by tidal processes, could result in a very different vertical distribution where the same species occupies both types of shore.

It is instructive to consider the responses of archiannelids of the genus *Protodrilus.* The species *P. hypoleucus* selects sands which are coarser than 250 μm diameter, *P. chaetifer* is limited to sands coarser than 500 μm diameter, *P. adhaerens* prefers sands in the 210–105 μm range and *P. symbioticus* prefers sand of 200 to 300 μm diameter. However, within these soil grade preferences, the populations may be very variable. The distribution of both *P. symbioticus* and *P. hypoleucus* is related to the presence of an attractive film of micro-organisms upon the sand grains: this attractive agent may be destroyed by drying, heating, acid cleaning or treating with alcohol, formol or cetyltrimethyl-ammonium bromide. Not all micro-organisms are equally attractive to *Protodrilus*: when presented with autoclaved sand plus micro-organisms, the preference of *P. symbioticus* was *Phaeodactylum* + bacteria > *Phaeodactylum* > *Dunaliella* > *Chlorella*;

innocula of natural sand bacteria, soil bacteria and *Pseudomonas* sp. > *Corynebacterium erythrogenes*, *Flavobacterium* sp. and *Serratia marinorubra*. Similarly, the preference of *P. hypoleucus* was natural sand bacteria > *Corynebacterium erythrogenes* > *Flavobacterium* sp., *Serratia marinorubra*, sterile control. However, even when these desirable factors are present, *Protodrilus symbioticus* has a preference for, and moves towards, well oxygenated sands: a combination of this response and that towards strong light serve to maintain this animal at an optimal depth of 2–6 mm within the soil. Although this animal can tolerate temperatures from −3·7 to +34·3, it has a preference for 15°C. In summer it moves into the cooler, deeper soil layers, but in winter it migrates away from the colder temperatures, i.e. it migrates downwards on deep, sandy beaches and horizontally down the shore in shallow sand. This animal can tolerate a wide range of salinity, and is likely to respond only to extreme changes in the salinity[44–49].

Vertical Distribution

On rocky shores, the meiobenthos, which occupies an essentially two dimensional habitat, is either associated with algae (with or without deposited sediment), or such situations as where the rock surfaces themselves provide a suitable substratum, or with other animals or in rock pools. In England, three critical zones for meiofauna have been recognized on such shores as Tinside, Plymouth. These zones are:

1. Between M.H.W.S.T.L. and M.H.W.N.T.L., where several intertidal species reach their lower limits. Meiofaunal species characteristic of the zone were *Hyale nilssoni*, *Clunio marinus*, *Trichocladius* cf. *vitripennis* and *Lasaea rubra* which are the counterparts of *Littorina littorea*, *L. littoralis*, *Patella vulgata*, *Monodonta lineata*, *Chthamalus stellatus*, *Ascophyllum nodosum* and *Fucus vesiculosus*.

2. Between E.L.W.N.T.L. and M.T.L. which represents the upper limit for some infra-littoral species. This zone was characterized by the main colonization of the polychaete *Fabricia sabella* which is accompanied by *Jassa falcata*, *Apherusa jurinei*, *Amphiglena mediterranea*, *Grubea pusilla*, *Odontosyllis ctenosoma*, and *Platynereis dumerilii* which are counterparts of *Gibbula cineraria*, *Rhodymenia palmata*, *Gigartina stellata* and *Chondrus crispus*.

3. Between M.L.W.S.T.L. and M.L.W.N.T.L. which represents the upper limit for other infra-littoral species. This zone was characterized by *Pleonexes gammaroides*, *Rissoa parva*, *Oridia armandi*, several Polychaeta errantia and nematodes, e.g. *Prochromadorella paramucrodonta*, *Neochromadora poecilosomoides* and *Chromadora brevipapillata* which are counterparts of *Laminaria digitata*, *Himanthalia elongata*, *Pyura stolonifera*, *Verruca stroemia* and *Calliostoma zizyphinum*[50].

In the sedimentary shore, on the other hand, presenting as it does an essentially three dimensional habitat, the meiobenthos penetrates the substratum to varying depths depending upon the composition of the substratum and the environmental tolerances of the individual species. In intertidal estuarine areas, the meiofauna is normally most abundant in surface layers, i.e. *ca.* 1–2 cm in depth[11, 15, 29, 30, 51–53]. Penetration below this depends upon the nature of the substratum. In anaerobic or poorly oxygenated soils, the ostracods, copepods and other oxygen sensitive organisms are confined to a very superficial layer: in Southampton Water, 95% of the harpacticoid copepods lived in the surface 0–5 mm and few occurred below 1 cm[53]; at Whitstable, the copepods and ostracods tended to be confined to the surface 4 mm, but the nematodes extended to about 7 cm depth where the underlying impermeable London Clay prevented further penetration[29, 52]; in the Danish Waddensea, the bulk of the harpacticoid copepods and ostracods (i.e. ⩾88%) occurred in the surface 2 cm, and none lived below 8 cm depth; on the other hand, although 83% of the nematodes lived in the surface 2 cm, they extended to the 6 cm depth at least[30]. In the salt marshes of Georgia, nematodes were able to reach a depth of 12–14 cm in those situations where the macrofauna and currents prevented marked consolidation of the soil, but at the landward edge of the marsh, where the soils were compact or bound by roots, the nematodes penetrated to the 3–6 cm depth only. Here the mud was anaerobic at depths which exceeded a few millimetres[11], however, nematodes have a marked tolerance of anaerobic conditions[54, 55].

On steeper, better drained, and therefore better aerated beaches, a greater depth penetration by the meiobenthos occurs[40, 56–59]. Here the amount of pore water may become important; a factor to which the harpacticoid copepods are particularly sensitive, turbellarians are rather less so, and oligochaetes are unaffected[39]. In a tideless beach, the meiobenthos is found at depths to the water table which may be as deep as 52 cm below the surface[40]. On the Atlantic coast of France, the meiofauna is found to 1 m depth, with concentrations at 30 and 70 cm[56]. In a beach at Miami, Florida, the meiofauna from the low water to mid-beach levels were concentrated mainly in the surface 7–10 cm; at the M.H.W.L. the harpacticoids, annelids, turbellarians and nematodes were concentrated between 25 and 30 cm below the surface, i.e. to a depth about 10 cm above the water table which was reached by some of the meiofauna[58].

Depending upon the depth of substratum available, the meiofauna may migrate into the depths of the soil to escape extreme temperatures, either low or high[29, 44, 52, 56]. Nevertheless many of these animals can withstand prolonged freezing; individuals of the harpacticoid copepod *Platychelipus* can withstand −9°C, in sea ice, for 9 hr[53], and, in the Baltic area, a high proportion of the meiofauna, in the sand of an ice covered beach is alive[39]. The response of the meiobenthos to light is varied, and may, in some species, lead to a vertical migration. The best known example of this phenomenon is that performed by *Convoluta roscoffensis* which comes to the soil surface in a period of daytime tidal exposure in order that its symbionts may perform photosynthesis (p. 203). The archiannelid *Proto-drilus symbioticus* undertakes a migration in response to excessive amounts of light[45]. The harpacticoid copepods show a variety of response to light depending upon habit[39, 60], but although the species *Tachidius discipes*, *Microarthridion fallax*, and *Parathalestris intermedia* migrate downwards at low temperature, they do not show a change from positive to negative light response at low temperatures[60]. Heavy rainfall may induce a down-ward migration[58]; but this is not true of all situations[53]. Finally, some vertical migrations are related to the tidal movement[52, 61]; even under calm conditions ciliates, turbellarians, nematodes and harpacticoids migrate downwards as the surge zone and tide cross their habitat[61].

Horizontal Distribution

As with the macrofauna, a traverse from the high-water mark to low-water mark and below is characterized by a zonation of the meiofauna; this zonation is related, at least in part, to environmental conditions already discussed[3, 5, 15, 29, 52, 57–59, 62, 63].

In tideless seas, the introduction of saline waters to higher beach levels depends upon wave action, but there is a transition from the saline condi-tions at the waters edge to the fresh water in the land; a distance which may be only a few metres. Here the distribution of meiobenthic organisms in response to preferred environmental factors may be clear cut[39].

With increasing tidal height the areas exposed by the receding tide become increasingly wider, and at an increasingly lower gradient. In this situation the zone occupied by each species is greatly extended[5] (Fig. 9.6). In the Danish Waddensea, all groups are most abundant near the low-water mark[30], but on more exposed coasts the richest zone is about the M.T.L., and in these areas migration allows the selection of a preferred substratum[3, 57–59, 63].

In estuaries, however, the gradation in sediment grade, from the relatively coarse at L.W.M. and increasing fineness towards the H.W.M., may lead to the marked separation of related species. In Southampton Water, the relatively resistant species *Platychelipus littoralis* occurred towards the H.W.M., whereas *P. laophontoides* is concentrated near the low-water mark[53]. Here dissimilar tolerances of low salinity account for the differences in distribution, but the influences of changing soil grade upon the environmental parameters are important also.

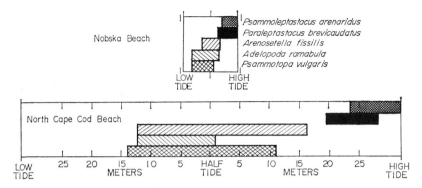

Fig. 9.6. Horizontal distribution of five common species of intertidal copepods in two sandy beaches near Woods Hole, Massachusetts. Each bar represents the horizontal extent of the median 80% of the total population. Data for the top 16 cm of sand (after Pennak[5]).

Seasonal Changes

Where studies have been carried on for upwards of a full year, the intertidal meiobenthos has been shown to undergo seasonal changes in abundance[3, 29, 30, 39, 52, 53, 55, 64–66].

At Whitstable, Kent, the meiofauna, as a whole, showed a single seasonal maximum which was strongly correlated with temperature, but the individual groups, viz. nematodes, harpacticoids and ostracods, showed some individual departure from this pattern[29] (Fig. 9.7). The harpacticoid copepods are of great interest because, although a single marine beach may contain many species, the preponderance of individuals are representatives of a few species only[5, 29, 52, 53, X]. At the fixed station, at Whitstable, Kent, three species only were of significance throughout the period of investigation. A further 16 species occurred intermittently, but chiefly at the time of the annual maximum in the summer and early autumn; this occurrence of "aberrants" was smaller in 1954 than in 1955, when the copepod numbers as a whole were greater[29]. Bearing in mind the relatively distinct harpacticoid meiofaunas recorded on other parts of the Whitstable shore, and what is known of other areas, it is reasonable to postulate that: a single marine beach may contain numerous species of harpacticoid copepods, forming, at the same or different levels on the shore, a number of minor communities, in each of which the preponderance of individuals are representatives of a few species only. The less abundant representatives in any one of the minor communities are made up from those species which are preponderant in other minor communities nearby. Their occurrence outside these communities is intermittent and dependent

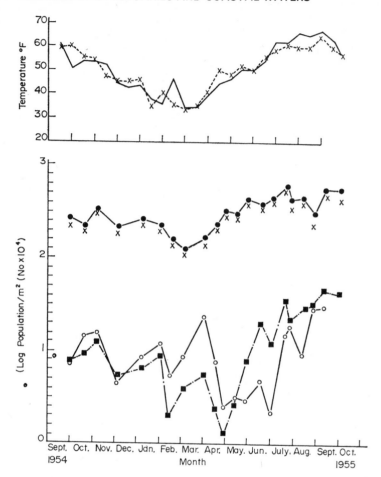

Fig. 9.7. The seasonal abundance of the meiofauna. ——— Mean air temperatures,
× ----- × Mean sea temperatures, ●———● Total Meiofauna, × nematodes,
○———○ ostracods, ■— · —■ harpacticoid copepods (after Perkins[29]).

upon large and successful broods which allow them to overflow into other
minor communities. In this case it would not be unreasonable to refer
to the few species typical of the individual situation as "autochthonous
residents" and the immigrants as "allochthonous" to that situation.

References

1. Moore, H. B. (1931). *J. mar. biol. Ass. U.K.* **17**, 325–328.
2. McIntyre, A. D. (1961). *J. mar. biol. Ass. U.K.* **41**, 599–616.
3. McIntyre, A. D. (1969). *Biol. Rev.* **44**, 245–290.

4. Wigley, R. L. and McIntyre, A. D. (1964). *Limnol. Oceanogr.* **9**, 485–493.
5. Pennak, R. W. (1950). *Annls Biol.* **27**, 449–480.
6. Harding, J. P. (1953). *Ann. Natal Mus.* **12**, 359–365.
7. Harding, J. P. (1955). *National Institute of Sciences of India, Bulletin No. 7: Symposium on Organic Evolution* 104–106.
8. Stout, J. D. (1963). *Tuatara* **11**, 57–64.
9. Nielsen, L. B. (1963). *Natura jutl.* **12**, 195–211.
10. Crosby, L. H. and Wood, E. J. F. (1959). *Trans. R. Soc. N.Z.* **86**, 1–58.
11. Teal, J. M. and Wieser, W. (1966). *Limnol. Oceanogr.* **11**, 217–222.
12. Schurumans-Stekhoven, J. H. (1935). *In* "Die Tierwelt der Nord und Ostsee" (Grimpe and Wagler, eds), Tiel 5b and c, Akademische Verlagsgesellschaft, m.b.H., Leipzig.
13. Goodey, T. (1951). "Soil and Freshwater Nematodes", Methuen, London.
14. Riemann, F. (1966). *Arch. Hydrobiol.* **31** (Suppl.), 279.
15. MacCoy, C. V. (1966). *Verh. int. Verein. theor. angew. Limnol.* **16**, 1471–1475
16. Perkins, E. J. (1960). *J. Ecol.* **48**, 725–728.
17. Capstick, C. K. (1957). *J. Anim. Ecol.* **26**, 295–315.
18. Capstick, C. K. (1959). *J. Anim. Ecol.* **28**, 189–210.
19. Gerlach, S. A. (1953). *Acta zool. fenn.* **73**, 1–32.
20. Everard, C. O. R. (1960). *Ann. Mag. nat. Hist.* Ser. 13, **3**, 53–59.
21. Jansson, B.-O. (1968). *Ophelia* 5 ,1–71.
22. Barker, D. (1963). *J. mar. biol. Ass. U.K.* **43**, 785–795.
23. Boaden, P. J. S. (1963). *J. mar. biol. Ass. U.K.* **43**, 239–250.
24. Laakso, M. (1969). *Ann. zool. fenn.* **6**, 98–111.
25. Jansson, B.-O. (1962). *Oikos* **13**, 293–305.
26. Wieser, W. (1953). *Ark. Zool.* Ser. 2, **4**, 439–484.
27. Wieser, W. (1960). *Limnol. Oceanogr.* **5**, 121–137.
28. Perkins, E. J. (1958). *Ann. Mag. nat. Hist.*, Ser. 13, **1**, 64–77.
29. Perkins, E. J. (1958). "The Microfauna of the Shore", Ph.D. thesis. University of London.
30. Smidt, E. L. B. (1951). *Meddr Kommn. Danm. fisk.-og Havunders* **11**, 1–151.
31. Boaden, P. J. S. (1963). *J. mar. biol. Ass. U.K.* **43**, 79–96.
32. Pennak, R. W. and Zinn, D. J. (1943). *Smithson. misc. Collns* **103** (9), 1–11.
32a. Delamare-Deboutteville, C. (1953). *Vie Milieu.* **4**, 321–380.
33. Delamare-Deboutteville, C. (1954). *Rassegna Speleologica Italiana.* **6**, 119–122.
34. Delamare-Deboutteville, C. (1954). *Vie Milieu.* **5**, 310–329.
35. Delamare-Deboutteville, C. (1954). *Vie Milieu* **5**, 408–451.
36. Delamare-Deboutteville, C. (1954). *Publnes Inst. Biol. apl., Barcelona* **17**, 119–129.
37. Delamare-Deboutteville, C., Gerlach, S. A. and Siewing, R. (1954). *Vie Milieu* **5**, 373–407.
38. Remane, A. (1933). *Wiss. Meeresunters. Abt. Kiel.* N.F. **21**, 161–222.
39. Jansson, B.-O. (1968). *Akademisk Avhandling.* 1968, 16 pp.
40. Fenchel, T., Jansson, B.-O. and von Thun, W. (1967). *Ophelia* **4**, 227–243.
41. Wieser, W. (1960). *Limnol. Oceanogr.* **5**, 121–137.
42. Warwick, R. M. and Buchanan, J. B. (1970). *J. mar. biol. Ass. U.K.* **50**, 129–146.
43. Wieser, W. (1959). *Limnol. Oceanogr.* **4**, 181–194.
44. Gray, J. S. (1965). *J. Anim. Ecol.* **34**, 455–461.
45. Gray, J. S. (1966). *J. Anim. Ecol.* **35**, 55–64.

46. Gray, J. S. (1966). *Veroff. Inst. Meeresforsch. Bremerh.* 2, 105–110.
47. Gray, J. S. (1966). *J. Anim. Ecol.* 35, 435–442.
48. Gray, J. S. (1966). *J. mar. biol. Ass. U.K.* 46, 627–645.
49. Gray, J. S. (1967). *J. exp. mar. Biol. Ecol.* 1, 47–54.
50. Wieser, W. (1952). *J. mar. biol. Ass. U.K.* 31, 145–174.
51. Rees, C. B. (1940). *J. mar. biol. Ass. U.K.* 24, 185–199.
52. Perkins, E. J. (1958). *Nature, Lond.* 181, 791.
53. Barnett, P. R. O. (1968). *Int. Revue. ges. Hydrobiol. Hydrogr.* 53, 177–209.
54. Moore, H. B. (1931). *J. mar. biol. Ass. U.K.* 17, 325–358.
55. Wieser, W. and Kanwisher, J. (1961). *Limnol. Oceanogr.* 6, 262–270.
56. Renaud-Debyser, J. (1963). *Vie Milieu* 15 (Suppl.), 1–157.
57. Ganapati, P. N. and Rao, C. G. (1962). *J. mar. biol. Ass. India* 4, 44–57.
58. Bush, L. F. (1966). *Bull. mar. Sci. Gulf Caribb.* 16, 58–75.
59. McIntyre, A. D. (1968). *J. Zool. Lond.* 156, 377–392.
60. Perkins, E. J. (1965). *Trans. J. Proc. Dumfries. Galloway nat. Hist. Antiq. Soc.* Ser. 3, 42, 6–13.
61. Boaden, P. J. S. (1968). *Sarsia* 34, 125–136.
62. Nicholls, A. G. (1935). *J. mar. biol. Ass. U.K.* 20, 379–405.
63. Govindankutty, A. G. and Nair, B. N. (1966). *Hydrobiologia* 28, 101–122.
64. De Zio, S. and Grimaldi, P. (1966). *Veroff. Inst. Meeresforsch. Bremerh.* 2, 87–94.
65. Hopper, B. E. and Meyers, S. P. (1967). *Mar. Biol.* 1, 85–96.
66. Muus, B. J. (1967). *Meddr. Kommn. Dann. Fisk. og-Havunders.* 5, 1–316.
X. Perkins, E. J. Previously unpublished work.

10

Macrobenthos of Rocky Shores

Rocky shores provide a wealth of stable surfaces for the attachment of sessile and sedentary organisms, unlike the shifting, unstable environment of sandy/muddy shores which are hospitable only to the infauna as permanent residents. In general, rocky shore populations are only evident in the outer areas of coastal plain estuaries, and decline upstream in a manner dependent upon the availability of suitable surfaces as well as individual salinity tolerances. In fiordic estuaries, the development of sandy/muddy shores is usually confined to the head of the fiord, and the rocky and stony shores downstream of this are colonized by normal hard ground biota. The division into the inhabitants of rocky and sedimentary shores is to some degree a convenience; subjects which are pertinent to both will be treated where most appropriate and so that the topic is a unity, rather than divided under these two headings.

Zonation

Both rocky and sedimentary shores are characterized by a zonation of the inhabitants, although it is only in the former that this phenomenon is overt. In temperate latitudes the essential features of zonation are those of Fig. 10.1[1], and these features are apparent globally, even though the component species may differ and some zonal colonizations may be absent when conditions are adverse[2-8].

The distribution of intertidal plants and animals is influenced by certain *critical levels*. These levels may change, from shore to shore, in relation to local environmental variations, e.g. surf action, rock configuration and illumination, but four regions of the shore are more critical than the others (compare with the meiobenthos, p. 232):

1. From, or just below, M.L.W.S.T. to M.L.W.N.T., where most intertidal plants and animals reach their lower limits. Here two sets of species may be distinguished (a) those for which the neighbourhood of of M.L.W.S.T. is critical, e.g. *Patella vulgata, Balanus balanoides, Littorina littoralis, L. littorea, Gibbula umbilicalis, Thais (Nucella) lapillus* and *Fucus serratus*, and (b) those which reach their lower limits

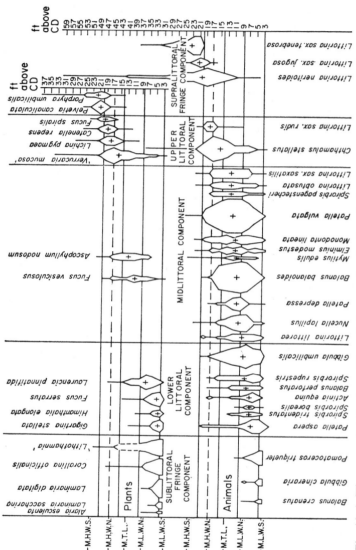

Fig. 10.1. The mean distribution of littoral plants and animals at Dale, arranged to show the variations in patterns of distribution within the five major components of the littoral population. The centre of abundance of each species is shown by a cross (after Moyse and Nelson Smith[1]).

ca. M.L.W.N., e.g. *Littorina neritoides, L. saxatilis, Monodonta lineata Chthamalus stellatus, Lichina pygmaea, Ascophyllum nodosum,* and *Fucus vesiculosus.*

2. Those sub-littoral species which penetrate the littoral region successfully, generally only reach M.L.W.N.T., e.g. *Laminaria digitata, Gibbula cineraria, Rhodymenia palmata* and *Gigartina stellata.*

3. The levels between E.H.W.N.T. and M.H.W.S.T. are critical for those intertidal species which reach their upper limits there. Nevertheless some species normally limited to the region M.H.W.N.T.–M.H.W.S.T. may be extended to higher levels under the influence of spray or splash: within this group sub-group (a), e.g. *Patella vulgata, P. depressa, Lichina pygmaea, Pelvetia canaliculata, Chthamalus stellatus,* and *Fucus spiralis* extends higher up the shore than sub-group (b), e.g. *Osilinus lineatus, Balanus balanoides, Littorina littorea, L. littoralis, Ascophyllum nodosum, Fucus vesiculosus, Gibbula umbilicalis, Thais (Nucella) lapillus, Laurencia, Fucus serratus, Mytilus* and *Sabellaria alveolata.*

4. The least critical levels on the shore are those between E.L.W.N.T. and E.H.W.N.T.[9]

Such critical differences may be noted in Florida[10, 11] and Jamaica[10, 12], where the tidal range is 1–3 ft and the vegetational zones are compressed, but sharply delineated, and each may occupy only one inch of vertical height. Similar critical differences may be observed in the Solway marshes where *Pelvetia canaliculata* and *Fucus spiralis* colonize stones which protrude a few inches above the general level of the salt marsh[X]. In each case, the fluctuations in water level resulting from marked changes in hydrographic and meteorological conditions must exceed such critical limits which must therefore represent a response to a long term average[10].

In the clear, saline waters of outer estuarine areas the zonation from green through brown to red algae develops to a considerable diversity of species. The change from brown to red is a response to the change in the quality of the light which penetrates to the depth at which the algae live (p. 18). However, the red algae may be green, brown, purple or blackish in colour. They are adapted to live in the reduced light intensities of the infra-littoral, but *Gigartina* and *Porphyra*, which habitually colonize the shore, have lost their red colour; *Corallina*, which colonizes rock pools retains its red colour. In unshaded conditions, *Gracilaria verrucosa* reaches its maximum growth rate at 60 cm below the surface of the sea. In light intensities of the order of full daylight, the phycoerythrin pigment disappears, but is regenerated upon return to lower light intensities. Bleached plants grow more slowly than unbleached plants, when both are in low light intensities or at intensities which would cause bleaching: in the latter case, the growth rate of the unbleached plant is reduced however. The assimilation rate of unbleached *Rhodymenia* and *Chondrus* is about twice that of bleached plants. Although the loss of phycoerythrin reduces the growth rate, *Gracilaria* is not placed in serious difficulty because this happens naturally, and rapidly, in high summer illumination. Despite bleaching and a reduced growth rate, the fully bleached plants reach a maximum size in late summer. Nevertheless, sporelings of *Gracilaria* are irreversibly harmed by exposure to full daylight for 4–5 hr, which possibly explains its absence from pools at higher shore levels[13].

The ability of the phycoerythrin pigment to absorb green light is exploited by sporelings of the red algae in two ways: (1) the permanently submerged species, e.g. *Brongniartella byssoides*, use it as an accessory pigment in photosynthesis; and (2) in species which have colonized intertidal habitats, e.g. *Plumaria elegans*, it protects the sporelings in the early stages of growth[14,15].

Curiously, in the clear waters of Malta, the upper 15 m of sub-littoral zone is dominated by brown algae, and green algae are then the most important down to at least 75 m: red algae are important only in shade communities down to this depth. In this situation, the light intensity rather than its spectral composition seems to be the major factor in controlling the vertical distribution of algae[16].

In contrast to the red algae which live in relatively sheltered conditions, *Pelvetia canaliculata* and *Fucus spiralis* have an essentially terrestrial mode of existence during the neap phase of the tidal cycle. At such times, these algae may be desiccated by a hot sun or a cold wind, they may be ice-covered, or at times of heavy persistent rainfall come to live in essentially freshwater conditions.

The zonation of fucoids in the Solway Firth and at Port Erin, Isle of Man, is very similar. *Pelvetia canaliculata* occurs at E.H.W.S.T. and extends to M.H.W.S.T. The *Fucus spiralis* zone is about M.H.W.N.T. and, in the Solway, corresponds to the level of *Spartina* found at the edge of developing marshes. Below the *Fucus spiralis* zone, the *Ascophyllum nodosum* and *Fucus vesiculosus* zones extend to M.L.W.N.T. Unlike Port Erin, *Ascophyllum* does not extend to lower levels than *F. vesiculosus* in the Solway Firth where both tend to remain in discrete zomes[17, 18].

In the rocky, outer areas of coastal plain estuaries, and far upstream in fiordic estuaries of high latitudes, the laminarians, or kelps, dominate the infra-littoral flora. Until World War II, these interesting plants were little studied in British waters. However, economic need changed this situation[19]: an industry to harvest and process these algae was developed and continues to the present time. The principal species found in British waters are *Laminaria hyperborea*, *L. digitata*, *L. saccharina* and *Saccorhiza poly-*

schides, (*Alaria esculenta* is common only on very exposed shores), which may occur in such dense populations that they may be appropriately called *Laminaria* forests[20]. Of all the sub-littoral species, *L. hyperborea* is the most important, both economically and in its effect on other algae[20]. In Scotland and the Isle of Man, *L. digitata* dominates over a narrow zone, and *L. saccharina* and *L. hyperborea* may compete, particularly on unstable bottoms, but the species may become dominant in different conditions, Table 10.1[20].

Table 10.1. Conditions favouring dominance by the species of *Laminaria* (after Kain[20])

	Sheltered	Medium	Exposed
Unstable bottoms (gravel and stones)	*L. saccharina*	*L. saccharina* *L. hyperborea*	Neither
Stable bottoms (boulders and rock)	*L. saccharina*	*L. hyperborea*	*L. hyperborea*

The *Laminaria* spp. and *Saccorhiza* are all able to grow above a minimum irradiance of 1–2 μg cal/cm^2 sec (1 ft candle $\equiv$ *ca.* 1 μg cal/cm^2 sec.) and the compensation point is likely to equal this value[21–23]. In comparison, the minimum irradiance required for growth of *Macrocystis*, at 15°C, is 21–23 μg cal/cm^2 sec, i.e. ten times that of the laminarians. The compensation point of Antarctic phytoplankton is *ca.* 5 ft candles, and ice micro-algae 2·5 ft candles. Antarctic macro-algae grow actively at less than 13 ft candles. There is, at present, no evidence to suggest that heterotrophic growth by these species of Laminariaceae occurs at very low irradiances[23]. Light, unlike temperature, is probably not a differentiating factor in the competition between *Laminaria* spp. and *Saccorhiza*. The more rapid growth, and shorter life span (*ca.* 1 yr) of *Saccorhiza* may be considered that of an opportunist species.

Compared to the Isle of Man and the Channel Islands, the Scottish *L. hyperborea* forest is taller, and is a result of a greater longevity rather than enhanced growth. At a given age, the weight of *L. hyperborea* stipe depends on local conditions rather than latitude. In Manx waters, the weights and lengths of stipes, within age-groups greater than three, decrease with depth, i.e. with light intensity. The weight of newly produced frond is proportional to the weight of the stipe when this exceeds 20 g, but is relatively greater on smaller stipes. The decrease of frond weight with depth is not wholly dependant upon the resources of the stipe. The growth rate of *L. hyperborea* decreases with depth, but downward extension is controlled by those factors which influence the establishment or removal of plants rather than their rate of growth[21].

Saccorhiza is usually epilithic, however, at Port Erin, Isle of Man and

Lough Ine, Eire, it grows attached only to small stones or shells. This species is confined to the sub-littoral, such that it is only at E.L.S.W.T. that the highest plants, on the shore, are exposed. It has been found to depths of 24 m and may occur between *Alaria sulcata* or *L. digitata* and *L. hyperborea* on some coasts; on others, viz. Port Erin and the coast adjacent to Lough Ine, it is always found below the *L. hyperborea* zone[25]. In the whirlpool area of Lough Ine, *Laminaria saccharina* was very abundant near the surface, however it was replaced by *L. hyperborea* at depths between 6 and 12 m. At depths greater than 12 m a large proportion of the algal biomass was due to *Dictyota dichotoma, Dictyopteris membranacea, Calliblepharis ciliata* and *Delesseria sanguinea*[26].

Latitude may have an important influence upon zonation and its component species. *Calliostoma zizyphinum*, the painted top-shell occurs on the lower shore in S. Devon, but in the Solway Firth, *ca.* 4°15′ further north, it occurs only in association with *Delesseria sanguinea* below

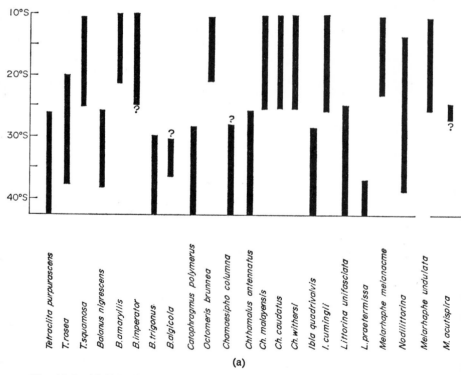

(a)

Fig. 10.2. (a) Distribution of littorinids and barnacles on east Australian coast (after Guiler[7]).

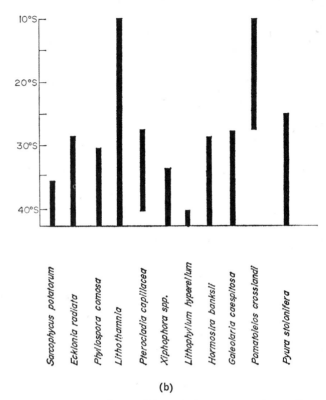

(b)

Fig. 10.2. (b) Distribution of serpulids and algae on east Australian coast (after Guiler[7]).

E.L.M.S.T.[27]. With a coastline which extends through 10° of latitude, the effect upon the components of zonation is seen clearly on the Australian east coast, Fig. 10.2[7].

Pronounced belts may be evident on some shores, e.g. barnacles or mussels in Britain, oysters, *Crassostrea amasa*, in Australia[7] and oysters in British Columbia[X]. Such zonation may be a consequence of biological as well as physical factors. It may be deduced that the occurrence of a sessile or sedentary prey species, in a particular zone or belt, is the consequence of difference in the ability of predator and prey species not only to survive, but grow and reproduce in the environmental conditions of the particular zone. Above this level, conditions are too rigorous for the successful survival of the prey, while below, its predators can give full rein to their appetites. The mussel, *Mytilus*, is apparently an example; normally it lives in a zone between the middle and lower shore, and may

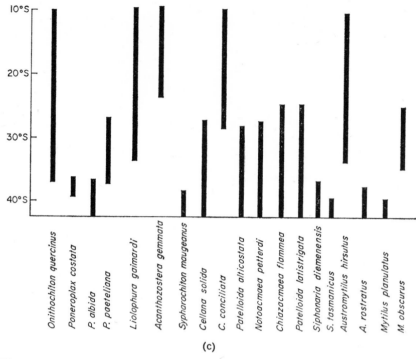

(c)

Fig. 10.2. (c). Distribution of mussels, belt-forming patelloids on east Australian coast (after Guiler[7]).

often occur only in a narrow band in the middle shore. It has been found that transplantation of more adult mussels, which are less vulnerable to attack by crabs and flounders than the newly settled juveniles, to more favourable situations below the low-water mark leads to an enhanced rate of growth and an improved condition. Perhaps freedom from the attacks of natural enemies may explain the success of mussels in colonizing and fouling the cooling water systems of large power stations and other marine installations[28]. It seems likely that *Carcinus* may also control the distribution of the dog-whelk *Thais* (*Nucella*), the top-shell *Gibbula cineraria* and the sea urchin *Paracentrotus lividus; Portunus puber* may influence the distribution of the last two species[29-31].

Large changes may be induced by single events. In 1967, a large spatfall of mussels, *Mytilus*, at Silloth, on the Solway Firth, grew well throughout the summer, to yield individuals in good condition and having thin shells. In early December, this promising mussel bed was wiped out, in a few days, by a flock of oyster catchers[X]. In the Gulf of St. Lawrence, mass mortalities of the scallop *Pecten magellanicus*, may result from unexplained concentrations of starfish[31]. In Curlew Bay, Lough Ine, the sea urchin *Paracentrotus lividus* is widespread on clean rock and shell gravel. Here it denudes the rock of algae by grazing them as fast as they grow, these algal growths are principally composed of *Enteromorpha clathrata*, but include *Chylocladia verticillata*, *Polysiphonia* spp., *Ceramium* spp. and *Ectocarpus confervoides*. Elsewhere, in the absence of *P. lividus*, these algae are abundant and provide shelter for the gastroped molluscs *Rissoa parva* and *Bittium reticulatum*, and the amphipod *Caprella acanthifera*[33]. In Maltese waters, the browsing of echinoderms is important in the upper 15 m[16]. Littoral algae are taken by crustacea also[34, 35, X]. Removal of the invertebrate grazers can lead to remarkable changes in the balance of algae on a shore: such a change occurred as a consequence of the Torrey Canyon incident in which annihilation of these grazers resulted in an unusually heavy development of *Enteromorpha*[36].

Zonation should not be regarded as a static system. On the coast of Ghana, the algae *Hypnea*, *Chaetomorpha*, *Padina* and *Sargassum* all show seasonal shifts both in distribution and quality[37]. Similar seasonal and annual shifts of the algal populations occurred on the Visakhapatnam coast of India, where many algae formed distinct bands only at times characteristic of the individual species[38]. On both coasts, the tidal behaviour associated with seasonal changes in the mean sea level and seasonal changes in sand levels on the beach have an important influence[37, 38]. On European shores, similar changes affect the non-fucoid algae, of which *Enteromorpha* is a notable example. At the time of summer neap tides, populations of this alga which flourished at levels above the M.H.W.M.N.T., become scorched by the sun, die and float away on the succeeding spring tides[X].

The movements of the vagile epifauna are more pronounced than the attached species and marked seasonal migrations are undertaken by many species. In autumn, as the days shorten and temperatures decline, adult *Carcinus* move off the shore to winter just below the low-water mark, returning once more with the warmer days of the following spring[31, 39-41]. This movement is apparently temperature controlled[40]; juveniles tend to remain on the shore, throughout the winter, only moving off it in the coldest weather conditions. Little migration on and off the shore, with the tide, occurs during winter[31, 39-41]. During the summer, onshore migrations may be extensive; individuals both remain on the shore and move to and fro with the tide. Movements on shore may extend up in to the *Littorina saxatilis*/*Pelvetia canaliculata* zone; at these times, larger individuals of *Carcinus* prey upon *L. saxatilis*, and smaller crabs shelter among the *Pelvetia* during the exposure period[41].

9

The dog-whelk, *Thais* (*Nucella*) *lapillus* becomes inactive at 3°C [37a], and in winter shore dwellers tend to creep into sheltered rock crevices, and hibernate [37a, 42]. At low temperatures, adhesion and contact with surfaces is lost; many individuals fall to the seabed [37a]. On the shores of the Gareloch, where the surfaces colonized by *Thais* (*Nucella*) *lapillus* may not be vertical, a positive downshore migration takes place in autumnal temperatures which are in excess of 3°C [X]. In winter, loss of adhesion and subsequent movement by wave action may account for changes in the vertical distribution of *Littorina* [43]; in the Gareloch, *L. littorea*, like *Thais*, apparently makes a positive downshore migration with the onset of winter [X]. Littoral fish tend to move offshore in the winter months [44]. Some species invade the shore by means of pelagic larvae rather than by migration, e.g. the nudibranch molluscs, such as *Aeolidia papillosa* and *Doto coronata*. The former species belongs to a group which exploits a stable food supply and produces only one, occasionally two, generations per year. The latter are opportunists which exploit fluctuating food supplies and produce several generations per year [45].

At Whitstable, Kent, the pelagic larvae of *Littorina littorea* settled out from the plankton, at levels below the low water mark; the juveniles were transported upshore by hydrographic forces to the level occupied by the adults: some shoreward movement of the young *L. littorea*, over the stony beach, occurred by active migration, rather than by passive transport [46]. On this shore they tended to remain in the same position for many weeks once a home position was achieved [47]. *L. saxatilis*, on the other hand, is viviparous, i.e. it bears its young alive, and a marked upshore migration of juveniles was not evident at Whitstable [48]; tagged adults, moved little from the site of release.

In the spring and summer of 1968, a total of 8000 *L. littorea, L. saxatilis* and *Thais* (*Nucella*) *lapillus* were tagged and released on the shores of the Gareloch. All showed large and rapid residual movements, i.e. up to 12 m per week, with a greater movement in the longer term, i.e. > 25 m. The movement from the release point was radial, and some of the *Nucella* were recovered in pots from below the E.L.W.M.S.T. A repeat experiment performed with *L. littorea* during the months October to December 1970, with residual movements > 65 m confirmed this result. In the Solway Firth, *L. littorea* is often found far out on the sands; here, unlike the Gareloch, passive transport by hydrographic forces rather than an active migration seems more probable [X]. However, both series of observations suggest that there is much to learn of shore population migration and recruitment, both by active migration and passive water transport.

Predator/prey relationships and the locomotory activities of the individual species, are only two aspects of the manner in which zonation is achieved. The ability of the individual organism to respond to, and survive those features of its physical and chemical environment, other than a response to salinity changes, are equally important.

Environmental Tolerance
Resistance to Desiccation

Possibly one of the greatest hazards to animals living on rocky and stony shores, is the liability to desiccation.

On exposure to the air, the barnacles, *Balanus balanoides* and *Chthamalus stellatus*, which live at higher shore levels, form a *micropylar opening*, or *pneumostome*, in the opercular valves. This pneumostome permits access of air to the mantle cavity, but controls water loss. The sub-littoral *B. crenatus* does not form a pneumostome. The 50% lethal exposure time (i.e. L.E.$_{50}$), is *ca.* 14 hr, 4 days and 5 days for *B. crenatus*, *B. balanoides* and *Chthamalus* respectively, and is consistent with the position occupied by each species on the shore. All three species have the ability to accumulate lactic acid under anaerobic conditions[49]. The air retained in the mantle cavity of *Balanus balanoides* may be renewed periodically either by diffusion or by active pumping through the pneumostome. Since all the exposed animals are inactive, the respiratory rate in damp air is lower than in water.

The talitrid amphipods *Talitrus saltator*, *Orchestia gammarella* and *Talorchestia deshayesii* live among stone and shingle at the E.H.W.M. All are very susceptible to desiccation, but in unsaturated air and in the absence of food *T. saltator* survives by far the longest: the availability of moist food increases the survival times significantly[51].

The sphaeromid isopods *Campecopea hirsuta* and *Dynamene bidentata* show an interesting difference in response to desiccating conditions which is related to their normal environment. Both occur intertidally, but the former inhabits the empty shell of *Chthamalus* or more preferably the lichen *Lichina pygmaea*, whereas the latter lives at lower intertidal levels. On exposure to dry air, at *ca.* 18°C, *D. bidentata* loses 50% of its initial weight in 6 hr, whereas *C. hirsuta* does not lose more than 20% of its initial weight in 12 hr. The efficiency of *C. hirsuta* in resisting desiccation is related to its ability to roll its body into a tightly closed ball, thus protecting the whole of its sensitive ventral surface against water loss[52].

In the Pacific, the starfish *Pisaster ochraceus* remains attached to inter-tidal rock surfaces throughout periods of exposure, whereas *P. giganteus* and *P. brevispinus* tend not to. Upon exposure to the air *P. ochraceus* loses relatively less body water than the other two species, presumably due to a difference in the structure of the body wall; this species also reattaches to surfaces more readily than the others[53].

At temperatures $\leqslant 30°C$, death of the gastropod mollusc *Thais* (*Nucella*) *lapillus* ensues more rapidly with decreasing humidity. Above 30°C, the median lethal time decreases slightly with decreasing humidity, and may be a function of the cooling effect of high evaporation at these temperatures. The operculum of this species is considered to have little importance as a mechanism for resisting desiccation, and its chief function may be that of defence against predation by crabs[54]. The movement of air, affects the results of desiccation experiments significantly[55, 56]; indeed, the provision of shelter greatly extends the survival of winkles, dog-whelks and shore crabs[28].

The effect of desiccation upon the intertidal limpets *Patella aspera* and *P. vulgata* depends upon the species and the shore level occupied. The median lethal water loss in *P. aspera*, low level *P. vulgata* and high level *P. vulgata* is 30–35%, 50–55% and 60–65% respectively. Water loss was shown to be well below the lethal level, even on a hot summer's day. Possibly the upper level of distribution is set by an interplay between loss of water and the time required to compensate this loss during the period of immersion[57].

In general, the resistance to desiccation and the position occupied in the littoral zonation are correlated. Species which are found entirely or mainly above M.T.L., e.g. *Littorina saxatilis, Neobisium maritimum, Campecopea hirsuta* and *Anurida maritima*, are usually more resistant to desiccation than those forms found entirely or mainly below M.T.L., e.g. *Tanais chevreuxi, Dynamene bidentata* and *Porcellana platycheles*. Those species which live in the outer regions of crevices, e.g. *Campecopea hirsuta, Littorina saxatilis, Lasaea rubra* and *Eulalia viridis*, are more resistant than those from the inner regions, e.g. *Ovatella myosotis, Dynamene bidentata* and *Porcellana platycheles*. Temporary inhabitants of crevices, e.g. *L. saxatilis, Anurida maritima*, and *Eulalia viridis*, are usually more resistant than species permanently living in crevices, e.g. *Dynamene bidentata, Porcellana platycheles, Tanais chevreuxi* and *Ovatella myosotis*[56]. Individuals of the lamellibranch mollusc *Lasaea rubra* taken from higher zonation levels are more resistant to desiccation than those from lower zonation levels[56, 59].

The survival of the shore crab, *Carcinus*, exposed to desiccation, is reduced by a low humidity and high temperature and is prolonged by a high humidity and a low temperature. Survival is also dependent upon size and the provision of shelter[28].

Turbulence

It is convenient to consider the effects of wave action and currents under the heading of turbulence since in practice they frequently work in concert, i.e. those situations subject to turbulent exposure are often subject to strong action of both waves and currents, whereas where turbulent exposure is weak, both these forces are often weak also.

At Plymouth, swells and refracted waves are an important influence upon the distribution of the biota upon the breakwater. Here, most plants, with the exception of *Lichina*, showed their greatest abundance and widest zone on the north face, particularly at its eastern end, which is probably the most sheltered point on the breakwater. However, *Himanthalia elongata* and *Rhodymenia palmata* were different upon the two sides. Most of the animals, notably *Chthamalus stellatus* and *Patella intermedia*, were favoured by wave action; their numbers were greater and the zone wider on the south side, especially at its western end, which was the most wave beaten part of the whole breakwater.

Those organisms to which wave action is unfavourable, i.e. their zone is narrowed and their numbers reduced, may be listed in order of decreasing sensitivity thus: *Mondonta lineata*, *Gibbula umbilicalis*, *Pelvetia canaliculata*, *Fucus spiralis*, *Laminaria saccharina*, *Fucus serratus*, *Laminaria digitata* and *Fucus vesiculosus*. Conversely, those organisms for which wave actions was apparently favourable, as indicated by increasing numbers and zone width, were: *Porphyra umbilicalis*, *Lichina pygmaea*, *Chthamalus stellatus*, *Patella intermedia*, *P. aspera*, *Littorina neritoides* and *L. saxatilis*[59].

The fucoid algae are a complex group[60] of considerable interest because they are easily accessible for study, and hybrids between the species *Fucus spiralis*, *F. vesiculosus* and *F. serratus* occur regularly. *F. spiralis*, and the hybrids which approach it, are limited to the top of the shore; lower down only the hybrids approaching *F. vesiculosus* persist[61]. Unlike *Pelvetia canaliculata*, these species are found on both sides of the Atlantic[3, 4]. The zonation of *F. vesiculosus* and *F. serratus* is affected by latitude: it is lower on the coast of Devon than on the Manx or Argyll coasts. In the first year of life the average rate of elongation of *F. vesiculosus* is 0·48 cm/week on the coast of Devon, 0·45 cm/week on the Manx coast and 0·68 cm/week on the Argyll coast. The average rate of elongation of *F. serratus* is 0·49 cm/week on the Devon coast, 0·68 cm/week on the Manx coast and 0·85 cm/week on the Argyll coast. Growth in the second and subsequent years shows a rhythm induced by alternating emphasis on receptacle formation and frond extension. The rate of growth is enhanced in shelter from rough water[17] and the vesiculation of the thallus of *F. vesiculosus* is directly related to the degree of exposure of the habitat to severe wave action[61]: the extreme form for very sheltered localities is *F. vesiculosus* var. *vadorum* Aresch and for exposed situations, *F. vesiculosus* var. *evesiculosus* Cotton[61]. (*F. vesiculosus* var. *evesiculosus* is more properly *F. vesiculosus* f. *linearis* (Huds)[60].) At Lough Ine, in Ireland, a population of *Fucus* has the texture and serration of *F. serratus*, the spiral twisting and round and rimmed receptacle of *F. spiralis* and with somewhat irregular vesicular inflations. At Lundy Island plants possess the vesicles of *F. vesiculosus* and the serrated fronds characteristic of *F. serratus*[61]. *Ascophyllum nodosum* is also influenced by its exposure to wave action and in severe conditions, it is reduced to a few short, straps of frond. In the Solway Firth, the association between *Ascophyllum nodosum* and *Polysiphonia lanosa* is most pronounced in areas of moderate exposure, viz. Southerness Point; *Ascophyllum* appears without *Polysiphonia* in the shelter of Urr Water and similar situations[X], a condition similar to that found in the Bay of Fundy and its tributary inlets[3].

Comparison of a large form of *Fucus vesiculosus*, from areas subject to wave action, and a small form, from quiet waters, revealed that both showed similar seasonal changes in respiration intensity, which reached a minimum in summer: in both, growth was maximal in summer[62].

The algae of the intertidal zone are particularly subject to storm damage. In rough weather, they may be torn free and deposited in sheltered salt marsh and *Zostera* area[17, 63, X].

In sheltered areas may be found sterile forms of *P. canaliculata, F. spiralis, F. vesiculosus, F. serratus, F. ceranoides* and *Ascophyllum*, which reproduce vegetatively. The species *F. vesiculosus, A. nodosum, F. serratus* exist in a free living form. In reduced salinity *Ascophyllum nodosum* undergoes a change of form; it becomes considerably branched and may lose its bladder. Although these *ecad forms* are generally free living, they may remain attached in the early stages of development, but only rarely after the densely tufted stage has been reached. These algae lie free on a marsh, or draped across vegetation or become embedded in mud. In the Baltic, these sterile free living forms of *Ascophyllum* lie on the sub-littoral seabed at depths of up to 10 m: they lie loose on mud or sandy mud in sheltered waters of reduced salinity. In Norway they lie on mud[64].

A study of the serpulid polychaetes around Dale Peninsula, Pembrokeshire, revealed that *Spirobis tridentatus* is the most tolerant of exposure, while on the more sheltered shores around the mouth of Milford Haven *S. rupestris* is abundant. Here, however, the situation is complicated by the differing ability of the various species to colonize the infra-littoral, and differing tolerance of low salinity: in the latter case *S. pagenstecheri* penetrates furthest to a rock reef at Landshipping Quay, *S. tridentatus* to a point two miles further downstream and all species are present at Garron Point some three miles downstream from Landshipping Quay. This penetration of the Daucleddau arm of Milford Haven, occurs in the same order as in the Bristol Channel and the Severn estuary: here, however, the upstream limits are reached over a distance of 100 miles instead of the three in Milford Haven[65].

Perhaps the finest observations upon the influence of turbulence have been made at Lough Ine where a marked gradation from exposure to shelter occurs. The influence of water currents upon other environmental conditions is summarized in Table 10.2[66]. Here the algal species could be classified on the basis of their preferences thus: *Porphyra umbilicalis*, *Alaria esculenta*, *Laminaria digitata*, *L. hyperborea*, *Saccorhiza polyschides*, *Lomentaria articulata* and *Laurencia pinnatifida* prefer open stations; *Fucus serratus*, *Leathesia difformis*, *Himanthalia elongata* and *Laminaria*

Table 10.2. The influence of water currents upon other environmental conditions (after Lilly *et al.*[66])

Unfavourable effects		Favourable effects
	Prevention of sedimentation	
Prevention of larval settlement. Dislodgement from substratum. Interference with feeding. Favourable effects upon competitors and predators.	←Strong current→	Prevention of local fluctuations in O_2 and CO_2 tensions. Distribution of larvae. Supply of planktonic food. Adverse effects on competitors and predators.
Local fluctuations in O_2 and CO_2 due to photosynthesis and respiration.	←Weak current	
Clogging of feeding and respiration mechanisms by suspended sediment. Smothering by settling of sediment. Low O_2; high CO_2, H_2S and other reducing substances. Favourable effects on competitors and predators.	←Sedimentation→	Organic debris for food. Bacterial food. Adverse effects on competitors and predators.

saccharina intermediate stations; and *Pelvetia canaliculata*, *Fucus spiralis*, *F. vesiculosus*, *Ascophyllum nodosum*, *Polysiphonia lanosa* and *Stilophora rhizodes* preferred stations remote from the open sea. *Ulva lactuca* was found in all parts of the area[67].

The fauna of boulders, in the Rapids area, differed depending upon the degree of shelter which they offered, Table 10.3[66]. In studies of the fauna of the algae, viz. *Saccorhiza polyschides*, *Laminaria digitata* and *L. hyperborea* in Lough Ine rapids, it was found that the polyzoa *Hippothoa hyalina* and *Tubulipora* sp. favoured weak currents, whereas the hydroid *Tubularia bellis*, the polyzoan *Membranipora membranacea*, and the crustacean *Jassa falcata* favoured moderate or strong currents. The limpet *Patina pellucida* favoured moderate currents in the Rapids, and moderate

Table 10.3. The distribution of the commoner animal species upon boulders in relation to shelter from current action in Lough Ine Rapids (after Lilly et al.[66])

Preference for tops	No obvious preference for tops or bottoms	Preference for bottoms
	Preference for main stream	
Tubularia bellis	Sertularia operculata†	
Patella vulgata*	Plumularia setacea†	
	Sagartia elegans†	
Patella aspera*	Metridium senile	
	Corynactis viridis§	
	Balanus crenatus	
	Costazia costazii	
	Hippothoa hyalina	
	Umbonula littoralis†	
	Mytilus edulis‡	
	No obvious preference for main stream or sheltered bays	
Acmaea virginea	Myxilla rosacea	Haliclona limbata
	Amphilectus fucorum	Leuconia nivea
	Halichondria bowerbankii	Hymedesmia bronstedi¶
	Pomatoceros triqueter	Spirorbis spp.
	Balanus balanus	Escharoides coccinius
	Verruca stroemia	Diastopora patina
	Scruparia chelata†	Schizomavella spp.
	Scrupocellaria reptans¶	Lichenopora hispida
	Crisia eburnea	
	Tubulipora plumosa	
	Musculus marmoratus	
	Hiatella arctica	
	Kellia suborbicularis	
	Monia squama	
	Heteranomia squamula	
	Didemnum maculosum	
	Preference for sheltered bays	
Acanthochitona	Haliclona indistincta	Serpula vermicularis
crinitus	Adocia cinerea	Bugula flabellata
	Mycale rotalis	Escharella variolosa
	Anthopleura ballii	Membraniporella nitida
	Caryophyllia smithi	
	Cellepora pumicosa	

* Confined to upper surfaces of large boulders in full current on Sill.
† Possibly prefer tops.
‡ Mainly on Sill.
§ Extends into some shelter, and prefers bottoms on Sill.
¶ Not on Sill.

wave action in the sea; nevertheless it was more plentiful at the seaward stations[68].

Further studies have revealed thirty-nine species of plants and animals living in association in the zone of rapid current. Of these, five species, viz. *Jassa falcata, Caprella acutifrons, Mytilus edulis, Electra pilosa* and *Alcyonidium hirsutum*, showed a clear preference for living on algae in these conditions. On the other hand, fifteen species, viz. *Sphacelaria pennata, Platynereis dumerilii, Grubea clavata, Oridia armandi, Pomatoceros, triqueter, Spirorbis corrugatus, Dexamine thea, Erichthonius brasiliensis, Corophium bonelli, Rissoa parva, Bittium reticulatum, Amphipholis squamata, Didemnum maculosum, Aetea truncata* and *Scrupocellaria reptans* preferred quiet water stations. The species *Callithamnion tetragonum, Eusyllis blomstrandii, Caprella acanthifera* and *Calliostoma zizyphinum* preferred situations with a moderate current. None of these species showed a clear preference for a particular alga as a substratum[69].

In the Lough Ine rapids, and especially on the Sill, *Sertularia operculata* occurred abundantly upon boulders situated in the strong current. Away from these situations, it declined in abundance and was absent from Codium Bay which lies just within the lough. Apart from the species which lived on the boulders with *S. operculata* (Table 10.3), the hydroid carried a characteristic flora and fauna. A small proportion only of its surface was affected by epiphytes and encrusting animals. These included attached diatoms of the genera *Grammatophora* (the most evident), *Licmophora, Striatella, Synedra, Cocconeis,* and *Navicula* (in mucilage tubes). Chlorophyta and Phaeophyta were represented by single species, viz. *Rhizoclonium* sp. and an *Ectocarpus*-like alga, respectively. No Cyanophyta were found. Rhodophyta were more numerous, but all were microscopic and included the species *Acrochaetium* sp., *Rhodochorton* sp. or spp., *Ceramium* spp.; an encrusting genus, possibly *Melobesia minutula* coated a considerable area of the *Sertularia* with a layer of red cells. Encrusting Polyzoa comprising the species, *Hippothoa hyalina, Costazia costazii, Tubulipora* sp., *Electra pilosa* and the tufted species *Scruparia chelata* and *Crisia eburnea,* covered ≤ 4% of the surface of the *Sertularia.* Of the free living animals which were associated with the *Sertularia* about 98% were the amphipods *Caprella acutifrons,* with some *C. acanthifera* and *Phthisica marina;* other amphipods of the species *Stenothoë monoculoides, Jassa falcata* and *Elasmopus rapax* occurred in small numbers in many samples. The isopod *Janira maculosa* occurred to a very variable degree. Mytilid spat sometimes occurred in large numbers (in the Solway Firth, large numbers of small amphipods and mussels occurred upon the hydroid *Abietinaria abietina,* which is taken by the Dab, *Limanda limanda,* apparently for this reason[70]). The gastropod *Rissoa parva* and the brittle star *Ophiothrix fragilis* occurred occasionally. These populations associated with *Sertularia* did not undergo any apparent seasonal cycle of change.

Upon transfer to the sheltered waters of Codium Bay, the *Sertularia*

9*

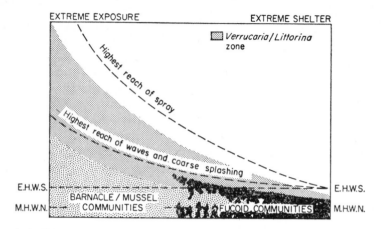

Fig. 10.3. Diagram showing the position of the *Verrucaria/Littorina* zone in relation to exposure (after Lewis[73]).

rapidly died. The number of living hydranths was reduced significantly by the end of the first week; although apparently dead, small portions of the coenosarc remained alive for up to 9 weeks. In the first week especially, the *Sertularia* became coated with an adhering sediment. The perisarc was colonized progressively by species absent or less abundant formerly: *Rhizoclonium* and *Ectocarpus* (?) increased, and *Enteromorpha*, *Cladophora*, *Sphacelaria*, *Lithosiphon* and *Leathesia* all appeared in abundance. The diatom *Navicula ramosissima* increased considerably, but *N. salinicola* disappeared after the second week; *Cocconeis scutellum* completely coated the perisarc of some of the transferred *Sertularia*. Of the associated animals, *Caprella acutifrons* declined after the first week, and the remainder were clogged with sediment or dead. In general, the less abundant animals declined, disappeared or occurred spasmodically as before: the amphipod *Dexamine spinosa* appeared; spat of the gastropods *Bittium reticulatum* and *Rissoa parva*, and the lamellibranchs *Pecten maximus*, *Chlamys* sp., *Cardium* sp. and an anomiid settled upon *Sertularia* in Codium Bay, whereas none were present in the Rapids[71].

On exposed shores, the zonation of plants and animals is more extended than on more sheltered shores[1, 72, 73] (Fig. 10.3); in tideless areas, wave action will produce a littoral zone, absent from more sheltered parts of such areas[73, 74]. In those areas of outer estuaries where longshore drift transports shingle and stones among the larger boulders, it is usual to find that the species which have a preference for living at the bottom of the boulders are absent; the constant abrasive action of the transported material renders the situation uninhabitable; on the bigger boulders, those levels, which are above the influence of the longshore transport of these coarse materials, are colonized by a normal flora and fauna.

Temperature

During the summer months, the algae *Enteromorpha*, *Ulva*, *Fucus*, *Porphyra*, *Chondrus* and *Griffithsia* are subject to wide semi-diurnal fluctuations of temperature, which in the normal environmental range have little influence upon their respiration[75]. Under extreme temperature conditions, i.e. when the sea and littoral algae become frozen, considerable damage may result from abrasion by ice[3, 4]. Even where such damage does not occur, photosynthesis and respiration may be depressed and the onset of reproduction can be delayed [18, 76, 77]. The rate of assimilation may be affected by temperature and light[78], while the ability of *F. vesiculosus* to absorb or evolve CO_2 is much reduced during tidal exposure[79].

High and low temperatures may be lethal to shore-dwelling animals which have a variable tolerance to these extremes depending upon species and phylum, e.g. barnacles are more tolerant than top shells, the geographical distribution of the species and their position in the shore zonation[80, 81] (Table 10.4). Least tolerance of high and low temperatures was shown by species found only in the infra-littoral fringe or below low water[81]. As with the ability to resist desiccation, the temperatures experienced on the shore are well within the tolerance limits of most species, and even exceptional extremes of temperature may have little direct influence on the distribution of adult intertidal animals. Nevertheless, in inshore waters and shallow seas, mass mortalities of fish and invertebrates may result from abnormally hot summers or cold winters[82–90]. In such a winter, geographical location may be important, and animals which inhabit promontories normally of a milder climate may be at particular risk[87]. Not all of the effects are directly due to temperature extremes, abrasion or entrapment in ice may inflict heavy casualties[3, 4, 77, 83]. On the other hand, in the Solway Firth, during the winter of 1962–63, the mortality of cockles, *Cardium edule*, in Auchencairn Bay was increased significantly by bird predation[77]. Mass mortalities may occur as a consequence of sudden, but non-lethal, increases or decreases in temperature: in the Gulf of St. Lawrence such changes caused debility in scallops, *Pecten magellanicus*, and exposed them to the attack of predators[32]. Oil spills and the development of severe anaerobic conditions produce similar results[88, 88a].

Populations may be eliminated from some areas[87] and widespread changes in distribution may result: the polychaete *Cirratulus cirratus* underwent little change in distribution as a result of the 1962–63 winter, whereas the related *Cirriformia tentaculata* virtually disappeared from the

Table 10.4.   The temperature tolerances of some intertidal invertebrates (after Evans[80], Southward[91])

Species	Time in hours to reach 50% mortality at temperatures (°C) of								Point of heat coma °C	Lethal point in °C 50%	Lethal point in °C 100%	Distributional status	Normal zonation
	−10	−5	0	30	35	37	40	50					
Chthamalus stellatus	12–24	72–120	8	—	—	—	29–30	$\frac{1}{2}$	43	52·5	53·7	Southern	
Elminius modestus	12–24	48–72	8	—	—	$7\frac{1}{2}$	$5\frac{1}{4}$	—	36–38	48·3	49·5	Australasian	
Balanus perforatus	<3	<22	8	—	—	—	3–5	$\frac{3}{4}$	38–40	45·5	47·0	Southern	
B. balanoides	12–24	120–190	8	—	—	—	—	—	35–37	44·3	45·3	Northern	
Patella depressa	—	—	—	—	—	—	$3\frac{1}{4}$–$3\frac{3}{4}$	—	37–38	43·3	—	Southern	Mid-littoral
P. vulgata	—	—	—	—	10	—	$2\frac{3}{4}$–3	—	37	42·8	—	Intermediate	Mid-littoral
P. aspera	—	—	—	—	$8\frac{1}{2}$–9	—	$\frac{3}{4}$–1	—	37	41·7	—	Southern	Lower mid-littoral to infra-littoral
Monodonta lineata	2–3	6–24	138–179	72–100	—	2	3	—	—	45·0	45·3	Southern	
Gibbula umbilicalis	2–3	16	30–79	24–72	—	—	$\frac{3}{4}$–1	—	—	41·8	42·0	Southern	
G. cineraria	2–3	2–3	12–30	$3\frac{3}{4}$	—	—	—	—	—	35·5	36·0	Northern	
Calliostoma zizyphinum	2–3	2–3	12–24	$3\frac{1}{2}$	—	—	—	—	33–34	34·5	34·8	Northern	
Littorina neritoides	—	—	—	—	—	—	14–15	—	38	46·3	—	Southern	Supra-littoral fringe to mid-littoral
L. littorea	—	—	—	—	—	—	$11\frac{1}{2}$–12	—	39	46·0	—	Intermediate	Mid-littoral
L. saxatilis	—	—	—	—	—	—	$9\frac{1}{2}$–10	—	37	45·0	—	Northern	Supra-littoral fringe to mid-littoral

south and east coast of England and from north and west Wales. This difference is a reflection of the temperature tolerances of the two species, e.g. after 24 hr at $-2°C$, 13 and 100% of *C. cirratus* and *C. tentaculata* respectively died in the laboratory[91]. Recovery from such events is of a variable duration. At Pornic on the coast of Brittany, the barnacle *Chthamalus stellatus* had recovered from the winter of 1962–63 by the summer of 1965[92]. The recovery of *Cirriformia tentaculata* is likely to be protracted, and if winter temperatures are returning to those existing before the mild period of 1900–1940, it may never recover the foothold which it lost in 1962–63[91]. Consideration of the mass mortalities which have afflicted the oyster industry of the River Blackwater, Essex, in this century, suggests that a recovery period of 20 years may be required[X].

In Weymouth and Poole Bays, many species survived the severe winter of 1962–63 without considerable loss. However, *Venus verrucosa* and *Venerupis rhomboides* suffered heavy mortalities, and it seems likely that the Scallop *Pecten maximus* was virtually exterminated. Such mortalities seem to have occurred in those places where the sea temperature fell to 3°C or below for a month or more. Curiously, species near their northern limits in this area were much less affected than *V. verrucosa*, *V. rhomboides* and *P. maximus* which, although distributed mainly to the southward of Britain, would not normally be considered to be near their extreme limits in this area[89].

Although, exceptional conditions can have far reaching consequences, it seems likely that evidence for a causal relation between temperature and the distribution of marine organisms must be sought in non-lethal terms such as debilitating effects, or combination with other factors[81], especially where an animal is not near the limits of its range. Interestingly, the instantaneous lethal temperatures of populations of *Elminius modestus* Darwin collected from habitats in New Zealand and Great Britain are similar[93]. Although balanids from various sources in Britain and Europe were able to adjust their activity-temperature responses to new temperature regimes, chthamalids from South Africa, Europe and Britain were unable to do so[93].

At non-lethal levels, temperature controls a wide variety of activities from feeding to reproduction. The rate at which the cirri of different barnacle species beat is a characteristic of the individual species, but with increasing temperature the cirral beat increases to an optimum and then declines. Species which have a southern origin, e.g. *Balanus perforatus*, *Chthamalus stellatus*, *Balanus amphitrite* and with the brackish water *B. eburneus* are active to a higher temperature than species of a northern origin, e.g. *Balanus balanus*, *B. balanoides* and the brackish water *B. improvisus*; the reverse is true also[94–96]. *Balanus crenatus*, a common inhabitant of sheltered waters, has a narrower temperature range than *B. balanoides*; the immigrant *Elminius modestus*, which is more abundant in sheltered waters also, has a wider range of temperature for cirral activity than *B. crenatus*[94]. However, many factors control cirral activity.

Within each species, the rate of cirral beat is dependent upon the shore level which the individual occupies: *C. stellatus*, *B. balanoides* and *E. modestus* from a low shore level have a more rapid beat than those from a high shore level[97]. Apparently, this phenomenon is as much related to a difference in growth rate (e.g. the growth rate of *B. balanoides* is dependent upon tidal level[98]) as to an adaptation to the temperature differences at the two levels[97].

The importance of adaptation to environmental conditions upon cirral beat may be illustrated by reference to the species *B. balanoides*, *B. balanus*, *B. crenatus*, *C. stellatus* and *Verruca stroemia*. The boreoarctic species *Verruca stroemia* may be found at the lowest levels of the intertidal zone, and occurs commonly on shallow banks and rough inshore grounds. It has a widespread distribution, viz. Red Sea, Adriatic, and Indian Ocean, but the major area of its distribution is along the eastern Atlantic seaboard from Spain to Finmark. It is absent from the Western Atlantic, but is found in the coastal water of Iceland, Greenland and the Faeroes. Although it is moderately euryhaline, it is absent from the Baltic. In the Clyde Sea area, it is widely distributed on suitable stony ground and upon crustaceans and molluscs. After settlement in spring, it grows rapidly, until the following winter when growth effectively ceases, but it is renewed in the following spring. Growth is directly related to environmental conditions, and is rapid when food is abundant. Measurement of the mean specific growth rates of this and other species under the partially controlled conditions of raft exposure (Table 10.5), revealed that those of *Verruca stroemia*, *Balanus balanoides*, *B. crenatus* and *B. balanus* are essentially similar, but that of *Chthamalus stellatus* is very different[99]. This is consistent with the difference in cirral beat in the optimum range of *C. stellatus* and *B.*

Table 10.5. The size and specific growth rate of acorn barnacles (after Barnes[99])

Species	Maximum size (mm)	Mean specific growth rate ($\times 100$) at half size
Balanus balanoides	25	1·0
B. crenatus	25	1·3
B. balanus	43	0·9
Chthamalus stellatus	10	0·1
Verruca stroemia	8	1·0

balanoides, the former being slower[94, 99]. The failure of *C. stellatus* to respond with increased growth, as did the other species when transferred to the raft may be an indication that this species requires a greater environmental stimulation for full metabolic activity[99]. Nevertheless, it must be recognized that this lower intrinsic metabolic rate of *C. stellatus* may be an adaptation to life at a high tidal level comparable with that of *Patella vulgata* (p. 262).

Increasing age and starvation may reduce the frequency of the cirral beat[95]. During the starvation of *B. balanoides*, carbohydrate reserves, if available, are consumed; if these reserves are not available then protein and lipids are utilized immediately food is withheld[100].

In still waters, the cirral beat of the species *Balanus balanoides*, *B. crenatus*, *B. improvisus*, *B. nubilis*, *B. perforatus* and *Elminius modestus* is characterized by a steady rhythm of short bursts of activity interspersed by inactive periods; in contrast, the surf- and current-loving *B. cariosus* and *Tetraclita squamosa* become totally inactive. In moving water, on the other hand, all species show a near continuous activity, but the current velocity required to induce this response depends on the species, habitat and previous history of the individual. Increased oxygen content at first has little effect but then reduces activity; a gradual reduction of oxygen content first leads to increased activity and then signs of distress attributable to anoxia are evinced.

A slight reduction of pH causes a reduction in the active periods, whereas a slight increase induces continuous activity; normal behaviour is resumed upon restoration to normal sea water. When the pH exceeds 9 or falls below 7, distress is evinced and cirral activity ceases. No consistent diurnal or tidal synchronized rhythms of cirral activity have been found[101].

In the species *Balanus balanoides* the rate of the cirral beat is related to the individual body weight. In any one animal, the cirral beat is not influenced by temperature between 7·5 and 10°C, but between 10 and 20°C it rises rapidly and then falls at temperatures in excess of 20°C. Temperature has little effect upon the amount of oxygen consumed per cirral beat per animal. However, the amount of oxygen consumed per cirral beat is dependent upon the size of the animal[102].

Over a temperature range 6–18°C, the oxygen consumption of the lobster *Homarus vulgaris* is directly proportional to the oxygen concentration; within specific limits of size and condition, the oxygen uptake of both sexes is the same. With increasing size the relative oxygen uptake in sea water decreases. In sea water of a constant tension, oxygen uptake increases with temperature. Under increased CO_2 tension, respiratory movements are retarded when the pH falls below 7·0 and are completely inhibited at 6·5. At a pH below 7·0, changes in CO_2 tension do not affect the rate of respiration. In the air, the oxygen uptake though at a low level, is directly proportional to temperature[103].

When the temperature of the gastropod molluscs *Thais* (*Nucella*) *lapillus*, *Littorina littorea*, *L. littoralis* and *L. saxatilis* is raised in air and when submerged, the oxygen uptake in air is consistently higher than in water up to the temperature at which heat death occurs. Once a snail has entered heat coma, two distinct respiratory patterns emerge depending upon treatment. In submerged snails, the respiratory rate falls until death ensues, whereas those kept in air respire at an increased rate initially, followed by a fall prior to death. In exposed and submerged snails, a heat coma ensues between 4 and 8, 6 and 7, 10 and 9, 6 and 5°C degrees below the heat death point in *L. saxatilis*, *L. littoralis*, *L. littorea* and *T. lapillus*, respectively. Therefore heat coma may act as a potential zonation controller. The generally higher temperature of onset of coma in exposed snails may be an additional safeguard against desiccation which is a severe risk once coma has been reached. The temperature tolerances of these molluscs varies with geographical distribution, those from Cardigan Bay having higher heat coma and lethal temperatures than those from the Firth of Forth[104].

The physiology of *Patella vulgata* and *P. aspera* is closely related to their normal habitat. The respiratory rates of individuals of *P. aspera* and high and low level *P. vulgata*, of equal weight, are essentially the same in January. In July, however, the respiratory rates of *P. aspera* and low level *P. vulgata* are higher than in January, but that of the high level *P. vulgata* is unchanged; a difference which may be attributable to the thermal acclimation of the high level *P. vulgata* and may be an adaptation to conserve energy in this more rigorous part of the environment[105, 106]. The low level population of *P. vulgata* is probably the normal form of this species: it is possible that an inability to perform a similar acclimation may be a factor which prevents *P. aspera* colonizing high shore level habitats[106].

However, the relatively low median water loss of *P. aspera*, viz. 30–35%, compared with 50–55% and 60–65% in low and high level *P. vulgata*[106], and the influence of increased wave exposure in facilitating colonization at high shore levels at Carrigathorna compared with the more sheltered Lough Ine[107] suggest that desiccation may be of particular importance in controlling the distribution of these species upon the shore. When water is dropped onto the apex of the shell of an exposed *P. vulgata*, a small amount, after running down the shell, passes beneath and induces a response which depends upon the salinity of the water. Sea water, or water of a high salinity, causes, the animal to lift the front of the shell, advance the head and extend the cephalic tentacles up to or beyond the edge of the shell, broadly in correspondence with the salinity of the water; fresh water, on the other hand, induces withdrawal and a firm clamp down of the shell. A greater response to splash and a greater tolerance of reduced salinity is evinced by limpets from a high tide level compared with those from a low tide level[108]. With respect to the composition of sea water, *P. vulgata* examined by the method described above showed a response little different from that to sea water in the range, pH 2·5–9·5, however at pH 10 the positive upward response was reduced and at 10·5 a strong contraction occurred. The nature of the response of *P. vulgata* to the major cations of sea water is modulated by the presence of calcium ions, but the response to NaCl, $MgCl_2$ and KCl alone is essentially the same as that to sea water; Na_2CO_3 and Na_2SO_4 induce the downward, clamping movements, of which the response to isotonic Na_2SO_4 is particularly powerful and prolonged[109].

Growth, Reproduction and Seasonal Change

Each of these aspects of the biology of the inhabitants of the rocky shore are closely related both with each other and with environmental parameters such as temperature. For this reason, although they could be considered under individual headings it seems more logical to consider them in a unified way.

Although the algal species *Pelvetia canaliculata*, *Fucus spiralis*, *F. vesiculosus* and *Ascophyllum nodosum* can penetrate far upstream in estuaries, the position occupied by individuals can influence their reproduction. In the Urr water, a tributary of the Solway Firth, the period of receptacle development of *P. canaliculata* and *F. spiralis* at Craigbrex was shortened compared with Rockcliffe, some 2 miles downstream; *F. vesiculosus* and *A. nodosum* were unaffected[18] (Fig. 10.4).

The laminarian *Saccorhiza polyschides* has a life history in which a large sporophytic phase alternates with a gametophyte phase which is microscopic. The growth of the sporophyte and its fruiting is completed in a single year; after the fruiting period, the stipeless bulb, which is incapable of further growth or of further fruiting, eventually decays and becomes detached[25]. In winter, the development of a new generation of sporophytes is apparently linked to the amount of light received by the plant and not to temperature or to day length. The latter is important only in so far as it controls the amount of light received, i.e. the product of light intensity and day length[110]. The form of the sporophyte stipe is related to the turbulence of the water: where the currents are weak the plant is broad, flimsy, curved and non-digitate, but where the current is strong the blade is long, tough, flat and very digitate. In those situations where the current is weak, but wave action is strong, the blades are short, flat, tough and with few digits[111].

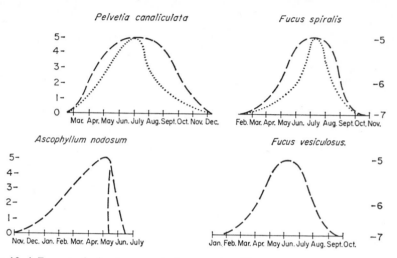

Fig. 10.4. Receptacle development in four species of brown seaweed from sites along the estuary of Urr water (1962–63 Season). Rockcliffe -----, Craigbrex..... (after Williams, Perkins and Gorman[18]).

An experimental comparison of the laminarians *S. polyschides* and *Chorda filum* showed that while the development of the sporophyte of the former was irreversibly inhibited at salinities below 9‰, *C. filum* could produce a sporophyte in a salinity of 5‰, though not at 4‰. These tolerances are consistent with the observation that *S. polyschides* is absent from the Baltic and other brackish water areas, but *C. filum* is recorded from the Gulf of Finland. Like animals, algae are sensitive to the ionic composition of the sea water, and an increased concentration of Ca permits the survival of *Delesseria sanguinea* and other red algae in salinities which would be fatal otherwise[112].

In any one year, the crop of *Laminaria hyperborea* is stable until the rough seas of winter dislodge the plants from the sea bed, and until the plants shed their old laminae in the spring. Over a period of years, the crop may be very variable, perhaps due to physical changes in the environment[113].

The sporophyte of *L. saccharina* reaches maturity in 8–12 months. Its longevity depends on the season of germination, bathymetric zone and habitat. Winter sporophytes rarely persist to maturity. Spring sporophytes form the bulk of all *L. saccharina* populations, except in the sub-littoral zone on very sheltered coasts where summer sporophytes may be equally numerous. On the British coast, the life span of a *L. saccharina* sporophyte does not exceed 3 yr. The sporophyte grows continuously throughout life, and at a rate which is dependent upon season; rapid growth occurs from January to June, and slow growth from July to December. Distal frond tissue is cast off continuously throughout the life of the sporophyte and at a rate which varies with the time of germination and with the habitat. Complete regeneration of the frond after cutting occurs in sporophytes up to the age of 1 yr. No regeneration occurs when the whole frond is removed. The fertility and longevity of the gametophyte, and hence of sporophyte production, depends on bathymetric zone and habitat: sporophytes develop at higher levels during winter, early spring, late summer and autumn, and at lower levels in spring, summer and autumn[19].

In southern Norway, *Laminaria digitata* grew most rapidly from February to April, and most slowly in late summer when the temperatures are high. Growth was reduced as temperatures approached 18°C, and stopped once it was exceeded. In autumn, growth recommenced as the temperatures decreased, and according to the fall in temperature became more rapid in late autumn and winter. In northern Norway, this species showed most growth from March to May, after which it decreased markedly. Here the decrease continued into the autumn and in the absence of light ceased in midwinter. In those years when the water temperature was low during the summer, growth was considerable[114].

The fresh weight of the British laminarians, *L. digitata* and *L. saccharina* changes seasonally, with maxima in spring and late summer. The chemical constitution of the fronds also varies seasonally. In spring, mannitol is at a minimum and laminarin is absent, but crude proteins, ash and alginic acid are at a maximum. The reverse is true in autumn.

The dry weight is at a minimum in spring and at a maximum in autumn. Both the weight of laminarin and the dry weight of these algae are greater in sheltered than in exposed areas. The greatest variation between samples occurs in those taken from the Scottish sea lochs. The observed differences between growth of these algae, in exposed and sheltered waters, suggests that growth may be restricted in the latter[115].

At three stations on the west coast of Scotland, the photosynthesis of *L. saccharina* and *L. hyperborea* occurs in two main bursts, spring and autumn, and is related to the availability of nutrients in relation to the thermocline and the autumn overturn[19, 116]; similar changes occur in the Barents Sea[117]. The changes in dry matter, ash, mannitol, laminarin, crude proteins, inorganic nitrogen and alginic acid content are related to this cycle[116]. In the actively growing region of the frond, which is proximal to the stipe, laminarin is absent in *L. saccharina*, the dry matter very low (10–12%) and is composed chiefly of mineral matter (36–40%) and mannitol (20–30%), proteins are minimal and alginic acid indetectable. A storage portion, which exists two-thirds of the way up the frond, contains reserves of laminarin and proteins. The distal portions represent tissue which has spored or is badly eroded. The concentrations of iodine and

Table 10.6. "Concentration factor", or ratio of trace-element content in algae (fresh weight) to trace element content in sea water (after Black and Mitchell[119])

Alga	Ni	Mo	Zn	V	Ti	Cr	Sr
Pelvetia canaliculata	700	8	1000	100	2000	300	20
Fucus spiralis	1000	15	—	300	10000	300	8
Ascophyllum nodosum	600	14	1400	100	1000	500	16
F. vesiculosus	900	4	1100	60	2000	400	18
F. serratus	600	3	600	20	200	100	11
Laminaria digitata							
Frond Atlantic Bridge	200	2	400	10	90	200	90
Stipe Atlantic Bridge	300	3	600	10	200	230	16
Frond Ardencaple Bay	200	2	1000	20	100	200	18
Stipe Ardencaple Bay	400	2	900	30	90	200	14

fucosterol are highest in the stipo-frondal region where they are probably required for growth. Such concentration gradients may not be present where nutrients are available throughout the year and where laminarin is absent[118].

The brown algae are all capable of extracting and concentrating trace elements from the sea[119] (Table 10.6). Perhaps the most remarkable is *Fucus spiralis* which contains 10 000 times more titanium than the surrounding sea water[119]. Studies with ^{65}Zn and *Laminaria digitata* and *Fucus vesiculosus* showed that this element is absorbed from the sea throughout the life of the plant, and that once absorbed little is lost[120, 121]; rapid losses occur only on death[120]. The affinity between divalent metals and alginate extracted from *Laminaria digitata* decreases in the order Pb > Cu > Cd > Ba > Sr > Ca > Co > Ni, Zn, Mn > Mg. However, zinc is absorbed by *L. digitata* in greater amounts than its affinity for alginate would imply, and although other divalent metals are largely bound by this means, some other factor is operative with zinc, possibly it is bound with fucoidin[120].

In populations of the barnacle *Balanus balanoides* density affects growth, in volume, to a significant degree, only when the individuals are closely packed, i.e. either touching or nearly touching. Even where densities reach 4 individuals/cm², the loss of cross-sectional area is compensated for by increase in height, and the effect of close packing is not in itself harmful[122]. In those situations where a particularly heavy spatfall has occurred, aberrant morphological forms may result. The influence of such overcrowding is perhaps best seen in the acorn barnacles which change in form from the normal flattened cone to an elongated tower life form. Growth is more rapid when these barnacles live in an uninterrupted current, but there is little evidence for a greater growth rate in the leading barnacles in the presence of a unidirectional current. In a constant food supply, these barnacles grow at a constant linear rate until the metabolic changes associated with the onset of maturity occur. In *Balanus balanus, B. balanoides* and *B. crenatus*, which mature slowly, the phase of linear growth may be interrupted by unfavourable environmental conditions; then growth in their second and third season is usual. On the other hand, the growth rate of *Elminius modestus* is reduced once it has reached a diameter of 6–7 mm, whatever the season. Barnacles parasitized by *Hemioniscus balani* grow more slowly than individuals which are unparasitized[122].

When grown upon a raft, *Chthamalus stellatus* grows much more slowly than *B. balanoides* which reaches its maximum size within 6 months of settlement. Under these conditions of continuous immersion, *C. stellatus* grows more rapidly than it does at either high or low shore levels, but the difference in growth is not so marked as in *B. balanoides*. This increased growth is a function of the greatly increased opportunity for feeding, especially where the growth panels are kept free of algal and other growth. On the shore, algae interfere little with *Chthamalus* and the relatively calm conditions of life on the raft may be offset by the lack of violent water in wave exposed positions which it normally favours.

Both *B. balanoides* and *Chthamalus* are at the limit of their range in the British Isles: *Chthamalus* is a tropical-Lusitanian form whereas *B. balanoides* is a boreo-arctic form. Changing climatic conditions have produced a seesaw of dominance on the south coast of England (p. 17). At the extreme southern limit of *B. balanoides*, where conditions are marginal for its breeding, *Chthamalus* forms a major community, even at lower shore levels. Further north, where the breeding of *B. balanoides* is more successful, the slow growing autumn settlement of *Chthamalus* is rapidly overgrown by the fast growing spat of *B. balanoides* which settle in the succeeding spring. In these waters, *Chthamalus* as a major community is restricted to an upper shore level, usually above that of *B. balanoides*. Presumably its survival here stems from its greater ability to withstand desiccation[124].

Many barnacles are cross fertilizing hermaphrodites. Both *Balanus crenatus* and *Elminius* will copulate within the temperature range 4–15°C[94]. Indeed, the breeding of *Elminius* is characterized by extreme eurythermy; it can breed at any season provided that the water temperature does not fall below 6°C and the food supply is adequate. This type of breeding is well suited to life in shallow estuaries and upon sheltered coasts in temperate latitudes and accounts for the success of this species in competition with the indigenous British species *B. balanoides*[125].

Although the breeding of *Elminius modestus* is strongly eurythermal, other species, e.g. *Chthamalus stellatus*, *Balanus amphitrite* and *B. perforatus*, breed over a limited temperature range and require a minimum temperature before breeding can commence. The boreo-arctic species *Balanus balanus* and *B. balanoides* produce a single annual brood; in the former, ovarian development continues throughout the early winter and copulation does not occur until mid-winter, there is little retention of embryos before the naupliar release in the spring.

The fertilization of *B. balanoides* takes place in autumn at a time which depends upon the environmental conditions. Fertilization is advanced by the falling temperatures of autumn and by reduced nutrition, and an increase in the age of the animal; it is inhibited by constant light of even a moderate intensity[126]. *B. balanoides* will not reproduce when the temperature exceeds 10°C. Indeed, a period of 4–6 weeks with less than 12 hr of light per day, and a temperature below the critical value of 10°C is essential for the maturation of the gametes and for reproduction to occur. There is some evidence that total darkness may increase the temperature limit, and that a lower temperature may allow an increase in the critical light factor. There is in any case a north–south gradient of reproduction; an example of adaptation to life in boreo-arctic waters[127].

Characteristic of the racial differences between *B. balanoides* inhabiting the coasts of Europe and the north eastern United States is that the latter are fertilized 1½–3 weeks in advance of the former. The subsequent development of the embryos and release of the larvae is adapted to the individual locality, and the actual time of liberation is mediated by the hatching substance in response to prevailing food conditions: in British waters, this is usually in March coincident with the commencement of the vernal plankton increase[126].

The eggs of the boreo-arctic species of cirripedes are significantly larger than those of all others. In the species *Balanus balanoides* eggs are larger where winters are severe and summers relatively cold irrespective of latitude. The change in egg size of warm water or eurythermal species, e.g. *Chthamalus stellatus*, over a similar latitude range is smaller than for *B. balanoides*. Production of such a large egg may ensure the development of a large larvae, well equipped to deal with the food available, and thus ensure a good survival. Egg size is also related to size of parent; the volume of eggs from *B. balanoides* native to the New England States of the U.S., are 1·3–1·6 times as great as the volume of British specimens[128].

In all the species, *Balanus balanus*, *B. crenatus*, *B. perforatus*, *B. amphitrite* var. *denticulata*, *Verruca stroemia*, *Elminius modestus* and *Chthamalus stellatus*, the rate of development of excised egg masses increases 3–4 times for every 10°C rise in temperature within the range at which breeding could occur; at the upper end of the range, the rate of development increased more slowly. The winter breeding northern forms *B. balanus* and *B. balanoides* could not withstand temperatures > 15°C, the intermediate forms, *Verruca stroemia* and *B. crenatus*, which breed in spring and early summer, could withstand temperatures up to 23°C. The southern midsummer breeding forms, *B. perforatus*, *B. amphitrite* and *Chthamalus stellatus*, withstood 30°C. *Elminius modestus*, which breeds all the year round, had a range of tolerance from 3 to 30°C[129, 130].

Well nourished individuals of *B. perforatus* and *C. stellatus* commenced breeding after 2–3 weeks at 15–16°C, *B. amphitrite* required 17–18°C, whereas *E. modestus* bred at 8–9°C. The moulting rate of well fed individuals of *C. stellatus* and *B. amphitrite* increased steadily in temperatures from 7 to 28°C, those of *B. perforatus* showed an increased rate from 7 to 24°C and *E. modestus* from 4 to 23°C. In all species, the moulting rate dropped markedly at temperatures of the order of 29–31°C[131].

In the absence of an adequate food supply, the ovaries of *B. balanoides* regress, but on the resumption of normal feeding, even after a period of five months, ovarian development is renewed, and the total feeding time to maturation shows little variation in fed and starved animals[132]. Clearly *B. balanoides*, like *Carcinus*[28], has a pronounced ability to withstand and recover from the effects of starvation.

Like the barnacles, the ascidians of Scottish waters breed in a manner related to geographical distribution. The northern boreo-arctic species *Pelonaia cornata* breeds only for a short period in January–February, and it seems likely that winter temperatures are too high for this species to breed further south. Boreo-arctic species centred in the boreal, but extending further north, e.g. *Dendrodoa grossularia*, may breed from May to December, but south boreal species, e.g. *Aplidium nordmanni*, breed for a shorter period in summer which suggests that the northern boundary is fixed by summer temperatures too low for breeding further north[133].

The barnacle *Balanus balanoides* shows a marked seasonal variation in its feeding rate, with peaks of feeding activity from March to May and in October. The onset of breeding in November is accompanied by a sharp decline in feeding, which practically ceases in late November and December. Food intake is greater at higher shore levels, particularly at temperatures exceeding 18°C; in contrast, the feeding of barnacles from lower shore levels is smaller and carried on at a constant rate[134]. Such a cycle of feeding corresponds to the seasonal changes of body weight and composition of this species[135].

In British waters, limpets of the genus *Patella* are represented by three species, viz. *P. vulgata* L., *P. intermedia* Jeffrys and *P. aspera* Lamarck. All three species are present on the coasts of northern Brittany, southern England and west Wales. *P. vulgata* and *P. aspera* only are present on the coasts of the Isle of Man and Ireland. In general, *P. vulgata* inhabits drier places on the shore than *P. aspera*[107]. In Lough Ine, *P. vulgata* extends over, and is confined to, the whole intertidal range, whereas *P. aspera* is confined to the sub-littoral fringe and very shallow sub-littoral. On the nearby shore, at Carrigathorna, which is on the open Atlantic coast, the range of *P. vulgata* extends from the M.T.L. to 1 m above H.W.S.T., and *P. aspera* extends over the whole intertidal range and from M.T.L. down provides almost the whole limpet population. Within Lough Ine both species have more depressed shells at low tide levels than at high tide levels, but in the more exposed Sill area, *P. aspera* is taller than in the more sheltered waters within the lough[107]: this is an example of the well known phenomenon that where a limpet spends a large part of its time holding fast to the substratum, an elevated form of shell results.

At Lough Ine too, the dog-whelk, *Thais* (*Nucella*) *lapillus*, taken from the open coast at Carrigathorna has a wider shell aperture, a more lightly constructed shell, a heavier body, a larger foot and a tighter grip upon the substratum, than those which live in the Rapids area. In a comparative test, performed at Carrigathorna, upon individuals from both sources, those from Carrigathorna were much more successful in maintaining a foothold in rough weather. Conversely, when both types were transferred to sheltered positions at Barloge Creek, where there were many crabs, those of the sheltered type survived the crab predation much better than those from the open coast[136].

The shape of the shell of the mussel, *Mytilus edulis*, is influenced by environmental factors also. Here the shell morphology is influenced by growth rate and population density, but shells from sheltered piers have higher, narrower and thinner shells than those collected from shores experiencing heavy seas. From comparable areas of exposure one would expect that shells from an area of high density would have a greater length: height ratio, but this is only true in some cases. Population density is not of unique importance as a factor which determines shell form, for it is modified significantly by differing growth rates which even within the same habitat are high variable (cf. *Carcinus* and *Thais*). Nevertheless population density probably exerts its influence through physical compression[137]. In south western Britain, the ranges of *M. edulis* and *M. galloprovincialis* overlap; although both species could be distinguished in the Camel estuary, both responded to environmental influences in the same way[138].

The shore gastropods *Littorina littorea*, *Gibbula umbilicalis* and *Thais (Nucella) lapillus* show most shell growth during the summer months and seasonal changes are determined largely by the water temperature[37]. The winter months are a period of inactivity and reduced growth[47, 139, 140] although growth may be affected by the maturation of the gonad[139]. Individuals of *L. littorea* at Whitstable, Kent, only became active above 8°C[47].

On the shores of the Gareloch, populations of *Thais*, and *L. littorea* to a lesser extent, are characterized by significant numbers of growth static forms. Furthermore, even the individuals which are growing actively do so at widely varying rates. This characteristic is shared by *Carcinus maenas*: juvenile crabs of uniform size, *ca.* 4 mm, introduced to tanks at Chapelcross in the early summer, grew well on the adequate food supply provided. However, by the end of the season these crabs ranged in size from 15 to 30 mm carapace width[X].

The dog-whelk, *Thais (Nucella) lapillus*, becomes active at 3°C and begins to spawn at 9–10°C. Feeding and crawling rates are at a maximum *ca.* 20°C; both are reduced at 25°C and the animal enters a coma at 27–28°C[37]. The more southerly whelk tingle, *Urosalpinx cinerea*, hibernates when water temperatures fall below 7°C and does not begin to feed until the temperature reaches 9–10°C[142]. It commences spawning at 12–13°C[143]; feeding reaches a maximum at 25°C and ceases at

30°C[37, 142]. The limpet *Patella vulgata* breeds during the winter months. The release of gonads is slow and the recovery period is prolonged. Similar behaviour is evinced by the top-shells *Monodonta lineata* and *Gibbula umbilicalis*, but their recovery period may be shorter than that of *P. vulgata*. The spawning of intertidal mussels, *Mytilus*, is more rapid than these gastropods. Both *Ostrea* and *Pecten* may produce two batches of gametes before autumn: the female of the former species produces only one batch of eggs, but a high proportion of these females switch rapidly to the male phase and commence sperm production immediately after spawning[144]. The developed gonads of *Patella* store fat, but show no appreciable glycogen reserves. Nevertheless, at the time of spawning, i.e. November–January, the blood glucose and tissue glycogen concentrations decline sharply. Unlike oysters, the glycogen levels remain low throughout the winter period[145].

The mussel, *Mytilus*, occurs widely and often abundantly around the British Isles. Depending upon situation, it may be the subject of a fishery, or a nuisance as a fouling agent (p. 370). In British waters, the gonads of *M. edulis* begin to ripen once the sea temperature has risen above *ca*. 7·0°C, and spawning follows a rapid rise in temperature to 11·5–13·0°C. Normally, the onset of spawning is coincident with a period of spring tides and of predominantly sunny weather. After spawning, the spent mussels enter a "*neuter*", or resting, stage which is followed by a redevelopment of the gonad in July or August. Apparently there is no correlation between nutritional condition and the ripening of the gonads, or indeed with the subsequent spawning: similar results have been recorded in Europe and N. America[146]. There is no correlation between spawning and temperature in Passamaquoddy Bay, New Brunswick[147], but definite evidence of a relationship between spawning and temperature is found in *M. californianus*[148-151]. The spawning in *M. edulis* in Passamaquoddy Bay may be related to the lunar cycle, but the mechanism is very obscure[146]. In the Bay of Arcachon, Les Landes, France, the spring spawning period lasts for twelve weeks, i.e. from the end of March to early June. Here spawning is discontinuous, with maxima at the time of the new and full moon. A second spawning occurs from December and January. Although at least some of the spawning is subject to hormonal control, the second spawning period is apparently halted by falling temperatures in February, and gametogensis is halted at temperatures below 7·5°C[152]. However, in Morecambe Bay, newly settled *Mytilus* spat has often been observed in February and March with further spatfalls in the late summer. Taking the larval life to be of about three weeks duration, this suggests a spawning at low temperatures[153].

Many attempts have been made to induce the spawning of *M. edulis* on a systematic basis; most were only a qualified success. However, Riley of the Fisheries Laboratory, Lowestoft, has succeeded by means of the following method: Suitable mussels are collected before inshore temperatures reach 8°C; they are maintained, in trays, in running sea water, at a temperature of 6–7°C, and are fed upon algal or yeast cultures. Some 2–3 weeks before they are required, they are raised to 10°C, and are disturbed as little as possible. When they are required to spawn, 10 are carefully removed, placed in a plastic bucket and shaken for 4–5 min. They are then returned to well aerated sea water, kept at 10°C: normally spawning occurs in males in 1 hr and females in 3–4 hr. One male only is required, more cloud the sea water unnecessarily, and should therefore be removed. The eggs collect below the exhalent siphon, and fertilized eggs may be transferred by pipette to clean, aerated sea water. A second spawning will take place overnight.

On the following day, strongly phototactic, free-swimming ciliated larvae are present in the water. They may survive without food for 7 days at 10–15°C, and become veligers, but under these conditions none survive more than 7 days. This spawning may be induced from early April to August or September: the larvae produced may be used to feed the larvae of herring, sand eels, dab, and pleuronectid hybrids with small yolk sacs. Indeed, mussel larvae are taken by young brill in nature.

References

1. Moyse, J. and Nelson-Smith, A. (1963). *Fld Stud.* **1** (5), 1–31.
2. Guiler, E. R. (1952). *Pap. Proc. R. Soc. Tasm.* **86**, 1–11.
3. Stephenson, T. A. and Anne (1954). *J. Ecol.* **42**, 14–45.
4. Stephenson, T. A. and Anne (1954). *J. Ecol.* **42**, 46–70.
5. Lewis, J. R. (1955). *J. Ecol.* **43**, 270–290.
6. Crisp, D. J. and Southward, A. J. (1958). *J. mar. biol. Ass. U.K.* **37**, 157–208.
7. Guiler, E. R. (1960). *J. Ecol.* **48**, 1–28.
8. Berry, A. J. (1964). *Malay. Nat. J.* **18**, 81–103.
9. Evans, R. G. (1947). *J. mar. biol. Ass. U.K.* **27**, 173–218.
10. Dalby, D. H. (1968). *Fld Stud.* **2** (Suppl), 31–38.
11. Kurz, H. and Wagner, K. (1957). *Fla St. Univ. Stud.* No. 24.
12. Chapman, V. J. (1946). *Ecology* **27**, 91–93.
13. Jones, W. E. (1959). *J. mar. biol. Ass. U.K.* **38**, 153–167.
14. Boney, A. D. and Corner, E. D. S. (1962). *J. mar. biol. Ass. U.K.* **42**, 65–92.
15. Boney, A. D. and Corner, E. D. S. (1963). *J. mar. biol. Ass. U.K.* **43**, 319–325.
16. Larkum, A. W. D., Drew, E. A. and Crossett, R. N. (1967). *J. Ecol.* **55**, 361–371.
17. Knight, M. and Parke, M. W. (1950). *J. mar. biol. Ass. U.K.* **29**, 439–514.
18. Williams, B. R. H., Perkins, E. J. and Gorman, J. (1965). H.M.S.O., U.K.A.E.A., P.G. Report 650 (CC).
19. Parke, M. W. (1948). *J. mar. biol. Ass. U.K.* **27**, 651–709.
20. Kain, J. M. (1962). *J. mar. biol. Ass. U.K.* **43**, 377–385.

21. Kain, J. M. (1964). *J. mar. biol. Ass. U.K.* **44**, 415–433.
22. Kain, J. M. (1965). *J. mar. biol. Ass. U.K.* **45**, 129–143.
23. Kain, J. M. (1969). *J. mar. biol. Ass. U.K.* **49**, 455–473.
24. Kain, J. M. (1963). *J. mar. biol. Ass. U.K.* **43**, 129–151.
25. Norton, T. A. and Burrows, E. M. (1969). *Br. phycol. J.* **4**, 19–53.
26. Larkum, A. W. D. and Norton, T. A. (1968). *Br. phycol. Bull.* **3**, 601–602.
27. Perkins, E. J. (1968). *Trans. J. Proc. Dumfries Galloway Nat. hist. Antiq. Soc.*, Ser. 3, **45**, 14–43.
28. Perkins, E. J. (1967). *Trans. J. Proc. Dumfries. Galloway Nat. hist. Antiq. Soc.*, Ser. 3, **44**, 47–56.
29. Muntz, L., Ebling, F. J. and Kitching, J. A. (1965). *J. Anim. Ecol.* **34**, 315–329.
30. Ebling, F. J., Hawkins, A. D., Kitching, J. A., Muntz, L. and Pratt, V. M. (1966). *J. Anim. Ecol.* **35**, 559–566.
31. Crothers, J. H. (1968). *Fld Stud.* **2**, 579–614.
32. Dickie, L. M. and Medcof, J. C. (1963). *J. Fish. Res. Bd. Can.* **20**, 452–482.
33. Kitching, J. A. and Ebling, F. J. (1961). *J. Anim. Ecol.* **30**, 373–383.
34. Nicholls, A. G. (1931). *J. mar. biol. Ass. U.K.* **17**, 675–708.
35. Naylor, E. (1955). *J. mar. biol. Ass. U.K.* **34**, 347–355.
36. Smith, J. E. (1968). "Torrey Canyon Pollution and Marine Life", University Press, Cambridge.
37. Lawson, G. W. (1957). *J. Ecol.* **45**, 831–860.
37a. Largen, M. J. (1967). *J. Anim. Ecol.* **36**, 207–214.
38. Umamaheswararo, M. and Sreeramulu, T. (1964). *J. Ecol.* **52**, 595–616.
39. Edwards, R. L. (1958). *J. Anim. Ecol.* **27**, 37–45.
40. Naylor, E. (1962). *J. Anim. Ecol.* **31**, 601–609.
41. Perkins, E. J. and Penfound, J. M. (1971). *Spectrum. British Science News*, No. 84, 7–8.
42. Moore, H. B. (1936). *J. mar. biol. Ass. U.K.* **21**, 61–89.
43. Gowanloch, J. N. and Hayes, F. R. (1926). *Contr. Can. Biol. Fish.* **3**, 135–165.
44. Gibson, R. N. (1967). *J. Anim. Ecol.* **36**, 215–234.
45. Miller, M. C. (1962). *J. Anim. Ecol.* **31**, 545–569.
46. Smith, J. E. and Newell, G. E. (1955). *J. Anim. Ecol.* **24**, 35–56.
47. Newell, G. E. (1958). *J. mar. biol. Ass. U.K.* **37**, 229–239.
48. Berry, A. J. (1961). *J. Anim. Ecol.* **30**, 27–45.
49. Barnes, H., Finlayson, D. M. and Piatigorsky, J. (1963). *J. Anim. Ecol.* **32**, 233–252.
50. Grainger, F. and Newell, G. E. (1965). *J. mar. biol. Ass. U.K.* **45**, 469–479.
51. Williamson, D. I. (1951). *J. mar. biol. Ass. U.K.* **30**, 73–90.
52. Wieser, W. (1963). *J. mar. biol. Ass. U.K.* **43**, 97–112.
53. Landenberger, D. E. (1969). *Physiol. Zoöl.* **42**, 220–230.
54. Gibson, J. S. (1970). *J. Anim. Ecol.* **39**, 159–168.
55. Ramsay, J. A. (1935). *J. exp. Biol.* **12**, 355–372.
56. Kensler, C. B. (1967). *J. Anim. Ecol.* **36**, 391–406.
57. Davies, P. Spencer (1969). *J. mar. biol. Ass. U.K.* **49**, 291–304.
58. Morton, J. E., Boney, A. D. and Corner, E. D. (1957). *J. mar. biol. Ass. U.K.* **36**, 383–405.
59. Southward, A. J. and Orton, J. H. (1954). *J. mar. biol. Ass. U.K.* **33**, 1–19.
60. Powell, H. T. (1957). *J. mar. biol. Ass. U.K.* **36**, 407–432.
61. Burrows, E. M. and Lodge, S. (1951). *J. mar. biol. Ass. U.K.* **30**, 161–176.
62. Rathsack-Kunzenbach, R. (1960, 1961). *Ber. dt. bot. Gesell.* **73** (11), 37–38.

63. MacNae, W. (1957). *J. Ecol.* **45**, 113–131, 361–387.
64. Gibb, D. C. (1957). *J. Ecol.* **45**, 49–83.
65. Nelson-Smith, A. and Gee, J. M. (1966). *Fld Studies.* **2**, 331–357.
66. Lilly, S. J., Sloane, J. F., Bassindale, R., Ebling, F. J. and Kitching, J. A. (1953). *J. Anim. Ecol.* **22**, 87–122.
67. Ebling, F. J., Sleigh, M. A., Sloane, J. F. and Kitching, J. A. (1960). *J. Ecol.* **48**, 29–53.
68. Sloane, J. F., Ebling, F. J., Kitching, J. A. and Lilly, S. J. (1957). *J. Anim. Ecol.* **26**, 197–211.
69. Sloane, J. F., Bassindale, R., Davenport, E., Ebling, F. J. and Kitching, J. A. (1961). *J. Ecol.* **49**, 353–368.
70. Williams, B. R. H., Perkins, E. J. and Hinde, A. (1965). H.M.S.O., U.K.A.E.A., P. G. Report 611 (CC).
71. Round, F. E., Sloane, J. F., Ebling, F. J. and Kitching, J. A. (1961). *J. Ecol.* **49**, 617–629.
72. Lewis, J. R. (1964). "The Ecology of Rocky Shores", English University Press, London.
73. Lewis, J. R. (1968). *Sarsia* **34**, 13–36.
74. Lewis, J. R. (1965). *Bot. goth.* **3**, 129–143.
75. Newell, R. C. and Pye, V. I. (1968). *J. mar. biol. Ass. U.K.* **48**, 341–348.
76. Tikhovskaya, Z. P. (1960). *Botan. Zhurnal.* **45**, 1147–1160.
77. Williams, B. R. H., Perkins, E. J. and Bailey, M. (1963). *Trans. J. Proc. Dumfries. Galloway Nat. hist. Antiq. Soc.* Ser. 3, **41**, 30–44.
78. Hyde, M. B. (1938). *J. Ecol.* **26**, 118.
79. Bidwell, R. G. S. and Craigie, J. S. (1963). *Can. J. Bot.* **41**, 179–182.
80. Evans, R. G. (1948). *J. Anim. Ecol.* **17**, 165–173.
81. Southward, A. J. (1958). *J. mar. biol. Ass. U.K.* **37**, 49–66.
82. Patterson, A. H. (1907). "Wildlife on a Norfolk Estuary", Methuen, London, 352 pp.
83. Smidt, E. L. B. (1944). *Folia geogr. dan.* **2** (3), 1–35.
84. Russell, F. S. and Yonge, C. M. (1947). "The Seas", Warne, London and New York.
85. Yonge, C. M. (1949). "The Sea Shore", Collins, London. New Naturalist Series No. 12.
86. Stephen, A. C. (1953). "Life on Sandy Shores", Essays in Marine Biology, being the Richard Elmhurst Memorial Lectures. 50–71. Oliver and Boyd, Edinburgh.
87. Crisp, D. J. (ed.). (1964). *J. Anim. Ecol.* **33**, 165–210.
88. Perkins, E. J. and Abbott, O. J. (1972). *Mar. Poll. Bull.* **3**, 70–72.
88a. Logan, J. W. M. and Perkins, E. J. (1973). Unpublished work.
89. Holme, N. A. (1967). *J. mar. biol. Ass. U.K.* **47**, 397–405.
90. Kristensen, I. (1957). *Levende Nat.* **60**, 93–95.
91. George, J. D. (1968). *J. Anim. Ecol.* **37**, 321–337.
92. Barnes, H. and Barnes, M. (1966). *Cah. Biol. mar.* **7**, 247–249.
93. Ritz, D. A. and Forster, B. A. (1968). *J. mar. biol. Ass. U.K.* **48**, 545–559.
94. Southward, A. J. (1955). *J. mar. biol. Ass. U.K.* **34**, 403–422.
95. Southward, A. J. (1957). *J. mar. biol. Ass. U.K.* **36**, 323–334.
96. Southward, A. J. (1962). *J. mar. biol. Ass. U.K.* **42**, 163–167.
97. Southward, A. J. (1955). *J. mar. biol. Ass. U.K.* **34**, 423–433.
98. Barnes, H. and Powell, H. T. (1953). *J. mar. biol. Ass. U.K.* **32**, 107–127.

99. Barnes, H. (1958). *J. mar. biol. Ass. U.K.* **37**, 427–433.
100. Barnes, H., Barnes, M. and Finlayson, D. M. (1963). *J. mar. biol. Ass. U.K.* **43**, 213–223.
101. Southward, A. J. and Crisp, D. J. (1965). *J. mar. biol. Ass. U.K.* **45**, 161–185.
102. Newell, R. C. and Northcroft, H. R. (1965). *J. mar. biol. Ass. U.K.* **45**, 387–403.
103. Thomas, H. J. (1954). *J. exp. Biol.* **31**, 228–251.
104. Sandison, E. E. (1967). *J. exp. mar. Biol. Ecol.* **1**, 271–281.
105. Davies, P. Spencer (1966). *J. mar. biol. Ass. U.K.* **46**, 647–658.
106. Davies, P. Spencer (1967). *J. mar. biol. Ass. U.K.* **47**, 61–74.
107. Ebling, F. J., Sloane, J. F., Kitching, J. A. and Davies, H. M. (1962). *J. Anim. Ecol.* **31**, 457–470.
108. Arnold, D. C. (1957). *J. mar. biol. Ass. U.K.* **36**, 121–128.
109. Arnold, D. C. (1959). *J. mar. biol. Ass. U.K.* **38**, 569–580.
110. Norton, T. A. and Burrows, E. M. (1968). *Proc. int. Seaweed Sym.* **6**, 287–296.
111. Norton, T. A. (1969). *J. mar. biol. Ass. U.K.* **49**, 1025–1045.
112. Norton, T. A. and South, G. R. (1969). *Oikos* **20**, 320–326.
113. Walker, F. T. and Richardson, W. D. (1955). *J. Ecol.* **43**, 26–37.
114. Sundene, O. (1964). *Nytt. Mag. Bot.* **11**, 83–107.
115. Black, W. A. P. (1950). *J. mar. biol. Ass. U.K.* **29**, 45–72.
116. Black, W. A. P. and Dewar, E. T. (1949). *J. mar. biol. Ass. U.K.* **28**, 673–699.
117. Tikhovskaya, Z. P. (1940). *C.r. Acad. Sci., U.R.S.S.* **29**, 120–124.
118. Black, W. A. P. (1954). *J. mar. biol. Ass. U.K.* **33**, 49–60.
119. Black, W. A. P. and Mitchell, R. L. (1952). *J. mar. biol. Ass. U.K.* **30**, 575–584.
120. Bryan, G. W. (1969). *J. mar. biol. Ass. U.K.* **49**, 225–243.
121. Gutnecht, J. (1963). *Limnol. Oceanogr.* **8**, 31–38.
122. Crisp, D. J. (1960). *J. Anim. Ecol.* **29**, 95–116.
123. Caspers, H. (1964). *Mitt. hamb. zool. Mus. Inst.* **1**, 13.
124. Barnes, H. (1956). *J. mar. biol. Ass. U.K.* **35**, 355–361.
125. Crisp, D. J. and Davies, P. A. (1955). *J. mar. biol. Ass. U.K.* **34**, 357–380.
126. Crisp, D. J. (1964). *J. mar. biol. Ass. U.K.* **44**, 33–45.
127. Barnes, H. (1963). *J. mar. biol. Ass. U.K.* **43**, 717–727.
128. Barnes, H. and Barnes, M. (1965). *J. Anim. Ecol.* **34**, 391–402.
129. Crisp, D. J. (1959). *J. Anim. Ecol.* **28**, 119–132.
130. Patel, B. and Crisp, D. J. (1960). *Physiol Zool.* **33**, 104–119.
131. Patel, B. and Crisp, D. J. (1960). *J. mar. biol. Ass. U.K.* **39**, 667–680.
132. Barnes, H. and Barnes, M. (1967). *J. exp. mar. Biol. Ecol.* **1**, 1–6.
133. Millar, R. H. (1958). *J. mar. biol. Ass. U.K.* **37**, 649–652.
134. Ritz, D. A. and Crisp, D. J. (1970). *J. mar. biol. Ass. U.K.* **50**, 223–240.
135. Barnes, H., Barnes, M., and Finlayson, D. M. (1963). *J. mar. biol. Ass. U.K.* **43**, 185–211.
136. Kitching, J. A., Muntz, L., and Ebling, F. J. (1966). *J. Anim. Ecol.* **35**, 113–126.
137. Seed, R. (1968). *J. mar. biol. Ass. U.K.* **48**, 561–584.
138. Lewis, J. R. and Seed, R. (1969). *Cah. Biol. mar.* **10**, 231–253.
139. Williams, E. E. (1964). *J. Anim. Ecol.* **33**, 413–432.
140. Williams, E. E. (1964). *J. Anim. Ecol.* **33**, 433–442.
141. Kristensen, I. (1957). *Archs. néerl. Zool.* **12**, 351–453.
142. Hancock, D. A. (1959). *Fishery Invest. Lond.* **22**, 1–66.
143. Cole, H. A. (1942). *J. mar. biol. Ass. U.K.* **25**, 477–508.

144. Orton, J. H., Southward, A. J. and Dodd, J. M. (1956). *J. mar. biol. Ass. U.K.* **35**, 149–176.
145. Barry, R. J. C. and Munday, K. A. (1959). *J. mar. biol. Ass. U.K.* **38**, 81–95.
146. Chipperfield, P. N. J. (1953). *J. mar. biol. Ass. U.K.* **32**, 449–476.
147. Battle, H. (1932). *Contrib. Canadian Biol.*, *N.S.* **7**, No. 20, 255–276.
148. Young, R. T. (1946). *Ecology* **27**, 354–363.
149. Young, R. T. (1942). *Ecology* **23**, 490–492.
150. Coe, W. R. and Fox, D. L. (1942). *J. exp. Zool.* **90**, 1–30.
151. Wheldon, W. F. (1936). *Univ. Calif. Pub. Zool.* **41**, 35–44.
152. Lubet, P. (1957). *Ann. biol.* **33**, 19–29.
153. Waugh, G. D. (1960). *Proc. malac. Soc., Lond.* **34**, 113–122.
 X. Perkins, E. J. Formerly unpublished work.

11

Macrobenthos of Sedimentary Beaches

In general, the rocky headlands and shores of outer estuaries give way to shores of sedimentary materials further upstream where the water is shallower. Except for isolated pinnacles and reefs, the infra-littoral substratum is composed of diverse sediments.

Before considering beaches composed of finer sediments it is worthwhile considering those of an intermediate nature, that is those which are composed predominantly of materials ranging from 0·5 to 64 mm on the Udden-Wentworth scale (i.e. coarse sand, very coarse sand, granules, and pebbles) and which are bound together with varying amounts of fine sediments, viz. *the bound shingles*.

Bound shingle occurs widely in those situations where wave and current action prevent the deposition of the more easily transported fine sands. In coastal plain estuaries, these materials may occur as an interface between a shingle upper shore and a middle and lower shore, composed of the finer grades, e.g. at Whitstable, Kent; as isolated patches of "scars" in large tracts of sandy shore, e.g. the Solway Firth; as patches of sea bed, e.g. The Stone, River Blackwater (so called because of this type of bottom). These materials occur on the middle and lower shore and the upper infra-littoral, in most situations, in the fiordic Scottish sea lochs, where the rocky shore does not drop straight into the sea. At upper levels on these fiordic shores, the finer sedimentary materials may have been removed by wave action and, like the shingle beach of a coastal plain estuary, the movement of this shingle, when the shore is covered by the tide, may be inimical to living organisms. Away from this extreme situation, the superficial inhabitants of the littoral bound shingles are essentially those of rocky shores, viz. the algae, *Pelvetia canaliculata, Fucus spiralis, Ascophyllum nodosum, Fucus vesiculosus, Gigartina stellata, Ulva lactuca, Enteromorpha, Cladophora*; sponges, *Halichondria panicea*; the gastropod molluscs, *Littorina saxatilis, L. littorea, L. littoralis, Thais (Nucella) lapillus, Acmaea testudinalis, A. virginea*; lamellibranch molluscs, *Mytilus edulis*; barnacles, *Chthamalus stellatus, Balanus balanoides, Elminius modestus*; starfish, *Asterias rubens, Henricia sanguinolenta*; and many others. However, unlike the rocky shores, the bound shingles of both fiordic and coastal plain estuaries are inhabited by an infauna characterized by the lamellibranch molluscs *Modiolus modiolus*, the horse mussel, *Venerupis pullastra*, pullet shell, *Mya arenaria* and *M. truncata*, the soft-shelled clams, and *Hiatella artica*.

Below the L.W.M. the clean, stony, gravelly grounds are inhabited by epifaunal populations which are characterized by the hydroid *Abietinaria abietina*, the lamellibranch *Modiolus modiolus*, *Chlamys opercularis*, and the sea-squirt *Dendrodoa grossularia* (this small orange coloured ascidian is known to the Essex oyster fisherman as "pock" and is notorious for the irritating fluid which it squirts when handled), the sea urchins *Psamme-chinus miliaris*, *Echinus esculentus* and the polyzoan *Flustra foliacea*. On these grounds too, the bright yellow boring sponge *Cliona celata* may be found, and where the water is clear red algae such as *Delesseria sanguinea*. In coastal plain estuaries, at least, these grounds may be important to skate as egg-laying areas[X]. Essentially similar communities are recorded in N. America[1X].

The muddy gravels which occur at greater depths, and fringe the bound shingles in fiordic estuaries, are characterized by the beautiful pelican foot shell, *Aporrhais pes-pelecani*; in some areas, e.g. the outer Solway Firth adjacent to Wigtown Bay, this species is so abundant that it forms the principal component of the dead shells stranded at the high water mark[X].

The fine sedimentary deposits of estuaries represent a transition between the land and the very fine offshore sediments found at depths greater than 30 m. The faunas and floras of mud and sand are often treated as belonging to distinct and separate systems. However, the differentiation between the two sedimentary types is rarely clear cut for the sediments of coastal plain estuaries tend to become finer with increasing distance from the open sea, and, on the individual shore, with distance moved from low-water mark towards high-water mark. At the mouths of such estuaries finer sediments accumulate at depths where wave action is insufficient to raise them into suspension. In fiordic estuaries, fine and very fine sediments are found increasingly with the greater depths below the low-water mark: a function of the relative low waves which influence them. Clearly, then, there is a unity in the sedimentary distribution of estuaries, with coarser sediments grading into finer sediments depending upon the hydrography of the individual estuary. Equally clearly, these changes, in sediment composition and distribution, are paralleled by changes in the composition and abundance of the flora and fauna.

Zonation and Succession

The zonation of sedimentary shores is overt only at the higher levels where salt marshes develop in temperate estuaries and mangrove swamps develop

Table 11.1.   Zonation of the Solway Firth biota (after Perkins and Williams[19])

	Vegetation / Substrate	Sedimentary beach — Fauna: Inner Solway	Sedimentary beach — Fauna: Outer Solway	Rocky beach
Supra-littoral E.H.W.M.	Carr and woodland / Dune land			*Xanthoria, Ramalina*
	Phragmitetum, Juncetum, Festucetum / Shingle and coarse sand			*Littorina neritoides*
	General salt marsh community / Bare sand	*Corophium volutator, Hydrobia vulvae*	*Corophium volutator*	*Pelvetia, L. saxatilis, Chthamalus*
Shore	*Puccinellietum*	*Scrobicularia plana, Macoma balthica, Arenicola marina*	*Macoma balthica*	
	Salicornietum, Spartinetum / Bare and bound shingle patches			*Fucus spiralis, Balanus balanoides, Elminius, Ascophyllum, Mytilus*
	Zosteretum	*Macoma balthica, Arenicola marina*	*Macoma balthica, Arenicola marina, Cerastoderma* sp., *Tellina tenuis*	
	Bare muddy sand	*Arenicola marina, Cerastoderma* sp.		*Fucus vesiculosus, Mytilus*
L.W.M.	Bare sand	*Lanice conchilega*	*Lanice conchilega, Owenia fusiformis, Sabellaria* sp. (*Sabellaria* on south shore especially)	
Infra-littoral	Bound shingle / Coarse sand	Channel bottoms—medium, coarse sand. Barren except for fish and vagile epifauna; Bound shingle; *Modiolus modiolus, Abietinaria abietina, Dendrodoa grossularia*	Coarse, medium exposed sand; *Donax vittatus, Mactra corallina*	*Laminaria, Delesseria*
	Fine sand		Fine sand; *Pectinaria, Mactra, Spisula*	Bound shingle, coarse, medium sand; fine sand; muddy sand
	Muddy sand		Muddy sand "sheltered"; *Abra alba, Nucula sulcata*	

in tropical estuaries[2]. The general form of such a zonation in a coastal plain estuary is typified by that found in the Solway Firth[3] (Table 11.1).

Even on rocky shores, zonation cannot be considered as other than a dynamic equilibrium in which, at any given level or situation, long term changes may occur. Unlike the rocky shore, those composed of sediment are liable to, and indeed undergo, marked changes in level and composition depending upon the relative influence of accretion and erosion. Clearly, the zones of colonization observed upon such shores represent a response to conditions at any given level, but since the shore is itself undergoing changes in level, each zone represents a given stage in a succession typical of the particular coastal area. In these areas then, zonation cannot be considered in isolation but must be considered in relation to succession, and vice versa.

Temperate Beaches

Salt Marshes

Salt marshes and mangrove swamps develop in the sheltered waters at the head of an estuary, or in the protection of a sand or shingle bar. Here the deposits of fine material, transported and deposited by the tides, form areas of natural reclamation[2, 4]. Salt marshes and mangrove swamps are rarely juxtaposed, although this situation does occur near Sydney, on the coast of Australia[5]. As the accumulation of sediment proceeds, so the resulting sand and mud flats become colonized, first by algae and then by higher plants, which, in turn, influence the rate at which deposition takes place[2, 4].

The quality of the sediment, which accumulates, influences the plant colonization, and a lack of silt may be responsible for poverty of the primary zones of colonization: *Salicornia* spp. are not normally successful colonists of sandy soils[6] and where the sand to silt ratio is less than 15 : 85, then *Spartina townsendii* is a more successful colonist than *Puccinellia*[7]. On the other hand, the soil composition was less important than the consequences of tidal elevation in controlling plant distribution patterns in the salt marsh and mangrove swamps of Sydney[5, 8]. Nevertheless, the marshes on the east coast of Britain are characterized by thick, fairly compact mud based on sandy flats, whereas those on the coasts of Wales, N.W. England, and southern Scotland are much more sandy; the latter are rather poor in precursor species and are dominated by grasses[2, 3, 6].

Comparison of the type of marsh, and the composition of the seabed around Britain, shows that where marshes are dominated by grasses, i.e. they are relatively impoverished, then significant amounts of silt occur on the adjacent seabed. For example, the relatively enclosed nature of the

Irish Sea, with its short fetches which contrast with the open nature and long fetches of the North Sea, permits the nutrient rich silt to remain undisturbed on the seabed, the salt marshes are poor in this material and are impoverished[3]. In those areas of the Solway Firth where the silt content is relatively high, the marsh zonation and succession may be summarized thus:

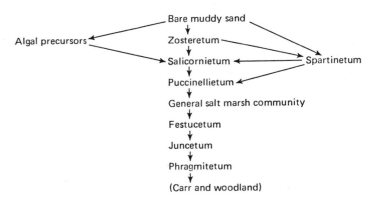

In areas where the silt content of the soil is reduced, the marsh zonation succession can be summarized thus:

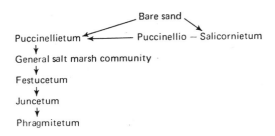

Broadly similar conditions are found in the Dovey estuary[9], at Findhorn on the Moray Firth, and on the shores of Loch Crinan[X].

Although the succession in the more silty areas of the Britian west coast marshes resembles that found in the rich marshes at Scolt Head, Norfolk[10, 11], and that of the Zwartkops estuary, South Africa[12], the qualitative and quantitative differences may be very considerable. Indeed, it is of interest to note that land reclamation in Morecambe Bay, has not been a success owing to lack of silt[13, 14], and the 15 000 acres (60 km²) of land, which might become available to agriculture, as a result of a barrage scheme in the Bay, is not likely to have the value which has been assumed[15].

The fiordic sea lochs of Scotland evince little salt marsh development, but where it does occur it is impoverished and confined to the head of the loch. Clearly, as in the coastal plain estuaries of western Great Britain, the fertilizing silt which could produce richer marshes is removed by the short, steep waves which influence these shores, and, since the fetches are short and the waves low, it comes to lie permanently at the bed of the fiord. Thus the factors which promote richness or impoverishment of the marginal salt marshes are shared equally by fiordic and coastal plain estuaries.

As plants colonize mud flats, providing obstructions which promote further deposition, the waters of the tide are increasingly confined, first within the clumps of colonizing plants, then between poorly-defined banks covered with plants, and finally within embryo and more developed creeks[2]. Within the creeks, filamentous algae and *Halimione portulacoides* may flourish[2-4, 10, 16]; in the more fertile, silty areas, where grazing is minimal, this low growing bushy plant may spread outwards from the creek edges, where drainage is good, to cover large areas of marsh and smother all other vegetation[2]. By straining out silt and building up banks, it may reduce drainage in adjacent areas[17]. Those soils, which occur in the neighbourhood of creeks, may be better drained and aerated than those further removed from the creek.

With continued accretion a marsh grows upwards until fewer tides cover the whole of it[2]. Marsh development occurs in two essential stages: (1) the *slobland stage*, which is covered by almost all tides and is characterized by *Zostera* spp., *Spartina townsendii* and *Salicornia* spp.; and (2) the *salting stage*, in which the tide is confined to the marsh creeks, except at spring tides, and is characterized by *Halimione portulacoides*, *Agropyron pungens* and *Aster tripolium*. During this latter stage pans develop[4].

In the richer marshes, the mature *Salicornietum strictae* is accompanied by a *Fucus limicola* association, dominated by *F. vesiculosus* ecad *caespitosus*, and in the *Asteretum tripolii* there may be a community composed almost entirely of *Pelvetia canaliculata* ecad *libera* and *Bostrychia scorpioides*[10]. Ecad forms of *Ascophyllum nodosum* occur in suitable marshes, and differ according to the level of the marsh occupied, viz. mud or muddy sand below high water neaps, or on turf, salt marsh and creeks above high water neaps[18]. Such algae are absent from the Solway Firth, although where stones or other suitable substrata for attachment occur in a marsh, *Fucus spiralis* is found at levels associated with *Spartina*, and *Pelvetia canaliculata* with the General Salt Marsh Community: in those shores from which *Salcornia* spp. are absent their place in the zonation is taken by *Spartina*. In shores of bound shingle, or of riven rock, *Spartina* has colonized the *F. spiralis* zone and has partially displaced this species[X].

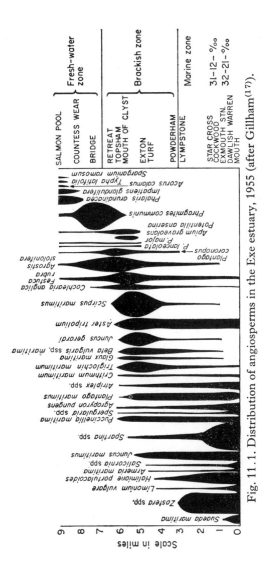

Fig. 11.1. Distribution of angiosperms in the Exe estuary, 1955 (after Gillham[17]).

In the River Exe estuary, the upstream penetration of the shore-dwelling angiosperms is controlled primarily by salinity (Fig. 11.1), although plant competition and mobility of the substratum are important additional factors[17]. In the Randers Fjord, Denmark, the marine and brackish plants penetrate upstream, to a salinity of 2‰, in the order *Zannichellia major* > *Ruppia spiralis* > *Zostera marina* > *Zostera nana*; the fresh water plants penetrate downstream, to a salinity of 18‰, in the order *Potamogeton pectinatus* > *Potamogeton perfoliatus* > *Myriophyllum spicatum*[19].

On the coast of Sweden, where the tides are small (range ≤ 150 cm), and the surface salinity varies between 1 and 30‰, three main successions can be distinguished. They are:

1. The *Scirpus maritimus—Alnus glutinosa* series which is characteristic of soft ground. The *Zostera marina, Ruppia spiralis—Zostera nana* association of the open water gives way to *Eleocharetum parvulae, Scirpetum maritimi, Caricetum paleaceae, Eleocharetum uniglumis, Agrostetosum stoloniferae, Juncetum gerardii* (with sub-associations), *Caricetum veltae, Blysmetum rufi, Caricetum mackenziei*. With communities belonging to *Bromion racemosi* and *Alneta glutinosa* this series reached an *oligohalobe* stage.
2. The *Salicornia strictissima—Rosa* series which is characteristic of sandy, gyffja bottoms. The *Salicornia* of lower levels is succeeded in turn by *Puccinellietum maritima* and *P. distantis, Juncetum geradii, Artemisietum maritimae, Sagineto maritimae—Cochlearietum danicae*, and with *Cynosurion cristati* communities. This series leads to land vegetation.
3. The *Salicornia europea—Empetrum* series which is characteristic of places with a salt accumulation at higher beach levels. Initial communities are extremely salt tolerant and transitional communities correspond to those of the preceding series. This series ends in poor, grazed heath communities[20].

Once a mud flat reaches a suitable level, the transformation into salt marsh may be rapid. On the coast of Anglesey, a high level flat of bare sand was transformed, via the pioneer *Puccinellia maritima* to *Juncus maritimus*, in 10 yr. Under the influence of a large tidal range the marshes of Bridgewater Bay, Somerset, are building up at a rate of 6 ft in 30 yr. In the past 150 yr, the loss of tidal area, in Poole Harbour, due to natural siltation and reclamation has increased to a rate 12 times that of the previous 6000 yr[15]. Estimates for the Wash vary from an addition of 45 000 acres since Roman times[2] to 80 000 acres (32 400 ha) since the seventeenth century (see also p. 149): here it is estimated that embankment could continue at a rate of 15 000 acres (6100 ha) per century[15]. In the Fal estuary, the succession from pioneer grasses to tidal woodland was completed in 100 yr[15]. The development of a marsh may be influenced by years which are good for seeding and propagation, bad winters such as that of 1962–63, the advent of species like *Spartina* or loss due to disease, such as affected *Zostera* on the Atlantic coasts of Europe and North America in the early years of the twentieth century. Reclamation works can lead to the formation of new marshland on the seaward side[21]. Some engineering works, e.g. viaducts and training walls, can lead to rapidly accelerated accretion[22], but others can lead to increased erosion[23]. In areas such as the Solway Firth, where channel shifts occur regularly, dramatic effects can result, even when the channel is a significant distance away from the marsh area, i.e. when these distances are measured in hundreds of yards: a shift towards H.W.M. will cause erosion, and away from the H.W.M. deposition[X].

Deposition of sediment may be defined either as *accretion*, i.e. the depth of sediment deposited in unit time, or as *true accretion*, i.e. the depth of sediment deposited in unit time minus the reduction in thickness due to all settlement factors[24]. A thin layer of distinctive material, spread over the sediment as a marker, can act as a reference point in subsequent sampling by core sampler or knife[2]. Markers used have included natural coloured sand, dyed sand, brick dust, iron filings and coal dust. Rapid accretion can be measured by means of deeply-set bamboo canes which are sufficiently slender not to interfere with the accretion process. Accretion does not proceed at the same continuous rate at all shore levels, and rates at comparable levels differ in different marshes. Mature marshes may have a static seaward limit[24]. In the mature marshes of the Dovey estuary the maximum rate of accretion occurred towards the lower or seaward limit of the marsh[25], whereas in the rapidly growing marsh in Bridgewater Bay, the maximum rate of accretion occurred in its upper or landward limits[24]. In the Solway Firth, using silt from the Irish Sea tagged with [106]Ru as an indicator, it was found that a maximum of accretion occurred around the accretion edge of M.H.W.M.N.T. and diminished to the seawards and landwards of this level. Deposition was high along creek margins also[3]. In this area, the amounts of sedimentary material available for deposition in any one situation are dependent upon internal reworking and the variable supplies of material from the Irish Sea[3].

A most potent force for change in salt marshes emerged in 1870, when a new, vigorous salt marsh grass was discovered on the shores of Southampton Water, Hampshire. Today it is one of the most widespread salt marsh plants in Europe, and has been transplanted to North America and Australasia; in all, this species now covers some 52 400–68 500 acres (21 000–27 000 ha) of the world's salt marshes[7]. It occurs in male-sterile and fertile forms, and since the fertile form has no name is best referred to as *Spartina* only[26]; it has a dwarf form in the Dovey estuary[27]. In

10*

favourable conditions, its growth may be extremely rapid, and it is an active colonizer of mud and sand. It is generally welcomed as an aid to stabilizing mud banks and silt entrapment, although in rare cases it may be a hindrance to navigation, e.g. the dock entrance at Sharpness, Bristol Channel[7, 28, 29]. In some areas it is regarded as being disadvantageous to amenity[7, 26] and in those areas where it replaces the food of wildfowl, viz. *Enteromorpha* and *Zostera*, it can limit the value of coastal nature reserves[30]. Nevertheless, there is considerable concern for those areas where salt marshes are deteriorating due to the "die-back" of *Spartina*[26, 31, 32]. Deliberate transplanting to areas where it became an important primary colonizer facilitated the spread of *Spartina* from its origin in Southampton Water[7, 26, 31, 33]. Although the seed sets undergo annual fluctuations, these sets are an important part of colonization, especially in pioneer zones, but have a decreasing influence upshore[26, 33]. In the Dee estuary, Cheshire, transplanted fragments survived well in the pioneer zone and the lower marsh pans, but not elsewhere[26]. In Poole Harbour, *Spartina* is limited to levels with a maximum depth of submergence of about 1 m[34]; the period of immersion in turbid waters, at neap tides, is apparently critical, as *S. anglica* at least can withstand total immersion in clear sea water for $4\frac{1}{2}$ months[35].

In the Dovey estuary, colonization by *Spartina* occurred rapidly, especially in the more muddy areas. In the absence of seeding, the rate of linear spread on a muddy substratum was about double that on sand. In this estuary, the established *Spartina* was invaded by other marsh species, viz. *Puccinellia maritima, Armeria maritima, Agrostis stolonifera, Spergularia salina, Aster tripolium*, and *Salicornia stricta*[36].

In the grazed sward of the Dovey, the *Spartina* slowly invaded established *Puccinellia, Armeria* and *Festuca* sward, though with uncertain effect[36]; a similar invasion is very marked in the Urr Water/Auchencairn Bay complex of the Solway Firth[23]. *Spartina* can exterminate turfy vegetation through shading effects[37].

All grass salt marshes are affected by grazing. A heavily grazed *Puccinellia* marsh can persist almost unchanged for 50 yr, but once it has reached a sufficiently high level, it can be transformed to a Red Fescue, *Festuca rubra*, or Sea Couch, *Agropyron pungens*, marsh in a decade by simply excluding grazing animals. Such changes are irreversible because the taller grasses accelerate siltation and raise the marsh above its original level[15]. *Spartina* marshes transform to *Phragmites* marsh or *Puccinellia* marsh when ungrazed or grazed, respectively[38]. Salinity is important here, for within the next 10–20 yr only those older *Spartina* marshes, which lie in mesohaline zones ($0 \cdot 3$–$1 \cdot 0\%_0$ chlorinity), are likely to be transformed into *Phragmites* marsh; temperate forms of *Phragmites* are limited to chlorinities $< 1\cdot2\%$ of soil solution[15, 34]. In Bridgewater Bay, the succession from *Spartina* to *Scirpus/Phragmites* began when the marsh was about 22 yr old, and in 12 yr was so rapid that 50% of the *Spartina* was replaced by the invading species along the 2 mile (1·6 km) landward edge of the marsh[39].

About 270 miles (9·8%) of the English and Welsh coast line are bordered by salt marsh. In Great Britain, these marshes cover at least 200 000 acres (80 970 ha) of which *Spartina* dominates some 30 000 acres (12 000 ha)[15]. These areas provide feed for wintering wildfowl, recreation for the wildfowler and bird watcher, and are a valuable agricultural resource. They are used for grazing throughout the world: *Spartina* may be used for silage[7]. In Britain, the value of good agricultural land was £260 per acre in 1968, but even prior to this date, enclosed marshland in southern England was sold for £60 per acre and its value for agricultural purposes could be reckoned at £100 per acre in 1968. In 1957, the value of grazing in the Wash was £15 per acre[15]. Raw materials for brooms, matting, paper and packing materials are derived from *Spartina*: in Florida up to 1260 tons (1280 m tons) of *S. spartinae* and *S. pectinata* may be collected in a good year and used in brush making[7]. The reed, *Phragmites*, may be used in basket making: at one site in southern England the annual rental, of such ground, was £4 per acre. In north-western England salting pastures are rented at 20–25p per acre per annum: here turf-cutting may yield an income of £80 per acre per annum[15].

Marshes, or *wetlands*, both fresh water and marine, are intrinsically a valuable resource, and are therefore worthy of proper management. Frequently they have been considered either as sites for "reclamation", or drainage or as convenient sites for toxic spoil heaps[40]. With rare exceptions reclamation schemes have not given favourable results. Holland, where the pressure for living space is acute, is one of these exceptions, but in others the results have been disastrous. In North America, where these activities have been carried out with typical efficiency, the results are visually horrifying and have an impact upon fish nursery grounds which is causing considerable concern. The magnitude of this problem will be appreciated when it is recalled that the salt marshes along the eastern coast of the United States were probably the most magnificent and extensive in the world, stretching, in the shelter of offshore bars and beaches and in coastal plain estuaries, from Florida to Maine[2].

Dunes, Shingle Beaches and Cliffs

Three other habitats marginal to the coast, viz. dunes, shingle beaches and cliffs, may be found either independently or in association with salt marshes.

Both dunes and shingle beaches are formed by wind, though not by the same direct processes. Dunes formed of blown sand are found in many places around the coast of Britain, although none of these reach a size comparable with those found along the French and Dutch coasts[9].

Initially, the fore dunes are colonized by an open growth of such species as *Agropyron junceiforme*, *Honkenya peploides* and *Glaucium flavum* which act as precursors for the vigorous marram grass, *Ammophila arenaria*, colonization of main dunes. The plant colonization of the main dunes shows a succession leading to the normal land vegetation, thus[9, 41, 42].

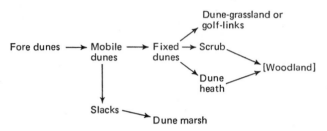

At Newborough Warren, Anglesey, there is a cyclic subseral association in the mobile slack between the dunes[42], thus:

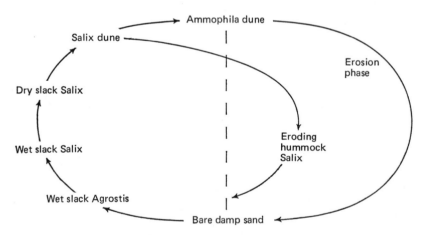

In this dune system, the plant community gradients from the *Festucetum* of the relatively nutrient rich wet "stable" grassland to the impoverished dry *Ammophila* dune followed the environmental gradient[43]. At Tentsmuir, Fife, the dune slack vegetation develops from a *Honkenya peploides–Juncus gerardii* to a *Filipendula ulmaria–Juncus effusus* type in a manner which depends on the water table depth and flooding frequency of the slacks[44]. Shingle beaches are characterized by a rather open vegetation of such species as *Glaucium flavum*, *Crambe maritima*, *Silene maritima*, *Honkenya peploides*, *Matricaria maritima* and *Rumex* spp. Studies upon the density of *Glaucium flavum* indicate that the sand/clay fraction ($\leqslant 2$ mm in diameter) is essential to plant growth here[45]; indeed, at those sites in Wigtownshire where sand is mixed with shingle or the shingle is of a finer nature, the dominant plant may be *Honkenya peploides*, which spreads as a mat over the surface of the shingle[41]. The principal source of water for these plants is precipitation, and the water table is inaccessible to most[45].

Clearly, stability is essential for successful colonization and the levels of a shingle beach which are more exposed to wave action may be too disturbed for plant survival[45]. Further inland, the open vegetation of the primary colonization may be replaced by a fine turf, which covers the shingle, and is similar to that upon dunes; species such as *Eryngium maritimum* (sea holly), gorse, sloe and Burnet Rose may be present[9, 41, X].

Exposed cliffs are colonized typically by such species as *Festuca ovina*, *Agrostis setacea*, *Armeria maritima*, *Plantago* spp., *Sedum anglicum*, *Erodium cicutarium*, *Scilla verna* and *Thymus serpyllum*[9]. In the Southwick area of the Solway Firth, the salt marsh development induced a succession from that typical of exposed cliffs to a woodland and scrub more typical of inland areas[X].

Sand and Mud Flats

The zonation of such shores may be demonstrated by reference to the Solway Firth (Table 11.1)[3, 23, 49, 50]. In the Inner Solway, the fauna may be ascribed to a *Macoma balthica* community. Passing downshore from the soils with a higher silt content, around the M.H.W.M.N.T., the animal succession was *Corophium volutator* and *Scrobicularia plana* to *Scrobicularia plana*/*Macoma balthica* to *Macoma balthica*/*Arenicola marina* to *Arenicola marina*/*Cardium edule* to *Bathyporeia* sp./stunted *Lanice conchilega* in the coarser sediments near to the L.W.M.

In the Outer Solway, the animal zonation from M.H.W.M.S.T. to E.L.W.M.S.T., followed the pattern *Corophium volutator* to *Macoma balthica*/*Arenicola marina* to *Arenicola marina*/*Cardium edule*/*Tellina tenuis* to *Tellina tenuis*/*Owenia fusiformis* and then, at and below E.L.W.M.S.T., to *Tellina fabula*, *Abra alba*, *Nucula sulcata*, *Donax vittatus* depending upon silt content, i.e. wave and current action. In the tributary estuaries, on the north side of this area, the fauna represented a microscosm of the Solway itself, rather than any one part of it. The middle banks had a poor, sparse fauna, but on the fringe banks the fauna tended to be richer with increasing distance from the channels and reached a

Table 11.2.   The principal species of the bottom fauna of the
N.E. Irish Sea (after Jones[51, 52])

(a) The fauna of rough ground:

*Alcyonium digitatum**	*Hyas araneus**
*Metridium senile**	*Chlamys opercularis**
*Lepidonotus squamatus**	*Buccinum undatum**
*Pomatoceros triqueter**	*Flustra foliacea**
Balanus porcatus	*Asterias rubens**
*Pandalus montagui**	*Ophiothrix fragilis**
Pandalina brevirostris	*Ophiura albida**
*Porcellana longicornis**	*Echinus esculentus**
*Eupagurus bernhardus**	*Psammechinus miliaris**

(b) The fine sand ground:

Infauna

Nephthys spp.*	*Tellina fabula**
Shenelais limicola	*Cultellus pellucidus**
Nucula turgida	*Ensis ensis*
*Venus striatula**	*Amphiura filiformis**
*Abra alba**	

Epifauna

*Aphrodite aculeata**	*Natica poliana*
Onuphis conchylega	*Buccinum undatum**
*Phascolion strombi**	*Philine aperta**
Crangon spp.*	*Asterias rubens**
*Eupagurus bernhardus**	*Astropecten irregularis**
*Corystes cassivelaunus**	*Ophiura texturata**
*Macropodia rostrata**	*Psammechinus miliaris**
*Aporrhais pes-pelecani**	*Solea lutea**

(c) Muddy sand ground:
(i) W. Cumberland coast

Infauna

Nephthys spp.*	*Mysella bidentata**
*Owenia fusiformis**	*Venus striatula**
Pectinaria spp.*	*Cultellus pellucidus**
Streblosoma bairdi	*Corbula gibba*
Phascolosoma procerum	*Turritella communis**
Diastylis spp.*	*Cylichna cylindracea*
Ampelisca spp.*	*Amphiura filiformis**
*Nucula turgida**	*Echinocardium cordatum**
Thyasira flexuosa	

Epifauna

*Aphrodite aculeata**	*Aporrhais pes-pelecani**
*Pomatoceros triqueter**	*Mangelia* sp.
*Phascolion strombi**	*Philine quadripartita**
Podoceropsis nitida	*Asterias rubens**
Schistomysis ornata	*Ophiura texturata**
*Eupagurus bernhardus**	*Solea lutea**
Anapagurus laevis	

(ii) West of the Isle of Man:
Dominant or characteristic species

Nephthys incisa	*Corbula gibba*
Abra prismatica	

Neither has been recorded from the Outer Solway Firth.

Influents or abundant species.
Of 21 species, 5 only are known from the Outer Solway Firth, viz.

Owenia fusiformis	*Turritella communis*
Pectinaria spp.	*Amphiura* spp.
Nucula sulcata	

Common species
Of 44 species, 7 only are known from the Outer Solway Firth, viz.

Aphrodite aculeata	*Chlamys opercularis*
Ampelisca spp.	*Ophiura albida*
Phascolion strombi	*Aporrhais pes-pelecani*
Pandalus montagui	

* Recorded from the Outer Solway Firth

maximum of abundance in the *Macoma balthica* zone[3, 23]. These faunas gradually merge with those typical of the deeper waters of the N.E. Irish Sea[51, 52], (Table 11.2).

Some species, e.g. *Nucula sulcata* an inhabitant of muddy sand deposits, ranged from depths exceeding 30 fm upstream to the shallows of Allonby Bay, i.e. almost to the Inner Solway. On the other hand, *Abra alba* which was characteristic of muddy sand deposits in the Outer Solway was replaced by *A. prismatica* at depths exceeding 10 fm; a difference which may be related to the greater salinity tolerance of *A. alba*[52]. A situation similar to that in the Solway Firth and N.E. Irish Sea, obtains on the Scottish east coast[53], and at the estuary-ocean boundary of the Columbia River, N. America[54].

It is of interest that the fauna of mud associations, at 90–110 m in Loch Nevis, a fiordic sea loch on the west coast of Scotland, and at a depth of 140 m, on the Fladen Ground, located 100 miles north east of Aberdeen, have many similarities. Both grounds are characterized by arenaceous Foraminifera and are dominated by Polychaeta. Some 23 species are common to both grounds, viz. *Pholoë minuta, Panthalis oerstedi, Ancistrosyllis groenlandica, Lumbrinereis impatiens, Skardaria fragmentata, Paraonis gracilis, Spiophanes kroyeri, Prionospio cirrifera, Rhodine loveni, Diplocirrus glaucus, Terebellides stroemi, Amphiura chiajei, Nucula tenuis, Cardium minimum, Thyasira ferruginea, T. croulinensis, Cyprina islandica, Abra nitida, Leucon nasica, Eudorella emarginata, Leptostylis villosa, Eriopisa elongata,* and *Calocaris macandreae.*

Although the fauna of Loch Nevis was similar to that of the fiords of N.E. Greenland and of Gullmar Fjord, it most closely resembled that of the Fladen ground, except that the macrobenthos weights for Fladen were 6·43 g/m² compared to 33·10 g/m² for Loch Nevis[55].

Although the foregoing paragraphs represent a general expression of conditions, clearly individual circumstances or events may introduce modifications. For example, *Cardium edule* is found in both the Inner and

Outer Solway, but it penetrates only a limited distance upstream in the inner area, i.e. to Annan, and then only as small individuals which do not grow large enough to form cockle beds capable of commercial exploitation. However, the status of *Cardium edule* is somewhat complex. It has now been assigned to the genus *Cerastoderma*[56] together with the closely related *C. lamarcki*. It seems probable that, in the past, these two species have been confused, and that records from areas with a salinity below 20‰ refer to *C. lamarcki*. In Danish waters, *C. edule* does not occur in waters of a salinity less than 20‰, whereas *C. lamarki* occurs in the salinity range 5–25‰. At salinities *ca.* 10‰, a third species, *C. exiguum*, is present. *C. lamarcki* may live among vegetation, and is not always buried in sand; young specimens may climb by the use of byssus threads in a manner similar to young *Mytilus*[57].

Another lamellibranch inhabitant of the Solway Firth, *Scrobicularia plana*, was badly affected by the severe winter of 1962–63 and until its recent recovery was absent from many areas. Both *Corophium volutator* and *Arenicola marina* are capable of, and do, undertake migrations which intermittently modify the zonation pattern. On the other hand, the tubiculous polychaetes *Owenia fusiformis* and *Lanice conchilega* are replaced by *Branchiomma vesiculosum* in the Wash[58] and Salcombe estuary[59]. The distribution of the infauna of the Solway Firth is typical of a general pattern for the coastal plain estuaries of N.W. Europe[58, 60–75], and many others. A similar zonation, though not necessarily including the same species, has been recorded from the Atlantic and Pacific coasts of North America[77–83], and the estuaries of South Africa[84], Tasmania[85] (Table 11.3) and New Zealand[86].

In British waters, the cockle *Cerastoderma* (*Cardium*) *edule* is the only member of this community to be worked commercially; *Mya arenaria* is toothsome, but not sufficiently abundant to support a fishery. The remaining species have apparently never been exploited commercially, presumably because of their small size. In contrast, a number of North American species are larger than those found living in sand flats of N.W. Europe, and these are exploited on a personal and commercial basis under the collective title of clams. The soft-shell clam, *Mya arenaria*, is taken on both the Atlantic and Pacific coasts. In the estuaries of Oregon alone, nine species, of a relatively large or large size, live intertidally: *Schizothaerus nuttalli*, gaper clam; *Clinocardium nuttalli*, cockle; *Mya arenaria*, soft-shell clam; *Saxidomus giganteus*, butter clam; *Protothaca staminea*, littleneck clam; *Macoma nasuta*, bent-nose clam; *Macoma secta*, sand clam; *Panopea generosa*, geoduck (pronounced gooeyduck); and *Tellina bodegensis*, bodega tellin. Not all these species are taken regularly by Man, but they range from 2 to 6 in. in length, and up to $6\frac{1}{2}$ lb in weight[87].

Table 11.3. The zonation at Pipe Clay Lagoon compared with that on other types of coast in South Tasmania (after Guiler[85])

	Lagoon		Semi-exposed coast	Exposed coast	Sheltered bay	% Exposure
Supra-littoral	*Arthrocnemum*		Lichens	Lichens	? Bare	100
Supra-littoral fringe	*Arthrocnemum* + *Salinator*	Upper shore	*Melaraphe*	*Melaraphe*	*Melaraphe*	70–100
Mid-littoral	*Bembicium* *Bittium*	Supra-zostera	*Bembicium* (local) Barnacle	Absent Barnacle	? local Barnacle + *Austrocochlea*	60–90 27–88
	Anapella *Zostera*	Zostera	*Galeolaria* Patelloid (*Brachyodontes* Patelloid, *Mytilus*	*Catophragmus* Patelloid	*Galeolaria* *Hormosira* and/or *Mytilus*	18–71 7–60
	Marcia			*Corallina*	*Corallina*	0–26
Infra-littoral fringe	*Austrocochlea*	Infra-zostera	*Laurencia* and *Lessonia*	*Sarcophycus*	*Cystophora*	0–2
Infra-littoral	*Pyura*	Lagoon bottom	?	*Macrocystis*	*Cystophora*	0

Benthic Communities or Associations

Under this heading, it is not intended to consider the close relationship between living organisms observed in parasitism, symbiosis and commensalism, but the more loose association of those free-living organisms which share the same habitat. In such situations, it is the ability to co-exist in the same environmental conditions, e.g. sediment of the same grade composition, that determines what component species shall comprise such an association. Nevertheless, it should be appreciated that the component species will influence each other by the trophic relationships which they bear to each other, or by the non-predatory relationships (p. 13), or by the mechanical effect by which one or more may tend to stabilize, or modify, the prevailing environmental conditions. That such community relationships are important was realized at least as long ago as 1881, by Semper[87a]. However, it was work carried out in the shallow coastal waters of Denmark[87b–87i], which led Petersen to characterize nine different associations, or communities, of benthos associated with the sedimentary sea bed. He defined the different faunas associated with differences in sediment quality and water depth as follows:

1. *shallow muds* inhabited by *Macoma communities*, widespread in distribution, and composed principally of *Macoma balthica, Mya arenaria* and *Cerastoderma (Cardium) edule*;

2. *shallow, muddy sands* frequently found in sheltered creeks and characteristically inhabited by *Syndosmya (i.e. Abra) communities* and composed principally of *Abra alba, Abra prismatica, Macoma calcarea* and *Astarte* spp.: *Echinocardium cordatum* may be found with this association;

3. *shallow sandy bottom*, on open coasts, inhabited by *Venus communities*, widespread in distribution, and composed principally of *Venus gallina, Tellina fabula* and *Montacuta ferruginosa*: *Echinocardium cordatum* frequently occurs in this association;

4. *sandy mud*, at intermediate depth, inhabited by *Echinocardium—filiformis communities*, and composed principally of *Echinocardium cordatum, Amphiura filiformis, Turritella communis* and *Nephtys* spp.;

5. *soft mud*, at greater depth than (4) inhabited by *Brissopsis—chiajei communities*, and composed principally of *Brissopsis lyrifera, Amphiura chiajei* and *Abra nitida*;

6. *soft mud*, at greater depths than (5) inhabited by *Brissopsis—sarsi communities*, and composed principally of *Brissopsis lyrifera, Ophiura sarsi* and *Abra nitida*;

7. *mud*, at depth in the Skagerrak, inhabited by *Amphilepsis—Pecten*

communities and composed principally of *Amphilepsis norvegica* and *Chlamys* (*Pecten*) *vitrea*;

8. *firm mud*, deep in the Kattegat, inhabited by *Haploops communities* and composed principally of *Haploops tubicola* and *Chlamys septemradiata*; and

9. *coarse sands*, at depth, inhabited by *Venus communities* which are widespread in occurrence and composed principally of *Venus gallina*, *Spatangus purpureus*, *Echinocardium flavescens* and *Spisula* spp.

This concept of community structure has proved most important to the appreciation of the manner in which biological systems work in the sea. In some senses, it could be regarded as an expression of the wider issues of zonation, since as we have seen there is a general tendency for all faunas, and floras, to exhibit this characteristic. However, it recognizes that an expression other than zonation is very necessary since the distribution of bottom type and its associated biota may be very complex and is frequently not amenable to treatment in such terms. Over the years, this important conclusion by Petersen lead to many studies in many parts of the world concerned with such relationships[51–55, 58, 60–83], and, in all, considerations of community have been found to be important. In the waters off north-east Europe and the Baltic, for example, such studies led to the conclusion that the benthic communities could be broadly categorized as Petersen had done, but that areas of seabed which experienced the same depth, range of temperature and of salinity and had the same type of substratum could be categorized by a particular community even though the individual situations are widely separated[87j]. These communities were categorized as follows:

1. *Communities which live in shallow, brackish waters.* Such communities are composed of individuals which are widely eurythermal (temperature limits 3–16 °C), and euryhaline (salinity limits 7–34 g/kg, and periodically exposed to salinities less than 23 g/kg); the upper limits of the distribution of these communities extend on to the shore.

(a) *Shallow Soft Bottom Community*
   (i) Boreal shallow sand association which inhabits relatively exposed coasts and is composed principally of *Arenicola marina*, *Nephtys caeca*, *Bathyporeia pelagica*, *Tellina tenuis* and *Donax vittatus*.
   (ii) Boreal shallow mud association which inhabits the more sheltered coasts and estuaries. This association is equivalent to the *Macoma* community of Petersen, and is composed principally of *Arenicola marina*, *Corophium volutator*, *Cerastoderma edule*, *Macoma balthica* and *Mya arenaria*.

(b) *Shallow Hard Bottom Community*

(i) Boreal shallow rock association which inhabits the shore and shallow waters and is composed principally of *Balanus balanoides, Patella vulgata, Littorina* spp. *Thais lapillus* and *Mytilus edulis.*

(ii) Boreal shallow vegetation association which inhabits the macrophytes of the shore and shallow waters and is composed principally of *Idotea* spp. *Hyale nilssoni, Hippolyte varians, Littorina littoralis, Lacuna vincta* and *Rissoa* spp.

2. *Communities which live in the offshore region.* Such communities are composed of individuals which are confined to depths below the E.L.W.M.S.T., and which although eurythermal (temperature limits 5–15°C) and euryhaline (salinity limits 23–35·5 g/kg) are less tolerant of such change than Group (1) above.

(a) *Offshore Soft Bottom Community.*

(i) Boreal offshore sand association which inhabits sandy substrata. This association is equivalent to the *Venus* community of Petersen, and is composed principally of *Sthenelais limicola, Nephtys* spp. *Ampelisca brevicornis, Bathyporeia guilliamsoniana, Dosinia lupinus, Venus striatula, Gari fervensis, Abra prismatica, Tellina fabula, Ensis ensis* and *Echinocardium cordatum.*

(ii) Boreal offshore muddy sand association which inhabits muddy sand in the offshore region, and with modification may be found in sheltered conditions and in estuaries. This association is equivalent to the *Echinocardium–filiformis* community of Petersen and is composed principally of *Nephtys incisa, Goniada maculata, Lumbrinereis impatiens, Diplocirrus glaucus, Scalibregma inflatum, Notomastus latericeus, Eumenia crassa, Owenia fusiformis, Pectinaria auricoma, Ampelisca spinipes, A. tenuicornis, Dentalium entalis, Turritella communis, Aporrhais pes-pelecani, Philine quadripartita, Nucula turgida, Cyprina islandica, Cardium ovale, Acanthocardia echinata, Dosinia lupinus, Spisula subtruncata, Abra alba, A. prismatica, A. nitida, Cultellus pellucidus, Corbula gibba, Amphiura filiformis, Ophiura texturata, Echinocardium cordatum, E. flavescens* and *Leptosynapta inhaerens.*

(iii) Boreal offshore mud association which occurs in soft mud. This association is equivalent to the *Brissopsis—chiajei* association of Petersen, and is composed principally of *Leanira tetragona, Glycera rouxi, Nephtys incisa, Lumbriconereis impatiens, Maldane sarsi, Notomastus latericeus, Eudorella emarginata, Calocaris macandreae, Nucula sulcata, N. tenuis, Abra nitida, A. prismatica, Amphiura chiajei* and *Brissopsis lyrifera.*

(b) *Offshore Hard Bottom Community*.
   (i) Boreal offshore gravel association which inhabits coarse deposits of sand, gravel, stones or shell at moderate depths. Epifaunal species are abundant, and the nestling species *Modiolus modiolus* may sometimes achieve great dominance. The community is composed principally of *Polygordius lacteus*, *Glycera lapidum*, *Serpula vermicularis*, *Potamilla* spp., *Crania anomala*, *Balanus porcatus*, *Galathea* spp. *Eupagurus* spp., *Hyas coarctatus*, *Buccinum undatum*, *Nucula hanleyi*, *Glycymeris glycymeris*, *Modiolus modiolus*, *Lima loscombi*, *Venus casina*, *V. fasciata*, *V. ovata*, *Venerupis rhomboides*, *Gari tellinella*, *Spisula elliptica*, *Asterias rubens*, *Ophiothrix fragilis*, *Ophiopholis aculeata*, *Echinus* spp. *Echinocyamus pusillus*, *Echinocardium flavescens*, and *Spatangus purpureus*.

3. *Communities which live in the deep sea*. Such communities are composed of stenothermal (temperature limits 3–7°C) and stenohaline (salinity limits 34–35·5 g/kg) species confined to depths below 70 m.
   (a) *Deep Soft Bottom Community*
   (i) Boreal deep mud association which occurs at greater depth than the offshore mud association and is equivalent to the *Brissopsis–sarsi/Amphilepsis–pecten* communities of Petersen. The principal species which comprise the community are *Glycera alba*, *Spiophanes kroyeri*,| *Chaetozone setosa*, *Maldane sarsi*, *Clymene praetermissa*, *Sternaspis scutata*, *Notomastus latericeus*, *Melinna cristata*, *Proclea graffi*, *Eriopisa elongata*, *Nucula tenuis*, *Nuculana pernula*, *Chlamys vitrea*, *Cardium minimum*, *Thyasira flexuosa*, *Abra nitida*, *Portlandia lucida*, *Ophiura sarsi*, *Amphilepsis norvegica* and *Brissopsis lyrifera*.

It is interesting to note that on the basis of this classification the bound shingle shores of many of the Scottish west coast sea lochs are inhabited by association (i) and (ii) of the shallow hard bottom community. At or about the low water mark, it is replaced by the boreal offshore gravel association characterized by *Eupagurus* spp., *Hyas* sp., *Buccinum undatum*, *Modiolus modiolus*, *Asterias rubens*, *Echinus esculentus* and *Psammechinus miliaris*. At a slightly greater depth, a muddy shingle gravel is inhabited by a boreal offshore muddy sand association in which *Chaetopterus variopedatus* and *Aporrhais pes-pelecani* may be particularly evident[X].

This concept of communities is, as we have seen, undeniably a valuable approach to the study of this environment. However, it does have certain notable deficiencies (the fate, alas, of so many attempts to bring order into the vast complexity of marine systems). For example, if we consider the shallow hard bottom community and its division into shallow rock and shallow vegetation associations, one must question if such a division can really be justified, or justified as it stands. True the species, *Balanus balanoides* and *Mytilus edulis* are dependent solely upon planktonic food sources, and so indirectly is their predator *Thais lapillus*; on the other hand, *Patella vulgata* and the *Littorina* spp. depend upon the macrophytic algae either attached or stranded as a source of food, or on those shores where such algae are absent upon the epilithic chlorophytes and lichens[X]. Moreover, it is these grazing species which control the development of macrophytic vegetation upon such shores[87k]. By the same token, one can discuss the communities of sandy and muddy shores and in the most general terms these are so described for the sand and mud flats of the Solway Firth (p. 291). However, the essential difficulty with all such classifications is that they ignore the relationship which the animal and plant distributions bear to each other and for some curious reason botanists and zoologists alike tend to treat the salt marshes as if they are not a part of the marine system. The salt marsh, as we have seen, is composed of a number of zones or levels of accretion which range from those levels which are covered only by the highest tides down to the mud colonized by the algal precursor species covered by most tides, each zone being characterized by a particular plant species. Each of these plants is accompanied by a number of animals which can inhabit the level occupied by the particular plant. In the Solway Firth, the following associations have been noted:

1. Grass marsh association, found particularly at the lower levels of the marsh, subject to accretion, and where pans are present, the faunistic component includes *Corophium volutator*, *Hydrobia ulvae*, *Macoma balthica* and *Scrobicularia plana*; in those areas where the pans remain full of water during the exposure period *Palaemonetes varians* may be present.

   In those regions where *Salicornia* spp. and *Spartina* are absent and the accreting edge of the grass marsh is preceded by *Puccinellia* and/or an algal precursor species then *Corophium volutator*, *Hydrobia ulvae* and *Macoma balthica* are present.

2. *Salicornia* association (and including the edges of a grass marsh colonized by dwarfed *Salicornia* spp.) in which the faunistic component includes *Corophium volutator*, *Hydrobia ulvae*, *Macoma balthica* and *Scrobicularia plana*.

3. *Spartina* association in which the faunistic component includes *Corophium volutator*, *Hydrobia ulvae*, *Cerastoderma edule*, *Macoma balthica* and *Scrobicularia plana*.

4. *Zostera noltii* association in which the faunistic component includes

Arenicola marina, *Corophium volutator*, *Hydrobia ulvae*, *Cerastoderma edule*, *Macoma balthica* and *Mya arenaria*: *Scrobicularia plana* may be present at the highest levels.

5. *Macoma* community, which typically occupies the substratum at levels below the *Zostera noltii* where it is present, but when associations 2–4 are absent it inhabits the bare substratum up to the edge of the grass marsh.

These salt marshes, and particularly the *Spartina* marsh are highly productive[87i, 87m], and in addition to the more typically marine species support an abundance of invertebrate animals, of which the insects are a particularly important component[16]. In particular, the diptera may be so abundant that they are considered to affect adjacent human populations adversely, and remedial measures are undertaken[87n]. These marshes are also major contributors of plant litter, i.e. detritus, to adjacent shores and inshore waters. Plant litter may be thrown up in strand-lines on rocky and sandy shores and at the highest levels of salt marshes. In a study upon wrack fauna[87o], it was found that the species composition was 34·3% schizophagous, i.e. species feeding upon decaying matter, 52% carnivorous, 9% phytophagous and 4·7% parasitic; in order of abundance the schizophagous animals comprised 94·99%, the carnivores 2·88% and the phytophagous species 1·36% of the population. Of the total of species present the insects were predominant. The communities which inhabit strand-lines are, however, more readily characterized by their crustacean component viz. *Orchestia communities* which are associated with litter on rocky, stony and salt marsh shores, or *Talitrus/Ocypode* communities on shores of sand and sand/shingle mixtures[87p].

The definition of communities, has other limitations of which the difficulty in making quantitative comparisons is perhaps the most obvious. In recent years, attempts to overcome this have been made by the use of diversity indices[87q, 87r]; however, the essential inability of such methods to discriminate between populations composed of small numbers of species of high individual abundance and those of a large numbers of species of low individual abundance, suggests that much further work is required to improve these techniques. Further difficulties in this approach are particularly evident in those situations where it is desired to study the epifauna of which many of the component species are vagile, and marked periodic changes occur both naturally and as a result of fishing.

Tropical and Sub-tropical Beaches

In the sub-tropics and tropics, the low growing salt marsh vegetation, of temperate latitudes, is replaced by swamps of mangrove. These swamps develop upon the sheltered, muddy shores produced by accretion, especially in the mouths of rivers. Such a colonization may occur upon coral and other offshore reefs, also. Elsewhere, the mangroves promote accretion both by trapping sedimentary materials and by their own breakdown: in Florida a shoal was transformed to swamp in 30 yr[46]. When colonizing coral reefs, the mangrove may be accompanied by such angiosperms as *Thalassia testudinum*, turtle grass, *Cymodocea manatorum*, manatee grass, and occasionally, *Spartina alterniflora*.

The pioneer species is generally *Rhizophora mangle*, common mangrove, which forms a forest 30 ft or more in height, and which at a higher shore level is succeeded by *Avicennia*, black mangrove[2, 46]. However, on coral reefs, *Avicennia* may be found to the seawards of *Rhizophora*, and around Sydney, Australia[5, 8], and at Inhaca Island, Moçambique[47], *Avicennia marina* is the primary colonizer. At the latter site, *Rhizophora* can only develop into a tree where the salinity is much below that of the open sea, and is therefore of importance along the margins of creeks and channels. The seedlings of the mangrove *Ceriops* and *Bruguiera* develop only in the shade of other trees; this growth encourages waterlogging which kills off the *Avicennia* whereupon *Ceriops* and *Bruguiera* become dominant[47]. In Florida, *Thalassia testudinum* and *Cymodocea manatorum* are associated with the roots of *Rhizophora mangle*; *Salicornia perennis* and *Spartina alterniflora* are associated with *Avicennia nitida*[46].

In other estuaries around Sydney, Australia, where the mangroves *Avicennia marina* and *Aegiceras corniculatum* form a primary colonization succeeded by salt-marsh plants, the plant zonation is closely associated with elevation, of the substratum, above mean sea level. Very small differences in topography are involved, and the relationship between tide, shore level and salinity is the controlling factor at the site inhabited[8] (Table 11.4).

In these estuaries, salinity and water logging influence germination and growth. *Suaeda*, *Arthrocnemum*, *Triglochin*, and *Juncus* germinated in 80%, 40%, 20% and 20% sea water respectively, but *Casuarina* failed to germinate under saline conditions. Upon transfer from sea water all germinated in fresh water. The plants from the mangrove and *Arthrocnemum* zones required sea water for satisfactory growth. In culture, the tolerance to sea water decreased in the order *Arthrocnemum*, *Suaeda* > *Avicennia*, *Aegiceras* > *Triglochin* > *Sporobolus* > *Juncus* > *Casuarina*. With the possible exception of *Casuarina* the maximum growth of all species occurred in 20% sea water in dilute nutrient solution, although *Juncus* and *Sporobolus* grew equally well in non-saline nutrient solution. All these species germinated when submerged under 4 mm depth of water, but when coverage was 5 cm the germination of *Suaeda* was reduced; all except *Suaeda* and *Casuarina* grew well in waterlogged soils[48].

Table 11.4. The probable influence of physiography on migration of mangrove and salt marsh species landward and seaward (after Clarke and Hannon[8])

Species	Factors which may limit migration seaward	Factors which may limit migration landward
Avicennia marina *Aegiceras corniculatum*	Tidal action limits seedling establishment	High salinity enduring for long periods or wide salinity variation; decreased flooding (lower humidity).
Arthrocnemum australasicum	Number of flooding tides may limit seedling establishment (tidal scour); length of submergence (photosynthesis)	Increased periods of exposure
Juncus maritimus	Number of flooding tides may limit seedling establishment (tidal scour); high salinity especially if of long duration	None known in connection with physiography
Casuarina glauca	Number of flooding tides may limit seedling establishment (tidal scour); high salinity, especially if endured for long periods; high water table caused by flooding	None known in connection with physiography

At Inhaca Island, Moçambique, the temperature of the air and of the mangrove water has a mean annual range between 18·7 and 26·9°C, and between 17 and 37°C respectively. Rain may fall at any season, but, even in summer, it is rarely heavy; the 5 yr annual average is 1100 mm. The salinity of the water in the pools and channels ranges between 12 and 42‰, and is restored to *ca.* 35‰ by each high spring tide. The oxygen concentration varies little from the 5·2 ml/litre characteristic of fully saturated water at this temperature. An oxidized surface layer is present in soil on the outer fringes of the swamp, but it is absent where the mangrove shade is dense. Here the soil is coloured by the hydrate sulphides of iron and upon disturbance may liberate bubbles of free H_2S[47].

As in the coastal plain estuary of temperate latitudes, a zonation of fauna may be demonstrated in the mangrove swamps of tropical and sub-tropical waters[88] (Fig. 11.2). Speciation in mangrove swamps, particularly of the crabs is complex, consequently it is now evident that comparisons formerly thought possible[47] no longer hold true. Professor W. MacNae who informed me of this was kind enough to supply Table 11.5, which compares the mangrove swamps of Malaya and Inhaca, Moçambique. A similar variety of habitat has been recognized elsewhere in Malaya[90], Jamaica[88], Brazil[89] and Venezuala[91, 92]. At the same time the degree of speciation in Malaysian swamps is much greater than in those of Moçambique; a consequence to be expected from the higher temperatures

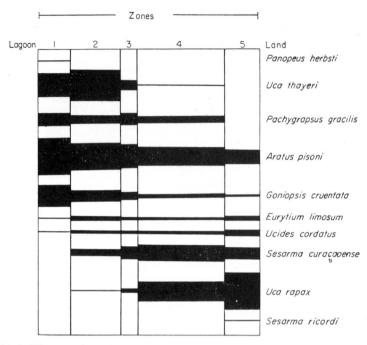

Fig. 11.2. The zonation of crabs in the Port Royal swamp. The width of the horizontal bars indicates the population density. Zones: 1, Fringe; 2, *Rhizophora*; 3, Transitional; 4, *Laguncularia* and *Avicennia*; 5, Back (after Warner[88]).

and greater extent of the Malaysian swamps. The animal association, with the mangrove trees, may be regarded as fortuitous and their distribution is controlled by five major factors, viz. the level of the water table, resistance to desiccation, need for protection from the sun, the degree of consolidation of the substratum and the availability of food in the upper layers of the substratum[47]. Mangrove crabs seek cover, from predation and desiccation, on the swamp floor; here they may dig burrows or inhabit communal "crab-runs" and the adults of one species, *Aratus pisoni*, climb trees. *Pachygrapsus gracilis* and *Sesarma curaçaoense* increase in mean size towards the upper swamp, whereas *Aratus pisoni*, adults and juveniles, are commonest in the lower swamp. Desiccation is thought to prevent the small crabs from penetrating the upper swamp. The aboreal habit, with its attendant problems of desiccation and the need for females to undertake periodic breeding movements to the water's edge, restricts *Aratus pisoni* to the lower swamp. The softer soil of the mid- and upper-swamp permits the extensive development of "crab-runs" by *Pachygrapsus gracilis* and *Sesarma curaçaoense*[88, 93].

In the shallow creeks of Inhaca Island, the goby, *Gobius nebulosus*, is the most common fish, and the pipefish *Syngnathoides biaculeatus* often occurs, but the estuarine fish *Ambassis safgha*, *Atherina afra*, *Therapon jarbua* and *T. theraps* only occur in small numbers. The most abundant shrimp is *Palaemon concinnus*, and prawns, viz. *Metapenaeus monoceras*, *Penaeus canaliculatus*, *P. indicus*, *P. monodon* and *P. semisulcatus*, are plentiful in season. The swimming crabs *Scylla serrata* and *Thalamita crenata* make burrows in the bank from which they pounce upon prey. Unlike temperate estuaries, polychaetes occur rarely in mangrove mud, and indeed only the species *Marphysa simplex* and *Dendronereis arborifera* have been found. Both species have well-developed gills and both are characteristic of oxygen depleted muds in estuaries[47].

At Inhaca Island, the mangrove swamps are replaced by extensive sand flats especially at lower tide levels; these flats are largely meadows of tropical marine angiosperms or "sea-grasses". Over these flats the salinity of the intertidal water ranged from 30·0 to 33·9‰, whereas the drainage pools in the mangrove swamp had a salinity as low as 12‰, but could rise to 40‰ upon drying out. The sand flats were influenced by the seepage of fresh water through dunes to the shore; similar seepages may be noted in the dune-backed shores of Britain.

These flats extend for some 4 km, from high water mark to low water mark, and the following zones have been recognized:

1. *A supra-littoral fringe* composed of two facies:
   (a) that on semi-exposed shores and dominated by the ocypodid or ghost crabs and talitrid amphipods; and
   (b) that of the sheltered shores on the landward edge of the mangrove swamp dominated by the crab *Sesarma* spp., fiddler crabs, *Uca* spp. and talitrid amphipods.
2. *An upper mid-littoral zone* composed of varying facies depending upon exposure: (a) the semi-exposed shores were dominated by the lamellibranch mollusc *Donax faba* and eurydicid isopods on a sandy slope with a community of tubiculous polychaetes on the flat at the base of the slope; and (b) on the sheltered shores this area was replaced by the lower levels of the mangrove swamp with various combinations between.
3. *A lower mid-littoral zone* composed of three facies: (a) a community, which lived on dry sand banks, dominated by the crab *Dotilla fenestrata*: (b) a community, which lived on waterlogged banks of muddy sand, dominated by the quadrangular crab *Macrophthalmus grandidieri*; and (c) a community, which lived in those areas, where water lay upon the surface, and of which the "sea grasses" *Halodule uninervis, H. wrightii* and *Zostera capensis* were the most conspicuous inhabitants.

Table 11.5. Distribution of mangrove fauna in the "high" mangrove of Malaya compared with that of the "low" mangrove at Inhaca, Moçambique (produced by the courtesy of Professor W. MacNae)

	Species in Malaya	Species at Inhaca
Landward fringe	(a) *Nypa* forest	Bare salt flat with upper margin of ferns and trees
	Various large sesarmas including *Sesarma mederi, S. chentongensis, S. singaporensis, S. versicolor* etc. (about 25 species occur but not altogether).	*Sesarma eulimena, S. ortmanni, S. meinerti*
	Thalassia anomala	
	Coenobita cavipes and other species of this genus.	*Coenobita rugosus, C. cavipes*
	Cardisoma carnifex uncommon—ground is too wet.	*Cardisoma carnifex*
	Littorina scabra, L. carnifera, L. melanostoma are all common on leaves of trees, at different heights of the tree at different tidal levels.	*Littorina scabra*
	At least 2 spp. of *Cerithidea: C. quadrata, C. obtusata.*	*Cerithidea decollata*
	Several ellobiids: *Cassidula auris felis, C. mustellina, Ellobium auris judae* and *E. auris midae* and several smaller forms.	
	(b) Bare sands flats at high springs	
	Nil, artificial	
Avicennia fringe (landward)	*Uca annulipes, U. lactea, U. triangularis, U. rosea* (*U. tetragonon* in S. Thailand)	*Uca inversa*
		Uca annulipes
	Thalassia anomala	*Sesarma longipes*

Shady muddy banks and flats

Uca rosea, U. manii
Metaplax elegans, Leipocten sordidulum,
Nannosesarma minutum, Sarmatium crassum,
Sesarma spp. juveniles of other species of
subgenera of *Sesarma. Clistocoeloma* spp., *Helice*
tridens, Ilyoplax spp., *Paracleistostoma* spp.,
Metopograpsus frontalis, Camptandrium spp.,
Cleistostoma spp., *Macrophthalmus* spp., *Thalassia*
anomala, Upogebia spp.
Ilyograpsus paludicola.
Terebralia palustris, T. sulcata, Telescopium
telescopium, T. mauritsi, Assiminea brevicula,
Salinator burmana, Cassidula spp.

Uca urvillei, U. gaimardi
Sesarma guttatum, Sesarma catinatum,
Ilyograpsus paludicola, Macrophthalmus
depressum, Paracleistostoma fossulum, Upogebia
africana, Sesarma smithii, Metopograpsus
thuknar, Terebralia palustris, Cassidula labrella,
Peronia peroni

Assiminea sp.

Channels

Scylla serrata
Periopthalmus chrysospilos, P. ?kalolo.
Periophthalmodon schlosseri
Boleophthalmus boddaerti
Scartelaos viridis
Penaeus spp., *Metapenaeus* spp.

Scylla serrata
Periophthalmus kalolo, P. sobrinus
Penaeus spp. *Metapenaeus monoceros*

Professr MacNae informs me that with only slight variations this scheme will apply to a wide range of Indo-Pacific and sub-tropical shores.

4. *An infra-littoral zone* dominated by: (a) plants of the genus *Cymodocea* when the area was permanently covered by standing water; or by (b) the starfish *Astropecten granulatus* when it was a sand bank.

Some 400 species have been recorded from these sand and mud flats in which, unlike the mangrove swamps, the polychaete worms are well represented. Here *Owenia fusiformis* inhabited an upper mid-littoral exposed to some wave action, and shell debris in the lower mid-littoral. In Britain, this polychaete worm characteristically inhabits coarser sediments found near the L.W.M. in estuarine channels (p. 291). As in the estuaries of N.W. Europe a starfish of the genus *Astropecten* characterized the fauna of clean, sandy sub-littoral sediments. In this area too, many sections of the upper mid-littoral, at which the transition from the flat to the slope above occurred, were marked by a green band of *Convoluta macnaei*; this species, like *C. roscoffensis*, of Brittany, contained symbiotic algae. Like *C. roscoffensis*, this species is positively phototactic and undertakes vertical migrations to the surface of the substratum; similarly, it reacts to vibration by disappearing into the sand[94].

As in areas like the Solway Firth, sediment movements at Inhaca Island have far-reaching consequences. At the time of exceptionally low tides tracts of the plant *Cymodocea* became exposed to the sun and died. The plant recovered rapidly in some areas, but, in others, its loss affected the tidal currents; a deposition of sand ensued and the chance of recovery was lost. The sand bank, which resulted, was eventually colonized by *Astropecten*. Loss of sand, in the mid-littoral zone colonized by *Dotilla*, permitted colonization by *Macrophthalmus* or *Halodule*, whereas areas where a shift of channel produced better drainage became more favourable to *Dotilla*. Elsewhere, an accumulation of sand at high shore levels favoured extended colonization by *Ocypode* spp.; removal of high level sand, down to the underlying sandstone, obliterated *Donax* in that area[94].

Environmental Tolerance

Response to Sedimentary Materials

The distribution of those animal species, which inhabit neritic sediments, is related to the composition of the sediment as well as to the temperature and salinity regime of the individual situation. In those shores which are exposed to wave and current action such that the median size of the sediment particle decreases towards the L.W.M., and bars may be formed, e.g. Brodick Bay, Firth of Clyde and the outermost shores of the Outer Solway

Firth, the populations are characterized by the lamellibranch mollusc *Venus fasciata* and the heart urchin *Echinocardium cordatum*.

In those areas, which are slightly more sheltered from wave action, the relatively well drained silt-free sediments, characterized by the lamellibranch *Tellina tenuis*, are associated with the strong currents produced by an asymmetrical tidal wave. Upstream and upshore of this level, tidal transport mechanisms are dominant and increasingly finer sediments are deposited. The colonization of this habitat is dependent upon an ability to live in association with sediments of increasing fineness, and the chemical and physical consequences which result. It must be emphasized again that, except in the innermost areas of extreme shelter, the sedimentary environment is a dynamic one. True the movements are not as dramatic or as obvious as those of the water, but they exist nevertheless, and survival requires an adaptation to these movements.

In coastal areas and outer estuaries a typical sediment movement, onshore in summer and offshore in winter, occurs (p. 141). Upstream, and with increasing distance from the open sea, such movements as shore building and reworking are influenced by equinoctial and subsequent tides, and by wave action. A shift of channel may induce a long term trend of erosion or of sedimentation. Isolated violent storms may have pronounced long term effects also. In the Solway Firth, changes in sediment distribution, on the south shore, were incipient in the early 1960s; however, in March 1967, a single violent storm triggered off large scale sediment movements, which were still evident in 1970, and which had a pronounced effect upon the infauna, particularly *Macoma balthica* and *Tellina tenuis*[X], (see p. 441). In the period 1958–1962, changes in sediment distribution adjacent to Bradwell Nuclear Power Station, River Blackwater, Essex, induced fluctuations of animal numbers and distribution[95].

In the Solway Firth, the channel of the tributary River Nith[96, X], in common with the rest of the estuary[97, 98, X] undergoes extensive movements. In September 1962, as a result of heavy rain, a meander on the east bank of Carse Gut reworked about 350 m³, i.e. *ca.* 500 tons dry weight, of sand in 48 hr. The rate of erosion was not constant, for about 50% of the total was reworked in the first 48 hr, and in the second 24 hr some 40 tons, i.e. < 10% of the total, of sand was reworked. As a result, approximately 380 living *Mya arenaria*, 5500 living *Macoma balthica*, 11 000 living *Cardium edule* and 2000 living *Retusa obtusa* were reworked from this sediment, and transported downstream. In essence then, the living fauna, dead shells and valves at all depths down to the channel floor were carried into the channel-floor shell beds, by the reworking of the flats which the lateral migration of channels brought about. However, the tidal currents which reworked the sediment concentrated the living fauna, and dead shells and valves, into the surface layers or shell beds[96].

The sediment movements of the Nith estuary probably had little direct effect upon the shells of the molluscs which lived there[96]. However, sediment movements can bring about abrasion of such shells: abrasion is greatest upon a gravelly, sand beach and least in the infra-littoral. The amount of erosion produced is proportional to the surface area per unit weight of shell, but grain size is an overriding factor. Little loss due to abrasion or solution resulted when the shells of *Mercenaria mercenaria*, *Mya arenaria* and *Aequipecten irradians* were exposed on a current-swept sand bottom for 1 yr: in this environment, the valves gained weight due to encrusting organisms[99]. Nevertheless, the effect of extensive sediment movement upon the newly settled spat of bivalve mollusc and other larvae remains unknown, but may be considerable.

The large scale movement of the sedimentary bed load of estuaries is, of course, a gross situation. However, all estuarine waters carry a variable sediment burden (p. 34) which ranges, for example, from the highly turbid waters of the Mississippi to the clear waters of the sea lochs of the west coast of Scotland. Bivalve molluscs, both larval and adult, are affected by such turbidity (p. 514). The adults are able to remove, from the water, large amounts of fine sediment which are deposited as pseudo-faeces[100] and which contribute to the build-up of sediments (p. 131). Where the sedimentary turbidity is excessive, these molluscs cease pumping, close the shell and die. The mortality of bivalve molluscs, e.g. *Crassostrea virginica* and *Mytilus edulis*, exposed for long periods in large quantities of water containing heavy concentrations of turbidity-creating substances may be due, in part, to clogging of the gills, but it may have been advanced by the high rate of respiration induced in the closed animal maintained at the relatively high temperature of 20°C. The effect of high turbidity may depend upon the natural habitat of the animal investigated. For example, the response of oysters from the clear waters of Long Island Sound will differ from those from the very turbid waters of North and South Carolina, even although this environment may adversely affect these molluscs[100]. The Japanese oyster, *Crassostrea gigas*, evinces local racial adaptations to environmental conditions[101].

Although the high concentrations, considered above, apply to estuaries, populations of *Mercenaria mercenaria*, a suspension feeder, grew 24% faster in sand than in an adjacent plot of sandy mud containing much organic matter[102, 103]. On the whole, one may expect high concentrations of silt and clay to affect suspension feeders more adversely than deposit feeders, since the latter are normally accustomed to sort their food from deposits containing a high proportion of non-edible material[103].

Shelter from excessive bottom disturbance is a necessary condition for the survival for many forms. The densest populations usually occur in the more sheltered areas, but in assessing the effect of water or current disturbance both bottom contour and depth must be taken into account[103]. Rich populations occur in bays, not only because of the shelter they provide, but also because of wave divergence and the presence of eddies, which appear to favour the settlement of larvae and the deposition of food in the form of detritus[103, 104].

Within limits, many common benthonic species appear to be tolerant of a fairly wide range of sediment grades, indeed, if they were not, they would not be distributed so widely[103]. Within the constraints imposed upon sediment distribution and redistribution by the hydrography and the degree of stability intrinsic to the individual estuary, the species of fauna and flora present exhibit an individual choice of substratum inhabited.

Polychaetes in the Sedimentary Environment

A wide range of sediment choice is exhibited by the polychaete worms. In Plymouth Sound, the sandy deposits at M.L.W.M. are dominated by populations of *Exogone hebes* and *Aricidea minuta*; at 5 m depth they are joined, as dominants, by *Scoloplos armiger*, *Magelona* spp., *Chaetozone setosa* and *Myriochele* sp. In the deposits with a high silt/clay content the dominant species at M.L.W.M. are *Caulleriella caput-esocis* and *Tharyx* spp.; at 5 m depth they are joined, as dominants, by *Syllidia armata*, *Sphaerosyllis* spp., *Nephtys hombergi* and *Cossura longocirrata*[105].

Polychaetes belonging to the genus *Nephtys* are of widespread occurrence in the sediments of outer estuaries and on open coasts. In the British Isles, the distribution of *Nephtys hombergi*, *N. cirrosa* and *N. caeca*, particularly in the North and East, is related to the distribution of warm water from the Gulf Stream. On all the coasts of Scotland the Arctic *N. caeca* occurs rarely, whereas *N. cirrosa* is abundant, while on the eastern coasts of the North Sea the reverse obtains[106] *N. cirrosa* is characteristic of clean sand beaches (i.e. with 1·8% of the soil particles < 0·125 mm), and *N. hombergi* of muddier beaches (i.e. 5·8% of soil particles smaller than 0·125 mm). Both species occur most abundantly below M.T.L., but are not restricted to any one zone; the upstream limit of penetration of *N. hombergi* is controlled by winter minima of salinity. *N. hombergi* is primarily an inhabitant of the sub-littoral, and occurs intertidally only in the lower reaches of estuaries or on extremely sheltered beaches. *N. cirrosa* occurs rarely in the sub-littoral, and inhabits all but the most exposed beaches[106, 107].

The distribution of *Nephtys californiensis* and *N. caecoides* in the shores of California, contrasts with that of the British species. On this coast, the beaches may be divided into three principal categories, thus:

1. Extremely exposed beaches with substrata too coarse for colonization by any burrowing animals.
2. Exposed beaches colonized by the burrowing anomuran crab *Emerita analoga*. Of these beaches the coarsest may contain *Nephtys californiensis* in very small numbers; beaches with intermediate substrata contain a moderate population (4–8/m²) of *N. californiensis*; fine substrata contain few if any *N. californiensis*. This species reaches an optimum in sediments which contain only a trace of the grade less than 0·15 mm, and have a particle modal size of 0·3–0·6 mm.
3. The muddy substrata of sheltered beaches, in bays and estuaries, have a varied and abundant fauna. They are inhabited by *N. caecoides* which is most abundant in soils which contain *ca.* 2% of grades < 0·1 mm. The intertidal distribution of both species is related to the availability of a suitable substratum, and the ecological isolation of the Californian species, unlike those from Britain, is complete[108].

In Southampton Water, studies upon the polychaete *Cirriformia tentaculata*, an inhabitant of muddy sand, showed that, while it was less abundant at H.W.M. than further downshore, temperature, salinity, oxygen content and water content had little influence upon its distribution, which was correlated primarily with soil grade and organic content. However, wave action, resulting from gales, had an important influence, primarily by the removal of young worms which lived near the surface of affected beaches. The adults which lived at a greater depth were unaffected by the wave induced erosion or by the deposition of a layer of sand from an adjacent beach[109].

The cosmopolitan polychaete, *Fabricia sabella*, lives in coastal waters which contain large amounts of silt upon which it feeds; in these situations the *Fabricia* populations may exceed 1·5 × 10⁶/m². In sheltered areas, *Fabricia* colonizes the silt trapped by the filamentous alga *Rhodochorton* but in less sheltered areas where the silt is removed by wave action, *Fabricia* cannot maintain itself. However, upon exposed headlands, large populations live in silt trapped between seed-mussels and in the plates of dead barnacles[110].

The important role which the nature of the substratum plays in the distribution of estuarine animals has long been recognized, and particularly with reference to the polychaete worms[111–113]. The pelagic larva of *Ophelia* is able, under favourable conditions, to settle and metamorphose immediately upon contact with a substratum suitable for adult life. Failure to make contact with a suitable sediment delays metamorphosis until one more suitable is found; in its absence the larva eventually dies, either without metamorphosing or attempting to do so with greater or lesser success[113]. In a classic study, D. P. Wilson was able to show that attractive and repellent factors existed upon sand, that the attractive factor was either micro-organisms living upon clean sand grains, or the organic products of these organisms, and that only sediment particles in this condition were likely to attract these larvae[114-116].

Around, and for some depth below the L.W.M., in the exposed outer areas and in the more turbulent situations in the middle reaches of temperate estuaries, massive colonies of tubiculous polychaetes of the genus *Sabellaria* create large hummocks, either with rocks as a base or spread over the sand from some relatively secure base, and consolidate large areas of otherwise shifting sand. In the Solway Firth such colonies, known as "the Ross" extend for many miles along the south shore and here occupy a zone colonized by *Lanice conchilega* and *Owenia fusiformis* elsewhere[23]. The capacity of a *Sabellaria* colony for gathering food and building materials is proportional to the surface area of the colony[117]. It is, therefore, a distinct advantage if its pelagic larvae settle in close relationship with established colonies, and with each other. As with barnacle cyprids (p. 366), and other polychaete larvae, the *Sabellaria* larvae have a searching phase prior to metamorphosis upon a suitable substratum. This metamorphosis is not greatly influenced by physical factors, and the strongest stimulus to settlement is received by the larvae when they make contact with the tubes of adults of their own species, or with their remains, or with the tubes of recently settled young. The essential factor, in this settlement, is the cement secreted by adults and young alike and with which the tubes are constructed; direct contact with this substance is essential for it cannot be detected from a distance. This metamorphosis-inducing substance in the cement of sandy tubes and in the secreted primary mucoid tubes is insoluble in water and is unaffected by drying. Larvae which had been aged in clean, glass vessels became progressively less capable of secreting a fully formed mucoid tube when placed with adult-tube sand; eventually they become unable to secrete even a partial tube. Later they may undergo an imperfect metamorphosis or die without metamorphosing. However, before the last irreversible stages are reached, some larvae successfully undertake metamorphosis, upon which, the other larvae, that make contact with them, quickly do likewise. The ability of some larvae to metamorphose when adult colonies are absent is clearly an advantage to the creation of new colonies[117]. The larvae of both *Sabellaria spinulosa* and *S. alveolata* are able to survive initial periods of starvation and, provided the periods are not too long, can complete development apparently unaffected. Both detect, by direct contact, the tubes of their own adults, which can act as releasers inducing settlement. The settling larvae can distinguish the tubes of the adults of their own species, but *S. alveolata* may, under some circumstances, choose the tubes of *S. spinulosa* instead of those of its own species[118, 119]; the larvae of *S. alveolata* are little influenced by the mineralogical character of the tube walls[118].

The sediments of the middle and upper reaches of an estuary may be colonized by a number of species, e.g. *Nereis diversicolor*, *Nephtys* spp., *Pygospio elegans*, *Scoloplos armiger* and *Cirratulus cirratus*, with a mode of larval life and settlement which shows some gradation between that of *Ophelia* and that of *Arenicola* (see below), may indeed be more highly evolved than that of the latter species. *Nephtys* has a planktonic larvae[119a]; that of *Pygospio* is planktonic, but it retains a photopositive response while it remains in water of a high salinity and thus ensures its settlement in water of a reduced salinity where it becomes negatively phototropic immediately prior to settlement[119b]. The means by which gametes are liberated and fertilization takes place is uncertain for *Nereis diversicolor*, however, the larvae are essentially bottom dwellers although some have been taken from the plankton[119c]. The larvae of *Cirratulus cirratus* are associated with the gills of the adult and no planktonic larva is known[119c]. *Scoloplos armiger* lays its eggs in cocoons from which the larvae hatch in 10–24 days depending upon temperature[119d]; upon escape from the cocoon the larvae, which are well developed, immediately burrow into the substratum, a response which results in part from a negatively phototactic response[119e]: the larvae may have a pelagic phase[119f].

The lug-worm, *Arenicola*, so abundant in the sandy, middle levels of estuarine shores and in areas like the Wadden-sea, does not have a pelagic larvae. No larvae occur in the plankton, but fertilized eggs and larvae are apparently carried to higher shore levels by the tide, i.e. as a demersal phase which is transported like sediment[120]. A concentration of juvenile *Arenicola*, at higher shore levels, is observed in the Solway Firth also[X]. The juvenile worms leave their burrows and by active swimming undertake the colonization of the lower shore[120]. The adults too swim actively[75, 121] and the rapid recolonization of shores denuded by bait diggers is commonly acknowledged.

A healthy adult lugworm dug out of, and replaced on, the sand at once begins to burrow; a process normally completed in a few minutes[122]. In a vigorously burrowing worm two groups of events alternate. In the first, waves of swelling run headwards along the trunk, at a rate of one every 5 or 6 sec. These burrowing waves are accompanied by slight internal pressure changes. As each wave approaches the front end of the worm, a new event occurs. A sudden relaxation in the musculature of the first few segments is accompanied by a strong contraction in the rest of the trunk. In consequence, a peak of extremely high pressure is produced, and this thrusts the distended anterior end sideways against the surrounding sand. Pressure peaks in the coelom of 60–80 cm occur commonly, and values of 110 cm have been recorded. The pressure drops after a second or two and the anterior end narrows once more; a further violent lateral movement is repeated when the next of the travelling waves reaches the anterior end of the trunk a few seconds later. In the interval between two such heaves, the head end lengthens and narrows and performs movements of a different kind. By extrusion, the proboscis rasps and thrusts its way forwards into the sand. Concurrently, the anterior segments obtain purchase upon the sand by a process known as "flanging". Flanging results when certain muscular belts, derived from the parapodia encircling the first few segments, become inflated and elevated. Normally, these belts lie flush with the general body wall, but they can be raised suddenly as prominent flanges.

These two phases generally occur in a regular alternation in a vigorously burrowing *Arenicola*, but modifications may be evinced. For example, once the worm has burrowed deep into the firmer, deeper substrata, the number of lateral heaves may be reduced to one for every second or third of the trunk waves; consequently, the rasping and flanging between heaves is increased. An *Arenicola* which is commencing to burrow into soft surface layers makes no sideways thrusts[122–124]. The forces which result from the muscular contractions described are insufficient to allow straightforward burrowing by *Arenicola*, and its ability to burrow depends upon the thixotropic properties of the sand in which the animal lives[125, 126]. In the Zwartkops estuary, South Africa, the burrowing crabs *Cleistostoma edwardsi* and *C. algöense* are influenced by soil thixotropy[12]. At Whitstable, Kent, the main factor which controls the local distribution of *Arenicola* is the depth of thixotropic sand which overlays the compact clay substratum[121].

For much of its life, *Arenicola* inhabits a cylindrical, U-shaped burrow with walls which are firmed by impregnation with its own secretions and by pressure from its body. It may creep to and fro within this tube, but it nearly always faces the same way. It irrigates the tube, almost continuously, with a stream of sea water which is propelled by waves of muscular contraction. During irrigation, which is nearly always towards the head, the side and ventral surfaces of the worm are pressed firmly against the tube wall and the swellings which become apparent occlude the space (containing the gills) between the dorsal surface of the worm and the tube. These waves progress along the body like a piston. This activity is rhythmic. If the experimental conditions are arranged so that the *Arenicola* can circulate water, in its burrow, without obtaining a supply of oxygen, then, although the irrigation cycles persist with their timing unchanged, their vigour is substantially reduced. When aerated water is admitted once more, the rate of irrigation is increased significantly[122, 127].

At the time of high tide, the lugworm obtains an adequate supply of oxygen by circulating water in its burrow. However, when the tide recedes, the surface may be dry, and stagnant water stand in part of its burrow, circulation is not then possible and a serious lack of oxygen may result. At such times, the worm creeps backwards to the surface of the water, where air drawn between its dorsal surface and the tube comes into contact with the gills, i.e. aerial respiration ensues; this process is an adaptation of the irrigation cycle, it is rhythmic and is sufficient to support the worm for a protracted period[127].

At those times when the surface water is unacceptable to *Arenicola*, viz. surface pools are heated to 26·2°C in summer, or subjected to hard frost or heavy rain, it desists from irrigating its burrow[127]. At such times, and in situations where the black sulphide containing layer is well developed, the animal must be protected against the toxic effects of sulphide. The blood of *Arenicola* catalyses the oxidation of sulphide; an activity which is associated with the respiratory pigments. Upon treatment with sulphide, sulphaemoglobin is not produced from *Arenicola* haemoglobin, but its oxyhaemoglobin reacts to produce a brown pigment. This brown pigment may be useless as a respiratory pigment, but it is a very active catalyst of sulphide oxidation; it will therefore tend to keep the blood sulphide concentrations at a low level. Evidently, an excess of sulphide will lead to the formation of more brown pigment from the oxyhaemoglobin: in this sense, the oxidation of sulphide in the blood will be autocatalytic. Observation shows that *Arenicola* is remarkably tolerant of sulphide, even in conditions when it cannot protect itself by irrigation of the burrow[128].

Arenicola may be remarkably tolerant of sulphide, but curiously, since it is present in all soils, it has a rather low tolerance of Fe_2O_3. It is repelled by, and will not burrow in, soil containing 2·0% Fe_2O_3 in the amorphous state and a solution of 0·021% Fe_2O_3 in suspension in sea water forms an envelope with its body mucus, prevents contact with the surrounding aerated water and thus kills the *Arenicola*[128a]; this is an interesting result since marine soils may have a total iron content greatly in excess of 2·0%[128b]. Suspensions of $CaCO_3$, $MgCO_3$, Stourbridge clay, kaolin and kieselguhr have the same effect as suspensions of Fe_2O_3[128a].

During periods when the environment is anaerobic *Arenicola* draws upon the glycogen in the body wall, but conserves its energy by becoming quiescent. This metabolism of glycogen yields products other than lactic acid, therefore, although an oxygen debt is avoided, resources may be wasted if normal activity is continued. Indeed, even if true glycolysis existed in the lugworm, it might be that the less efficient respiratory mechanism (i.e. compared with the vertebrate) would be unable to clear the debt within a reasonable period. In contrast, *Owenia fusiformis*, when exposed to anaerobic conditions, does not draw upon the large reserve of glycogen held by the coelomic cells. Nevertheless, this species can survive without oxygen for long periods, apparently by becoming quiescent. Neither species mobilizes their reserves of oil at such times[129]. The onuphid polychaetes *Hyalinoecia tubicola* and *Diopatra cuprea* inhabit the infra-littoral and littoral respectively. Both species ventilate their gills by means of an irrigation current induced by muscular contractions. As in *Arenicola*[122, 127], these contractions are controlled by an endogenous pacemaker upon which environmental oxygen deficiency, or CO_2 excess, has little influence. The intertidal *Diopatra cuprea* may endure oxygen depletion and perform anaerobic respiration about the time of low water. At such times the cessation of activity reduces its oxygen consumption to 31–37% of the active value. *Hyalinoecia tubicola* can withstand many hours of extremely low oxygen concentration and later respond to full aeration with greatly increased irrigation[130].

Comparison of the two polychaetes, *Chaetopterus variopedatus* and *Nereis diversicolor*, revealed an interesting difference in response to adverse conditions. Like *Arenicola marina*, *N. diversicolor* lives intertidally where the extremes of surface water temperature and salinity may be sufficient to injure the worm[127]. In contrast, *Chaetopterus* may be found occasionally at the L.W.M., but is found more typically at greater depths where such adverse conditions are unlikely to, or only rarely, occur. Both species have a regular irrigation pattern when maintained in well aerated water. However, if the supply of well aerated water is shut off, *Chaetopterus* responds by an increased rate of irrigation, i.e. a response to irritating conditions rather than one which is specifically respiratory: *Nereis* responds first by long spells of irrigation which alternate with equally long or longer periods of rest, and then by a gradual decrease and finally a total cessation of irrigation. On return to more normal conditions, *Chaetopterus* responds either by a great decrease or total cessation of irrigation, but *Nereis* responds by a prompt return to irrigation, so that initially the amounts of water pumped are much greater than normal[131]. The large polychaete *Nereis virens*, which lives near the L.W.M., responds to a decreased oxygen concentration by an increased duration of irrigation; eventually rhythmicity is replaced by virtually continuous activity[130, 132].

Arenicola marina is a cold-water species which is widely distributed in sedimentary shores. In European waters, two "tail-less' lugworms, viz. *Arenicola ecaudata* and *A. branchialis*, live in gravel and beneath stones. The former is known to occur from Iceland to Northern Spain, and the latter from the West of Scotland to Morocco, the Mediterranean and the Black Sea. Neither occurs on the western side of the Atlantic, and both are much more local in distribution than *A. marina*[133].

Crustacea in the Sedimentary Environment

The amphipod *Corophium volutator* is an important constituent of the shore zonation in the inner areas of estuaries, and is normally found associated with the finer sediments at the head of the shore. It has a salinity preference in the range 10–30‰, a tolerance from 2–50‰, but below 10‰ the animals are under increasing osmotic stress. In the field, they are scarce below 5‰, and absent below 2‰. Above 20‰ *C. volutator* is almost isosmotic with the medium, but below 20‰ the blood is maintained hyperosmotic to the medium[134, 135]. Over the salinity range, 1–35‰, *C. volutator* maintains Na^+, K^+, Ca^{2+} and Cl^- at a higher concentration and Mg^{2+} ions at a lower concentration than the medium[134]. These results suggest that *Corophium* migrates from low salinities[134, 135].

C. volutator normally inhabits the finer sediments, whereas *C. arenarium* normally inhabits the more sandy situations. *C. volutator* has been shown, by experiment, to prefer fine to coarse sand, whereas *C. arenarium* is indifferent to particle size. The discrimination of particle size by *C. volutator* may be due to one of three factors: (a) an appreciation of particle size; (2) an appreciation of the increased content of organic material and micro-organisms on fine sand grains and (3) the difficulty of constructing permanent burrows in sand of coarser grade[136].

Sedimentary particles are coated with populations of micro-organisms which may be attractive to the larvae of the polychaete *Ophelia bicornis* (p. 312). In the case of *Corophium volutator* and *C. arenarium*, the factor which attracts these species may be removed by boiling the sediment with acid which destroys the film; the effect is the same even when all traces of the acid are removed. In contrast, treatment with salts which cause the bacteria to detach from the sediment particles does not necessarily render the particles unattractive to *Corophium*. Possibly some of the materials secreted by the micro-organisms remain on the particles after the treatment, and are at least partly responsible for rendering the sand attractive to *Corophium*[137].

In the River Ythan estuary, Scotland, *C. volutator* became abundant when the salinity exceeded 5‰, but below this the low salinity overcame the effect of the substratum[138].

Within the limits of survival, i.e. 2–50‰ salinity, moulting occurs in the salinity range 2·6–46‰, but is most frequent between 5 and 20‰. Maximum growth occurs at a salinity of 15·4‰, but is only slightly less over the range 4·4–30·6‰, and it is reduced progressively below a salinity of 4‰[139]. Temperature influences *C. volutator* to the extent that it burrows more deeply into the mud when the temperature falls below 4°C (compare meiobenthos p. 234); the onset of breeding is coincident with a rise in temperature to 7°C[138]. *C. volutator* may swim actively, in the overlying waters, around the time of high water; this rhythmic swimming activity reaches a maximum in the early ebb and is independent of temperature. This rhythm is probably entrained by tidal changes in hydrostatic pressure, rather than by variations in the intensity of wave action or of water currents[140]. At Whitstable, Kent, however, *C. volutator* emerged from its burrows, during the period of exposure, to wander freely on the surface of the sand beneath water held by the sand ripple system[141].

Like the polchaete worms, crustacea which inhabit the intertidal sedi-

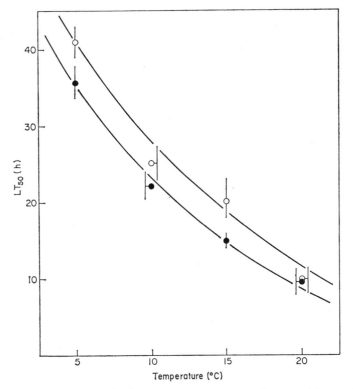

Fig. 11.3. The relationship between temperature and anaerobic survival in *Corophium volutator*. Menai Strait population, ○, and *C. arenarium* Red Wharf Bay population, ●. Vertical lines indicate 5% fiducial limits about LT_{50} values derived from probit regression analysis (after Gamble[142]).

ments are at some risk due to anaerobosis during the period of tidal exposure. *Corophium volutator* and *C. arenarium* have a similar response to anaerobic conditions (Fig. 11.3). The effects increase with increasing temperature, and juveniles are more susceptible than adults. Both the *Corophium* spp. and *Tanais chevreuxi*, which has a semi-sessile, crevice

II*

dwelling existence, are rapidly paralysed under anaerobic conditions. This may result from either a build up of toxic anaerobic products within the animal, or a quiescence comparable with that of the polychaetes *Owenia fusiformis* and *Arenicola marina* (p. 316). After exposure, the recovery time is proportional to the duration of the anaerobosis. *T. chevreuxi* is more resistant to anaerobosis than *Carcinus maenas* which is more resistant than the *Corophium* spp., but all three are more resistant than *Gammarus oceanicus*, *Idotea baltica* and *Crangon crangon* which are all highly mobile. Some animals, e.g. *Homarus vulgaris*, may acclimatize to low oxygen tensions[142].

The production of faeces by the isopod *Dynamene bidentata* fed upon *Fucus serratus*, in the laboratory, depends upon the size of the animals and apparently upon the degree to which aerobic conditions are maintained in the experimental vessels. Specimens kept in unchanged water will feed more actively at day than at night. However, if the water is renewed continuously or is aerated, the night feeding rises to about the level of that during the day. Such results can be explained by the assumption that a lack of oxygen, which results at night during the cessation of photosynthesis by the algae, inhibits the feeding activity of the isopods[143].

In contrast to *Corophium volutator*, the amphipod *Pectenogammarus planicrurus* is confined to a few scattered shingle beaches around the British coast. Its ability to live in these situations is related primarily to the size of the interstitial voids through which it passes without injury; at soil grades < 3·35 mm, this ability is much reduced. Although exposure, currents and turbidity influence the composition of the shore, neither they nor salinity have a direct effect upon this interesting species[144].

Although the estuarine distribution of species in the amphipod genus *Bathyporeia* is related to the salinity regime of the habitat, the distribution of this and other amphipod genera is related to the sediment distribution. Thus, the intertidal sand habitat may be classified according to the predominant soil grade and its inhabitants: "*Bathyporeia*" sand, 90–200 μm; "*Ampelisca brevicornis*" sand, 70–100 and 200–400 μm; and "*Urothoe brevicornis*" sand, 200–400 μm[145]. In the shallow waters of the Irish Sea and its tributary estuaries, the sediment is largely derived from the action of glaciers in the Pleistocene Ice Age (p. 154) and is composed principally of the fine sand grades (i.e. from 62·5 to 250 μm). Here the species of *Bathyporeia* predominate, a result consistent with the classification above[146]; the dominance ranking of the individual species is given by Table 11.6.

In the Dovey estuary, Wales, *Bathyporeia pilosa* occupies distinct zones at high and low shore levels; the high level population does not occupy such a high level in the summer as in the winter months. *B. pelagica*, which occurs

Table 11.6. Over-all dominance ranking
of the amphipoda in the shallow-water sand
habitat of the Irish Sea (after Fincham[146])

Bathyporeia elegans	42·3%
Bathyporeia nana	25·9%
Bathyporeia guilliamsoniana	14·6%
Perioculodes longimanus	5·5%
Ampelisca brevicornis	5·0%
Pontocrates arenarius	3·2%

both in the Dovey estuary, and in the adjacent open coast, migrates away from the estuary in the spring and returns to it in the early autumn[147].

Clearly, although the sand of the beach may be undergoing considerable movement (p. 151) animals like the *Bathyporeia* spp., which are able to leave it readily, can avoid the worst effects of such movements. Indeed, at Blyth, Northumberland, many of the small Crustacea, viz. the amphipods, *Bathyporeia elegans*, *Nototropis falcatus*, *N. swammerdami*, *Perioculodes longimanus*, *Hippomedon denticulatus*, and the cumaceans *Diastylis bradyi*, *Diastylis* spp., and *Pseudocuma cercaria*, leave the sediment at night and swim actively in the plankton. These benthic forms were caught more readily in April and July. On this part of the coast, considerable amounts of sedimentary material are removed from the shore by winter gales, and are returned to it later. The effect of these movements is uncertain, but may result in the offshore migration of these species during the winter[148].

The isopod *Eurydice pulchra* occurs in the outer areas of estuaries where it may be seen swimming swiftly in the rising tide. Although stranded in marsh pools, it may prefer open water and will burrow into sand[16]. It may spend most of its time buried in sand, and emerge to lead a pelagic life at night[149]. It also occurs in the sandy shores of open coasts[12, 72, 74, 75, 150–162]. It penetrates estuaries to the 27–28‰ isohaline in the summer and autumn, and to the 22–23‰ isohaline in winter[71, 159], and like the vagile epifauna its penetration may be greatest in summer[74]. In the Bristol Channel and Severn estuary, its distribution coincides with *Eurydice affinis* in the middle and outer areas, but *E. pulchra* does penetrate further upstream. Curiously, *E. pulchra* apparently prefers a coarser sediment than *E. affinis*[159, 161], but in the Burry estuary, it is found in association with *Corophium volutator*[140, 159]. In the Dee estuary, Cheshire, it occurred in association with *Corophium volutator, Haustorius arenarius, Bathyporeia pelagica* and *Macoma balthica* at Heswall[74], and with *Platynereis dumerilii, Haustorius arenarius* and *Macoma balthica* at Hilbre Island[75]. In the Exe estuary, it occurred on transects with many species which can be considered estuarine inhabitants, e.g. *Arenicola marina* and *Macoma balthica*[72]. This species has been defined as being more typically an inhabitant of "marine shores"[162], but this seems to be inadequate in the light of the foregoing sentences, and does not recognize its ability to colonize the wide variety of situations which are estuarine and to cohabit with species adapted to live in this environment. In the Dovey estuary, low numbers of *E. pulchra* are present in the shore during winter and summer, and conversely high numbers are found in spring and autumn. Moreover there is a significant difference in the pattern of the intertidal distribution at spring and neap tides respectively. At spring tides, during the spring and autumn, these animals are abundant at all levels from M.H.W.S.T. to below M.T.L., with about 70% above the M.H.W.N.T. At neap tides, these animals fall to below the M.H.W.N.T., and only *ca*. 3% are found above this level. The pattern above M.H.W.N.T. is maintained for about 3 days, on both day and night tides, at the time of the springs, but, with the fall in tidal height, the numbers above this level fall to that observed at the neap tides[160] (Fig. 11.4).

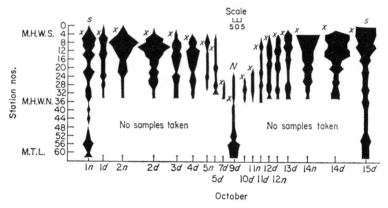

Fig. 11.4. *E. pulchra*. Variation in the distribution of animals along the transect with change in tidal height, 1–15 October 1966. *n*, Night-time tide; *d*, day-time tide; *x*, observed tidal height; *S*, spring tide; *N*, neap tide (after Fish[160]).

In Firemore Bay, Loch Ewe, *Eurydice* is tychopelagic and is commonly found only in impermanent sand[163]. In the outer Solway Firth its presence is related, at least to some degree, to those shores in which sand movements occur frequently: a characteristic which is shared by *Haustorius arenarius*[X].

The mysid crustaceans may be considered to be planktonic, but in estuaries, where great numbers are often associated with the seabed, it is convenient to consider them in the present context.

Schistomysis spiritus is the commonest species of mysid crustacean which occurs in sandy bays around Britain; indeed, at Kames Bay, Millport and

Gaineamh Smuagh, a bay in Loch Ewe, Wester Ross, it dominates the pelagic fauna. It occurs in shoals which make considerable vertical and horizontal movements. The relative degree of shoaling of the total population undergoes diurnal changes which are related to tidal movement. The largest catches are obtained at the tides' edge, at the time of low water; at the time of high tide, the population is dispersed between low and high-water marks, in summer and in calm conditions. This dispersed population is aggregated by the tide as it ebbs down the beach (a movement which may be compared with that of plaice and flounder in the Solway Firth, p. 402). Severe wave action causes the main population of *S. spiritus* to move from the water's edge, out to the seawards of the breakers. At night, *S. spiritus* tends to move away from the bottom, and although the population becomes distributed throughout the total depth available, shoals may form when it is out of contact with the bottom. However, this behaviour is influenced by the state of maturity of the individuals, the intensity of daylight, the intensity of moonlight and the intensity of wave action[164]. In the Clyde Sea area, *Praunus inermis* is one of the most common species; it is occasionally found over sand at the L.W.M., but is more usually found among *Laminaria* spp. and other algae below L.W.M.S.T.[165].

Species of *Leptomysis*, viz. *L. gracilis* and *L. lingvura* occur commonly in the western Scottish sea lochs. The former is found at depths of 50–200 m and may extend beyond this, but *L. lingvura* is a common inhabitant below the L.W.M. of sandy bays, and is only occasionally found at 50 m depth; *L. gracilis* is much more abundant than *L. lingvura*, which occurs only in small numbers[166].

Like other animals, individual species of mysid differ in their ability to penetrate regions of low salinity; the ability of European species to penetrate coastal estuaries is broadly in the order *Neomysis integer*, *Praunus flexuosus*, *Mesopodopsis slabberi*, *Gastrosaccus spinifer* and *Leptomysis gracilis*. *Neomysis integer*, which may be found with such typical inhabitants of fresh water as frogs, is normally most abundant in the waters of inner estuaries and is rarely found in the open sea: most typically it is found, in large shoals, at the tide's edge as it rises over sand banks with the flood. It is seen as a dense, overt band, sometimes a metre wide, in which the mysids are so crowded that it is difficult to visualize the advantage gained by such behaviour, unless some food material floats off the substratum when it is covered by the rising tide[16, 49]. *Praunus flexuosus* extends into water with a salinity as low as $0.3\%_{00}$[62], and *Leptomysis gracilis* lives in the mouths of estuaries, usually in water ranging from 12 to 530 m depth[16, 166]. Characteristically, *Neomysis integer* occurs, in shoals, in the inner Solway Firth, on the other hand *Gastrosaccus spinifer* was normally taken in sediment samples from the ebb-flood channel system of the Outer Firth[49].

In the Brisbane River estuary, Australia, *Gastrosaccus dakini* penetrates further upstream than *Rhopalophthalmus brisbanensis*. Most of the juveniles of *G. dakini* and *R. brisbanensis* are found in the salinity range 2–16‰ and 7–17‰ respectively. Both have a seasonal movement upstream which is greatest in October and is then succeeded by a downstream migration[167].

The species *Mysidopsis gibbosa* is caught, in large numbers, in the sandy bays of the Scottish west coast[168]. Three species of *Mysidopsis* are known from the Atlantic and Gulf of Mexico coasts of the U.S.A. *M. bigelowi* occurs from New England to Louisiana, *M. furca* is known only from the type locality in S. Carolina and *M. almyra* lives in the brackish waters of Florida and Louisiana where it is found in association with the copepod *Acartia tonsa*[169].

In general, the mysids are sensitive to a lowered oxygen tension. In the Tyne area, especially in summer, where pollution produced a depleted oxygen content, *Neomysis integer* occurred only rarely[170]. Similarly, *Mysis relicta*, an inhabitant of the fresh water lakes of Europe and North America and the reduced salinity of the Baltic, rarely occurs in fresh water with an oxygen tension of less than 4 p.p.m., although, in the Baltic, it can survive at 1·6 p.p.m. O_2[16].

The penaeid shrimps of the South Atlantic and Gulf coasts of the U.S.A. form the basis of one of the most valuable fisheries of that country. The fishery is based entirely upon three species which show a range of environmental adaptation; these species are *Penaeus setiferus*, common white shrimp, *P. aztecus*, brown shrimp, and *P. duorarum*; of these, the first was originally the more important, but the last two are now of increasing importance.

P. setiferus, unlike the other species, is diurnal in habit, and it lives in shallow, coastal waters upon terrigenous, silt bottoms. The adults spawn at sea over the entire range of abundance, and normally within the 10–12 fm contour. The juveniles have an optimum salinity of < 10‰, and are better adapted to the Louisiana than to the Texas nursery grounds. *P. aztecus* is found on grounds similar to those inhabited by *P. setiferus*, but normally lives outside the 10 fm contour. The optimum salinity for juvenile development is probably > 10‰, and the Texas nursery grounds are more suitable than those of Louisiana. *P. duorarum* usually occurs on bottoms of coral sand and shell; the adults normally spawn in water exceeding 28 fm depth, and the optimal salinity for the nursery areas is > 20‰. Nevertheless, the salinity of the nursery grounds is highly variable, e.g. those of North Carolina range from "nearly fresh to 37‰, and development of larvae is good even at salinities very different from the optimum". In general, the juveniles, of all three species, are found furthest inshore, moving to deeper more saline water with increasing age, and as sexual maturity approaches; growth is rapid and maturity may be reached in six months[171].

Molluscs in the Sedimentary Environment

Individual species of estuarine molluscs differ in their ability to penetrate an estuary, and also live in preferred zones upon the shore. Nevertheless, the range of distribution of the principal shore inhabitants may, and indeed does, overlap. Here a fundamental difference between sedimentary and rocky shores, or beaches, becomes apparent. For the rocky shore is essentially a two dimensional environment, whereas the sedimentary shore is three dimensional; here, although the inhabitants all require access to the surface, the bulk of their bodies lie at different depths, rather like the inhabitants of a block of flats. Although this concept may be applied to the population of a sedimentary beach, as a whole[72], the principal is best demonstrated by reference to the molluscs. In the inshore mud flats, at the mouth of the River Nith, a tributary of the Solway Firth, *Hydrobia ulvae*, in the surface layers, may reach a population of 15 000 per m² in the summer, falling to 6000–7000 per m² in spring, probably as a result of removal by winter storms. *Macoma balthica* reaches a density exceeding 100 per m², at 2–5 cm depth in the more silty sand, but the density declines rapidly as the grain size of the sediment increases. *Scrobicularia plana* lives in mud and silt at a depth of 8–12 cm; it frequently occurs below *Macoma*, but at much lower population densities. *Mya arenaria*, with a preference for silt and sand, lives at depths of 20–30 cm. Therefore, it is possible, and it indeed happens here, that the surface layers are occupied by *Hydrobia ulvae* and lamellibranch spat, while beneath live *Macoma balthica*, *Scrobicularia plana* and *Mya arenaria* at successively deeper levels[172].

The thin shelled lamellibranch, *Tellina tenuis* is characteristic of the more silt free sediments of outer estuarine areas (p. 291). In the Exe estuary, it lives at a depth ranging between 5 and 10 cm coming nearer to the surface at the time of high tide. Because this animal can live over a range of depth within the substratum, it would be possible for individuals to live in very close proximity without bodily interference. Nevertheless, in the Exe, moderately dense populations were uniformly distributed and exhibited a significant degree of "over-dispersion". A distribution, presumably brought about by the activities of the inhalent siphon delineating a "territory". Moderately dense populations, enlarged artificially, at first, continued to show this dispersion, but a still further increase led to its break down. Very dense, but patchy, populations in other areas, e.g. 7588 per m² in Kames Bay, did not show this phenomenon and indicated that the size of "territory" does not limit density[173].

Unlike *T. tenuis*, adults of the species *Scrobicularia plana*, *Mulinia lateralis*, *Mya arenaria*, *Cardium edule* and *Macoma balthica* generally have a random distribution and do not display territorial behaviour; however, the spat of *Mya arenaria* tend to aggregate after settlement, and populations of the adult may show aggregations which are related to the type of substratum. The general distribution of *Scrobicularia plana* is influenced by height on the shore and salinity, but the patchy distribution of this species cannot be correlated with these factors or the type of substratum[174].

At Whitstable, Kent, the density of *Macoma balthica* and *Hydrobia ulvae* was inversely related to the median particle diameter of the sediment[175]. However, the estuarine distribution of *Hydrobia ulvae* is not a function of sediment grade alone. In the River Crouch estuary, Essex, the survival of *H. ulvae* at salinities of $< 7\%_0$ is strongly temperature dependent, and increases with decreasing temperature; few animals are found at salinities below $2 \cdot 8\%_0$. In this estuary, *H. ulvae* is restricted to increasingly higher shore levels as the distance from the open sea increases. This arises because the isohalines of minimal salinity rise higher, in the intertidal zone, with distance moved upstream. In the Crouch, therefore, the population is restricted to a narrowing zone between high water springs (the level above which the effects of desiccation prevent further penetration of large populations[74]) and the $2 \cdot 8\%_0$ isohaline. Within this zone, populations are influenced by the effect of salinity upon activity. At a low salinity, activity is first reduced, and then the animal retires within its shell which is closed by the operculum: a response evoked by a salinity *ca.* $2 \cdot 1\%_0$. If the salinity is increased, the animal at first becomes more active, but once that to which it is acclimated is exceeded it retires within its shell.

During the exposure period, water may persist in ripples, on the shore, and here the *Hydrobia* climbs the ripples and then floats attached to the surface film. In consequence, they may be transported by the tide; they may be carried to less favourable conditions, but upon contact with a lowered salinity, they shut the shell, sink and thus avoid being transported to an area where the salinity is lethal. In those areas where the salinity is normally low, the floating reaction is not evoked and the animals maintain their position by crawling and burrowing. In these positions too, surface layers of water may not be so pronounced as at lower shore levels, further downstream, where the salinity is normally higher[176].

The salinity tolerance of *H. ulvae* may vary from place to place, and and may be dependent upon local physiological races: it can tolerate $1\%_0$, $1 \cdot 3\%_0$ and $1 \cdot 75\%_0$ at Randers Fjord, the River Adur, Shoreham, Sussex and at Burton Marsh, Dee estuary, Cheshire, respectively. Nevertheless, despite individual variations, it is reasonable to conclude that because the isohaline contours in estuaries become progressively higher in the intertidal zone as the estuary is ascended, the lower limit of the population of *H. ulvae*, in the intertidal zone, also becomes higher in the upper reaches of an estuary. Furthermore, it seems probable that, in general, there will be an increase in the number of deposit feeding species as an estuary is ascended, and that they will in turn be restricted to higher and higher zones upstream. This probably explains why, at Whitstable, *Scrobicularia plana* is found down to low water, but in the upper reaches of the Gwendraeth estuary, Wales, the population is confined to much higher levels in the intertidal zone[176, 177]. This is true for these species and for *Corophium volutator* in the Solway Firth and its tributary estuaries[49]. It seems significant that in 1938, during a period of low rainfall, the distribution of *H. ulvae* extended in an upstream direction in the estuaries of the Tamar, Tavy and the Yealm[70]. Furthermore, in the ordinary marine habitat, the most euryhaline species are to be found in the *upper* intertidal zones, but paradoxically, in the upper reaches of estuaries, the more euryhaline species are found *low* in the intertidal zone while the less euryhaline species are found *higher* in the intertidal zone. In other words, the vertical zonation of estuarine species, according to their osmotic adaptability, is the reverse of the zonation of such animals in the more fully marine habitat[176].

The lamellibranch *Scrobicularia plana* is well adapted for life in deep muddy sediments which may be referred to as *"Scrobicularia-butter"*. An adult *Scrobicularia* may live at depths of 30 cm in black anaerobic mud; its oxygen supply is derived entirely from the water above the soil surface (leakage from the animal may produce an oxidized layer around it). Access to the surface, is gained by means of the long siphons and an animal with a shell 4 cm long, may extend its inhalant siphon to a length of 28 cm; in the Gwendraeth estuary, the average length of shell was 30 mm, but individuals up to 54 mm length were found[177]. In this estuary, the population density reached 500 per m²; some specimens lived 16–18 yr and reached a length of 54 mm[177] (Table 11.7)[178].

The inhalent siphon of *Scrobicularia* actively sucks up deposits from the surface of the mud. This material is transported by the inhalent current to the ctenidia which select the food material and reject the remainder which is transported by ciliary currents to the base of the exhalent siphon, whence it is violently expelled. Larval settlement, in that part of the Gwendraeth estuary investigated, was generally unsuccessful: in January 1955, < 2% of the shells were smaller than 10 mm. This poor survival of spat was attributed to adult feeding which rendered the area unsuitable for settlement[177]. Subsurface deposits rich in sulphur bacteria and the like may be utilized as a source of food by *Scrobicularia*. However, unlike *Macoma balthica* which undertakes frequent migrations, and *Tellina tenuis*

Table 11.7. Life span of the lamellibranchs (after
Segestråle[178]). +means that the age given may be exceeded

Species	Life-span (years)
Arca sp.	2+
Unio crassus	12
U. pictorum	8–10
U. tumidus	8–10
Margaritifera margaritifera	60; 70–80; 50–150
Megalonaias gigantea	54+
Quadrula ebena	20–50
Anodonta cygnea	20–30
Pisidium pusillum	3
Sphaerium (Cyclas) corneum	3–4; 3/4–5/4
Sph. solidum	3/4–5/4
Sph. rivicola	1½–2; 3
Dreissena polymorpha	"several"
Chama sp.	50+
Cardium edule	3–4 (normally); 4; 7–9; 8+
C. corbis	7; 16
C. fasciatum	2
Tridacna gigas	100+
Saxidomus giganteus	20+
Mercenaria mercenaria	17; 40
V. (Tivela) stultorum	20; 26
Paphia undulata	2
Cumingia tellinoides	4
Abra (Syndosmya) alba	1 (generally)
A. nitida	1 (generally)
Amphidesma ventricosum	7
Macoma balthica	7–8 at 3 m depth
	25 at 25 m depth
Tellina tenuis	5–6+ ; 7+
Scrobicularia plana	16–18
Siliqua patula	14–16; 15; 11–13
Cultellus pellucidus	2
Mya arenaria	10–12; 19
M. truncata	8+
Teredo sp.	2
Meleagrina vulgaris (Ceylon)	7; 7–8
Meleagrina sp. (California)	4+
Ostrea edulis	7–12
O. virginica	6
O. crassissima	30+
Mytilus edulis	8–10
M. variabilis	5
Pecten maximus	18–22
P. grandis	10
Aequipecten irradians	ca. 2
Pecten yessoensis	7+
Notovola meridionalis	11 (maximum)
Placuna sp.	5

which migrates horizontally in response to depletion of its food supply, *Scrobicularia* rarely undertakes horizontal migrations[179]

Evidently, by drawing water into itself, for feeding and respiration, *Scrobicularia* is vulnerable to the changes in salinity of the overlying water even though its immediate surroundings may be stable, with respect to this parameter. Indeed *S. plana* can tolerate almost marine conditions, salinities *ca.* 20‰ and 2‰ for short periods. However, to a dilution of 50% sea water the osmotic pressure of the blood is not significantly different from that of the external medium, but at 30% sea water it osmoregulates. Open animals equilibrate to 80% and 60% sea water in 4–5 hr and 5–6 hr respectively. Curiously, however, estuarine lamellibranchs which dwell below the low-water mark may experience a longer immersion at low salinity than *S. plana*[180]. However, this species is intolerant of permanent immersion, and at St. John's Lake, Tamar estuary, this factor, in combination with an unsuitable substratum, viz. mainly gravel, prevented a permanent settlement of *Scrobicularia* although adequate spat were available[70, 180]. In contrast, the extensive beds of *Mytilus edulis* on Neille Point, in the River Tamar, lay in 12–18 in. of water at low tide, and here the salinity fell to 2·8% in wet weather[62, 180].

In basic structure, *Macoma balthica* is similar to *Scrobicularia plana*, but it is hardier and did not suffer so severely in the winter of 1962–63 as *S. plana* which is presumably near to its northern limit in the northern British Isles; *M. balthica* lives in the Gulfs of Finland and Bothnia, where the sea freezes for several months in the year. *Macoma* is normally associated with fine deposits which carry high densities of micro-organisms[175], an observation consistent with the conclusion that the abundance of *Macoma* is related to the amount of food available[69]. Unlike *Scrobicularia*, *Macoma* lives in a superficial layer of the substratum which it leaves readily, and makes a looped track in the soil surface. These movements are composed of a movement towards the sun followed by a movement away from it: under heavy cloud, in summer, the movements still occur, but are random[181]. The growth of *Macoma* is related to the environmental salinity, and is slower in reduced salinities[178, 182]. However, wide local variations in size are typical of this species, a possible result of its dependence upon detritus from the substratum as a source of food[178].

Outside the Baltic, *M. balthica* is confined to shallow waters, but in this sea it lives at a considerable depth (up to 200 m). The contrast between the two areas may result from an absence of echinoderms in the Baltic; outside, it, they may wipe out *Macoma* below the L.W.M.[183]. However, even in the deep water areas of the Baltic, the periodic failures of recruitment are apparently not due to adverse physical factors in the environment, but result from a dense population of the amphipod *Pontoporeia affinis* which devours the newly settled larvae[183, 184]. Once established, however, *Macoma* may live for 25 yr at 35 m depth and 7–8 yr in 3 m depth in the Baltic[178] (Table 11.7).

The tellin, *Tellina tenuis*, inhabits sands with a reduced silt content in the outer areas of estuaries. In 1965, in a sand bank at the mouth of the Dee estuary, *T. tenuis* and *M. balthica* were distributed in an inverse relationship to each other[75], but in the Exe estuary[72] and in the Solway Firth[X], *T. tenuis* tended to occur at lower shore levels than *M. balthica*. At Firemore Bay, Loch Ewe, the populations of *T. tenuis* may have as many as 8 yr-classes and the most abundant populations were found on the more sheltered beaches. Recruitment may be influenced by major large scale environmental factors including dispersal in adverse currents. In the isolated pocket beaches of the north coast of Scotland, in particular, larval dispersal, and loss, must be rapid and vigorous, and the chances of successful settlement must be very low. Apparently, such exposure is a major influence in the production of sporadic and haphazard recruitment in organisms which have a long stage in the meroplankton. It is implicit, even where chances of recruitment are good, that after this long, pelagic, larval life, i.e. 4–6 weeks, spat will not necessarily settle where they were spawned. Indeed, at Loch Ewe, in 1968, although conditions were favourable, i.e. with reduced predation, to a spawning at Firemore, a successful settlement occurred only in the more sheltered parts of the loch at Aultbea and Inverasdale. In contrast, the more regular recruitment of the Clyde Sea beaches is probably explicable in terms of a richer environment which allows *T. tenuis* to withstand cropping of the siphons and achieve reproduction[163].

The razor-shells may be found in exposed beaches and in outer areas of estuaries. The species *Ensis siliqua*, *E. ensis*, and *E. arcuatus*, which are apparently confined to European waters, are all found at L.W.S.T. on the beach, and in shallow water offshore. *E. siliqua* is restricted within the depth range to *ca.* 20 m, but the other species may be found at depths below this. All three species require some shelter from wave action although *E. siliqua* is the more resistant of the three species: all burrow nearly vertically into the sand.

On wave exposed beaches, *E. siliqua* may be the only lamellibranch present, but, in more sheltered areas, it is accompanied by *E. ensis*, together with such species as the lamellibranchs *Venus striatula*, *Donax vittatus*, *Tellina fabula*, *Gari fervensis*, *Mactra corallina cinerea*, *Lutraria lutraria*, the masked or burrowing crab *Corystes cassivellaunus* and the heart urchin *Echinocardium cordatum*.

Ensis arcuatus inhabits a coarser grade of soil in company with the lamellibranchs *Dosinia exoleta*, *Tellina donacina* and the heart urchin

Spatangus purpureus. Ensis spp. do not occur in water of reduced salinity, although its absence from estuaries may sometimes be due to a lack of deposits of a suitable grade. The genus is also intolerant of reducing conditions which may be another reason for its absence from all except the outer areas of estuaries. Individuals of both *E. arcuatus* and *E. ensis* from Jersey and Finistère are more tolerant to wave exposure than those from S.W. England. Each of the species of *Ensis* has a planktonic larva which may be reared to metamorphosis in one month[185]. The razor-shell, *Pharus legumen* is a little known member of the British marine fauna, formerly ascribed to the Solenacea, but it more properly belongs to the Tellinacea. It inhabits a restricted zone on the margin of the lower shore and shallow sub-littoral. Because it was so little known, it was formerly considered to be a species confined to the south and west of England[186]. However, in the Solway Firth, it occurs in sufficient numbers to be taken as food by plaice, *Pleuronectes platessa* and dab, *Limanda limanda*[49]. It is also taken in the Clyde[187]. According to the Burgh Museum at Kirkaldy on the Firth of Forth, numbers of this species may be thrown upon the shore by gales.

The soft-shelled clams, *Mya*, are resident in the deeper layers of sediment, but, unlike *Scrobicularia*, gain access to the surface with fused siphons. *Mya arenaria* is a characteristic inhabitant of stiff sands and muds in coastal plain estuaries; *Mya truncata* occurs frequently in the more stony sediments of fiordic estuaries, in company with *Venerupis pullastra*, *V. decussata* and *Hiatella arctica*: the two species of *Mya* may occur together in the latter environment[X]. *M. arenaria* may penetrate far up estuaries; however, its rate of feeding declines as the salinity decreases from 31‰ and at 4‰ it ceases feeding. Both in the field and in the laboratory, small clams are less tolerant of a low salinity than the adults: clams of 2–4 mm length die in 30–40 hr, whereas those with a shell length exceeding 20 mm survive more than 50 hr in fresh water. The tolerance limit is *ca.* 4‰, but above this salinity the clams can withstand a sudden change of 18‰ in a few minutes[188]. The larvae of *Mya* settle after two weeks in the plankton and, as with other lamellibranchs, the size at metamorphosis is very variable[189]. A newly settled *Mya* spat may vary from 200 to 300 μm in shell length; it attaches itself to the substratum by means of a byssus which may be retained until a length of 7 mm is attained. However, once the byssus is lost, the *Mya* juvenile adopts the adult mode of life with the shell buried in a vertical position and the siphons reaching to the surface of the soil. Growth is rapid, and the animal may reach a length of 30 mm by its first winter, and adults may live for 8 yr and attain a shell length of 15 cm[190].

Mercenaria (Venus) mercenaria, the quahog, or hard-shelled clam, or littleneck clam or cherrystone clam is widely distributed along the Atlantic coast of North America, where it is an important commercial species. It has been introduced to California and to Britain. In Chincoteague Bay, Maryland, the distribution of this animal correlates, in decreasing order of abundance, with the following bottom types: shell, sand and sand-mud mixtures, and mud. Consistent with this preference, the higher densities of clams are correlated with stronger currents. Winter temperatures may limit its distribution at the northern end of its range. In areas where the salinity values from January to June fall below 20‰ and may reach 13‰, the quahog is scarce, however few stations experienced a prolonged exposure to salinities < 21‰ and the most productive beds are located in relatively saline inlets.

The oysters, *Crassostrea* and *Ostrea*, are essentially marine forms which are able to tolerate reduced salinities: the former species has a salinity tolerance of 12‰, whereas the latter does not occur in those situations where the salinity falls below 20‰ for prolonged periods. The left valve of *Crassostrea* is relatively deep compared with that of *Ostrea*: an adaptation which enables it to survive in more silty conditions than *Ostrea* can tolerate[16]. In the Foveaux Strait, New Zealand, where the sediment types range from fine sand to coarse pebble gravel, the distribution of the oyster beds (*Ostrea sinuata*) is correlated closely with the distribution of medium to fine sandy pebble gravel[192].

Growth and Reproduction

In general, the growth and reproduction of neritic organisms is temperature dependent, although temperature is not the sole factor controlling these activities. Temperature may act as a trigger to spawning, but this may be modulated by the tidal cycle, brightness of light or chemical factors present in the water. On the other hand, grazing of the siphons, by predators, may prevent spawning in the bivalve mollusc, *Tellina tenuis*[163]. Nevertheless, the literature contains many references to the effect of temperature upon this activity and since the effect of many industries, and of the high energy society in general, is to influence this parameter it seems worthy of particular attention below. The attributes of examples drawn from individual phyla and species will be considered in systematic order.

Polychaete Worms

The breeding of *Arenicola* is somewhat variable and depends upon the environment of the adult. For example, at Whitstable, Kent, the breeding season is sharply defined, and extends for 14 days between the new and full moon spring tides in the second half of October. Spawning begins slowly, reaches a maximum at the intervening neap tides and then declines. The worms spawn for the first time at 2 yr old[120]. At St. Andrews, the breeding occurs during the second fortnight of October, but it begins during neap tides, and reaches a peak at the following springs; the littoral population at Booterstown, Dublin, begins to spawn at the spring during the first 2 weeks of November. Unlike the *Arenicola* which live at Kames Bay, Millport, and which spawn only in autumn, those at Fairlie Sands on the coast of Ayrshire spawn both in spring and autumn; however, the spring spawners live at the level of the laminarian zone though they are not of the laminarian variety[193], and this difference in spawning time may result from a difference in temperature regime at different shore levels. At St. Andrews and Dublin the breeding season is preceded by the first sharp fall in air temperatures to approximately the autumnal minimum: it was first thought that temperature initiated spawning[193], however it was later shown that the maturation of eggs and spawning is subject to hormonal control[193a]. In the population at Whitstable, germ cells are not detected in the body cavity, from November to June, but from August until the end of October 98% of the adult worms are ripe[120]. This is a marked contrast to *Nereis virens*, from Southend, Essex, in which, after a period of relative quiescence, the oocytes in the body cavity grow rapidly, between September and December, and then increase little until early May when spawning occurs (possibly on a spring tide)[194]. Once spawning has occurred 40% of the adult *Arenicola* die[120]; similarly, spent adults of *Nereis virens* may be washed up[194].

The fertilized eggs of *Arenicola* may either be retained in the burrow[193a], or washed from it and come to lie on the surface of the sand[120]. The larva hatches 4–5 days after fertilization, and usually the opaque trochophore rests attached to the substratum.

A. marina is a cold-water species which breeds in the cold months of the year. *A. ecaudata*, on the other hand, spawns all the year round, and *A. branchialis* is a winter-spawning species. This lack of a clear relationship between geographical distribution and time of spawning, i.e. the uncertain effect of temperature limitation, appears to be a characteristic of the *Arenicola* spp. as a whole, but more information is needed[133].

Like *Arenicola marina*, other shore-dwelling polychaetes have larvae which do not enter the plankton. The ragworm, *Nereis diversicolor*, is one of the commonest of all shore polychaetes in the British Isles and N.W. Europe. In 1949, at Chalkwell, Essex, on the Thames estuary, spawning occurred over a limited period centred round the third lunar quarter in February, and followed a sharp rise in temperature. In general, spawning occurs in early spring, but the period is variable (Table 11.8). The larvae develop in the substratum, and none are truly pelagic. Development is slow, the larvae become active at 7 weeks and assume the adult mode of life at 10 weeks when the worm is 4 mm in length[195].

Table 11.8. The breeding period of *Nereis diversicolor* (after Dales[195])

Author	Locality	Jan.	Feb.	Mar.	Apr.	May	June	July	Aug.	Sept.	Oct.	Nov.	Dec.
Dales	Chalkwell, England	O	O,S	O,L	O,L	—	—	—	—	O	O	O	O
McIntosh	St. Andrews, Scotland	O	O	O,S	O	O	O	O	—	O	O	O	O
Thamdrup	Skallingen	—	S	—	—	L	L	L	L	L	L	L	—
Just	Roscoff	S	S	—	L	L	L	L	L	L	S	S	S
Smidt	Copenhagen	—	S	S	S	S	S	S	S	S	S	S	S
Herpin	Cherbourg	S	—	—	O	—	—	L	L	—	—	—	—
Nunn	Plymouth	—	—	L	L	L	L	L	L	—	—	—	—
Hofker	Zuidersee	—	—	—	O	—	L	—	—	—	—	—	—
Thorson	Øresund	—	—	—	O	—	O	—	—	—	—	—	—
Blegvad	Nyborg	—	—	—	—	—	—	—	—	—	—	—	—
Menthal	B. of Pillau	S	S	S	S	—	—	—	—	—	—	—	—
Durchon	Luc-sur-Mer	—	—	—	—	—	—	—	—	—	—	—	—

O = Oocytes in the coelom;   S = spawning;   L = occurrence of larvae

Table 11.9.   Summary of the Spawning Periods for *Scoloplos armiger* from British and European waters (after Gibbs[196])

Area	Spawning Period	Authority
Denmark, Germany	March to May	Schulze[1961] Mau[196i] Thamdrup[196m] Linke[196g] Thorson[119c] Smidt[73]
Holland	End of February to mid-April	Ritzema Bos[196k] Horst[196f] De Groot[196c] Delsman[196d]
East coast of Scotland	February to April	Cunningham and Ramage[196b]
Isle of Man	March to April	Hornell[196e]
Isle of Man (Port St. Mary)	November to February	Southward[155]
North Wales	December to March	Lyster[196h]
Whitstable	Mid-February to March	Newell[196j] Anderson[119d] Chapman[196a] Gibbs[196]
Solway Firth	March to May	Perkins[23]

Scoloplos armiger normally inhabits the sediments of outer estuarine areas, but it can tolerate low salinities: in the Baltic it lives in water of 7·72‰ salinity. It generally spawns in spring (Table 11.9), at a time which is variable, but upon spring tides which follow a rise in temperature: in the Thames estuary when the sea temperature rises above 5°C. The eggs are retained in conspicuous cocoons and a single female may produce more than one cocoon. Polychaetes which produce cocoons have larvae which may or may not be pelagic (Table 11.10). The formation of a cocoon by *Scoloplos* is considered to be an adaptation to facilitate maintenance of the non-pelagic larvae in conditions favourable to the adult. The adult, although tolerant of a wide variety of soil grades, is restricted to fairly sheltered shores[196].

At Whitstable, Kent, large numbers of *Polydora ciliata* make burrows in the London Clay exposed at L.W.M.S.T.; therefore, on this shore, it occupies a situation comparable with *Lanice conchilega*, *Owenia fusiformis*, *Sabellaria alveolata* and *Branchiomma vesiculosum* in other estuaries. On this shore, this cosmopolitan species makes burrows in clay kept free of loose superficial material, and is not normally found in mussel beds or where there is an accumulation of silt; in this habitat, which is inhospitable to most other species, the *Polydora* is apparently free from direct competi-

Table 11.10. The reproductive attributes of some polychaete species having a cocoon forming habit (after Gibbs[196])

Species	Size of egg (μm)	Period spent in cocoon (days)	Stage at emergence	Later development	Authority
Phyllodoce maculata (L.)	120	2–3	Trochophore	Pelagic	McIntosh[196q]; Thorson[119c]
Micronereis variegata (Claparède)	240	21–28	6-chaetiger	Non-pelagic	Rullier[196u]
Marphysa borradailei Pillai	170	—	Late trochophore	6–12 h pelagic	Borradaile[196p]; Pillai[196s]
Lumbrinereis latreilli (Aud. and M. Edw.)	300	—	4–7 chaetiger	Non-pelagic	Okuda[196r]
Heteromastus filiformis (Claparède)	110	2–3	Trochophore	3–4 weeks pelagic	Rasmussen[196t]
Spio filicornis (O. F. Müller)	300 × 200	4	3-chaetiger	Long pelagic	Okuda[196r]
Axiothella mucosa (Andrews)	260 × 200	11–14	10-chaetiger	Non-pelagic	Bookhout and Horn[196o]
Lanassa nuda Moore	170	7	4-chaetiger	Non-pelagic	Okuda[196r]
Chone teres Bush	200	—	4-chaetiger	Short pelagic	Okuda[196r]
Haploscoloplos kerguelensis (McIntosh)	200	3	8-chaetiger	1–2 days pelagic	Okuda[196r]
Haploscoloplos fragilis (Verrill)	150	6	Late trochophore	4 days pelagic	Anderson[196n]
Scoloplos armiger (O. F. Müller)	250	18–26	9-chaetiger	Non-pelagic	Gibbs[196]

tion by these species. Breeding is promoted by a rise in temperature and takes place in March and early April at Whitstable, in March at Plymouth, and in May at St. Andrews. In Gullmar Fjord it has two spawning periods, viz. early May to June and the middle of September to December. On spawning, the eggs are laid into a string of capsules which are attached to the walls of the tube. The larvae take about a week to hatch and are released to the plankton at the 3-chaetiger stage. They spend some 6 weeks in the plankton before metamorphosis, and have grown to 17 or 18 segments[197].

Large populations of *Cirriformia tentaculata* inhabit the mud flats at Hamble, Southampton Water. Here, it breeds throughout the warmer months, i.e. late March to September; this protracted period of breeding contrasts with the short, well defined breeding season of most polchaetes, e.g. *Palola viridis*, *Nereis diversicolor*, *Arenicola marina* and *Tharyx marioni*. Evidently the breeding period is related to temperature. Growth is optimal at 20°C, slow at 8°C and ceases at 6°C; it does not grow in the colder winter months[198].

Although the precise nature of the stimuli associated with lunar periodicity in spawning is uncertain, polychaetes other than *Arenicola* evince this behaviour. Best known is the Palolo worm, *Palola viridis*, from the Samoan Islands. This species swarms during the last quarter of the moon in October and November. At these times, the posterior end of the worm packed with gametes becomes free and swarms at the surface of the sea, at the time of low tide, in the early morning of several days; death follows the release of gametes. *Platynereis dumerilii*, an inhabitant of outer estuaries in N.W. Europe, also occurs in the Pacific Northwest of the U.S.A. Here, at Friday Harbour, it swarms just after high water, one or two nights before the full moon from June to August. Here too, *Nereis procera* swarms a few days before and after, but not during the dark of the moon. Many other sexually active polychaetes are taken at night at Friday Harbour, viz. *Nereis vexillosa*, *N. brandti*, *Anaitides maculata*, *Autolytus varians*, *Flabelligera infundibularis*, *Podarke pugettensis*, *Armandia brevis* and *Ceratonereis* sp., but with these species no correlation with the phases of the moon has been demonstrated. It is of interest that colonies of the hydroid *Hydractinia echinata* maintained under a continuous illumination do not spawn. However, if these colonies are placed in darkness for at least an hour, and then exposed to light, spawning follows: the males in 50 min., the females 5 min. later. A similar response is evinced by *Hydractinia epiconcha*, a Japanese species and at Friday Harbour by the medusae—*Halistaura cellularia*, *Aequorea aequorea*, *Phialidium hemisphericum*. *P. gregarium*, *Aegena rosae*, *Leuckartiara* sp., *Sarsia tubulosa* *Aglantha digitale* and *Gonionemus vertens*. Here too, the ctenophores *Pleurobrachia bachei* and *Bolinopsis microptera* spawn shortly after sunrise[199].

Crustaceans

In the Dovey estuary, the isopod *Eurydice pulchra* requires about 20 months to reach maturity. It breeds in the period June–August; it has one brood per year and lives about 2 yr[160]. In contrast, at Arcachon the population attains sexual maturity in 8 months, has an annual reproductive cycle and premature individuals can procreate in the year in which they emerged from the marsupium[158]; presumably this is a consequence of the temperature difference between the two areas[160]. Indeed, in Swansea Bay, where the winter temperatures are significantly higher than in the Dovey estuary, early broods mature before the winter of their first year, and the breeding season extends from April to August[162].

The population of the amphipod *Bathyporeia pilosa* which inhabits the Dovey estuary reproduces throughout the year with a peak in spring and autumn. From November to January the high shore population reproduces more actively than the low shore population. *B. pelagica*, also, has a peak of reproduction in the spring and autumn, but it does not reproduce from November to February: possibly a temperature factor is involved[147].

Although temperature and light intensity may control breeding, paradoxes can occur. In the shallow waters of the Baltic, the amphipods *Pontoporeia affinis* and *P. femorata* breed only in the cold season. With increasing depth this becomes less pronounced, and reproduction in some part of the population occurs in the warm season. This may occur at depths as shallow as 60 m[200].

In north west Europe the mysids *Praunus inermis*, *Leptomysis gracilis*, *L. lingvura*, *Schistomysis spiritus*, *Mysidopsis gibbosa*, *M. didelphys* and *Neomysis integer* tend to breed throughout the year. An overwintering population breeding in spring, to give a brood which breeds in summer. In turn, the summer brood breeds in autumn to give an overwintering population[164–166, 168, 200]. However, specific differences and geographical distribution may influence this pattern. *Schistomysis ornata* and *Mysidopsis didelphys* unlike *M. gibbosa*, for example, have a main breeding period in autumn and early winter[168]. *Praunus inermis* which breeds throughout the year, in the Clyde and at Port Erin, where the sea temperature lies in the range 7–10·9 °C, breeds only from April to October in Nyborg Fjord, Denmark, where the sea temperature lies in the range 1·5–7·6 °C, the low winter temperature probably prevents the Danish mysids from breeding[165]. Similarly, *P. flexuosus* breeds throughout the year at Roscoff, but stops breeding in winter in Danish waters[16]. *Schistomysis spiritus* evinces a pronounced autumnal peak of breeding in the Clyde, but not in Loch Ewe where the numbers decrease sharply after September[164]. In the Kiel Canal, *Neomysis integer* breeds from April to September[201], however, at Plymouth, it breeds from February to November[202].

The species *Erythrops serrata* and *E. elegans* have overwintering populations which in the Clyde and Loch Ewe produce a spring generation
which grows and matures to produce a summer generation. This summer
generation, together with some remnants of the spring generation, comprise the overwintering population: the individual life span of each group
is about eight months, but this is reduced to 4–6 months in winter[203].
Unlike the species described, *Mysis relicta*, which inhabits various fresh
water lakes of Europe and North America, breeds principally in the
winter[16].

Molluscs

Breeding of the oyster, *Ostrea edulis*, is confined to the summer months.
Spawning takes place when the water temperatures exceed 15°C[204–208];
successful larval growth can take place only above 16°C[204], and a temperature of 17–18°C may be necessary[208–210]. Nevertheless, on the east
coast of England spawning rarely occurs below 16°C and successful larval
development has not been observed below 18°C[207, 208]. In the waters
near Naples, *O. edulis* commences breeding in April, when the water
temperature exceeds 15°C. In the Bay of Vigo, on the Atlantic coast of
Spain, larvae of the oyster may be present in the water from March until
December, and, in contrast to oysters from Naples and N.W. Europe,
these larvae seem to develop normally at 13–14°C. Even more curious,
larval oysters, in the Norwegian oyster polls, appear in abundance when
the water temperatures exceed 25°C; a temperature which results from
the insulating effects of a superficial layer of fresh water, beneath which the
oysters dwell in saline water[207].

Most individuals of *Ostrea* start life as males, but change to females once
their sperm is liberated; the female may then change back to male after
liberating eggs. This cycle of change may be repeated several times in the
life of an individual oyster; the rapidity of the change is dependent upon
the sea water temperature[206]. Unlike the mussel which freely emits both
male and female sexual products into the surrounding water, the male
oyster, *Ostrea*, liberates sperm to the surrounding water, whence it passes
to the female and fertilizes the eggs which are retained within the inhalent
cavity; *Crassostrea* more nearly resembles *Mytilus* in that the eggs are
ejected from the inhalent cavity by powerful, rhythmic contractions of the
adductor muscles.

Ostrea edulis may be induced to spawn in the laboratory. Stock brought
into the laboratory in the period January–March, and introduced to water
maintained at 21°C, show signs of shell shoot 10–14 days later, and the
first liberation of larvae occurs after 30–40 days. In North Wales, the

growth of *Ostrea edulis* is confined to the period April–October when the sea water temperatures exceed 12°C[211]. Both temperature and salinity affect the ciliary activity of *Crassostrea virginica*; these parameters may also influence its condition[212].

The European oyster, *O. edulis*, was introduced to the eastern sea-board of North America in 1949, here breeding oysters were found from the beginning of June until the end of August, at Milford Harbour, Connecticut. However, at Boothbay Harbour, Maine, where the summer is shorter and the temperatures lower, the breeding season is correspondingly shorter[213]. Here it rather resembles *Crassostrea virginica* which in general spawns at 20°C, although in a few northern areas it may spawn at 17°C and in Florida spawns only at temperatures above 25°C. This is not a simple relationship between latitude and minimum temperature required for spawning, for at the mouth of the Delaware River, bay oysters spawn at 25°C, whereas open coast oysters spawn at 20°C.

After 1 yr of adjustment, Chesapeake Bay oysters transplanted to Santa Rosa Sound, Florida, commenced spawning and reached mass spawning at a temperature 5°C above that which initiated similar behaviour in the parent stock. The spawning period was increased from the $3\frac{1}{2}$ months, normal in parent stock, to 5 months. Evidently, the reproductive activity of oysters is regulated primarily by changing temperature levels rather than a hereditary response to a specific temperature threshold. Furthermore, at Santa Rosa Sound, spawning is initiated following a 5°C rise within a 30 day period, and the actual onset of spawning has occurred at temperatures ranging from 18 to 27°C, and varies in time from March to May. Clearly, in this area at least, the initiation of oyster spawning reveals nothing about the time of year or the water temperature[214].

Once the larvae are released to the surrounding water, their survival depends to a marked degree upon the amount and quality of food available. In this connection, it has been shown that in estuaries of the United States from Florida to New England, tropical storms are favourable to the survival

Table 11.11.   The effect of a tropical storm upon the settlement of sedentary animals (after Butler[214])

| Organism | Incidence of organisms per cm² per week | | Increase |
	1 week before storm	1 week after storm	
Barnacle	0·3	170·6	568X
Oyster	1·1	73·0	67X
Mussel	0·1	3·2	32X
Bryozoa	0·1	2·1	21X
Plankton ml/100 l	0·9	5·0	5X

of oyster, barnacle, mussel and bryozoan larvae, and of the general plankton (Table 11.11). Such observations suggest that the high waters induced by the storms caused some trace elements or simple nutrients to be washed out from the adjacent marshlands, and that the enrichment which results is favourable to increased phytoplankton production and therefore to improved brood survival[214].

The cockle, *Cerastoderma* (*Cardium*) *edule* normally inhabits slightly muddy sand, but it can also live in coarse gravels and stiff clays. Its siphons are short, and in consequence it lives near to the soil surface, where during the exposure period it can be felt by the sensitive human foot. Like the mussel it breeds at a low temperature, and in British waters spawns from March throughout the summer. The fertilized eggs give rise to planktonic veliger larvae which settle in 2–3 weeks[208]; once settled, it may in favourable areas such as the Burry Inlet, S. Wales, live 10 yr or more[215]. The size at settlement is very variable, and the length at metamorphosis ranges from 150–345 μm[189]. At the time of settlement, particularly upon beds rich in adults, the larvae run the risk of passing into the inhalent siphon of the adult. Once inside the adult, they, like other unwanted materials, become coated in mucus and are returned to the environment where, unable to extract themselves from the mucus, they die[216].

Escallops, *Pecten maximus*, which may live in the extreme outer areas of estuaries such as the Solway Firth, as well as in the open sea, spawn twice each year, viz. in April–May and in August–September. In Manx waters, the first most intense spawning takes place at a temperature of 7·2°C[217], a reflection, perhaps, of its adaptation to life in the open sea. The second spawning, like that of oysters, may be the result of its ability to recover from the first spawning, and for the gonads to develop once more before the temperatures fall too low[208, 217]. Although scallops of the species *Pecten magellanicus*, from the Bay of Fundy and Georges Bank, *Notovola meridionalis* from Tasmanian waters and *Pecten maximus* from European waters undertake local movements, which may be very active, they do not undertake more extensive migrations[218].

The soft shelled clam, *Mya arenaria*, spawns when the sea temperature rises to 10–15°C. In consequence, the time and duration of spawning varies from locality to locality. In Canada, spawning begins in June, but further south in the U.S.A., it may begin in April. In Maryland, *Mya* spawns in two periods, viz May–June and September–October. Similar to the oyster and escallop, these two periods represent separate periods of gamete maturation, of which the second takes place at a higher temperature than the first[219]. Settlement occurs after a pelagic larval life of about 2 weeks, but although a single *Mya* female may produce three million eggs, large numbers of larvae may be lost from an estuary with a short flushing time (p. 75). This animal matures in 5 yr, but only about 1% of settled spat survive to this stage[190].

Venerupis pullustra is typically found beneath flattish, angular stones covering pockets of gravel about L.W.M. or in the shallow waters of the sub-littoral[220, 221, X]. The majority of these animals may be attached, by a byssus, to some solid object such as a stone or shell[220], or the byssus may be absent, as in the Isefjord, Denmark[221], the Gareloch and Loch Long[X]. At Keppel Pier, Isle of Cumbrae, Firth of Clyde, female *V. pullustra* require a minimum of 16°C for spawning, but the spawning period of about two months varies from year to year within the period mid-May to the end of August[222, 223].

Venus striatula occurs in a range of sediment grades, from fine to muddy sand, around and below the L.W.M. in outer estuaries such as the Solway Firth and its tributary estuaries[12, 51, 52]; it may also inhabit sand bars[X]. In both situations it may be associated with the heart urchin *Echinocardium cordatum*[12, 51, 52, X]. Spawning occurs only at temperatures above 11°C, and takes place in both sexes, at intervals throughout the spawning period, i.e. from late May–early July. However, it reaches a peak of intensity near the middle of the spawning period, and then falls off. Spawning may be induced by a rise of temperature or by the stimulation of sexual products. "Epidemic spawning" may be initiated by either males or females. A short recuperative stage follows spawning and is succeeded by a rapid development of gametocytes which continues to the following May. Other members of the Veneridae spawn as follows: *V. ovata*, May; *V. casina*, May; *V. fasciata*, February, March, April and July, *Venerupis decussata*, July; and *Gafrarium minimum*, June. The larva of *V. striatula* is a planktonic veliger which settles at a size of 200–220 μm: some settle when only 190 μm long and a few remain planktonic to a length of 280 μm (compare *Mya arenaria*, newly settled shell length 200–320 μm[189]). After spawning, the eggs develop normally at temperatures between 10·0 and 18·0°C, but development is slower at the lower temperature. At 5°C, the eggs are not injured and division occurs at a much reduced rate; at 20°C development is impaired and a large proportion of larvae are abnormal. Larval growth is increased with increasing temperature in the range 5–16°C, growth at 20°C is similar to that at 16°C, and a maximum rate of growth probably occurs about 18°C. At temperatures above 20°C the growth of larvae is retarded[223].

The hard-shell clam, *Mercenaria mercenaria*, spawns naturally only when the water temperature is $\geqslant 24°C$[224]. In the laboratory, however, clams subjected to a 20°C increase from winter conditions, spawned at 20·6 and 21·0°C in 3 weeks or less[225]. At Chincoteague Bay, Maryland, the temperature of the adjacent Atlantic Ocean rarely exceeds 24°C, and then

only in late summer. *M. mercenaria* occurs only in the inlets of Chincoteague Bay, while further north it occurs only in the more sheltered and
shallow bays and rivers; therefore, the high minimum temperature of
spawning may limit the distribution of this species[191]. Once spawned,
however, larval development depends upon temperature: the time to
metamorphosis is 7 and 16–24 days at 30 and 18°C respectively[226].

The gastropod slipper limpet, *Crepidula fornicata*, is a competitor of
the oyster even on its native grounds in the eastern U.S.A. However, in
the 1880s, it was inadvertently carried, along with shipments of oysters,
to England and the Pacific coast of the U.S.A. In both new habitats it
multiplied rapidly and became a dangerous competitor of the native
oysters, viz. *O. edulis* in England and *O. lurida* on the U.S. Pacific
coast[213].

In Britain, it has spread to most of the oyster beds on the south-east
and south coasts. Here it spawns once the sea temperature has risen to,
and exceeded, 10°C. Spawning is irregularly periodic either at, or immediately after, neap tides. In the Essex estuaries, developing embryos
are found from April until early September, and most females spawn at
least twice in a season. *Crepidula fornicata* forms chains of individuals, in
which the youngest member of the chain is male, the older members
female: it is therefore a *protandric hermaphrodite*. On the completion of the
functional male phase, the penis degenerates and cells are sloughed off
from the brown tip. As the penis becomes smaller, the width of the
gonoduct increases and the inner walls become folded longitudinally. The
distal part of the gonoduct develops into a prominent uterus, with folded
walls, into which a number of seminal receptacles open. Concurrently, the
penis becomes greatly reduced and is finally represented by a small brown
scar. The gonad, which even in the male phase contains small ova,
develops rapidly as oogenesis takes place and it becomes bright yellow
when fully functional[227].

Echinoderms

Like other groups, the spatangoid echinoderms exhibit a range of habit.
Spatangus purpureus, the purple heart urchin, characteristically inhabits
the coarser offshore deposits, especially clean shell gravel, but, in the Channel Islands and the Isles of Scilly, it may be found in shell gravel beaches
exposed at E.L.W.S.T. *Echinocardium cordatum*, on the other hand, has
been found in a wider variety of deposits than *S. purpureus*, viz. from mud
to gravel plus shells; its distribution ranges from some depth below the
low-water mark to the lower shore, where it inhabits sediments which
range from coarse to muddy sand. Other species, viz *E. flavescens* and *E.*

pennatifidum occur, below the L.W.M., in sediments ranging from muddy sand to clean shell gravel[228].

In the coastal waters of Northumberland, *E. cordatum* occurs most abundantly in the finer offshore sediments which, if abundance were the sole criterion, would be taken to be the ecological optimum. However, the growth rate of these animals is poor, and the population never reaches sexual maturity. In the clean sand of the intertidal habitat, on the other hand, *E. cordatum* is less abundant, but the growth rate of the population is four times faster than that of the fine, organic rich, offshore deposits; moreover, animals of this population became mature and bred in their third year, and thereafter annually. This population commences breeding 1 year later than the faster growing populations of the Clyde; the precise figure for the longevity of this North Sea population is uncertain, but must exceed 10 yr[229].

The poor growth and lack of reproduction in the organic rich, finer offshore deposits, may be due to the food supply being of a qualitatively unsuitable nature. However, one of the main factors contributing to this effect may be the inability of the animals, in these circumstances, to ingest sufficient quantities of sediment. Indeed, the presence of populations in such an unsuitable habitat is of great interest. It does not represent a chance occurrence, since on the Northumberland coast, at least, stocks of *E. cordatum* have persisted in these deposits. Because *E. cordatum* is found associated with several different bottom communities, it would appear that the larvae are not highly discriminatory and it is probable that the degree of discrimination in "larval choice" becomes diminished with the age of the larvae. The scarcity of near-optimum habitats on the Northumberland coast may force larvae to postpone settlement until they are eventually forced to metamorphose on a bottom which is near to the environmental limits for growth and survival. Indeed, the offshore distribution is within discrete and quantitatively concentric patches which bear no relationship to the sediment or to any other environmental factors. These patches remain stable in position and shape from the time of settlement and are difficult to explain without postulating that the larvae were aggregated at the time of settlement and while still planktonic[229]. This difference in the biology of *E. cordatum* living in an optimal and an unfavourable environment has important consequences related to ecological studies of pollution (p. 485), and the work of Buchanan must be regarded as one of the most important investigations carried out in recent years.

The time of spawning of the starfish *Asterias rubens* varies from April in southern England[230] to early June on the Atlantic coast of Canada[231]. On the Essex oyster beds spawning commences when the water temperatures reach 15°C[232]; in south western England it spawns from April onwards[230]. In contrast, *Solaster papposus*, a predator of *A. rubens*, commences spawning at a temperature of 9°C, i.e. in February as opposed to May for *A. rubens*. In Essex, the young *Asterias* settled in maximum density on the cultivated oyster beds, and also on the *Solaster* ground further offshore. Young *Solaster* settled mainly in the mid-river region, possibly a gregariousness of settling near parents as with other invertebrates (p. 313). Both young and adult *Solaster* preyed on *Asterias*, particularly the juveniles; the reverse occurs only to a limited degree. The predation of *Solaster* upon *Asterias*, combined with its own discrete settling behaviour, is probably responsible for the almost complete lack of overlap in the distribution of the two species in this area[232]. *Asterias* (*Marthasterias?*) *vulgaris*, common on the Atlantic coast of Canada, spawns in early June[231]. *Asterias forbesi* found along the U.S. Atlantic coast, begins to spawn in July at the northern end of its range; in Long Island Sound spawning commences within a few days of the sea water temperature reaching 15°C, i.e. about the middle of June, and continues intermittently until early September. Once spawning is completed the gonads are resorbed, a process which is usually complete by October. Active development of new sex cells begins, gametogenesis is rapid in November and December, by January and February many starfish are found with gonads of a normal size.

Survival of the larvae depends upon the incidence of disease, quality and quantity of food available and the presence of various ectocrines in the water. Both the amount and duration of settlement is very variable, the former depends upon the factors defined in the foregoing sentence, while the latter, which takes place in waves, occurred in 1 day in 1945, but took 91 days in 1961, and the average duration of this period is about 52 days. There is no apparent relationship between the abundance of adults in the pre-spawning period and the intensity of larval settlement[233].

References

1. Dexter, R. W. (1944). *Ecology* **25**, 352–359.
2. Steers, J. A. (1959). *Endeavour* April 1959, 75–82.
3. Perkins, E. J. and Williams, B. R. H. (1966). H.M.S.O., U.K.A.E.A., P. G. Report 587 (CC).
4. King, C. A. M. (1959). "Beaches and Coasts", Edward Arnold, London.
5. Clarke, L. D. and Hannon, N.J. (1967). *J. Ecol.* **55**, 753–771.
6. Chapman, V. J. (1960). "Salt Marshes and Salt Deserts of the World", Hill, London. Plant Science Monographs.
7. Ranwell, D. S. (1967). *J. appl. Ecol.* **4**, 239–256.
8. Clarke, L. D. and Hannon, N. J. (1969). *J. Ecol.* **57**, 213–234.
9. Hepburn, I. (1962). "Flowers of the Coast", Collins, London. New Naturalist Series No. 24.
10. Chapman, V. J. (1959). *J. Ecol.* **47**, 619–639.

11. Steers, J. A. (1959). "Scolt Head Island. The Physiography and Evolution of the Island", Heffer, Cambridge.
12. MacNae, W. (1957). *J. Ecol.* **45**, 113–131.
13. Steers, J. A. (1946). "The Coastline of England and Wales", University Press, Cambridge.
14. Steers, J. A. (1962). "The Sea Coast", Collins, London. New Naturalist Series, No. 25.
15. Ranwell, D. S. (1968). "Coastal Marshes in Perspective", University of Strathclyde, Regional Studies Group Bulletin, No. 9, 26 pp.
16. Green, J. (1968). "The biology of Estuarine Animals", Sidgwick and Jackson, London.
17. Gillham, M. E. (1957). *J. Ecol.* **45**, 735–756.
18. Gibb, D. C. (1957). *J. Ecol.* **45**, 49–84.
19. Mathiesen, H. and Nielsen, J. (1956). *Bot. Tidsskr.* **53**, 1–34.
20. Gillner, V. (1960). *Acta phytogeogr. suec.* **43**, 1–98.
21. Kestner, F. J. T. (1962). *Geogrl. J.* **128**, 457–478.
22. Marshall, J. R. (1962). *Scott. geogr. Mag.* **78**, 81–99.
23. Perkins, E. J. (1968–71). *Trans. J. Proc. Dumfries. Galloway Nat. hist. Antiq. Soc.*, Ser. 3, **45**: 15–43; **46**: 1–26; **48**: 12–68.
24. Ranwell, D. S. (1964). *J. Ecol.* **52**, 79–94.
25. Richards, F. J. (1934). *Ann. Bot.* **48**, 225–259.
26. Taylor, M. C. and Burrows, E. M. (1968). *J. Ecol.* **56**, 795–809.
27. Chater, E. H. (1965). *J. Ecol.* **53**, 789–797.
28. Bird, E. C. F. and Ranwell, D. S. (1964). *J. Ecol.* **52**, 355–366.
29. Hubbard, J. C. E. and Stebbings, R. E. (1968). *J. Ecol.* **56**, 707–722.
30. Ranwell, D. S. (1967). *I.U.C.N. Publications*: New Series No. 9 (1967): 27–37.
31. Goodman, P. J., Braybrooks, E. M. and Lambert, J. M. (1959). *J. Ecol.* **47**, 651–677.
32. Goodman, P. J. (1960). *J. Ecol.* **48**, 711–724.
33. Hubbard, J. C. E. (1965). *J. Ecol.* **53**, 799–813.
34. Ranwell, D. S., Bird, E. C. F., Hubbard, J. C. E., and Stebbings, R. E. (1964). *J. Ecol.* **52**, 627–641.
35. Hubbard, J. C. E. (1969). *J. Ecol.* **57**, 755–804.
36. Chater, E. H. and Jones, H. (1957). *J. Ecol.* **45**, 157–167.
37. Oliver, F. W. (1928). *J. Minist. Agric. Fish.* **35**, 709–721.
38. Ranwell, D. S. (1961). *J. Ecol.* **49**, 325–340.
39. Ranwell, D. S. (1964). *J. Ecol.* **52**, 95–105.
40. Hoffman, L. (1970). *New Scient.* **46** (697), 120–122.
41. Sutherland, D. (1925). *Scott. geogr. Mag.* **41**, 1–23.
42. Ranwell, D. S. (1960). *J. Ecol.* **48**, 117–141.
43. Chandapillai, M. M. (1970). *J. Ecol.* **58**, 193–201.
44. Crawford, R. M. M. and Wishart, D. (1966). *J. Ecol.* **54**, 729–743.
45. Scott, G. M. (1963). *J. Ecol.* **51**, 517–527.
46. Davis, J. H. (1940). *Publs. Carnegie Inst.* **517**, 303–412.
47. MacNae, W. and Kalk, M. (1962). *J. Ecol.* **50**, 19–34.
48. Clarke, L. D. and Hannon, N. J. (1970). *J. Ecol.* **58**, 351–369.
49. Perkins, E. J., Williams, B. R. H., Bailey, M., Hinde, A. and Gorman, J. (1963–1966). H.M.S.O., U.K.A.E.A., P.G. Reports: 12 parts.
50. Perkins, E. J. and Williams, B. R. H. (1964). U.K.A.E.A. Chapelcross, H.P. and S. Note No. 2: 12 parts.

51. Jones, N. S. (1952). *J. Anim. Ecol.* **21**, 182–205.
52. Jones, N. S. (1956). *J. Anim. Ecol.* **25**, 217–252.
53. McIntyre, A. D. (1958). *Mar. Res.* 1958. No. 1, 1–17.
54. King, A. R. and McCauley, J. E. (1969). *In* "Ecological Studies of Radio-Activity in the Columbia River Estuary and Adjacent Pacific Ocean", Prog. Rept. 69–9: 186, Dept. of Oceanography, Oregon State Univ. Corvallis.
55. McIntyre, A. D. (1961). *J. mar. biol. Ass. U.K.* **41**, 599–616.
56. Tebble, N. (1966). "British Bivalve Sea Shells", British Museum (Natural History), London.
57. Peterson, G. H. (1958). *Medd. Kommn. Danm. Fisk.-og Havunders.*, N.S. **2** (22), 1–31.
58. Kindle, E. M. (1930). *U.S. Nat. Res. Comm. Reprint and Circ.* No. 92, 14.
59. Nicol, J. A. C. (1950). *J. mar. biol. Ass. U.K.* **29**, 303–320.
60. Skertchley, S. B. J. (1877). "The Geology of the Fenland", Memoirs of the Geological Survey of England and Wales, London.
61. Vaillant, L. (1891). *Annls Sci. nat.*, Ser. 7 (Zool.). **12**, 289–305.
62. Percival, E. (1929). *J. mar. biol. Ass. U.K.* **16**, 81–108.
63. Alexander, W. B., Southgate, B. A. and Bassindale, R. (1935). *D.S.I.R. Water Pollution Research Tech. Paper* No. 5, 1–171.
63a. Smith, J. E. (1932). *J. mar. biol. Ass. U.K.* **18**, 243–278.
64. Fraser, J. M. (1935). *Proc. Trans. Lpool biol. Soc.* 1935, 48–65.
65. Nicol, E. A. T. (1935). *J. mar. biol. Ass. U.K.* **20**, 203–261.
66. Sparck, R. (1935). *J. Cons. perm. int. Explor. Mer.* **10**, 3–19.
67. Bassindale, R. (1938). *J. mar. biol. Ass. U.K.* **23**, 83–98.
68. Hantzschel, W. (1939). *In* "Recent Marine Sediments" (P. D. Trask, ed.), pp. 195–205, Am. Assoc. of Petroleum Geologists, Tulsa, Oklahoma; Thomas Murby, London.
69. Beanland, F. L. (1940). *J. mar. biol. Ass. U.K.* **24**, 589–611.
70. Spooner, G. M. and Moore, H. B. (1940). *J. mar. biol. Ass. U.K.* **24**, 283–329.
71. Bassindale, R. (1943). *J. Ecol.* **31**, 1–29.
72. Holme, N. A. (1949). *J. mar. biol. Ass. U.K.* **28**, 189–232.
73. Smidt, E. L. B. (1951). *Meddr. Kommn. Danm. Fisk.-og Havunders.* **11**, 1–151.
74. Stopford, S. C. (1951). *J. Anim. Ecol.* **20**, 103–122.
75. Perkins, E. J. (1956). *Ann. Mag. nat. Hist.* Ser. 12., **9**, 112–128.
76. McIntyre, A. D. and Eleftheriou, A. (1968). *J. mar. biol. Ass. U.K.* **48**, 113–142.
77. MacGinitie, G. E. (1935). *Am. Midl. Nat.* **16**, 629–765.
77a. MacGinitie, G. E. (1939). *Am. Midl. Nat.* **21**, 28–55.
78. Sanders, H. L. (1956). *Bull. Bingham oceanogr. Coll.* **15**, 345–414.
78a. Parker, R. H. (1956). *Bull. Am. Ass. Petrol. Geol.* **40**, 295–376.
79. Hedgpeth, J. W. (1957). *Mem. geol. Soc. Am.* **67**, 1, 587–608.
80. Stickney, A. P. and Stringer, L. D. (1957). *Ecology* **38**, 111–122.
81. Stickney, A. P. (1959). *U.S. Dept. of the Interior, Fish and Wildlife Service. Special Scientific Report.* No. 309, 1–21.
82. Sanders, H. L. (1960). *Limnol. Oceanogr.* **5**, 138–155.
83. Sanders, H. L., Goudsmit, E. M., Mills, E. L. and Hampson, G. E. (1962). *Limnol. Oceanogr.* **7**, 63–79.
84. Day, J. H. (1950). *Trans. R. Soc. Afr.* **33**, 53–91.
85. Guiler, E. R. (1951). *Pap. Proc. R. Soc. Tasmania.* 1950, 29–52.

86. Morton, J. E. and Miller, M. (1968). "The New Zealand Sea Shore", Collins, London and Auckland.
87. Marriage, L. D. (1958). "The Bay Clams of Oregon—Their Economic Importance, Relative Abundance and General Distribution", Fish Commission of Oregon, Contribution No. 2, Condensed as Educational Bulletin No. 2, 29 pp.
87a. Semper, K. G. (1881). "Animal Life as Affected by the Natural Conditions of Existence", Appleton, New York.
87b. Petersen, C. J. (1913). *Rep. Dan. biol. Stn.* **21**, 1–44.
87c. Petersen, C. J. (1914). *Rep. Dan. biol. Stn.* **22**, 89–96.
87d. Petersen, C. J. (1915). *Rep. Dan. biol. Stn.* **23**, 1–28.
87e. Petersen, C. J. (1918). *Rep. Dan. biol. Stn.* **25**, 1–57.
87f. Petersen, C. J. (1918a). *Rep. Dan. biol. Stn.* **26**, 1–62.
87g. Petersen, C. J. and Boysen Jensen, P. (1911). *Rep. Dan. biol. Stn.* **20**, 1–76.
87h. Boysen Jensen, P. (1914). *Rep. Dan. biol. Stn.* **22**, 1–39.
87i. Boysen Jensen, P. (1919). *Rep. Dan. biol. Stn.* **27**, 1–44.
87j. Jones, N. S. (1950). *Biol. Rev.* **25**, 283–313.
87k. Lewis, J. R. (1970). *In* "F.A.O. Technical Conference on Marine Pollution and its Effects on Living Resources and Fishing", F.I.R.: MP/70/E-22. Rome, 9–18 December 1970.
87l. Teal, J. M. and Kanwisher, J. (1961). *Limnol. Oceanogr.* **6**, 388–399.
87m. Teal, J. M. (1962). *Ecology* **43**, 614–624.
87n. Ranwell, D. S. (1972). "Ecology of Salt Marshes and Sand Dunes", Chapman and Hall, London.
87o. Backlund, H. O. (1945). *Opuscula Entomologica. Suppl.* (1945). 1–236.
87p. Perkins, E. J. (1973). *In* "The Biology of Plant Litter Decomposition", (C. H. Dickinson and G. J. F. Pugh, eds), Chap. 24. Academic Press, London.
87q. Margalef, R. (1956). *Investigación pesq.* **3**, 99.
87r. Warwick, R. M. and Buchanan, J. B. (1970). *J. mar. biol. Ass. U.K.* **50**, 129–146.
88. Warner, G. F. (1969). *J. Anim. Ecol.* **38**, 379–389.
89. Dansereau, P. (1947). *Revue can. Biol.* **6**, 448–477.
90. Berry, A. J. (1963). *Bull. natn. Mus. St. Singapore* **32**, 90–98.
91. Rodriguez, G. (1959). *Bull. mar. Sci. Gulf Caribb.* **9**, 237–280.
92. Rodriguez, G. (1963). *Bull. mar. Sci. Gulf Caribb.* **13**, 197–218.
93. Hartnoll, R. G. (1965). *Proc. Linn. Soc., Lond.* **176**, 113–147.
94. MacNae, W. and Kalk, M. (1962). *J. Anim. Ecol.* **31**, 93–128.
95. Key, D. (1965). *M.A.F.F., Bradwell Investigations, Report* No. 2, 8 pp.
96. Wilson, J. B. (1967). *Scott. J. Geol.* **3**, 329–371.
97. Abernethy, J. (1862). *Proc. Instn civ. Engrs* **21**, 309–344.
98. Marshall, J. R. (1962). *Scott. geogr. Mag.* **78**, 81–99.
99. Driscoll, E. G. (1967). *J. sedim. Petrol.* **37**, 1117–1123.
100. Loosanoff, V. L. (1961). *In* "Proc. Gulf Caribb. Fish. Inst., 14th Annual Session, 1961" (J. B. Higman, ed.), pp. 80–95, Institute of Marine Sciences of the Univ. of Miami.
101. Imai, T. and Sakai, S. (1961). *Tohoku J. agric. Res.* **12**, 125–171.
102. Pratt, D. M. (1953). *J. mar. Res.* **12**, 60–74.
103. Holme, N. A. (1961). *J. mar. biol. Ass. U.K.* **41**, 398–461.
104. Orton, J. H. (1937). *Nature, Lond.* **140**, 505–506.
105. Gibbs. P. E. (1969). *J. mar. biol. Ass. U.K.* **49**, 311–326.

106. Clark, R. B., Alder, J. R. and MacIntyre, A. D. (1962). *J. Anim. Ecol.* **31**, 359–372.
107. Clark, R. B. and Haderlie, E. C. (1960). *J. Anim. Ecol.* **29**, 117–147.
108. Clark, R. B. and Haderlie, E. C. (1962). *J. Anim. Ecol.* **31**, 339–357.
109. George, J. D. (1964). *J. mar. biol. Ass. U.K.* **44**, 383–388.
110. Lewis, D. B. (1968). *J. Linn. Soc. (Zool.).* **47**, 515–526.
111. Day, J. H. and Wilson, D. P. (1934). *J. mar. biol. Ass. U.K.* **19**, 655–662.
112. Wilson, D. P. (1937). *J. mar. biol. Ass. U.K.* **22**, 227–243.
113. Wilson, D. P. (1948). *J. mar. biol. Ass. U.K.* **27**, 723–760.
114. Wilson, D. P. (1953). *J. mar. biol. Ass. U.K.* **31**, 413–436.
115. Wilson, D. P. (1953). *J. mar. biol. Ass. U.K.* **32**, 209–233.
116. Wilson, D. P. (1954). *J. mar. biol. Ass. U.K.* **33**, 361–380.
117. Wilson, D. P. (1968). *J. mar. biol. Ass. U.K.* **48**, 387–435.
118. Wilson, D. P. (1970). *J. mar. biol. Ass. U.K.* **50**, 1–31.
119. Wilson, D. P. (1970). *J. mar. biol. Ass. U.K.* **50**, 33–52.
119a. Wilson, D. P. (1936), *J. mar. biol. Ass. U.K.* **21**, 305–310.
119b. Thorson, G. (1957). *Mem. geol. Soc. Am.* 67, **1**, 480.
119c. Thorson, G. (1946). *Meddr. Kommn. Danm. Fisk.-og Havunders. Serie Plankton* **4**, 1–523.
119d. Anderson, D. T. (1959). *Q. Jl microsc. Sci.* **100**, 86–166.
119e. Upton, B. (1953). *Br. J. Anim. Behav.* **1**, 87.
119f. Banse, K. (1955). *Kieler Meeresforsch* **11**, 2.
120. Newell, G. E. (1948). *J. mar. biol. Ass. U.K.* **27**, 554–580.
121. Chapman, G. and Newell, G. E. (1949). *J. mar. biol. Ass. U.K.* **28**, 627–634.
122. Wells, G. P. (1969). *Proc. Challenger Soc.* **4** (1), 36–50.
123. Trueman, E. R. (1966). *J. exp. Biol.* **44**, 93.
124. Trueman, E. R. (1966). *Biol. Bull. mar. biol. Lab. Woods Hole.* **131**, 369.
125. Chapman, G. and Newell, G. E. (1947). *Proc. R. Soc.* B. **134**, 431–455.
126. Chapman, G. (1949). *J. mar. biol. Ass. U.K.* **28**, 123–140.
127. Wells, G. P. (1949). *J. mar. biol. Ass. U.K.* **28**, 447–464.
128. Patel, S. and Spencer, C. P. (1963). *J. mar. biol. Ass. U.K.* **43**, 167–175.
128a. Reid, D. M. (1930). *J. mar. biol. Ass. U.K.* **16**, 109–115.
128b. Perkins, E. J., Gilchrist, J. R. S., Abbot, O. J., and Halcrow, W. (1973). *Mar. Pollut. Bull.* **4**, 59–61.
129. Dales, R. P. (1958). *J. mar. biol. Ass. U.K.* **37**, 521–529.
130. Dales, R. P., Mangum, C. P., and Tichy, J. C. (1970). *J. mar. biol. Ass. U.K.* **50**, 365–380.
131. Wells, G. P. and Dales, R. P. (1951). *J. mar. biol. Ass. U.K.* **29**, 661–680.
132. Lindroth, A. (1938). *Zool. Bidr. Upps.* **16**, 397–497.
133. Southward, E. C. and A. J. (1958). *J. mar. biol. Ass. U.K.* **37**, 267–286.
134. McLusky, D. S. (1968). *J. mar. biol. Ass. U.K.* **48**, 769–781.
135. McLusky, D. S. (1970). *J. mar. biol. Ass. U.K.* **50**, 747–752.
136. Meadows, P. S. (1964). *J. Anim. Ecol.* **33**, 387–394.
137. Meadows, P. S. (1964). *J. exp. Biol.* **41**, 499–511.
138. McLusky, D. S. (1968). *J. mar. biol. Ass. U.K.* **48**, 443–454.
139. McLusky, D. S. (1967). *J. mar. biol. Ass. U.K.* **47**, 607–617.
140. Morgan, E. (1965). *J. Anim. Ecol.* **34**, 731–746.
141. Perkins, E. J. (1965). *Trans. J. Proc. Dumfries. Galloway Nat. hist. Antiq. Soc.* Ser. 3, **42**, 6–13.
142. Gamble, J. C. (1970). *J. mar. biol. Ass. U.K.* **50**, 657–671.

143. Weiser, W. (1962). *J. mar. biol. Ass. U.K.* **42**, 665–682.
144. Morgan, E. (1970). *J. mar. biol. Ass. U.K.* **50**, 769–785.
145. Toulmond, A. (1964). *Cah. Biol. mar.* **5**, 319–342.
146. Fincham, A. A. (1969). *J. mar. biol. Ass. U.K.* **49**, 1003–1024.
147. Fish, J. D. and Preece, G. S. (1970). *J. mar. biol. Ass. U.K.* **50**, 475–488.
148. Bossanyi, J. (1957). *J. Anim. Ecol.* **26**, 353–368.
149. Soika, A. G. (1955). *Vie Milieu* **6**, 38–52.
150. Elmhirst, R. (1931). *Proc. R. Soc. Edinb.* **51**, 169–175.
151. Pirrie, M. E. Bruce, J. R. and Moore, H. B. (1932). *J. mar. biol. Ass. U.K.* **18**, 279–296.
152. Rees, C. B. (1939). *Proc. R. Ir. Acad.* **45**, 215–229.
153. Brady, F. (1942). *J. Anim. Ecol.* **12**, 1–27.
154. Watkin, E. E. (1942). *Trans. R. Soc. Edinb.* **60**, 543–561.
155. Southward, A. J. (1953). *Proc. Trans. Lpool biol. Soc.* **59**, 51–71.
156. Colman, J. S. and Segrove, F. (1955). *J. Anim. Ecol.* **24**, 426–444.
157. Colman, J. S. and Segrove, F. (1955). *J. Anim. Ecol.* **24**, 445–462.
158. Salvat, B. (1966). *Act. Soc. limn. Bordeaux*, 103 Série A L, 1–77.
159. Jones, D. A. and Naylor, E. (1967). *J. mar. biol. Ass. U.K.* **47**, 373–382.
160. Fish, S. (1970). *J. mar. biol. Ass. U.K.* **50**, 753–768.
161. Jones, D. A. (1970). *J. Anim. Ecol.* **39**, 455–472.
162. Jones, D. A. (1970). *J. mar. biol. Ass. U.K.* **50**, 635–655.
163. McIntyre, A. D. (1970). *J. mar. biol. Ass. U.K.* **50**, 561–575.
164. Mauchline, J. (1967). *J. mar. biol. Ass. U.K.* **47**, 383–396.
165. Mauchline, J. (1965). *J. mar. Ass. U.K.* **45**, 663–671.
166. Mauchline, J. (1969). *J. mar. biol. Ass. U.K.* **49**, 379–389.
167. Hodge, D. (1963). *Pap. Dep. Zool. Univ. Qd.* **2**, 91–104.
168. Mauchline, J. (1970). *J. mar. biol. Ass. U.K.* **50**, 381–396.
169. Bowman, T. E. (1964). *Tualane Stud. Zool* **12**, 15–18.
170. Jørgensen, O. M. (1929). *Proc. Univ. Durham Phil. Soc.* **8**, 41–54.
171. Broad, A. C. (1962). *In* "Biological Problems in Water Pollution. 3rd Seminar August 13–17, 1962", U.S. Dept. of Health, Education and Welfare, Public Health Service Publ. No. 999-WP-25. pp. 86–91.
172. Wilson, J. B. (1963). *Trans. J. Proc. Dumfries. Galloway Nat. hist. Antiq. Soc.* Ser. 3, **40**, 98–101.
173. Holme, N. A. (1950). *J. mar. biol. Ass. U.K.* **29**, 267–280.
174. Hughes, R. N. (1970). *J. Anim. Ecol.* **39**, 333–356.
175. Newell, R. (1965). *Proc. zool. Soc., Lond.* **144**, 24–45.
176. Newell, R. C. (1964). *Proc. zool. Soc., Lond.* **142**, 85–106.
177. Green, J. (1957). *J. mar. biol. Ass. U.K.* **36**, 41–47.
178. Segerstråle, S. G. (1960). *Commentat. biol.* **23**, 1–72.
179. Hughes, R. N. (1969). *J. mar. biol. Ass. U.K.* **49**, 805–823.
180. Freeman, R. F. H. and Rigler, F. H. (1957). *J. mar. biol. Ass. U.K.* **36**, 553–567.
181. Brafield, A. E. and Newell, G. E. (1961). *J. mar. biol. Ass. U.K.* **41**, 81–87.
182. Vogel, K. (1959). *Neues Jb. Geol. Paläont.* **109**.
183. Segerstråle, S. G. (1965). *In* "Botanica Gothoburgensia III. Proc. 5th Marine Biological Symposium". Goteborg 1965. pp. 195–204.
184. Segerstråle, S. G. (1962). *Commentat. biol.* **24** (7), 1–26.
185. Holme, N. A. (1954). *J. mar. biol. Ass. U.K.* **33**, 145–172.
186. Yonge, C. M. (1959). *J. mar. biol. Ass. U.K.* **38**, 277–290.

187. Halcrow, W. (1971). Personal communication.
188. Matthiessen, G. C. (1960). *Limnol. Oceanogr.* **5**, 291–300.
189. Jørgensen, C. B. (1946). *Meddr. Kommn. Danm. Fisk.-og Havunders, Serie Plankton.* **4**, 277–311.
190. Ayers, J. C. (1956). *Limnol. Oceanogr.* **1**, 26–34.
191. Wells, H. W. (1957). *Ecology* **38**, 123–128.
192. Cullen, D. J. (1962). *N.Z. Jl. Geol. Geophys.* **5**, 271–275.
193. Howie, D. I. D. (1959). *J. mar. biol. Ass. U.K.* **38**, 395–406.
193a. Howie, D. I. D. and McClenaghan, C. M. (1965). *Gen. Comp. Endocrinol.* **5**, 40–44.
193b. Howie, D. I. D. (1961). *J. mar. biol. Ass. U.K.* **41**, 771–783.
194. Brafield, A. and Chapman, G. (1967). *J. mar. biol. Ass. U.K.* **47**, 619–627.
195. Dales, R. P. (1950). *J. mar. biol. Ass. U.K.* **29**, 321–360.
196. Gibbs, P. E. (1968). *J. mar. biol. Ass. U.K.* **48**, 225–254.
196a. Chapman, G. (1965). *Biol. Bull. mar. biol. Lab., Woods Hole* **128**, 123–140.
196b. Cunningham, J. T. and Ramage, G. A. (1888). *Trans. R. Soc. Edinb.* **33**, 635–684.
196c. De Groot, G. J. (1970). "Eifurchung und Keimblattbildung bei *Scoloplos armiger*". Diss. Leiden, 74 pp.
196d. Delsman, H. C. (1916). *Tijdschr. ned. dierk.Vereen.*, Ser. 2, **14**, 383–394.
196e. Hornell, J. (1891). *Proc. Trans. Lpool biol. Soc.* **5**. 223–268.
196f. Horst, R. (1880). *Tijdschr. ned. dierk.Vereen.* Ser. 2, **5**, 121–130.
196g. Linke, O. (1939). *Helgoländer wiss Meeresunters.* **1**, 201–348.
196h. Lyster, I. H. J. (1965). *J. Anim. Ecol.* **34**, 517–527.
196i. Mau, W. (1882). *Z. wiss. Zool.* **36**, 389–432.
196j. Newell, G. E. (1954). *Ann. Mag. nat. Hist.* Ser 12, **7**, 321–350.
196k. Ritzema Bos, J. (1874). *Tidjschr. ned. dierk. Vereen.* **1**, 58–61.
196l. Schultze, M. S. (1856). *Abh. naturforsch. Ges. Halle* **3**, 213–221.
196m. Thamdrup, H. M. (1935). *Meddr. kommn. Danm. Fisk.-og Havunders Serie: Fiskeri*, **10**, 125 pp.
196n. Anderson, D. T. (1961). *Q. Jl. microsc. Sci.* **102**, 257–272.
196o. Bookhout, C. G. and Horn, E. C. (1949). *J. Morph.* **84**, 145–184.
196p. Borradaile, L. A. (1901). *Proc. Zool. Soc., Lond.* **2**, 714–720.
196q. McIntosh, W. C. (1869). *Ann. mag. nat. Hist.* Ser. 4, **4**, 104–107.
196r. Okuda, S. (1946). *J. Fac. Sci. Hokkaido Univ.* Ser. Zool. **9**, 115–219.
196s. Pillai, T. G. (1958). *Ceylon J. Sci. (Biol. Sci.)* **1**, 94–106.
196t. Rasmussen, E. (1956). *Biol. Meddr.* **23** (1), 84 pp.
196u. Rullier, F. (1954). *Archs Zool. exp. gén.* **91**, 195–233.
197. Dorsett, D. A. (1961). *J. mar. biol. Ass. U.K.* **41**, 383–396.
198. George, J. D. (1964). *J. mar. biol. Ass. U.K.* **44**, 47–65.
199. Fernald, R. L. (1959). *In* "Mar. Biol. Colloquium 1959. Oregon State University, Corvallis", pp. 51–58.
200. Segerstråle, S. G. (1967). *J. exp. mar. Biol. Ecol.* **1**, 55–64.
201. Kinne, O. (1955). *Biol. Zbl.* **74**, 160–202.
202. Tattersall, W. M. and Tattersall, O. S. (1951). "The British Mysidacea", Ray Society, London.
203. Mauchline, J. (1968). *J. mar. biol. Ass. U.K.* **48**, 455–464.
204. Orton, J. H. (1920). *J. mar. biol. Ass. U.K.* **12**, 339–366.
205. Orton, J. H. (1931). *J. mar. biol. Ass. U.K.* **17**, 301–313.
206. Orton, J. H. (1936). *Mém. Mus. r. Hist. nat. Belg.* (II. Ser.). **3**, 997–1056.

207. Korringa, P. (1957). *Année biol.* **30**, 1–17.
208. Waugh, G. D. (1960). *Proc. malac. Soc. Lond.* **34**, 113–122.
209. Cole, H. A. (1936). *Fish. Invest., Lond.* Ser. 2. **15**.
210. Cole, H. A. (1939). *Fish. Invest., Lond.* Ser. 2. **16**.
211. Walne, P. R. (1958). *J. mar. biol. Ass. U.K.* **37**, 591–602.
212. Fingerman, M. (1959). *Comml. Fish. Rev.* **21**, 10–11.
213. Loosanoff, V. L. (1955). *Science, N.Y.* **121**, 119–121.
214. Butler, P. A. (1962). *In* "Biological Problems in Water Pollution. 3rd Seminar August 13–17, 1962", U.S. Dept. of Health, Education and Welfare, Public Health Service Publ. No. 999-WP-25, pp. 92–104.
215. Hancock, D. A. (1965). *Fish. Invest. Lond.* Ser. 2, **24**, 40 pp.
216. Kristensen, I. (1957). *Archs. néerl. Zool.* **12**, 351–453.
217. Mason, J. (1958). *J. mar. biol. Ass. U.K.* **37**, 653–671.
218. Baird, R. H. (1966). *J. mar. biol. Ass. U.K.* **46**, 33–47.
219. Pfitzenmeyer, H. T. (1962). *Chesapeake Sci.* **3**, 114–120.
220. Quayle, D. B. (1949). *Proc. malac. Soc., Lond.* **28**, 31–37.
221. Rasmussen, E. (1958). *Oikos* **9**, 77–93.
222. Quayle, D. B. (1952). *Trans. R. Soc. Edinb.* **62**, 255–297.
223. Ansell, A. D. (1961). *J. mar. biol. Ass. U.K.* **41**, 191–216.
224. Loosanoff, V. L. (1937). *Ecology* **18**, 506–515.
225. Loosanoff, V. L. and Davis, H. C. (1950). *Biol. Bull. mar. biol. Lab. Woods Hole.* **98**, 60–65.
226. Loosanoff, V. L. and Davis, H. C. (1963). *Adv. mar. Biol.* **1**, 1–136.
227. Chipperfield, P. N. J. (1951). *J. mar. biol. Ass. U.K.* **30**, 49–72.
228. Gage, J. (1966). *J. mar. biol. Ass. U.K.* **46**, 49–70.
229. Buchanan, J. B. (1966). *J. mar. biol. Ass. U.K.* **46**, 97–111.
230. Vevers, H. G. (1949). *J. mar. biol. Ass. U.K.* **23**, 165–187.
231. Smith, G. F. M. (1940). *J. Fish. Res. Bd. Canada.* **5**, 84–103.
232. Hancock, D. A. (1958). *J. mar. biol. Ass. U.K.* **37**, 565–589.
233. Loosanoff, V. L. (1964). *Biol. Bull. mar. biol. Lab. Woods Hole.* **126**, 423–439.
  X. Perkins, E. J. Previously unpublished work.

12

Fouling and Boring Organisms

This chapter, the first to deal with the interactions of estuarine organisms and mankind, is concerned essentially with that part of the biota which has taken advantage of man's activities to extend its range. In essence, fouling organisms are those species which normally dwell attached to rock surfaces, but which colonize any available man-made surface, e.g. ships' bottoms, harbour walls and civil engineering works in general. Such a habit is relatively unspecialized, with little specialization shown by the organisms which adopt it. However, boring organisms can probably be regarded as the most specialized of that group of animals which colonize crevices to a varying degree. Life in crevices evinces increasing specialization depending upon the nature and depth of the crevice inhabited[1-3]. In general, the boring species have undergone a modification in structure fitting them for their specialized habitat[4]; however, some species, e.g. the polychaete worm, *Polydora ciliata*, normally regarded as a characteristic borer, may in suitable circumstances behave as a fouling species[5].

Fouling Organisms

Any organism which is capable of colonizing a surface is capable of becoming a fouling organism, depending upon the opportunity offered for settlement. Such organisms range from bacteria, *Desulphovibrio* to the mussel, *Mytilus*, but it should be appreciated that "fouling" is a term of approbrium which reflects the "harm" which man considers these species to cause relative to his interests; it has no biological meaning other than indicating this interaction. In any given area, the population of fouling organisms may be composed of hundreds of species representing bacteria, algae, protozoans, coelenterates, polychaetes, bryozoans, molluscs, crustaceans, and tunicates. However, in many parts of the world a relatively few genera or species preponderate: these include the diatom *Navicula grevillei*; the green algae *Enteromorpha* sp. and *Ulva lactuca*; hydroids, especially *Tubularia larynx*; serpulid polychaetes, especially *Hydroides norvegica*; bryozoans, especially *Bugula* spp.; molluscs, especially *Mytilus*; crustaceans, especially the acorn barnacles, *Balanus* spp. and *Elminius modestus*, and ascidians, e.g. *Botryllus schlosseri*[5-27].

The ubiquitous distribution of many fouling organisms, e.g. *Entermorpha intestinalis*, *Bugula neritina*, *B. avicularia*, and *Botryllus schlosseri*, may be due to their translocation on ships[23, 24, 28], although the location of the initial establishment of a species in a new area may be difficult to trace[24]. Nevertheless, the introduction of the barnacle *Elminius modestus* to Britain and its subsequent spread to Europe is particularly well documented, and the history of the spread of the bryozoan *Watersipora cucullata* in Australian and New Zealand waters is being followed in the same way[29].

Essentially, fouling is a phenomenon of ports, with respect to ships[24], and inshore waters, with respect to civil engineering structures. In ship fouling, the influence of the duration of the stay was very evident during World War II when convoy escort ships, which spent only short periods in harbour, were much less troubled by fouling problems than other vessels.

The colonies of fouling organisms may be studied by examination of ships' bottoms when in dry dock, by exposing settlement plates on rafts and tracing the history of the colonization, or by examination of power station culverts, dock gates and the like. During the examination of a ship's bottom a detailed questionnaire is completed with information which describes the distribution and approximate quantitative rating of the predominant organisms upon each of the following: wind and water plates, side plates, bottom plates and keel, condenser and fire main intakes, rudder, propeller-blades and boss. In addition, information is usually sought regarding the previous painting history, the state of steel and paint surfaces, and details of the ship's movements during the period between successive dockings. Unfortunately, or otherwise, such studies provide information only upon fouling in the presence of antifouling paints[24]. Various rating systems have been devised to describe the quantitative assessment of fouling growth, especially in the presence of antifouling paints[30, 31]. Although assessments of fouling growths have been carried out on a world wide basis, and the intensity of fouling at any port is known, there are, however, few criteria for comparing different ports or regions[23]. Nevertheless, despite some individual variations, it can be shown that different ports and types of traffic do influence the nature of ship fouling[24].

In New Zealand waters, ships in the general coastal trade are fouled characteristically by the algae *Enteromorpha* sp., *Ulva lactuca*; hydroids *Tubularia larynx*, *Phialella quadrata*; serpulid polychaetes, *Hydroides norvegica*, *Galeolaria hystrix*; the bryozoans *Bugula neritina*, *B. flabellata*, *Cryptosula pallasiana*; molluscs, *Mytilus edulis planulatus*, *Ostrea* sp. cf. *O. sinuata*; and barnacles, *Elminius modestus*, *Balanus amphitrite cirratus*.

Those ships which include ports near Auckland in their itinerary are characterized by the presence of *Balanus amphitrite, B. trigonis* and a relatively high incidence of *Bugula neritina, Hydroides, Tubularia* and oysters, whereas ships confined to more southern waters carry algae, *Bugula flabellata, Cryptosula, Galeolaria* and mussels; *Elminius modestus* was frequently the most common species in both groups. Ships engaged in traffic between New Zealand and East and South Australian ports are fouled by *Balanus amphitrite cirratus, Hydroides norvegica, Bugula neritina, Tubularia larynx, Elminius modestus* and the encrusting bryozoan *Watersipora cucullata*. In contrast, ships which sail into the tropical waters of the Pacific are colonized by the hydroid *Pennaria* (?) sp., a bryozoan *Schizoporella* sp., the pedunculate barnacles *Lepas anatifera, L. anserifera* and the acorn barnacles *Balanus amphitrite denticulata, B. auricoma, Chthamalus hembeli*, and those operating in Antarctic waters are colonized by *Conchoderma auritum* and *C. virgatum* only [24].

Investigations of settlement upon plates or panels suspended from rafts have been particularly informative with respect to the mode of development of fouling communities. Such communities undergo a succession of change from the original, clear surface to a climax community of *Mytilus* [8] (Fig. 12.1a), a succession which may be compared with that upon rocks in the Mediterranean [32, 33] (Fig. 12.1b). Although the sequence of colonization upon settlement plates may be completed in a few months or a year, the

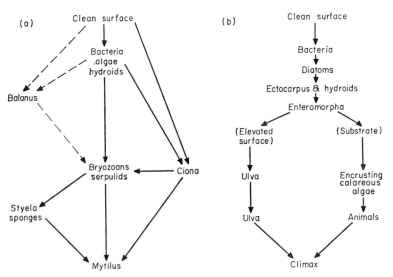

Fig. 12.1. Successional sequences observed in studies of marine fouling (after Hedgpeth [33]).

colonization of rocks, from cleared surface to climax, may take much longer to complete, i.e. months or years[34–40].

By studies in which test panels are suspended from rafts for limited periods only, it can be shown that individual species of fouling organisms have a characteristic period in which settlement takes place consequent upon the seasonal rhythm of reproduction[18, 23, 25, 41, 42] (Fig. 12.2). Such

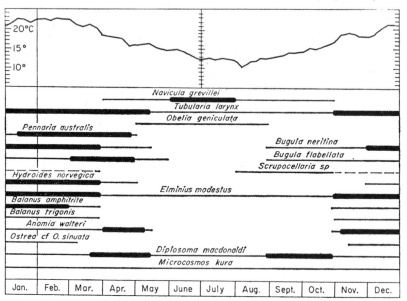

Fig. 12.2. The annual record of settlement of fouling organisms at Auckland. Sea surface temperatures are plotted as means of six-day periods from Skerman (after Skerman[23]).

seasonal changes are most marked in those areas in which marked changes of sea temperature occur, but some species may reproduce throughout the year, e.g. *Tubularia* and *Elminius* in New Zealand waters, and thus remain qualitatively important during the winter[23, 24]. Evidently the qualitative nature of a fouling settlement will depend upon the season at which the plate was first exposed[8, 18, 23, 24].

The colonization of test panels and buoys exposed in different parts of an estuary or harbour may show pronounced intra- and inter-specific differences[19, 23, 41, 43]. The southern part of Biscayne Bay, Florida, is in free communication with the sea and fouling is light; in contrast, fouling is heavy in the more enclosed northern waters of the bay, which are polluted and in which the strong tidal currents transport considerable amounts of suspended detritus[43].

In the port of Auckland, New Zealand, the heaviest barnacle and hydroid settlement occurred at Kauri Point where serpulids, bryozoans, oysters and ascidians were not uncommon. On the other hand, at Devonport, ca. 4 miles downstream, there were few barnacles and a high incidence of the serpulid *Hydroides* and bryozoans, especially *Bugula neritina*. The intensity of winter fouling was greatest at the North Head, the most seaward station, where most species were represented[23].

Upon prolonged immersion of the panels the composition of the populations changes. Barnacles and hydroids which succeeded the primary microbiological slime do not survive, but are replaced successively by bryozoans, ascidians and *Mytilus*, although seasonal factors which influence the liberation and survival of larvae may introduce variations into the sequence[8, 23, 25].

The bryozoan colonization provides a favourable base upon which *Mytilus* larvae may become attached. The ascidian *Styela* which apparently settles only upon bryozoans, may become established before *Mytilus*, and a *Styela* community, including sponges, may follow the bryozoan stage which dies from lack of food and oxygen as the *Styela* increases. Subsequently *Mytilus* overgrows the *Styela* community which dies also[8, 23].

The distribution of fouling organisms changes with increasing depth, although the water-line belt of algae probably represents the only regional difference in distribution on ships hulls[24, 44]. Nevertheless, a heavier growth may occur on bottom areas away from the light. Moreover, barnacles, particularly *Elminius*, may concentrate by plate landings, seawater intake vents and edges of bilge keels. *Mytilus*, a "small" mussel, usually appears nearer the water line than populations of the larger *Perna*; hydroids, especially *Tubularia*, are characteristically attached to the poorly illuminated areas below bilge keels and under the stern counter, and bryozoans, *Bugula* spp., are recorded more frequently towards the bows and at higher levels on the ships' sides[24].

In sea lochs of the Scottish west coast, a characteristic zonation with depth develops upon the ropes attached to mooring buoys: the diatoms or laminarians which live near the surface, are replaced by *Mytilus* which at some depth is replaced by the tunicates *Ciona* and *Ascidiella*[X]. No similar pronounced zonation was observed in panels immersed to 7 ft below E.L.W.S. at Lyttleton, New Zealand[18]. However, in Monterey Bay, California, the alga *Ulva linza* was dominant upon panels at 1 ft depth exposed for both long and short periods; it was accompanied by much less abundant growths of *Enteromorpha tubulosa*, *Giffordia sandriana*, *Macrocystis pyrifera* and *Polysiphonia* sp., together with a variety of animals. At 50 ft depth, extensive algal growths were rare: the red algae

Desmarestia herbacea and *D. munda* accompanied by the brown *Nito-phyllum northii* and the encrusting red alga *Bossiella corymbifera* occurred only at this depth. Here, *Balanus crenatus*, which showed little settlement on the shallow panel, was the dominant fouling animal: curiously, few *Mytilus* were found[25]. The barnacle *Elminius modestus* settles over a wide range of conditions from the supra littoral fringe to a 17 m depth in the harbours of southern England. On continuously immersed structures, *Elminius* settlements increase to a depth of 4–5 m; apparently, the depth of settlement is controlled by a low optimum light intensity[45].

In the relationship between data from experimental exposure of settlement panels and the ship-fouling problem, the species of greatest importance will be drawn largely from the primary colonizers of newly immersed surfaces, rather than those species which appear only in the advanced stages of immersion. However, there are some important exceptions to this generalization. At Auckland, *Ostrea* and *Scrupocellaria* are important fouling organisms, but they do not colonize newly immersed panels readily. The serpulid polychaete *Hydroides norvegica* is responsible for heavy fouling in Auckland harbour, but it is not abundant upon test panels where it meets considerable competition from the barnacles. Moreover, the severe depletion of barnacles which takes place on test panels in summer, may not arise with barnacle fouling on ships' hulls[23].

The Settlement of Sedentary Organisms

Fouling, as we have seen, is a term indicating man's response to those sedentary organisms which have an adverse influence on his activities. However, it is more rational to consider the settlement of these organisms as a problem of general interest rather than to specify this behaviour as a function of fouling alone.

Before considering the effects of surfaces upon the settlement of individual species and the succession which occurs once fouling has commenced it should be appreciated that water velocities have a considerable bearing upon the manner and quality of settlement, a consideration which is relevant to boring organisms also. The speeds above which the settlement of sedentary animal species cannot occur is summarized in Table 12.1: speeds above 2 kt are sufficient to prevent animal settlement, but a wide range of critical speeds has been recorded for some species[22, 41, 46–50]. Such discrepancies probably arise because the speed at which the main body of water moves may be measured with relative ease; near a surface, however, small dead spots and back eddies introduce considerable difficulty, and the velocity here may be much less than that of the main body of water. In contrast to the sedentary animals, diatoms

Table 12.1. The speed of sea water which prevents settlement by sedentary and boring organisms (after Smith[46, 47]; Doochin and Smith[49], Houghton[50] and Crisp[51])

Species	Habit	Speed in knots
Chthamalus fragilis	Sedentary barnacle	1·0
Balanus balanoides	Sedentary barnacle	1·0–2·0
B. amphitrite	Sedentary barnacle	0·5–2·0
B. improvisus	Sedentary barnacle	1·1–1·8
B. eburneus	Sedentary barnacle	0·4–0·8
Elminius modestus	Sedentary barnacle	1·0–2·0
Limnoria sp.	Sedentary isopod	1·5–1·9
Crassostrea sp.	Sedentary oyster	1·4
Mytilus edulis	Sedentary mussel	5·0
Teredo pedicellata	Boring shipworm	1·4–1·8
Schizoporella unicornis	Sedentary bryozoan	1·4

are able to attach themselves to a surface at much greater speeds: a diatom felt, i.e. primary film, may develop successfully upon a disc the periphery which is rotating at 11·2 kt[16].

It is of interest to consider the problems which face a larva which is trying to settle in the cooling water intake culvert of a power station. Here the flow velocity in the body of the water decreases as the walls are approached and as the drag of this solid boundary brings to rest the water in immediate contact with it. The velocity gradient across the larva imposes upon it a spinning force which increases with proximity to the boundary, and settlement becomes difficult. Conduits have a characteristic "entry length", thus the fluid shear near the walls, and hence the turning moment upon the larvae, decreases throughout the entry length, but remains steady thereafter. Therefore, the chance of settlement increases along the length of the culvert, and is demonstrated by an increase of fouling which occurs with distance from the mouth of the intake (Fig. 12.3). Clearly any roughening of the surface, e.g. expansion joints, or other animals already settled, which reduces this shear facilitates settlement[16a].

Although the speed of a body of water may limit the settlement of sedentary animals, sub-critical speed may have a stimulatory effect, e.g. barnacle cyprids are stimulated to settle at velocities of 4·8 cm/sec[51]. Once settled, however, currents faster than the critical speed, which prevents settlement, may stimulate growth in barnacles[51] and *Tubularia* grows abundantly in swift currents[23]. On the other hand, bryozoans at Beaufort, North Carolina, settle most freely at lower current speeds[41, 52]: of these species *Schizoporella unicornis* cannot settle in current velocities greater than 1·4 kt, and its maximum growth occurs in a current of 0·2 kt[49].

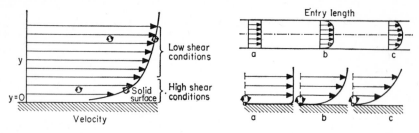

Fig. 12.3. Velocity profiles show that the turning moment exerted upon settling larvae increases, shown by thicker arrow heads, as the distance from the surface of the conduit walls decreases (left). There are also further changes as they enter the conduit (right). As the settling stages move further down the conduit, the turning moment decreases and their chances of orientating themselves correctly for settling increases correspondingly. This is confirmed by increase in fouling along the intake (after Board and Collins[16a]).

Water quality plays an important role in the settlement of sedentary organisms (p. 367), however evidence of the influence of the silt transported by water is conflicting. At Kauri Point, Auckland, a particularly heavy settlement of barnacles developed upon silt-laden films on test panels, and although this material is considered to be unfavourable for oyster settlement and growth, similar films from Auckland Harbour had to be washed off before the young oysters could be counted. Here, however, the effect of long term immersion of the panels was unknown[23]. At Beaufort, North Carolina, the settlement of bryozoans was prevented on those panels upon which a significant accretion of silt was allowed to occur[41]. Once settlement has occurred, the fouling organism may promote the accretion of silt and lead ultimately to its own death[23, 41]. The serpulid *Spirorbis vitreus* has a requirement for clear water, for irrespective of nutritional requirements its breeding season makes it particularly susceptible to damage by abrasion[53]. Although, in general, these parameters influence the settlement of sedentary organisms, succession and its related problems are regulated by many additional factors.

The formation of a primary film of bacteria or diatoms upon a surface is an important aid to colonization by higher organisms[6, 8, 9, 12, 14, 54, 55]. On the coast of California, bacteria are the most important micro-organisms in primary film formation[6, 9], whereas in Sydney Harbour, algae and diatoms are more important than bacteria on non-toxic surfaces, though probably not on toxic ones[22, 54]. Bacteria may promote the colonization of submerged surfaces in a number of ways, (1) by affording the planktonic larval stages of fouling organisms a foothold or otherwise mechanically facilitating their attachment; (2) by discolouring bright or glazed surfaces (bright, light-reflecting surfaces are colonized less readily than dark or discoloured ones); (3) by serving as a source of food (barnacles, mussels, and tunicates ingest and are nourished by bacterial food and; (4) by increasing the alkalinity of the film-surface interface, the deposition of calcareous cements may be favoured. The elaboration of ammonia from the decomposition of proteinaceous materials, the reduction of nitrates or nitrites, or the utilization of organic acids are bacterial processes which tend to increase the alkalinity, (5) by influencing the potential of the surface upon which they are growing, bacteria might expedite the attraction and attachment of fouling organisms; (6) by increasing the concentration of plant nutrients at the expense of the accumulating organic matter, which bacteria mineralize, bacterial activity tends to favour the growth of algae; and (7) where a toxic surface is present, as in the presence of an antifouling paint, the bacterial film may have a protective function for barnacles. It has been shown that the particulate and dissolved organic matter which accumulates on solid surfaces both precedes and explains the development of bacterial films upon them. Indeed, many bacteria of an innately sessile or periphytic habit prefer to grow only on solid surfaces[6].

Although bacteria are important in primary film formation it is considered that diatoms are pre-eminent because: (1) apart from bacteria, they are the most numerous representative of the microflora able to fulfil this function; (2) they are widely distributed throughout the world; (3) they are present at all seasons of the year; (4) they are oxygen producers and provide a rich source of nutriment for the bacteria; (5) they affect the hydrogen ion concentration of the slime; and (6) they are highly resistant to copper and mercury which are used widely in antifouling paints[12, 22]. Nevertheless, the denser settlements of diatoms occur in the presence of a bacterial film[8, 56].

Diatom colonizations of test plates may include 50 or more species[15, 22], though a relatively few species predominante generally[12, 15, 23]. Like the other sedentary organisms, the diatoms show a species succession which depends upon the duration of the plate exposure. In Loch Sween, glass slides, exposed for a short period, were colonized by species of active, motile diatoms, principally *Amphora laevis*, *A. macilenta*, *A. marina*, *Navicula pygmaea*, *N. retusa*, *Nitzschia closterium*, and other species of *Navicula* and *Nitzschia*, which occurred throughout the year. Eventually, the slides became colonized by a second group of diatoms, which were not so active as the primary diatom colonizers, and unlike this group were not found living freely in the mud, but attached to weeds and larger shell fragments. Dense colonies, composed principally of *Amphora coffeaeformis*, *Navicula bahusiensis* var. *bahusiensis* and *Cocconeis scutellum* (the last often becoming dominant), spread out from the point of infection for as long as the slide was exposed. The species *Navicula bahusiensis* var. *arctica*,

N. ramosissima, Amphora acutiscula and a number of very small forms including species of *Navicula* or *Fragilaria* occurred less regularly or were less important. The species *Achnanthes affinis* and *A. stroemii*, although active, apparently belonged to this group. In this group as a whole, colonization appeared to be relatively accidental: the density of the colony was affected by the situation of the slide, the time and intensity of the infection (i.e. the proximity and activity of neighbouring colonies), length of exposure of the slide, and physical and biological factors influencing the colony. Where a slide was left long enough, the colonization probably represented fairly accurately the kind of population on large natural surfaces.

In the last stage of the diatom invasion, the slides often carried large quantities of mud, detritus, faecal material with empty frustules, worm tubes and other foreign matter and were well colonized by encrusting forms. The largest diatoms found on these slides, viz. *Pleurosigma*, and the heavy chains of *Melosira sulcata*, which are normally mud dwelling species, occurred only with mud and in small numbers on the slides. This category included the species *Amphora costata, A. proteus, A. truncata, Diploneis* spp., *Gyrosigma rectum, Navicula cyprinus, N. palpebralis, N. scopulorum* var. *belgica, Nitzschia sigma, N. spathulata, Pleurosigma* spp. and *Tropidoneis* spp., together with a few other large unidentified species of *Navicula* and *Nitzschia*. In this loch there was no evidence that depth has much effect on the composition of the diatom flora, at depths between 1 and 10 m; a seasonal decrease of colonizing activity occurred in winter, but the range of species present remained fairly constant[15].

In Chichester Harbour, England, *Navicula grevillei* is one of the most important fouling organisms. Here settlement was at a maximum in winter temperatures ranging from 4 to 7°C, when the colonies reached a maximum length of 12 cm, but were rarely found below 15 cm depth[12]. At Auckland and Lyttleton Harbours, New Zealand, this species grew to 3 cm and 5–6 cm length respectively. This difference compared to Chichester Harbour, may be a function of depth, since the New Zealand plates were kept at a greater depth than 15 cm[18, 23], but in the absence of water transparency data, from the two areas, a firm conclusion cannot be reached.

In Newport Harbour, California, the colonial diatom *Licmophora* succeeded the small sedentary diatoms, and was accompanied, in some experiments, by the multicellar algae, notably *Ectocarpus granulosoides; Enteromorpha* sp., *Lophosiphonia villum* and *Pterosiphonia bipinnata* were noted frequently. In other experiments, these genera, and *Cladophora* and *Scytophonia* were present, but did not appear in quantity until after the diatom increase. At Port Lyttleton, *Enteromorpha* sp. was represented by a high density of minute filaments on the monthly panels of August,

September and November, but grew little and was not present in a significant quantity upon plates subjected to prolonged immersion. In contrast, all the locally based fishing and pleasure craft, and the inter-Island passenger ferry are heavily fouled by *Enteromorpha* at the water line during the summer months[18].

The swarmers of *Enteromorpha* are positively phototactic, and able to remain motile for 8 days. Although the period of motility is greatly reduced when these swarmers are subjected to intermittent lighting, darkness encourages settlement only in the long, but not the short, term[57-59]. The motile period is at a maximum in the optimum temperature range 17·5–27·5°C[57]. Optimum settlement occurs at 23°C, but is only slightly less at 18 and 30°C, and may be appreciable even at 4 and 38°C. Salinity has a greater influence, settlement is greatest at 26‰, but is poor at 34‰ in comparison with that in the range 17 to 25‰. Measurable settlement takes place at 7‰, and although lysis may be significant at 2‰ successful settlement can occur[58].

The hydroid *Tubularia larynx* has a cosmopolitan distribution, and occurs commonly on test panels, buoys, wrecks and ships. In Auckland Harbour, its populations were in general confined to short term panels[23]. Nevertheless, in some power stations this animal can be important as a source of trouble (p. 371). It reproduces by means of an actinula larva which is discharged by movements of the gonophore. Once liberated, it is apparently unable to swim and has little ability to select a site for settlement. At settlement, the larva becomes attached to a surface by the discharge of nematocysts on tentacles which point towards the aboral pole; settlement is considered to be complete when the larvae are attached by the aboral pole and the tentacles are free. Temperature has little effect upon the settlement of the larvae, but a greater settlement occurs either in total darkness, or in alternating dark and light than under continuous illumination. Because the initial attachment can occur only when the tentacle tips touch an object, the actinula can settle successfully only when the nematocyst threads can withstand the strains imposed by a water current, but once such an attachment has occurred, a permanent attachment will follow irrespective of whether the locality is ultimately suitable for growth of the organism or not[60]. Such settlement behaviour explains the gregariousness of *T. larynx*[10]. Exposure of the actinula larvae of *Tubularia* to hypotonic sea water accelerates their attachment to surfaces[10, 23]. Similarly, spatfalls of the acorn barnacles, particularly *Elminius modestus* are especially heavy after a period of heavy rainfall.

Once permanent attachment has occurred growth is rapid. At Millport, Firth of Clyde, both settlement and growth generally reach a maximum during August and September. In 1945, a heavy settlement of early July grew until the beginning of August, when the colonies became moribund, and autotomy of the polyps occurred. Regeneration did not take place until the middle of September and then the main mass of *Tubularia* settlement persisted until the end of October. The sequence of activity and dormancy of the colonies is apparently under the control of endogenous, rather than exogenous, factors. Once a settlement of *T. larynx* is established other organisms soon become associated with it or prey upon it. Many small organisms, including young *Mytilus*, diatoms *Licmophora* spp. and *Striatella* sp. and the suctorian protozoa *Ephelota* sp. and *Acineta* sp. find this hydroid a suitable substratum for settlement. The holotrich protozoan *Choenia* sp. and the nudibranch molluscs, *Cratena aurantia* and *Dendronotus frondosus*, prey upon it. Indeed the numbers of *C. aurantia* are such that it may play an important role in the reduction of the polyps which takes place at the end of the growing season[10]. At Auckland, New Zealand, settlement occurs throughout the year, but it is greatest from November to May, and much reduced in the months of winter and early spring[23]. Further south, at Port Lyttleton, South Island, N.Z., *Tubularia larnyx* is replaced by another hydroid *Phialella quadrata* which is an important fouling organism during the spring and early summer[18]. The association of bacteria and algae favours the settlement of *T. larynx* and other hydroids, but the whole of this group is replaced by one of three associations, viz. bryozoans and serpulids, balanoid barnacles, or the ascidian *Ciona*[8, 23].

The serpulid polychaetes secrete calcareous tubes by means of which they attach themselves to a variety of substrata depending upon the individual species, many of which have a world-wide distribution[53, 61]. The spirorbid *Spirorbis borealis* is characteristically an inhabitant of fucoid algae, upon which it may occur in large numbers. Its larvae are liberated mainly about the time of the moon's quadrature. Upon liberation, these larvae are positively phototactic for up to 2 hr, after which they swim at random; for a further period, of up to 2 hr, they visit a large number of different surfaces, crawling upon them, and swimming off again. Once a suitable surface is chosen they secrete an initial semi-transparent tube, and metamorphose. These larvae will settle upon a glass or stone surface only if it is colonized already by a bacterial film, but they have a strong preference for *Fucus*. They are so markedly gregarious, that concentrations of recently settled individuals can reach 10–20 per cm^2 [62]. When a larva of *S. borealis* is freshly liberated it shows a marked gregariousness; however, after it has been kept swimming for 3 hr less choice is exercised and larvae settle rapidly both in isolation and gregariously. After swimming for 8 hr the rate of settling declines and many larvae are apparently too weak to settle successfully[63]. On the coast of Wales, at least, *Spirorbis borealis* occurs most abundantly upon *Fucus serratus*. Studies of the settlement of its larvae upon Tufnol panels, treated with an extract of *F. serratus*, revealed that the larvae selected extract treated surfaces in significantly greater numbers than the controls, indicating that settlement is a response to some chemical substance derived from the alga, rather than to changes in the physical properties of the surface. This reaction results from the ability of the larvae to detect the active substance as an adsorbed layer, rather than as a result of a concentration gradient in the water. The precise mechanism of the response is unknown, but compared with larvae in contact with untreated surfaces, those in contact with the substance undergo a curtailment of the photopositive and crawling behaviour evinced prior to settlement[64]. Another spirobid, *S. rupestris*, normally lives on rocks encrusted with *Lithophyllum polymorphum*. The larvae of this polychaete settle very readily upon *L. polymorphum*, but not on other coralline algae or upon fucoids[65, 66]. They will settle rapidly on slates only if they have been first soaked in an aqueous extract of *L. polymorphum*; this active agent will pass a dialysing membrane of average pore diameter 24 Å, but its effectiveness is reduced by treatment with heat and formalin[66].

The serpulid *Mercierella enigmatica* has a cosmopolitan distribution within a range bounded in the north by the 60°F July isotherm, and in the south by the 70°F January isotherm. It occurs in ports and estuaries and while it can tolerate salinities from 1 to 33‰, it does not grow well or become mature at salinities $< 5‰$ or $> 30‰$. *M. enigmatica* collected from Kuramo Water, Lagos Harbour, at the end of November and at a salinity of 7‰, spawned and gave rise to larvae which settled in salinities of 6–15‰ three weeks later. Others collected at a salinity of 21‰ in mid-February, spawned and gave rise to larvae which settled in a salinity range of 1·5–15‰, 6 weeks later. This limit of 1·5‰ is the lower limit for settlement under laboratory conditions. In contrast *Hydroides uncinata* settled only when the high tide salinity exceeded 30‰ and the low tide salinity exceeded 20‰. The best growth, in Lagos Harbour occurred in a salinity ranging from 20–30‰, but below 20‰ the populations died. The upper limit was unknown, but growth was poor at a salinity which fluctuated between 30 and 33‰[67].

The acorn barnacles are of great interest because the mechanism of their settlement has been so widely studied. The adult of the barnacle releases nauplius larvae at a time of year which is characteristic of the individual species. This larva spends a prolonged life in the plankton at the end of which, in the period before settlement, it becomes a cyprid larva which seeks a suitable surface for attachment, settles and undergoes a metamorphosis to the adult form[68–70].

The currents of their environment transport the nauplius and cyprid larvae in a manner consistent with their form and behaviour (p. 89) and in those areas in which the current is too strong, the cyprid is unable to settle (p. 359). Once settled and metamorphosed however, the barnacles *Balanus balanoides*, *B. crenatus*, *B. amphitrite*, *B. improvisus* and *Elminius modestus* are all capable of rotating their shells so that the carinae point away from the current source and the cirral net is facing the current: a similar orientation is observed in populations of *Coronula diadema* which take up the same orientation in respect of water flow past the whole host upon which they are found. The cypris usually orientates towards the light, and *B. balanoides* cyprids attach in greater numbers during the day than at night, but the orientation is modified by water current during growth[72-74]. The cyprids of sub-littoral and littoral species alike settle and begin to metamorphose when they are entirely submerged; the myriads of cyprids, often seen in the littoral, may appear to have been forced to settle when stranded in the intertidal zone, by the retreating tide, but this is not so[75].

During settlement, a cypris responds to surface texture[76-79] and to surface contour[80]. In the latter case, the tendency of the cyprid larvae of *Balanus balanoides*, *B. crenatus* and *Elminius modestus* to settle in grooves is described as a *rugophilic* tendency; indeed cyprids can orientate along grooves which are either much smaller or considerably larger than themselves[80]. Both *Balanus balanoides* and *Elminius modestus* which are epizoic upon the shells of *Littorina* exhibit a pattern of orientation in which the carinae are directed upwards when the shell is in a resting position. This orientation is considered to result from the response of the settling larvae to light from above, superimposed upon the thigmotropic response to the grooves in the shell. Such an orientation may be an advantage to the barnacle since the cirral net then opposes the current of water created by the *Littorina* when moving[81]. The stalked barnacle *Pollicipes polymerus*, which often occurs in dense masses in the mid-littoral of the U.S. Pacific coast, orientates itself so that the cirral net faces the rush of water which results when waves have broken over rocks and boulders. Here unlike the permanent rotation by means of which the acorn barnacles respond to a water current, *Pollicipes* rotates the whole capitulum by means of the muscles of the peduncle[82].

The acorn barnacles exhibit gregariousness in settlement to a marked degree[75, 83]. Where smooth glass panels are used as settlement plates, *Elminius* settles more readily if the individual plate is already colonized by recently settled barnacles, or if a barnacled microscope slide is attached to the glass plate. Moreover, settlement is always greater on those plates which are exposed near rocks colonized by barnacles than on similar plates on areas of bare mud 50–100 m away[86]. When settling near other barnacles, cyprids apparently make no specific recognition of spat of their own species[42, 84], but barnacles of other families have no appreciable stimulatory effect[84]. Cyprids may respond positively to bases of dead barnacles, but an abundance of older barnacles may make some cyprids leave a surface without metamorphosing[42].

However, the cyprid larvae in making a gregarious settlement require to make physical contact with settled barnacles or their bases, which contain an insoluble quinone-tanned protein, formed from anthropodin chemically cross linked with polyphenols, to which they react by a tactile chemical sense[85-87].

The pedunculate barnacles *Lepas anatifera* and *L. australis*, unlike the acorn barnacles, are attracted to less brightly illuminated situations at settlement. However, the cyprid larva, like those of the acorn barnacles, exhibits gregariousness and rugophilic behaviour in choosing a suitable site for settlement[20].

The settlement of most benthic animals is highly variable. In some seasons, the liberation and development of early stage *Balanus balanoides* nauplii is followed by a more or less catastrophic reduction in the number of cyprids. In other years, tremendous, widespread and successful settlement may occur, e.g. the 1969 settlement in many British waters. The results of a similar, widespread and successful oyster settlement in British Columbia is still evident some years after the event. In the Firth of Clyde, failure years for *B. balanoides* are related to abnormalities in the spring diatom outburst and the resultant failure of food for the developing nauplii: factors such as a lowered viability of larval input, increased predation, adverse physical conditions, adverse water quality or biotic factors related to the diatom increase are not apparently the direct causative factor[88]. Once a barnacle has settled much depends upon the site chosen for settlement, the species *B. balanoides*, *B. crenatus* and *E. modestus* settle equally well when permanently submerged in water of 2–4 ft depth. Once metamorphosed, however, *B. balanoides* suffers a heavy mortality, *E. modestus* a less heavy mortality and *B. crenatus* the least mortality for the three species. The heavy mortality of *B. balanoides* is probably due to interspecific competition with *B. crenatus*, *B. balanoides*, when it settles sublittorally, is at a disadvantage compared with this species because it has slow metamorphosis, lower growth rate and higher mortality, while its membranous base may make it vulnerable to leverage by the calcareous base of *B. crenatus*. Moreover, on the shore, ecological observation suggests that *B. balanoides* can eliminate *Elminius modestus* and *Chthamalus stellatus* by these means[42].

The bryozoans, which represent a stage of colonization intermediate between the balanoid, algal and hydroid communities and the subclimax ascidian or climax communities, settle most readily at the lowest water speeds[41]. Most species, e.g. *Watersipora cucullata*, do not settle readily on a clean, unfilmed surface, but *Bugula flabellata* does[89, 90]. Although, infra-littoral species apparently do not exhibit a marked preference in their choice of substratum, shore species do, especially with respect to algae[91, 92] (Fig. 12·4).

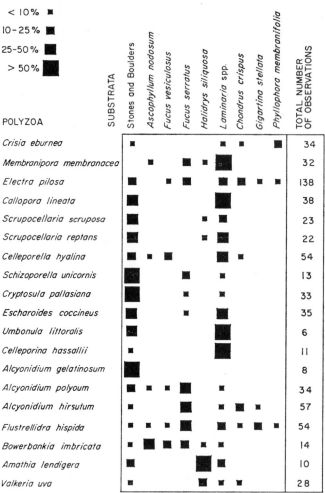

Fig. 12.4. The association of common polyzoans with stones and boulders and with the principal algae of the mid-littoral zone and the infra-littoral fringe. Observations concerning *Laminaria* refer to *L. digitata*, *L. hyperborea* and *L. saccharina*. *Membranipora membranacea* occurs mainly on the fronds of the first two species; *Electra pilosa* may also be found on fronds; *Celleporella hyalina* occurs on *Laminaria saccharina* fronds as well as on the rhizoids of the other species. The rest of the records are based on *Laminaria* holdfasts and refer principally to *L. digitata* and to some extent to *L. hyperborea* (after Ryland[92]).

Climax populations are composed largely of mussels, and some oysters may be present. The larva of the American oyster *Crassostrea* from Long Island Sound develops best at a temperature of 17·8°C and a salinity of 22·5‰. It is sensitive to the silt content of the water, amounts as low as 0·25 g/litre (i.e. 250 p.p.m.) and 0·5 g/litre (i.e. 500 p.p.m.) reduce egg survival to 73 and 31% respectively. At metamorphosis, a stage reached when the larva is 300 μm long, it drops to the sea bed, metamorphoses, and crawls over the bottom until it finds a surface free of dirt or silt when it becomes attached by its left valve. When offered the choice of old, clean oyster shells or artificial collectors made of plastic, cement, glass or resins, the oyster shells collect considerably more spat per unit area than the artificial surfaces[93].

The larva of *Ostrea edulis*, like so many other species, settles more readily when there is some shelter from the water current. In the Oosterschelde, its settlement is concentrated in the periods of slack water, but heavily silted surfaces are unfavourable to successful settlement and growth[94]. The larvae of *Ostrea* and *Crassostrea* exhibit gregariousness in settlement[95–98]. The larva of *Ostrea* is planktonic and when "full-grown", and eyed, metamorphoses. However, some are unable to perform metamorphosis, even though of a suitable size, a phenomenon which is apparently related to the illumination and dietary history of the larva prior to this stage[98, 99]. When they are ready to settle the larvae sink to the seabed where they crawl over the substratum by means of a ciliated and very mobile foot; after crawling for a period they usually swim off again. Such discriminatory behaviour indicates that they can delay attachment, at least to some extent, until a suitable substratum is found[100]: it has been suggested that copper will induce the metamorphosis of *Ostrea virginica*, but the evidence is conflicting[101–104]. The larvae of *Ostrea edulis* prefer surfaces filmed by diatom and bacterial slime[95], although Virginian oyster fishermen consider that *Crassostrea* which prefers clean surfaces to those covered in slime and detritus, settles more readily upon shell which has been cleaned previously[97]. The periostracum, outer layers of shell matrix, and body extract, all promote settlement and gregariousness in *Ostrea* and *Crassostrea*, but only when the larva has made contact with a surface which contains these materials[97, 98].

Like the oyster, the mussel, *Mytilus edulis*, has a free-swimming planktonic larva, a stage which has a duration of about 4 weeks. Settlement can occur at any time after the formation of the dissoconch, i.e. from 0·4 to 1·0 mm length, and is heaviest on floating structures in the region between the water line and a depth of 2 ft. On the beach, i.e. shore, intertidal structures and seabeds in estuarine areas, there is little persistent settlement above the M.T.L. and the heaviest settlements take place at and below

L.W.M.S.T.[105]. In the Menai Straits, N. Wales, the mature larvae of *Mytilus edulis* settle on filamentous substrates, viz. the red alga *Polysiphonia lanosa* (an epiphyte of *Ascophyllum nodosum*) and *Ceramium rubrum*, where it grows to a size ≤ 1·5 mm. These early plantigrades then re-enter the plankton, before finally settling on the adult beds, at a size between 0·9 and 1·5 mm. Such a mechanism may reduce competition between the very young and adult mussels[106].

The Effects of Fouling Growths

The effects of fouling growths are varied, depending upon situation, but the best known must surely be in connection with shipping. From ancient times mariners have been conscious of the influence of these growths upon the performance of their ships, and of the need to careen them periodically: a need which has coloured folk-lore and literature of the sea-faring nations. Indeed, much of the value of the Caribbean islands to the pirates who infested the area until the eighteenth century, must surely have been the discrete anchorages where such a vital operation could be carried out in security as well as the close proximity of rich sources of booty. Clearly, maritime business enterprises have always been subject to the influence of these varied, beautiful, but humble organisms.

All ships have a given fuel consumption for an optimum speed which depends on hull shape. Consequently fouling growths induce considerable losses of speed and significantly increased fuel consumption[24]. Figures quoted in the daily press suggest that on a large ship fouling may cost £10 000 per voyage in increased fuel bills. In those ports where it is customary to dock a ship with the same side always facing the landing stage, the uneven growth of *Enteromorpha*, which results, can make manoeuvering difficult[24]. This species is of particular economic importance in the operation of tankers[58]. Not all organisms affect the frictional resistance of a ship, but those like the encrusting bryozoans *Cryptosula* and *Watersipora* are of indirect importance because they provide a non-toxic substratum for the settlement of less desirable fouling species inherently more susceptible to antifouling toxins[24].

The mussel *Mytilus* affects a ship not only by increasing the frictional resistance to its passage, but by blocking the condenser intakes also. In power stations, which use a seawater cooling system, fouling organisms evidently affect the flow of cooling water in intake culverts. However, this is not the most important function of fouling here, for a large settlement of *Mytilus* between the fine rotating band or drum screens and the condensers can occur. When such a mussel, of a sufficient size, dies, the flesh rots, but the shell on falling off the culvert wall will pass to the condensers; it may become jammed in a condenser tube giving rise to *Impingment Corrosion*, and a puncture of the tube. The importance of such a problem will be appreciated when it is recalled that the modern power station boiler may be more than 100 ft high and is operated at temperatures in excess of 300 °C. At these temperatures and pressures, any impurity in the boiler feed water can result in rapid corrosion of the boiler tubes and possibly a disastrous explosion. Once impingement corrosion and a condenser tube puncture has occurred, that part of the system has to be shut down while the condenser tube is drawn and replaced: shut down of a large unit may cost a generating authority more than £2000 per day.

In some stations where conditions are particularly favourable for its growth, *Tubularia larynx* may form a vast mat-like growth on the walls of the intake culvert. Because of its fine, mat-like structure it sieves out large amounts of fine sediment. It grows larger in order to keep the hydranths free of the silt, but continues to filter out the fine sediment. Ultimately, the weight of sediment becomes more than the stolon structure, attached to the culvert wall, can bear. Large pieces of the *Tubularia*/sediment mat fall off into the culvert, roll along it, and become plastered over the end of the condenser tubes: a shut-down in order to clear the condenser end box follows.

Where breaks occur in the paint on a ship's hull, widespread corrosion may develop beneath the paint film. This corrosion is characterized by the presence of a black sludge-like deposit of iron sulphide. The process is anaerobic, and the sulphate reducing bacteria, *Desulphovibrio* spp., are implicated in the process[24] (see p. 119). Formerly, it was considered that the hydrogenase-containing strains of bacteria bring about the cathodic depolarization of the iron. More recently, it has been proposed that the principal contribution of the sulphate-reducers to the corrosion process is the formation of H_2S which, with perhaps other sulphur compounds, may act directly or indirectly to form corrosion cells with the metal[24, 107]. In those estuaries where the water quality is poor, a similar corrosion of iron work, exposed to the waters, may occur in the culvert systems of power stations.

Incrustation by fouling organisms such as the serpulid worm, *Mercierella enigmatica*, may prevent the closure of the sluice gates of a tidal harbour[108]. Mussels may give trouble at the sills of dock gates also. With the increasing consideration of fish farming in Britain comes the need to construct enclosures or other structures essential to the working of the farm. Here too fouling organisms are a nuisance and one of the costs to be borne by the enterprise. The fouling organisms increase the weight of the structure dramatically, especially when it is a netting. It is thus more liable to breakage and the flow of water through the meshes is impeded[26, 27].

Perhaps one of the most curious of all forms of fouling is that which occurs in the creeks and estuaries of Essex. Here, no doubt due to the tremendous enrichment which results from the release of sewage from London, a vast growth of *Enteromorpha* and other filamentous algae occurs, in the intertidal areas, in the summer months. From time to time, depending upon the tidal cycle, large mats of the algae are lifted off the mud flats and float away. Unfortunately, not all this material floats on the surface where it can be seen and avoided. Some floats at varying depths beneath the surface where it fouls trawls, dredges and other gear which may be broken by a large mat of algae. All such gear is expensive and troublesome to replace: local fishermen given these mats the unlovely, but highly descriptive name of "clodge".

Control of Fouling Organisms

A rough surface provides a good key for the settlement of fouling organisms[109, 110], conversely a smooth surface provides a poor key, and barnacles, for example, which settle upon a polished glass plate may be unable to maintain their hold. In practice, however, such polished surfaces, even if they could be prepared, would be unlikely to remain in an unpitted condition. Furthermore a surface which is unacceptable to some species, may not be unacceptable to others, e.g. fibrous glass surfaces which prevent the settlement of *Balanus crenatus* and *Pomatoceros triqueter* larvae are acceptable to *Tubularia larynx*[111]. In fish farming, a wide range of netting materials are theoretically available for use, but in practice the surfaces which these materials present for fouling organisms exhibit a wide degree of suitability (Table 12.2). Of the materials tested only the galvanized

Table 12.2. Increase in the weight of netting fabrics exposed to fouling activity (after Milne[26, 27]). (Reproduced by permission of the Controller of H.M. Stationery Office)

| Fabric | | Multiplication factor for weight increase | | |
Type	Wt kg/m²	July	Sept.	Nov.
Nylon	0·23	2	85	108
Ulstron	0·34	2	64	110
Courlene	0·20	2	85	126
Polythene—				
(a) standard	0·18	2	112	200
(b) cupra-proofed	0·18	1	44	94
Netlon	0·34	1	36	48
Plastabond	3·25	0·75	10	13
Galvanised—				
(a) chain link	2·03	0·3	0·5	2·75
(b) weldmesh	3·4	0·3	0·5	2·5

chain-link and weldmesh performed really well in discouraging fouling growth[26, 27]: a function, apparently, of the toxicity of the zinc. It is this relationship between surfaces, on ships and engineering works, and would-be colonizers which makes the problem of fouling difficult to combat: a statement of especial truth in connection with the cooling water culverts of power stations. Here it is impossible to obtain concrete which is smooth and free of bubble holes. These surfaces, combined with the bends, oints, and large chambers at screens, present an environment which is extremely favourable to the growth of fouling organisms. A situation which is in no way helped by the large flow of water which brings an excellent food supply to the inhabitants of the culvert.

Since surfaces, as such, offer a poor prospect for the control of fouling organisms most treatments depend upon chemical methods. Ships and those structures to which access can be gained esaily and frequently, if necessary, are protected by means of antifouling paints. Such paints normally have a mercury, copper or copper arsenical base[7, 22, 24]. In some specialized cases a layer of tar painted upon a zinc sheathing may be used[7].

Anti-fouling paints act by a leaching out of the toxic element which then poisons the settled organism[24, 112, 113], or poisons or disrupts the behaviour of the would-be settler[114]. Clearly, because of the leaching out of these toxic salts, the anti-fouling paint confers protection for a limited period only[59], and this relatively short useful life (although some may last for 1–2 years[118]) is a major reason for the lack of use of these paints in power station intake culverts.

Different organisms have a different response to the individual toxin, and to the different toxins present in antifouling paints. Some species of diatom are able to colonize a test plate soon after application of the paint (Table 12.3), and in this Australian situation mercury was more inhibitory

Table 12.3. Diatoms growing on test plates after 14 days exposure at Port Hacking, N.S.W. (after Crosby and Wood[22])

Species	Non-toxic controls	Mercury paint	Cold plastic MI. 143E	High leaching rate copper
Nitzschia longissima	+		+	+
N. closterium	+		+	
Amphora arcta	+	+	+	+
Licmophora abbreviata	+		+	+
L. flabellata	+		+	
Climacosphenia moniligera	+			
Grammatophora marina	+		+	
Cocconeis heteroidea	+			
Melosira moniliformis	+			
Coscinodiscus centralis	+		+	

than copper, although the actual paint concentrations are unknown[22]. However, in laboratory tests performed in England, the species *Amphora exigua, Amphiprora paludosa, Achnanthes longipes, Navicula grevillei, Nitzschia closterium* and *Cocconeis scutellum* were all more resistant to mercury salts than to copper salts[115]. The slime films which develop on the surface of an antifouling paint, which has copper as the toxic agent, rapidly acquire a concentration of copper many times that of a saturated solution of copper in sea water. Moreover, even when on a non-toxic panel, such a slime may rapidly remove and concentrate the copper from sea water[13, 23]. The hydroid *Tubularia* will survive exposure to a moderate concentration of antifouling toxins[116]; moreover, the settlement and metamorphosis of the actinula larva is accelerated after exposure to copper ions[10]. Evidence with respect to oysters, however, is conflicting[101–104]. Such observations indicate the essential problem of heavy metal toxicity both with respect to antifouling paints and pollution problems, viz. that whereas metals like Cu, Zn, Mo, for example, are toxic at relatively high concentrations, they are growth promoters and essential to life at very low concentrations (see also p. 553). Clearly, with old antifouling paint, once the failure point has been reached, growth of fouling organisms far from being retarded may well be encouraged.

Barnacles which have been growing in contact with an antifouling paint may exhibit a marked malformation of the basis and shell[24, 112, 113]. Like barnacles, *Tubularia* colonies attach directly to surfaces treated with antifouling paints, but the majority of the other hydroids appear as epizoites upon barnacles, tubeworms and encrusting bryozoans which separate them from the toxic layer, e.g. *Cryptosula* and *Watersipora* (p. 355).

Watersipora cucullata will settle readily on moderately toxic antifouling paints which contain copper, but it is sensitive to mercury. However, on copper containing paints, its rapid growth provides a non-toxic substratum for susceptible species[21, 29, 89]. Indeed, antifouling paints may alter the whole course of fouling by confining the more susceptible species to settlement upon the more resistant ones[24]. The toxic agents of antifouling paints are carried in a variety of paint matrices which include P.V.C. resin, copolymer vinyl acetates, polyisobutylene, glycerol phthalate resins and nitrile rubber. The bacterial slime will form on a surface exposed to sea water for 5 days, whether a copper containing antifouling paint is present or not. The addition of a bactericide to the paint matrix does not improve its antifouling performance, on the other hand bacterial accumulation of free copper will prevent the settlement of macrofouling organisms. The liberation of copper from an antifouling paint which contains resin or paraffin may be increased 2–4 times by bacterial action; further bacteria may use the paint matrix as a source of carbon. The leaching rate of toxin from an antifouling paint depends primarily upon character and concentration of the poison, pH, temperature and salinity: a decrease of pH having a marked effect. Sulphate-reducing bacteria, by producing H_2S, may transform cuprous oxide into sulphide which has little antifouling activity.

In recent years, copper has become very expensive and alternatives to its use in antifouling paints have been sought. In Japan and the U.S.S.R. arseno-organic poisons have been used with success. D.D.T. has been used, but is effective only against barnacles. At 1–5% presence of D.D.T. reduces the copper leaching rate by 50%, but in the presence of iron and copper, D.D.T. tends to breakdown and release HCl, simultaneously losing its toxic action[118].

In the difficult situation presented by the intake system to a ship's condensers, fouling can be controlled by the local generation of chlorine[117]. In those situations, where it is possible to run a ship into fresh or near to fresh water on frequent occasions, no fouling growths develop: in the Solway Firth area, many fishermen avoid the use of antifouling paints by these means[X].

In general, the control of fouling in power station intake culverts is achieved by the use of chlorine. These fouling organisms are relatively resistant to high doses of chlorine as a toxin[117, 119] (Table 12.4). Clearly,

Table 12.4. The time required to kill fouling organisms when treated with intermittent dosages of chlorine at various concentrations (after Turner, Reynolds and Redfield[119] and Lovegrove and Robinson[117])

Continuous treatment	Length of treatment to kill all test organisms (days)		
Species	10·0 p.p.m.	2·5 p.p.m.	1·0 p.p.m.
Anemones	4	6–8	15+
Mussels	5	5	12–15
Barnacles	3	4	5–7
Molgula	1	1	3
Bugula	1	1	—

Intermittent treatment	Survival in 10 p.p.m. chlorine (days)			
Species	1 hr/day	2 hr/day	4 hr/day	8 hr/day
Anemones	10+	10+	10+	10+
Mussels	10+	10+	10+	10+
Barnacles	10+	10+	10+	5
Molgula	10	5	5	5
Botryllus	8	3	2	2
Bugula	5	1	2	2

if this treatment were carried out, either intermittently or continuously to the point of success, i.e. mussel death, it would be very expensive. However, only the nuclear power stations and those conventional stations which are at the top of the efficiency league operate 24 hr a day, others operate more or less intermittently, and frequently at peak periods only. At the other end of the scale of chlorine treatment, the susceptibility of fouling organisms to a low residual doseage, i.e. 0·1 p.p.m. of chlorine is very variable[117] (Table 12.5). It is now considered in N. America that injection of chlorine

Table 12.5. The relative sensitivity of various major fouling organisms to the presence of residual chlorine at a concentration of 0·1 p.p.m. (after Lovegrove and Robinson[117])

	Plants	Animals
Very high resistance	Slime forming organisms e.g. bacteria diatoms, algae (*Ulothrix*)	
High resistance	*Entermorpha* (green) *Ectocarpus* (brown)	Hydroid *Tubularia* Cirripedes *Elminius*
Moderate resistance	*Polysiphonia* (red) *Ulva* (green)	Hydroid *Gonothyraea*
Low resistance	*Scytosiphon* (brown) *Laminaria* (brown) *Cutleria* (brown)	

at a continuous rate of 1·5–2·0 p.p.m., to give a residual of 0·5 p.p.m., is necessary to control and does control fouling growths in power station culverts. The Central Electricity Generating Board (of England and Wales) has found that injection of chlorine continuously to give a residual concentration of 0·5 p.p.m. is sufficient to control the common fouling organisms, viz. hydroids, barnacles and mussels. This highly successful method of chlorine injection does not have to be closely related to the chlorine demand arising from ammonia and other organic compounds which occur naturally in sea water. However, when H_2S, metallic sulphides and mercaptans are present, in the water, special care is required[120].

With regard to the corrosion of iron work that results from the action of sulphate-reducing bacteria, clearly a paint with an improved ability to key onto such metal work is needed. All conventional and epoxy paints can be punctured and subsequently rise to let in the bacteria which cause the corrosion. A new protective covering which has the desired qualities has been produced by the British Columbia Research Council, Vancouver. This covering has the advantage that it can be applied by brush or spray to a sand blasted surface while under water. It has an affinity for iron, and an ability to bind up residual water. It binds to an iron surface with a strength which is great and which increases with time. It has the further advantage that it will bend under compression and tension without cracking and corrosion does not creep beneath it[121].

Boring Organisms

Although very many marine animals burrow into the substratum, only a relatively few species have such an impact that they are considered to be a nuisance or pests. Clearly, the burrowing or boring habit ranges from such animals as *Macoma balthica* an abundant resident of estuarine sediments, but only of those sediments, to the piddock, *Pholas*, which can inhabit a range of substrata from stiff clay to a compact limestone. The borers, of economic importance, can be defined under two headings, viz. borers in wood, and borers in rock. It is convenient to discuss them under these headings, although it must be appreciated that rock borers can and do burrow into wood, e.g. *Pholas*[4], *Barnea candida*[122], *Zirphaea crispata*[X].

Wood-boring Animals

Wood-boring animals are drawn from two phyla, viz. the lamellibranch molluscs, e.g. *Teredo*, and the crustacean arthropods, e.g. *Limnoria*. The best known of the molluscs are the shipworms, often referred to collectively as teredo, composed of 66 species which belong to the genera *Teredo*, *Bankia*, *Zachsia* and *Bactronophorus*, and of which only the first two are of major importance[4, 123]. In addition to the highly evolved and aberrant shipworms, some other genera from the related family *Pholadidae* are characteristically wood-borers, viz. *Xylophaga* and *Martesia*[4, 124].

Although the genera are world wide in distribution, that of the individual species is more limited. In general, the distribution of the individual species is based on broad geographical areas related to temperature. *Teredo navalis* is primarily responsible for damage done by these animals on the Dutch coast, in the Thames estuary and elsewhere on the east coast of England. Further north on British coasts, *T. navalis* is replaced by *T. norvegica*, a larger species which ranges north to within the Arctic Circle on the Norwegian coast. However, *T. navalis* is apparently unique in that it is found on the Atlantic coast of N. America from the Panama Canal Zone to Newfoundland, and also in British Columbia on the western coast; it is also found in Australia. The distribution of *T. norvegica* outside Europe is not so widespread as *T. navalis*, but it has been recorded from Australia and New Zealand[4, 123].

It is of interest, that on both the eastern and western coasts of North America, the greatest number of species is found in the tropics. The number of species declines with distance moved northwards, thus: 23 species live in the Panama Canal Zone; 16 species live off the coast of

Florida; 10 species live off the coasts of Virginia, North Carolina, South Carolina and Georgia; 3 species live off the coasts of New Jersey, Delaware and Maryland; and from New York northwards only 2 species are found. Curiously, only 7 species live in the offshore waters of Texas, Louisiana and Mississippi. On the Pacific Coast, where the water is colder, due to upwelling, Californian waters have three species, and from Oregon northwards only two species are found[123]: along the whole of this coast the most important species is *Bankia setacea*[25, 123, 135].

Some species of shipworm are hermaphrodite[4], and all have a planktonic larva which resembles those of the other bivalve molluscs[4, 123]. These larvae, which may be as small as 50 μm in diameter, need to settle after a limited period of planktonic life: about 3–30 days depending upon species[123]. Test blocks require up to two months immersion in sea water before the teredo larvae will settle: this may be sufficient time to permit either the leaching out of resins and toxins, and the softening of the wood fibres, or the accumulation of an attractive film of bacteria and diatoms[126–128]; cellulolytic bacteria and fungi facilitate the entry of shipworms into wood, but since these animals cannot progress along a gradient of decreasing wood hardness, help to maintain them within the wood subsequently[128a]. After a longer period, six months say, fouling organisms, e.g. ascidians, may have settled so thickly that they prevent further settlement by shipworms[126, 127]. Settlement may be prevented by water current velocities which exceed 1·4–1·8 kt (Table 12.1).

Once a teredine larva has settled, it sheds its cilia and, by means of a large muscular foot, searches the surface of the wood for a suitable spot in which to commence boring. Once such a location has been found, it attaches by means of a byssus and starts to scrape away the fibres of the wood with the edges of its valves. At this stage, it covers itself with a protective case which is formed with particles of wood and other material stuck together. Under this cover, it rapidly disappears beneath the surface of the wood, leaving only a "pinhole" to mark its entry. This hole is enlarged only a little in later life, and such holes are so inconspicuous that a piece of timber may be riddled with teredines without any obvious signs of the infestation upon the wood surface[4, 123, 129]. Initial entry is at right angles to the surface of the wood, but the mollusc soon begins to burrow along the grain, but whatever the intensity of the infestation, the individuals do not breach the burrows of other shipworms, and pronounced avoiding action may be taken[4]. In Plymouth Sound where concurrent infestations of *Teredo navalis* and *T. norvegica* occurred, the former species lived at a more superficial depth in the wood than the latter[124].

The size of shipworms, at maturity, depends upon species, some reach a length of 6 in. only, whereas others such as *Bankia setacea* and *Teredo dilatata* can grow to 4 ft[4, 123]. The rate of growth and the size reached by the adult varies according to species, and probably food supply and temperature[4]. At Beaufort, N. Carolina, *Bankia gouldii* grew from 0·25 mm at settlement to a length of 100 mm in 36 weeks: at this site *Teredo dilatata* were considered to reach 4 ft in *ca.* 1 year[130]. At Plymouth, Devon, *Teredo navalis* excavated burrows 11 in. (280 mm) in length in 31 weeks; however, full-grown specimens of *T. navalis* may be 12–16 in. in length. A related Australasian species may reach a length *ca.* 5 ft 10 in[4].

On the coast of British Columbia, *Bankia setacea* penetrates wood to a depth of $\frac{1}{8}$ in. in the first month, to *ca.* $\frac{1}{2}$ in. by the end of the second month, and subsequent growth continues at a rate of 2–3 in. per month. It lives for at least two years, and attains a length of 3–4 ft[131]. Growth may be depressed when intense populations create crowded conditions[25], but when borers are so crowded, they exhaust the supply of wood, and being unable to migrate die in the first 6–8 months. Unlike *Bankia setacea*, *Teredo navalis* rarely exceeds a length of 6 in. or a life span of 4–5 months, in the waters of British Columbia[131]. Under crowded conditions in a 6 in. × 6 in. wood block, *Bankia australis* will grow rapidly to a length of 120 mm, at which point growth slackens off; in uncrowded conditions it may reach a length of 200 mm[126].

The teredine borers can live in a wide range of salinity, although at lower salinities their activity may be affected. *Teredo navalis* and *T. norvegica* may remain active to a salinity of 6‰, but 4–5‰ is lethal[4, 122]; however, *T. navalis* larvae require a salinity of 12‰ for development to proceed normally[123]. The great outbreaks of *T. navalis* in Holland in 1730–32, 1770, 1827, and 1858–59 are said to have followed periods of low rainfall, when the salinity of the Zuidersee and coastal waters was unusually high. This species also penetrates further upstream in San Francisco Bay during the summer and periods of extreme drought[4]. *Bankia setacea* maintains its normal boring action in salinities down to 12‰, but ceases boring at 9‰[123]. In laboratory studies, the adult of *B. setacea* exposed to salinities of 8 and 10‰ died after 6 and 10 days exposure respectively[132]. The larve of this species will develop to the veliger stage in salinities ranging from 16 to 40·4‰, but at 14‰ and less, the larvae fail to develop [133]. Some tropical species are said to live in fresh water[4]. *Teredo navalis* can live in waters with an oxygen concentration as low as 1 p.p.m. *Bankia setacea* can live in the presence of waters, which are heavily contaminated with waste sulphite effluent (see p. 541), and the levels of oxygen are essentially zero for a part of the year[123]. It is of interest that if the concentration of the potassium ion or the magnesium ion is increased by 1‰, *B. setacea* is rendered inactive[123]. In general, the presence of significant amounts of silt, sewage or factory waste will affect teredine populations adversely[4].

Teredo navalis is most active in the temperature ranges 15–20°C, is tolerant to 30°C and becomes inactive below 5°C, although it can withstand a period of freezing. *T. norvegica*, a more northern species, is adapted to lower temperatures[122]. The optimum temperature for the larvae of *T. navalis* ranges from 20 to 30°C, but temperatures < 11·7°C and > 35°C are lethal[123]. Breeding is related to temperature and in those areas in which the annual range of water temperature is fairly narrow it continues throughout the year, this is true even in more northerly locations such as on the Pacific coast of the U.S. around Seattle and Tacoma where the winter temperatures seldom drop below 7·2°C. The attack by *Bankia* ceases at 7·2°F in most locations, but a temperature from 12·8 to 18·2°C is required for it to recommence[123]. In San Francisco Bay, the breeding season of *B. setacea* extends from February to July, with a peak in April–May[4, 25].

The New Zealand species *Bankia australis* breeds throughout most of the year, but ceases when the water temperatures fall to 5°C. It recommences breeding about three months later when the temperature is rising and may have reached 10°C[126].

In British waters, *Teredo navalis* may breed at less than 1 in. in length, it requires a sea water temperature of 15°C for breeding to commence and an optimum is reached *ca.* 20°C[122]. In San Francisco Bay, *T. navalis* breeds from July to January. At Wood's Hole, Massachusetts, larvae are present in the water from early May to early October. On the coast of Holland its larvae have been found settling in wood from July to September[4, 124].

Teredo norvegica is able to breed all the year round in British waters, but in the colder Norwegian waters the most intense attack on timber occurs in June and July[122].

The range of depth distribution of the shipworms is uncertain. They may attack timber at least as high as the mean tide level[4, 122]. The wood piddock *Xylophaga* has been found in telegraph cable at a depth of 1500 m, *Bankia australis* has been recovered from a ship's wooden guardrail which had sunk to a depth of 70 fm[134] and *Teredo norvegica* was found in the Naples–Palermo telegraph cable raised from a depth of 3500 m[135].

It has been known since 1923, that the shipworms can digest wood[4, 134], although since then there has been some debate as to the necessity for wood in the teredine diet. As one might expect, these organisms have a specialized enzyme system, in order to deal with this food source[136–139]. However, as much as 11%, by volume, of the intestinal contents may be diatoms and other forms of skeletal material; moreover, teredine burrows are little larger than the mollusc itself, and bearing in mind normal conversion factors, and the fact that up to 50% of wood may pass through the digestive tract of the teredine unchanged, it does not seem to be a very good food supply for a fast growing animal[134]. Further, although neither *Bankia* nor *Teredo* have been recorded as rock-borers, they have been recorded from a variety of materials other than wood and including hempen rope, gutta percha (in telegraph cables), jute, manila cordage, cork, plastics and lead[134, 138]. Laboratory studies have shown that *Bankia setacea* can settle and grow on non-wood substrates based on agar gel. In these studies, wood was shown not to be essential to the shipworm, and non-cellulosic wood fractions were not necessarily desirable. Cellulose or wood flour were desirable additives to the culture media[138].

In the estuaries and creeks of Essex, *Teredo navalis* is abundant wherever suitable, water-logged wood is available for settlement. In general, the most intense infestation occurs near to the mud surrounding the stake or pile. This suggests that the rich supplies of microbenthonic food available, at this level, are important to the shipworm[X]; although it must not be forgotten that micro-organisms are probably most able to attack the wood at this level also (see p. 188).

Except for the shipworms, the gribble, *Limnoria*, is the most destructive animal which attacks wooden structures in the sea. This arthropod genus is composed of six species, of which the most widespread and destructive, *Limnoria lignorum*, is cosmopolitan and has been reported in Europe—from the Lofoten Islands to the Black Sea; North America—Atlantic and Pacific coasts; the Falkland Islands; South Africa; Australia and New Zealand. This small isopod may be accompanied by the amphipod *Chelura terebrans*, which together with *C. insulae* achieves a world-wide career of destruction. In warmer, brackish or fresh waters, a larger isopod *Sphaeroma* may be important: damage by this isopod has been reported from California, Florida, South Africa, India, Ceylon, Australia and New Zealand. In San Francisco Bay, *S. pentadon* attacked clay and friable rock in addition to wood[4].

Limnoria is much smaller than the shipworms, and has a length $ca.$ $\frac{1}{8}-\frac{1}{6}$ in. It burrows obliquely to the surface of the wood leaving a tunnel $ca.$ $\frac{1}{20}$ in. diameter, each tunnel may reach a length of $1\frac{1}{2}-2$ in., but the average depth of penetration rarely exceeds $\frac{1}{2}$ in. The tunnel diameters are uniform because with each moult and increase in size the animal begins a new one. Unlike those of the shipworms, the tunnels break into each other from all directions. Consequently, the superficial layers of wood are soon reduced to a friable, spongy mass which is easily broken by wave or mechanical action. Once the superficial layers have fallen away fresh galleries are constructed. Unlike the damage due to the teredines, with which they may occur, that due to *Limnoria* is overt.

The gribble, like most crustaceans, is monoecious, i.e. the sexes are separate, and the eggs are carried in the brood pouch of the female until hatching occurs. A single brood may vary in number from 8–29 young. On hatching the young resemble the adults, in all except size, and join the adults in their attack. The adults are apparently more vagrant than the juveniles[4].

Although *Limnoria* may occur near the high-water mark, it is most abundant between low water and the half-tide level[4]. There are three species in British waters, viz. the boreal *L. lignorum*, which is found on all coasts, is least abundant in the south and west, where the temperate *L. quadripunctata* is abundant. This species also occurs in the Isle of Man, Ireland and the Channel Isles, while *L. tripunctata* is found only on the coasts of central southern England and South Wales. Each of these three species can survive at shore levels equivalent to 60–65% exposure. However, in mixed populations, *L. tripunctata* dominates the upper level of *Limnoria* attack, *L. quadripunctata* dominates the middle zone and *L. lignorum* is found at the base of exposed piling or sub-littorally[139]. In the Clyde, *L. lignorum* characteristically attacks wood sub-littorally; that at the base of piling works is particularly severe at, or just above, the level of the substratum[X].

In Monterey Bay, California, the attack of *L. quadripunctata*, upon wood exposed sub-littorally, is not serious[25]. Further north in Alameda Harbour, San Francisco, *Limnoria* infestation of wood blocks was greatest near the surface, declined rapidly to 15 m, and remained nearly constant to the mud line at 35 m depth. In similar blocks, similarly exposed, at Elliot Bay, Seattle, *Bankia setacea* infestations increased progressively to the mud line. *Bankia* and *Limnoria* infestations of the test blocks are characteristic of the individual species: *Bankia* occurs predominantly upon the upper horizontal surface, and *Limnoria* occurs predominantly on the lower horizontal surface[128]. In the Mediterranean it has been found, with *Chelura*, in submarine telegraph cable at a depth of 290 fm[4].

Field data suggests that a salinity of 15–16‰ is limiting for *Limnoria*[4, 140], and that the gribble does not penetrate estuaries as far as *Teredo navalis*[4]. Laboratory studies have shown that boring by *Limnoria* decreased linearly with salinity and ceased altogether below 10‰. The limiting salinity varied inversely with temperature, and was 12 and 14‰ at 17·6 and 14·9°C respectively. A salinity of 6‰ is lethal in 65 days, but in hypersaline conditions survival is not reduced up to a value of 48‰[140]. *Limnoria* has a relatively high oxygen requirement and this is apparently the reason why it confines its boring activities to the superficial layers of the wood, for provision of an adequate oxygen supply at a greater depth would be difficult[4, 141]. Over a range of salinity from 6 to 36‰ *Limnoria* shosw no significant correlation between the rate of respiration and the dilution of the medium[141]; it is able to regulate its blood concentration to some degree, and in reduced salinities maintains the blood at a hyperosmotic value[142]. *Limnoria* has a limited ability to digest cellulose, and exists primarily upon the diatoms and bacteria from rotting wood[127].

Damage Due to Wood-boring Organisms

Wood-boring animals, especially the shipworms, are an ancient enemy of seafaring mankind, and as such attracted the attention of Greek and Roman writers. The shipworm has attacked the triremes of Athens, the galleys of Venice, Drake's Golden Hind and present day oyster smacks with equal impartiality[4, 122]. It caused particular concern during the revival of wooden boat building which followed World War I. During the drought years of the eighteenth and nineteenth centuries, the Dutch suffered particularly when the great teredo infestations of the dykes threatened the nation with disaster[4]. Today, wherever wooden structures are used in the salt waters of the world, these organisms still affect the activities of man and threaten his security.

With the introduction first of the all-steel and then of fibreglass hulls, the impact of the wood borers upon shipping has decreased dramatically, especially with the more advanced nations who are the prime users of these materials. Similarly, those nations which are rather poorly endowed with natural forests have replaced, or are tending to replace, wooden pilings with those of concrete. In countries like Britain, therefore, the marine borers are of a reduced importance, purely because wood is now used less widely than in earlier times.

However, in North America, and particularly on the Pacific coast, where forests are still of primary economic importance, the situation is completely different. Here wood is the building material par excellence, and most pilings and harbour structures are made of this material. Moreover, lumber and pulp manufacture is the major industry which means that large numbers of logs have to be transported and stored before processing. In British Columbia, the sea is a great highway and, since level ground is scarce, it is also used for storage. The actual loss of logs due to *Bankia setacea*, along the coast of British Columbia, is difficult to estimate, but is of the order of $500 000 to $1 000 000 per annum. Additional damage to standing booms and piles, A-frame floats and floating camps brings the annual loss to the logging industry to well in excess of $1 000 000. In the absence of treatment the effect of marine borers upon dry docks is very costly (p. 385)[125, 143].

13*

Methods of Control

Tests by the British Columbia Research Council have shown that there is no practical difference in susceptibility to attack or, once infested, in the rate of destruction of Douglas fir, hemlock, Western red cedar, yellow cedar (cypress), spruce, balsam fir, white pine, alder, birch, maple, oak and cottonwood[125]; while oak and teak are attacked as easily as the coniferous woods[4]. Some tropical woods have a natural resistance to attack and apparently owe their immunity to essential oils or alkaloids; such woods include the South American Greenheart (*Nectandra rodioei*); the Australian Jarrah (*Eucalyptus marginata*) and Turpentine (*Syncarpia laurifolia*); and the New Zealand Totara (*Podocarpus totara*). Others like the manbarklak resist attack supposedly because of their high silica content[4, 123]. However, no one species of wood has been found naturally resistant to all marine borers in all geographical locations. In one location a wood species may give extremely good service life; elsewhere it fails miserably[123].

The methods available to control wood-boring animals fall into two classes: (1) those methods which aim at protecting a surface so that borers cannot enter the wood; and (2) methods of impregnating the wood with substances which are either poisonous or at least distasteful to the would-be borer.

The earliest form of surface protection seems to have been to sheath the wood at risk with metal. Greek and Roman galleys were sheathed with lead, a method that was still in use in the seventeenth century. Copper sheathing was introduced in the eighteenth century and in more recent times, zinc and Muntz metal have also been used. These methods have been used upon boats in particular. Piles may be protected by these means, but casings of concrete or cement are more usual. In harbour works and piers, a covering of broad-heading iron nails ("scupper-nailing") may be used; here, the wood around each nail becomes impregnated with rust and confers protection. In other situations, surface applications of tar, pitch or suitable paints may be used. However, all of these methods are open to the objection that once a small unprotected space has been created, or if one is left during the "sheathing" process, then enough shipworm larvae to destroy the interior timber may gain ingress. The attacks of crustaceans are more readily controlled by these means, because the damage they cause is more localized and superficial than that caused by the shipworms[4].

Impregnation of the wood is normally based upon the use of creosote. However, *Limnoria tripunctata* is creosote resistant, and in western N. America, at least, this species causes considerable damage[128, 143], here pile barriers are frequently used to protect impregnated piles from attack[128]. To improve the performance of creosote impregnation, additions based upon organo-lead, organo-uranium or zinc based compounds, are made to the creosote[143].

A prerequisite of all impregnation methods is that they shall be performed adequately, but even then they have a limited life because the creosote is gradually washed out of the wood[4]. In the River Blackwater, Essex, a temporary wooden jetty was built in 1957 to facilitate the construction of a barrier wall for Bradwell Nuclear Power Station. These timbers were treated, nevertheless when they were drawn in 1960, some of the 15 in. sq. timbers had lost 76% of cross-section, and an average loss of 46% of cross-section had resulted from the attacks of *Teredo navalis*[122].

An essential difficulty of all control methods, especially those used in connection with the teredine borers, is that the toxins have a taste to which the boring organism reacts. The British Columbia Research Council have developed a method of treatment of sawmill lumber by sodium arsenite spray; structures such as roadcribbing, docks, dolphins and boats may be treated by first enrobing the structure with polythene sheet. Pilings are not easily treated by these means. Saw logs may be treated in ponds[123, 125, 129, 131 143–146]. In the development of this method, the likely effects of the sodium arsenite upon nearby oyster beds and other marine life was considered and investigated. Results indicated that no obvious adverse effect was likely to result from this treatment[147]. The economic aspects of this treatment are of interest. A large size Davis raft may contain timber to the value of \$75 000–\$80 000 and, in the absence of treatment, a loss of up to 10% may be sustained. The cost of chemical treatment is *ca.* \$400 dollars, so that the saving per raft as a result of treatment is good. Where flat rafts can be treated by a tidal ponding technique, the treatment cost for a raft 65 ft × 450 ft is \$75·00; in open water the same treatment would cost \$200. The cost of treating a dry dock is *ca.* \$200–\$300 per dock per treatment, and two or three treatments are required per year. Nevertheless, one dry dock operator has estimated that this protective scheme applied to two docks, amortized over 10 yr, will result in a maintenance saving of *ca.* \$1 500 000[125].

Another innovation to protect saw-mill logs was to establish monitoring stations for marine borers, so that attack between the surface and 20–30 ft depth could be assessed. This can be checked continuously by exposing blocks of Douglas fir in a vertical series to the desired depth. Use of this method, which permits forecasts to be made of larval attack, has eliminated the need for chemical treatment[123]. In British Columbia, the use of this method and of sodium arsenite has reduced the loss to the sawmill business, due to these animals, from \$1 000 000 to \$10 000 per annum.

Because harbour pilings are subject to continual abrasion by ships, and by logs and other dejecta in the waters of the harbour, they are particularly exposed to borer attack. Attack due to *Limnoria* is overt and easily assessed during an examination of a pile, but that due to the teredines is not readily assessed. Clearly, it is vital that load-bearing structures should be sound, and if they become unsound, that this should be recognized before they collapse. The inspection of such piling has been improved significantly by the introduction of non-destructive testing by sonic means. Three or four bankian channels cause essentially no loss in pile strength, and to downgrade or reject a pile on this basis would be misleading and costly, since the cost of replacement of the individual wharf pile is *ca.* $400. This high replacement cost combined with the even higher risk and financial loss arising through structural failure, makes the cost per day of inspection by sonic means, viz. $400–$500, with pile inspection at a rate of $50–$125 per day, very reasonable[123, 148, 149].

In parts of the world where wood in the sea, derived from a variety of sources is abundant, viz. trees living at the head of the shore falling into the sea during storms, massive shore strandings resulting from the break-up of Davis and other rafts, and waterlogged timber sinking in holding areas, then populations of wood-borers are difficult to eliminate. However, in those areas, like the British Isles, where wood supplies are limited and the use of timber in harbour structures is declining, much of the continued infestation by *Teredo*, in particular, is probably due to old, rotten, wooden hulks lying around in estuaries and creeks. In these situations, the impact of marine borers can be minimized by the removal and burning of these loci of infection. Indeed, as a matter of policy, such old hulks are not tolerated in the Old Port of Marseille, and here infestations of *Teredo norvegica* are rare[7].

Rock-boring Animals

The marine animals which bore into rock are more numerous and diverse than those which bore into wood, and include sponges, polychaete worms, crustacea, molluscs and echinoderms.

Boring sponges of the family Clionidae occur commonly in old mollusc shells, but are found in limestone rock also. An oyster shell may be attacked by *Cliona celata* which leaves holes about $\frac{1}{20}$ in. diameter on the surface and riddles the mass of the shell with branching and interconnecting tubes. When the shell has been demolished, the sponge may grow into a massive form, which, at its largest, is the well known Neptune's Cup sponge. When burrowing into limestone rock, or into limestone pebbles, the tubes created by *Cliona* are generally confined to a superficial depth[4].

The polychaete worm *Polydora* can live either as a fouling organism[5] or burrow into substrata ranging in hardness from that of London Clay to limestone, shale and sandstone[4, 150]. Although burrowing into softer materials may be achieved by abrasion with the uncinate chaetae of the fifth segment, this method seems to be improbable for use with limestone. Here, it has been suggested that the worm employs a sequestering agent linked with the biochemistry of mucus[127, 150]. When mature, the adults of *Polydora ciliata* lay eggs into a string of capsules, which are apparently secreted by the nephridia, and are attached to the wall of the tube. The larvae take a week to hatch and then spend some 6 weeks in the plankton before metamorphosis. At Whitstable, Kent, there are two larval settlements, one in late March and the other in May. Apparently, these two settlements result from two different populations which are derived from the North Sea and English Channel respectively and which overlap in distribution in this area[151]. In addition to burrowing in a variety of substrata, *Polydora* burrows into mollusc shells also[150, 152].

In some areas, the mussel, *Mytilus edulis*, is much affected by the parasitic copepod *Mytilicola intestinalis*. At Whitstable, where *ca.* 25% of the mussels are infested by *Polydora ciliata*, the number of *Mytilicola* is significantly less in a host in which this polychaete is present in the shell[152].

Although some barnacles may burrow into limestone and coral, the isopod *Sphaeroma quoyana* is the only crustacean known to have an economic effect upon rock structures. This species burrows into soft sandstone or claystone in Australia and New Zealand[4, 127].

Although the bivalve molluscs are characteristically burrowers into the sedimentary substrata, some can live attached to rock surfaces, e.g. *Mytilus*, *Crassostrea*, and others have adopted the boring habit. These last species are capable of boring even hard and compact rock, a habit which is characteristic of certain families such as the Pholadidae, but has been acquired by various species of unrelated families also.

The pholads are distributed throughout the world, and all burrow entirely by mechanical means. The boring movements are performed thus: the foot first takes a tight grip of the end of the burrow and draws the shell forward; the abrading surfaces engages the rock by diverging the anterior ends of the valves, a result of muscular action; and the shell is then rocked through 90°[4, 127]. By these means, the pholads can burrow into wood, peat, shale, micaschist, sandstone and limestone. In the case of *Pholas dactylus*, an animal of 5 or 6 in. length, the burrow created may be 1 ft deep[4].

The venerid, *Petricola pholadiformis*, was introduced from North America to the east and south-east coast of England in 1893. Superficially, this animal closely resembles *Barnea candida*, and other pholads, although the Veneridae and the Pholadidae are distinctly separate families[4, 153]. *Petricola* has been found burrowing in a range of substrate from London Clay to chalk. This species is found on the coast of West Africa, in addition to Britain and North America[153].

Although it has a thin shell, lacking spines or teeth, but covered by a thick periostracum, the Mediterranean Date-shell, *Lithophaga lithophaga*, is an accomplished borer. This member of Mytilidae, and closely allied to the edible mussel, creates its burrow by means of an acid secretion. This secretion is produced by glands which are situated in the mantle and are absent in the non-burrowing forms. Its boring is restricted to limestone and other calcareous rocks[4]. A similar habit is adopted by the New Zealand Date mussel *Zelithophaga truncata*[127].

The hiatellid, *Hiatella arctica*, is, perhaps, the commonest British rock-boring mollusc. This species has a range of habit from nestling among, and attached by, byssal threads to granitic, bound shingle near to the L.W.M. in Scottish west coast sea-lochs[X], to burrowing in hard compact limestone[4]. In New Zealand, *H. australis* has a limited boring ability, and lives in wood, algal holdfast, and hard clay in addition to rock; the young bivalve attaches, by a byssus, in a niche that it proceeds to enlarge[127].

Damage Due to Rock-boring Animals

The destructive action of rock-borers is less rapid than that of wood-borers. In general, engineers consider that little economic harm results from their activities, but they must facilitate erosion in coastal areas. Although engineers differ in their opinions, the breakwater at Plymouth is extensively burrowed by the sponge *Cliona celata*; the polychaetes, *Polydora ciliata*, *P. hoplura*, *Potamilla reniformis*, *Dodecaceria concharum*; and the molluscs *Hiatella arctica* and *Gastrochaena dubia*.

The crustacean, *Sphaeroma quoyana*, damaged the stonework at Wairoa Hawke's Bay, New Zealand. The brickwork of Calcutta Docks was damaged by the bivalve mollusc *Martesia fluminalis*, while *Pholadidea penita* destroyed concrete, of poor quality, at San Francisco and on the Alaskan coast. *Lithophaga aristata* is said to have damaged concrete at Panama.

The selection of stone for use in marine structures is likely to be influenced by considerations other than its resistance to boring organisms. Nevertheless, it would seem wise to avoid the soft and friable rocks which are suitable for the mechanical borers, e.g. *Pholas*, *Sphaeroma*, and the calcareous rocks which are susceptible to attack by *Lithophaga* or *Cliona*, for example. Clearly, whatever the ability of borers to penetrate concrete, the rubble derived from calcareous or friable rocks is not a suitable component for its preparation[4].

The polychaete *Polydora* can be a troublesome pest of oysters, in which the shell may be perforated by its galleries[150, 154–156]. By its boring action, this polychaete makes the oyster expend energy, which it otherwise would not, upon extra shell secretion: living space within the shell is more confined due to the formation of blisters in the nacreous layer[156]. From the point of view of the oyster fisherman this may be a relatively unimportant consideration. To him, it is important that the *Polydora* makes the shell much more friable, and liable to break when it is being opened. In these conditions, the oyster is a less desirable product for the market.

References

1. Morton, J. E. (1954). *J. mar. biol. Ass. U.K.* **33**, 187–224.
2. Morton, J. E., Boney, A. D. and Corner, E. D. S. (1957). *J. mar. biol. Ass. U.K.* **36**, 383–405.
3. Kensler, C. B. (1961). *J. Anim. Ecol.* **36**, 391–406.
4. Calman, W. T. (1936). "Marine Boring Animals Injurious to Submerged Structures", British Museum (Natural History), Economic Series, No. 10, 38 pp.
5. Persoone, G. (1965). *Helgoländer wiss. Meeresunters.* **12**, 444–448.
6. ZoBell, C. E. (1939). *Collecting Net.* **14** (5).
7. Berner, L. (1944). *Bull. Inst. océanogr.*, No. 858, pp. 44.
8. Scheer, B. T. (1946). *Biol. Bull. mar. biol. Lab., Woods Hole.* **90**, 244–264.
9. ZoBell, C. E. (1946). "Marine Microbiology", Chronica Botanica Co, Waltham, Mass.
10. Pyefinch, K. A. and Downing, F. S. (1949). *J. mar. biol. Ass. U.K.* **28**, 21–43.
11. Allen, F. E. and Wood, E. J. F. (1950). *Aust. J. mar. freshwat. Res.* **1**, 92–105.
12. Hendey, N. I. (1951). *Jl. R. microsc. Soc.* **71**, 1–86.
13. Woods Hole Oceanographic Institution (Comp.) (1952). "Marine Fouling and its Prevention", Prepared for Bureau of Ships, Navy Department, U.S. Naval Institute, Annapolis, Maryland: 388 pp.
14. Wood, E. J. F. (1953). *Aust. J. mar. freshwat. Res.* **4**, 160–200.
15. Smyth, J. C. (1955). *J. Ecol.* **43**, 149–171.
16. Wood, E. J. F. (1956). *Pacif. Sci.* **10**, 377–381.
16a. Board, P. A. and Collins, T. M. (1965). *Discovery* May 1965. 5 pp.
17. Barnard, J. L. (1958). *Calif. Fish Game* **44**, 161–170.
18. Skerman, T. M. (1958). *N.Z. Jl. Sci.* **1**, 224–257.
19. Skerman, T. M. (1958). *N.Z. Jl. Sci.* **1**, 402–411.
20. Skerman, T. M. (1958). *N.Z. Jl. Sci.* **1**, 383–390.
21. Wood, E. J. F. and Allen F. E. (1958). "Common Marine Fouling Organisms of Australian waters", Department of the Navy, Melbourne: 23 pp.
22. Crosby, L. H. and Wood, E. J. F. (1959). *Trans. R. Soc. N.Z.* **86**, 1–58.
23. Skerman, T. M. (1959). *N.Z. Jl. Sci.* **2**, 57–94.
24. Skerman, T. M. (1960). *N.Z. Jl. Sci.* **3**, 620–648.
25. Haderlie, E. C. (1968). *In* "Biodeterioration of Materials" (A. H. Walters and J. J. Elphick, eds), pp. 658–679, Elsevier, Amsterdam, London, New York.

26. Milne, P. H. (1970). *Mar. Res.*, **1970**. No. 1, 31 pp.
27. Milne, P. H. (1970). "Fish Farm Enclosures". *In* "World Fishing, Dec. 1969–July 1970" (H. S. Noel, ed.) pp. 20, Commercial Exhibitions and Publications, London.
28. Allen, F. E. (1953). *Aust. J. mar. freshwat. Res.* **4**, 307–316.
29. Skerman, T. M. (1960). *N.Z. Jl. Sci.* **3**, 615–619.
30. Bureau of Ships. (1944). (Unpublished) Docking report manual. Navy Department, Washington D.C.
31. Pyefinch, K. A. (1945). *J. Iron Steel Inst.* **152**, 229P–243P.
32. Huve, P. (1953). *C.r. hebd. Séanc. Acad. Sci., Paris.* **236**, 419–422.
33. Hedgpeth, J. W. (1957). *Mem. geol. Soc. Am.* **67**, 129–52.
34. Hewatt, W. G. (1935). *Ecology* **16**, 244–251.
35. Kitching, J. A. (1937). *J. Ecol.* **25**, 482–495.
36. Moore, H. B. (1939). *J. Anim. Ecol.* **8**, 29–38.
37. Moore, H. B. and Sproston, N. G. (1940). *J. Anim. Ecol.* **9**, 319–327.
38. Guiler, E. R. (1949). *Pap. Proc. R. Soc. Tasm.* **1949**, 135–201.
39. Ralph, P. M. and Hurley, D. E. (1952). *Zool. Publ. Victoria Univ. Coll., Wellington.* **19**, 22 pp.
40. Guiler, E. R. (1954). *Trans. Proc. R. Soc. Tasm.* **88**, 49–66.
41. Maturo, F. J. S. (1959). *Ecology* **40**, 116–127.
42. Meadows, P. S. (1969). *Hydrobiologia* **33**, 65–92.
43. Weiss, C. M. (1948). *Ecology* **29**, 153–172.
44. Pyefinch, K. A. (1950). *J. Anim. Ecol.* **19**, 29–35.
45. Houghton, D. R. and Stubbings, H. G. (1963). *J. Anim. Ecol.* **32**, 193–201.
46. Smith, F. G. W. (1946). *J. mar. biol. Ass. U.K.* **28**, 353–370.
47. Smith, F. G. W. (1946). *Biol. Bull. mar. biol. Lab., Woods Hole* **90**, 51–70.
48. Doochin, H. (1949). *Univ. of Miami, Marine Lab. Tech. Rept.* for 1949.
49. Doochin, H. and Smith, F. G. W. (1951). *Instrum. Pract.* **12**, 1304–1307.
50. Houghton, D. R. (1959). *Dock Harb. Auth.* Oct. 1959. 8 pp.
51. Crisp, D. J. (1955). *J. exp. Biol.* **32**, 569–590.
52. McDougall, K. D. (1943). *Ecol. Monogr.* **13**, 321–374.
53. Crisp, D. J., Bailey, J. H. and Knight-Jones, E. W. (1967). *J. mar. biol. Ass. U.K.* **47**, 511–521.
54. Wood, E. J. F. (1950). *Aust. J. mar. freshwat. Res.* **1**, 85.
55. Aleem, A. A. (1957). *Hydrobiologia* **11**, 40–58.
56. Coe, W. R. and Allen, W. E. (1937). *Bull Scripp's Instn Oceanogr. tech. Ser.* **4**, 101.
57. Jones, W. E. and Babb, M. S. (1968). *Br. phycol. Bull.* **3**, 525–528.
58. Christie, A. C. and Shaw, M. (1968). *Br. phycol. Bull.* **3**, 529–534.
59. Pearson, C. R. (1968). *In* "Biodeterioration of Materials" (A. H. Walters and J. J. Elphick, eds), pp. 610–616, Elsevier, Amsterdam, London, New York.
60. Hawes, F. B. (1958). *Ann. Mag. nat. Hist.* Ser. 13, **1**, 147–155.
61. Nelson-Smith, A. and Gee, J. M. (1966). *Fld Stud.* **2**, 331–357.
62. Knight-Jones, E. W. (1951). *J. mar. biol. Ass. U.K.* **30**, 201–222.
63. Knight-Jones, E. W. (1953). *J. mar. biol. Ass. U.K.* **32**, 337–345.
64. Williams, G. B. (1964). *J. mar. biol. Ass. U.K.* **44**, 397–414.
65. Gee, J. M. and Knight-Jones, E. W. (1962). *J. mar. biol. Ass. U.K.* **42**, 641–654.
66. Gee, J. M. (1965). *Anim. Behav.* **13**, 181–186.

67. Hill, M. B. (1967). *J. Anim. Ecol.* **36**, 303–321.
68. Visscher, J. P. (1928). *Biol. Bull. mar. biol. Lab., Woods Hole* **54**, 327–355.
69. Knight-Jones, E. W. and Crisp, D. J. (1953). *Nature, Lond.* **171**, 1109.
70. Walley, L. J. (1968). *Philos. Trans. R. Soc. B.* **256**, 807.
71. Pyefinch, K. A. (1949). "Marine Ecology". Wiley, New York.
72. Crisp, D. J. (1953). *J. Anim. Ecol.* **22**, 331–343.
73. Daniel, A. (1957). *Proc. zool. Soc. London.* **129**, 305–313.
74. Crisp, D. J. and Stubbings, H. G. (1957). *J. Anim. Ecol.* **26**, 179–196.
75. Crisp, D. J. and Knight-Jones, E. W. (1953). *J. Anim. Ecol.* **22**, 360–362.
76. Moore, H. B. and Kitching, J. A. (1939). *J. mar. biol. Ass. U.K.* **23**, 521–541.
77. Pope, E. C. (1945). *Rec. Aust. Mus.* **21**, 351–372.
78. Gregg, J. H. (1945). *Biol. Bull. mar. biol. Lab., Woods Hole.* **88**, 44–49.
79. Pyefinch, K. A. (1948). *J. mar. biol. Ass. U.K.* **27**, 464–503.
80. Crisp, D. J. and Barnes, H. (1954). *J. Anim. Ecol.* **23**, 142–162.
81. Roskell, J. (1962). *J. Anim. Ecol.* **31**, 263–271.
82. Barnes, H. and Reese, E. S. (1960). *J. Anim. Ecol.* **29**, 169–185.
83. Knight·Jones, E. W. and Stevenson, J. P. (1950). *J. mar. biol. Ass. U.K.* **29**, 281–297.
84. Knight-Jones, E. W. (1955). *Nature, Lond.* **174**, 266.
85. Knight-Jones, E. W. (1953). *J. exp. Biol.* **30**, 584–598.
86. Crisp, D. J. and Meadows, P. S. (1962). *Proc. R. Soc. B.* **156**, 500–520.
87. Crisp, D. J. and Meadows, P. S. (1963). *Proc. R. Soc. B.* **158**, 364–387.
88. Barnes, H. (1956). *J. Anim. Ecol.* **25**, 72–84.
89. Wisely, B. (1958). *Aust. J. mar. freshwat. Res.* **9**, 362–371.
90. Crisp, D. J. and Ryland, J. S. (1960). *Nature, Lond.* **185**, 119.
91. Ryland, J. S. (1959). *J. exp. Biol.* **36**, 613–631.
92. Ryland, J. S. (1962). *J. Anim. Ecol.* **31**, 331–338.
93. Loosanoff, V. L. (1965). *U.S. Dept. of the Interior, Fish and Wildlife Service.* Circular 205, 36 pp.
94. Korringa, P. (1952). *Qt. Rev. Biol.* **27**, 266–308; 339–365.
95. Cole, H. A. and Knight-Jones, E. W. (1949). *Fish Invest., Lond.* Ser. 2 **17** (3).
96. Knight-Jones, E. W. (1949). "Aspects of the Setting Behaviour of the Larvae of *Ostrea edulis* on Essex Oyster Beds", Report to the Special Scientific Meeting on Shellfish of the International Council for the Exploration of the Sea, Edinburgh, October, 1949.
97. Crisp, D. J. (1967). *J. Anim. Ecol.* **36**, 329–335.
98. Bayne, B. L. (1969). *J. mar. biol. Ass. U.K.* **49**, 327–356.
99. Walne, P. R. (1966). *Fish. Invest., London.* Ser. 2. **25** (4), 1–53.
100. Cole, H. A. and Knight-Jones, E. W. (1939). *J. Cons. perm. int. Explor. Mer* **14**, 85–105.
101. Prytherch, H. F. (1934). *Ecol. Monogr.* **4**, 49–107.
102. Korringa, P. A. (1940). *Archs. néerl. Zool.* **5**, 1–249.
103. MacGinitie, G. E. and MacGinitie, N. (1949). "Natural History of Marine Animals", McGraw-Hill, New York.
104. Lynch, W. F. (1958). *In* "15th Int. Congr. Zool.", Sect. 3, Paper 9, 3 pp.
105. Chipperfield, P. N. J. (1953). *J. mar. biol. Ass. U.K.* **32**, 449–476.
106. Bayne, B. L. (1964). *J. Anim. Ecol.* **33**, 513–523.
107. Starkey, R. L. (1953). *In* "Proc. 6th Int. Congr. Microbiol.", 347–349.
108. Tebble, N. (1953). *J. Instn munic. Engrs.* **1953**, 1–7.
109. Barnes, H., Crisp, D. J. and Powell, H. T. (1951). *J. Anim. Ecol.* **20**, 227–241.

110. Barnes, H. and Powell, H. T. (1953). *J. mar. biol. Ass. U.K.* **32**, 107–127.
111. Barnes, H. and Powell, H. T. (1950). *J. mar. biol. Ass. U.K.* **29**, 299–302.
112. Weiss, C. M. (1948). *Ecology* **29**, 116–119.
113. Stubbings, H. G. (1959). *Nature, Lond.* **183**, 1282.
114. Wisely, B. (1962). *Nature, Lond.* **193**, 543–544.
115. Stanbury, F. H. (1944). "Experiments in the Growth of Marine Plants with Special Reference to the Effects of Copper and Mercury Salts", Marine Corrosion Sub. Com. Rept. Iron and Steel Corr. Comm. M.S.
116. Harris, J. E. (1946). *J. Iron Steel Inst.* **2**,
117. Lovegrove, T. and Robinson, T. W. (1968). *In* "Biodeterioration of Materials" (A. H. Walters and J. J. Elphick, eds), pp. 617–631, Elsevier, Amsterdam, New York, London.
118. Dolgopolskaya, M. A. and Gurevich, E. S. (1968). *In* "Biodeterioration of Materials". (A. H. Walters and J. J. Elphick, eds), pp. 680–684, Elsevier, Amsterdam, London, New York.
119. Turner, H. J., Reynolds, D. M. and Redfield, A. C. (1948). *Ind. Engng. Chem.* **40**, 450–453.
120. Holmes, N. (1970). *Chemy Ind.* 1970, 1244–1247.
121. Clark, T. British Columbia Research Council. Personal communication.
122. Perkins, E. J. (1961). *C.E.R.L. Laboratory Note* No. RD/L/N.97/61: 8 pp.
123. Trussell, P. C. (1967). *Sea Frontiers.* **13**, 234–243.
124. Lebour, M. V. (1946). *J. mar. biol. Ass. U.K.* **26**, 381–389.
125. Trussell, P. C. (1959). *In* "Marine Boring and Fouling Organisms" (D. L. Ray, ed.), pp. 462–468, Friday Harbour Symposium, University of Washington Press, Seattle, 1959.
126. Hurley, D. E. (1959). *N.Z. Jl. Sci.* **2**, 323–338.
127· Morton, J. E. and Miller, M. (1968). "The New Zealand Seashore", Collins, London, Auckland.
128. Walden, C. C., Allen, I. V. F. and Trussell, P. C. (1967). *J. Fish. Res. Bd. Can.* **24**, 261–272.
128a Board, P. A. (1970). *J. Zool., Lond.* **161**, 193–201.
129. Trussell, P. C. Anastasiou, C. J. and Fulton, C. O. (1956). *Pulp Pap. Can.* **57**. 67–71.
130. Sigerfoos, C. P. (1908). *Bull. Bur. Fish., Wash.* **27**, 191–231.
131. Tru,ssell, P. C. (1962). *New Scient.* **13**, 432–435.
132. Trussell, P. C., Greer, B. A., and Le Brasseur, R. J. (1956). *Pulp Pap. Can.* **57**, 77–80.
133. Townsley, P. M., Richy, R. A. and Trussell, P. C. (1966). *Proc. natn. Shellfish. Ass.* **56**, 49–52.
134. Hurley, D. E. (1961). *N.Z. Jl. Sci.* **4**, 720–730.
135. Jona, E. (1913). *Atti. Soc. ital. Prog. Sci.* 6 (Genova, 1912), 263–292.
136. Townsley, P. M. and Richy, R. A. (1965). *Can. J. Zool.* **43**, 1011–1019.
137. Liu, D. and Townsley, P. M. (1968). *J. Fish. Res. Bd. Can.* **25**, 853–862.
138. Trussell, P. C., Walden, C. C. and Wai, E. (1968). *Material und Organismen.* **3**, 95–105.
139. Jones, L. T. (1963). *J. mar. biol. Ass. U.K.* **43**, 589–603.
140. Eltringham, S. K. (1961). *J. mar. biol. Ass. U.K.* **41**, 785–797.
141. Eltringham, S. K. (1965). *J. mar. biol. Ass. U.K.* **45**, 145–152.
142. Eltringham, S. K. (1964). *J. mar. biol. Ass. U.K.* **44**, 654–683.
143. Editor (1965). *New Scient.* Nov. 11, 1965: 414.

144. Trussell, P. C., Fulton, C. A., Anastasiou, C. J. and Gillespie, R. E. (1956). *Pulp Pap. Can.* **57**, 72–76.
145. Allen, I. V. F., Trussell, P. C. and Walden, C. C. (1966). *Materials Protection* **5** (6), 41–44.
146. Trussell, P. C., Walden, C. C. and Allen, I. V. F. (1966). *Materials Protection* **5** (5), 43–46.
147. Walden, C. C., Allen, I. V. F. and Trussell, P. C. (1966). *Materials Protection* **5** (7).
148. Walden, C. C. and Trussell, P. C. (1962). "Sonic Inspection of Marine Piles", Wood Preserving News, British Columbia Research Council, Reprint No. 179: 3 pp.
149. Walden, C. C. and Trussell, P. C. (1965). *Dock Harb. Auth.* **46** (535).
150. Dorsett, D. A. (1961). *J. mar. biol. Ass. U.K.* **41**, 577–590.
151. Dorsett, D. A. (1961). *J. mar. biol. Ass. U.K.* **41**, 383–396.
152. Williams, C. S. (1968). *J. Anim. Ecol.* **37**, 704–712.
153. Purchon, R. D. (1955). *J. mar. biol. Ass. U.K.* **34**, 257–278.
154. McIntosh, W. C. (1908). *Zoologist* Ser. 4, **12**, 41–60.
155. Lunz, G. R. (1941). *J. Elisha Mitchell Scient. Soc.* **57**, 273–283.
156. Korringa, P. (1951). *Arch. Zool.* **10**, 32–152.
X. Perkins, E. J. Previously unpublished work.

13

The Inshore Harvest

The production of an individual species, which forms the basis of an estuarine fishery, may not rival the great fisheries of the world, e.g. cod, on the Grand Banks of Newfoundland or around Bear Island, and the anchovietta of the Humboldt current, but estuarine fisheries are nevertheless very important. These fisheries are based upon a wide variety of vertebrate and invertebrate species; moreover, the valuable attached algae live only in coastal waters. Such fisheries are composed of those fish and invertebrates whose life is closely tied to that of the estuary and adjacent inshore waters, plus the extremely valuable catches of *anadromous* and *catadromous* species which merely use the estuary as a transit route to and from the feeding and spawning grounds. The anadromous fish, e.g. sea-lamprey, *Petromyzon*, sea-run sturgeons, *Acipenser*, and salmon, *Salmo* (Atlantic salmon), *Oncorhynchus* (Pacific salmon), ascend to fresh water to spawn, and the young descend from fresh water to the sea to feed; the catadromous fish, e.g. the eels, *Anguilla*, undertake migrations of the reverse order.

Seaweeds

Although the attached algae, i.e. seaweeds, inhabit a relatively insignificant part of the sea, they are of a considerable economic importance.

The red algae, especially, have been used as food for centuries. In Britain, dulse, *Rhodymenia palmata*, and Irish Moss, *Chondrus* and *Gigartina*, have long been utilized; *Porphyra*, of which 200 tons is harvested annually, is boiled to form laver bread, a paste which is fried and eaten with bacon by the people of South Wales. This last is of some interest, because *Porphyra* harvested in West Cumberland, and sold in South Wales, represents the critical food chain with respect to the authorization of the radioactive effluent from the Windscale works of the United Kingdom Atomic Energy Authority. In general though, the human consumption of seaweeds in western countries is insignificant compared with that in the countries of Asia: in Japan, 78 000 tons of *Porphyra* and 142 000 tons of *Laminaria* are harvested annually to produce the foods "Amanou" and "Kombu" respectively[1].

The red algae are a source of jelly-like materials of which the "agar" obtained from *Gelidium* spp. is best known. Agar is used as a culture medium for bacteria, and in the food, cosmetic, pharmaceutical and canning industries. Carrageenin from *Chondrus* and *Gigartina* is used in the food, textile and pharmaceutical industries, in particular in the preparation of chocolate, toothpastes and in book binding. Both the crude and prepared extracts of *Chondrus* and *Gigartina* are used in the "fining" of beer[1].

The calcareous rhodophytes *Lithothamnion* and *Lithophyllum* often occur in the outer areas of coastal plain, and fiordic estuaries, where they tend to be rather insignificant. Off the Atlantic coasts of France and Ireland, however, they occur abundantly and may give rise to beach sands containing over 90% of carbonate. Such *Lithothamnion* deposits may be used in agriculture and industry: in Brittany, the small town of Pontrieux alone produced 50 000 tons in 1966[2]. The calcareous algae are an important constituent of coral reefs[3].

Some minute species of red algae are parasitic on other algae, e.g. *Choreonema thuretii* upon *Jania* spp.[4], other species are endozoic and have been recorded from sponges, e.g. *Haliclona*; hydroids, e.g. *Campanularia*; polyzoa, e.g. *Alcyonidium*; and tunicates, e.g. *Styela*[5]. In some British coastal areas, seaweed, thrown ashore in storms, is used as fertilizer, e.g. in Ayrshire and Fife.

In western countries, a considerable industry was based on the kelp as a source of iodine and of alkali carbonates for the soap, bleach and glass industries. Subsequently, the use of alginates derived from kelp, laminarians, for a wide variety of food and industrial preparations (Table 13.1) led to a revival of the industry in Britain[1] (Table 13.2). Although there is a potential crop of *ca.* 1×10^6 tons p.a., British exploitation of this resource is confined to part-time workers in the Orkneys and Hebrides. In the United States, populations of the kelp, *Macrocystis*, are farmed on the coast of California and harvesting is mechanized[1].

Fish

The Transitory Migrants (i.e. Diadromous Species)

The economic importance of the anadromous and catadromous fish during their transient passage through an estuary depends much upon their legal status in the individual country. During their estuarine period, they may be subject to the effects of pollution, or the influence of engineering structures; they may be a source of food for estuarine inhabitants, particularly during their juvenile stages, but as adults their trophic effect is generally small. In estuaries then, their importance is economic.

In Britain, the Atlantic Salmon, *Salmo salar*, is taken commercially in

Table 13.1. Some of the manufactured products in which alginates
are used (after Johnston[1])

Pharmaceutical	*Textiles*
Aureomycin tablets	Size for cotton and rayon
Anti-acid tablets	Textile print pastes
Aspirin tablets	Plastic laundry starch
Calamine lotion	
Dental impression compounds	*Paper*
Toothpaste	Milk containers
Surgical jellies	Insulation board
Mineral oil emulsions	Food wrappers
	Greaseproof paper
Foods	Acoustic tiles
Bakery icings	
Salad dressings and creams	*Miscellaneous*
Frozen foods	Paints
Fruit syrups and concentrates	Polishes
Candy	Ceramic glazes
Milk puddings	Leather ware
Ice cream	Boiler compounds
Sherbet	Battery plate separators
Chocolate drinks	Waxes
Sterilized cream	Jointing cements
Cheese	
Rubber	
Latex creaming and thickening	
Foam rubber, cushions, etc.	
Rubber coatings	
Tyres	
Electrical insulation	

The above table gives some of the examples of products incorpora-
ting alginates (U.K. and U.S. market) but it is far from complete,
there are several hundred such products and the number is increasing
every day.

a few estuaries, e.g. in the Solway Firth, by stake and poke netting and in
the Tweed and Dee by seining. In the Solway Firth, drift netting, or
whammelling, had been carried on from mediaeval times, but was banned
around 1960, together with that which had developed off the Tweed. In
the Solway, a kind of hand netting, i.e. haaf netting, is still carried on[6,7, X].
Taken as a whole, however, this fishery is relatively unimportant, both
compared with that in other countries and the catch in British rivers. This
situation has arisen largely because of the British system of *riparian ownership*, i.e. a landowner whose land shares or contains a river has the right to
the fish in that river. The catches of eels, taken by long, cylindrical eel-
nets, are of little importance in Britain, but of a considerable importance
in Europe.

Table 13.2.   A brief history of brown seaweed utilization in Britain (after Johnston[1])

1746	Kelp burning in Highlands and Islands of Scotland, for carbonate obtained from the ash.
1760	Fall in the role of kelp as a supply of soda; competition from Spanish "barilla" following the removal of duty.
1800	Le Blanc process for soda manufacture developed.
1811	Iodine discovered in seaweed ash by Courtois, studied in detail by Clement and Desormes, then by Guy-Lussac (1813).
1820	Kelp (seaweed ash) used as source of iodine.
1863	British Seaweed Company founded by Charles Stanford, new extraction method using seaweed not ash; cheaper iodide discovered in nitrate deposits in Chile, exported in crude form (fully reported 1930), kelp industry protected by a price agreement until 1935.
1881–3	Stanford patents a method for extraction of alginate from seaweed; Algin Company founded by Stanford, but he died soon after, little progress.
1934	Cefoil Ltd. formed to develop manufacture of transparent paper from alginate; many difficulties in process and market.
1935	End of price agreement with Chile, end of kelp industry.
1939	World War 2—Ministry of Supply requirement for non-inflammable fibre for camouflage netting for guns, etc., four government factories built to produce suitable alginate fibres; at the end of the war Alginate Industries Ltd. formed, took over war-time factories; many alginate derivatives developed.
1951	Role of seaweed meal as animal feeding increased (a role realized back around 1920).
1956	ICI introduced a new range of cotton dyestuffs, which combine directly with the cotton; alginates used as thickeners.

Over last ten years—tremendous advances in many industrial uses for alginates, especially in cosmetics and food industries; many new factories opened but harvesting still a problem.

In contrast to the United Kingdom, North America has no system of riparian ownership. Here important commercial fisheries, in addition to the sport fisheries, exist in the estuaries and offshore areas. Such fisheries are important in fresh water also[8-12]. In Alaska alone, the number of sockeye salmon, *Oncorhynchus nerka*, returning to a single river may exceed 20×10^6 in a single year, and each fish may be worth $3 when canned[12]. In the State of Washington, salmon contribute $> \$100 \times 10^6$ annually to the economy of the state[9].

Oceanographic changes may influence the timing of migrations and the size of populations, but relatively less is known of these factors than those which are due to man on the spawning rivers. In general, man-made alterations in the environment cause disturbance and loss to the population: dams and diversions are particularly important[13]. Because of the influence of a progressively and rapidly degrading environment and of increased fishing efforts, states such as British Columbia, Oregon and Washington, have well developed schemes of improvement based upon large hatcheries and natural rearing[10, 14–18]. On the Big Qualicum and Fulton Rivers of British Columbia, engineering works, which control water flow and provide ideal gravel beds for spawning, have boosted egg-to-fry survival rate by 6–10 times over natural survival[17]. Commercial fisheries are carried out in North America by trolling, gill-netting and purse-seining[9, 17]; sport fishing is carried on with the use of flies, spinners, trolls, spoons and the like.

Non-diadromous Species

Those species which are more permanently associated with the estuary than the anadromes or catadromes may be categorized in one of four groups thus:

1. residents, which are species that spawn in the estuary and spend a significant portion of their lives there;
2. summer migrants, which are usually warm water species that "invade" estuary only during the warm seasons;
3. local marine forms, which are species indigenous to the local neritic waters, such as adjacent sounds, and are usually captured near the mouth of the estuary; and
4. fresh water fishes, which are species normally confined to inland waters[19–23] (Table 13.3).

In the Columbia River estuary, the bulk of the catch at all stations was made up of euryhaline species, and the greatest variety and number of fishes were taken in water of oligohaline to mesohaline salinity (i.e. 0·5–18‰), a distribution which was apparently related to large blooms of the copepod *Eurytemora hirundoides*[20]. In the Solway Firth, the species were more varied in the outer areas. Penetration of the inner area was most marked in the summer months: here, since salinities were generally in excess of 20‰, temperature was apparently important. Nevertheless, the effect of sedimentation upon salinity had resulted in a marked downstream movement of the upper limit of penetration of many species[22, 23].

In the Aransas River, Texas, 26 species of fish showed varying degrees of salinity tolerance. Of 9 species of freshwater fishes, viz. *Dorosoma petenense, Notropis lutrensis, Ictalurus furcatus, Lepomis megalotis, L. macrochirus, Gambusia affinis, Micropterus salmoides, Chaenobryttus gulosus* and *Cichlasoma cyanoguttatum*, which entered the saline portion of the river, the last 5 were found in water of salinity 17·4‰, to which they must therefore have at least a temporary tolerance. Eight species, viz. *Brevoortia*

Table 13.3. Classification of the fish of the Mystic River, Connecticut, according to habit (after Pearcy and Richards[19])

I. Residents	II. Summer migrants	III. Local marine	IV. Freshwater	V. Diadromous
Cyprinodon variegatus	Synodus foetens	Raja erinacea	Esox niger	Alosa pseudocharengus
Fundulus heteroclitus	Strongylura marina	Anchoa mitchilli (I?)		Osmerus mordax
F. majalis	Pomatomus saltatrix	Urophycis chuss		Anguilla rostrata
Microgadus tomcod	Alectis crinitus	U. regius		
Apeltes quadracus	Seriola zonata	Tautoga onitis (I?)		
Gasterosteus aculeatus	Bairdiella chrysura	Prionotus carolinus		
Pungitius pungitius	Leiostomus xanthurus	Myoxocephalus		
Syngnathus fuscus	Menticirrhus saxatilis	octodecemspinosus		
Tautogolabrus adspersus	Mullus auratus	Cyclopterus lumpus		
Gobiosoma bosci	Stenotomus chrysops	Liparis atlanticus		
G. ginsburgi (II?)	Sphyraena borealis	Paralichthys oblongus		
Myoxocephalus aeneus	Mugil cephalus	Scophthalmus aquosus		
Pholis gunnellus	Eutropus microstomus			
Menidia beryllina	Paralichthys dentatus			
M. menidia	Echeneis naucrates			
Pseudopleuronectes	Alutera schoepfi			
americanus	Balistes capriscus			
Opsanus tau	Sphaeroides maculatus			
	Chilomycterus schoepfi			

patronus, Anchoa mitchilli, Syngnathus scovelli, Bairdiella chrysura, Eucinostomus argenteus, Gobiosoma bosci, Trinectes maculatus and *Mugil curema* lived in all salinities between 0 and 35‰, but were not adapted to salinities > 35‰. The gar *Lepisosteus spatula* and the clupeid *Dorosoma cepedianum* are euryhaline, but have to return to fresh water to spawn. Seven species, viz. *Fundulus grandis, Lucania parva, Cyprinodon variegatus, Mollienisia latipinna, Menidia beryllina, Mugil cephalus* and *Dormitator maculatus* were fitted for life in a wide range of salinities[24].

In the Mystic River estuary, Connecticut, 97% of the total number of fish larvae taken were derived from demersal or non-buoyant eggs, and a greater number of species had demersal eggs in the estuary than in nearby neritic areas[19]; 5 species of flatfish—4 from the northern Pacific and 1 in the north-west Atlantic—are known to lay non-buoyant eggs[25]. Such demersal eggs may be an adaptation to enable the utilization of shallow, inshore breeding and nursery grounds, in temperate and arctic regions, during the winter, since they are less subject than buoyant eggs to offshore dispersal by currents or to the dangers of freezing.[25] The pleuronectids of British estuaries, notably the plaice, *Pleuronectes platessa*, and flounder, *Platichthys flesus*, migrate downstream to the sea to spawn and have pelagic eggs; in contrast, the pelagic herring, *Clupea harengus*, which enters the outer areas of estuaries, e.g. Clyde and Solway Firths, and the River Blackwater, to spawn during the spring has a demersal egg. This supports the contention above, although there is the difficulty that once hatched the pelagic larvae are subject to the circulation of the individual estuary (p. 88), and the salinities and temperatures which they encounter may have an important influence upon their distribution[26, 27]. In the Mystic River, pelagic eggs were common in the plankton collections from May to August, whereas peaks of larvae derived from demersal eggs occurred during March–April, 1959, and January 1960[19].

In the Tamar estuary[28, 29] and Solway Firth[22] flounder, *Platichthys flesus* fast from November to March; from March onward feeding, accompanied by an upstream migration, occurs, and the animal reaches a peak of condition about July. As the gonad develops so the type of food taken changes, i.e. from molluscs to polychaete worms, until by November the gonad fills the body cavity, and the fish move downstream and out of the estuary to spawn. Plaice show a similar pattern of behaviour[22]. Fish like the flounder which migrate down to the sea to spawn may or may not return to the same estuary once they have spawned[29].

The size of the brood of young fish and its success may vary markedly from year to year, and may depend directly upon environmental conditions, on the amount and size of food available, competition within the brood[30, 31], and competition between broods of different species[32, 33] and adult fecundity[34-41]; the last may be part of a density dependent population regulating mechanism[38].

Evidently, the differing success or failure of individual broods has an important influence upon a fishery, but the success or failure of a fishery may not depend solely upon differences in recruitment. Fishing itself has an influence and to achieve an optimum i.e. maximum sustained yield, a fishery must be managed so that it is neither *underfished* nor *over-fished*. A fishery which is under-fished is normally composed of very many under-sized individuals, which indicates a vigorous competition for food. On the other hand, a fishery which is over-fished is composed of small, undersized individuals which have not been allowed to grow to a mature size; in this state, the maintenance of catches, at a nearly constant level, requires an ever increasing expenditure of fishing effort, so that the economic return per unit of effort shows a consistent decline[42-45].

Estuarine areas are important as nursery grounds for the young of fish which spawn offshore; the young of plaice and flounder may be abundant in areas like the Solway Firth, Morecambe and Liverpool Bays. These areas are important feeding grounds of the adults also, and although many species of fish may inhabit an estuary, a natural segregation related to feeding habits occurs. For example, sole, *Solea solea*, feed on the more muddy grounds below the L.W.M. Plaice and flounder migrate on to the shore and feed when it is covered; the flounder usually migrates to a higher level (sometimes the *Corophium volutator* zone) than the plaice, and stays on the shore longer than the plaice. The dab, *Limanda limanda*, characteristically feeds on a wide variety of animals associated with hydroids from below the L.W.M.[22, 46-48].

As in offshore areas, pelagic fish, e.g. herring, are taken by drift netting, and demersal fish by trawling. Advantage is taken of the tidal on/offshore feeding migrations and of aggregation upon spawning grounds[19, 22, 49]. In the former case, for example, the area of the Solway Firth is such that plaice and flounder are too scattered for effective trawling when they are feeding upon the marginal banks; trawling is therefore restricted to the period about the time of low water, when the fish are confined to the narrow channels[22].

Inshore fisheries, both for fin-fish and shellfish, are very valuable, but, from official statistics, it is difficult to separate the contribution made by each group; moreover, it is even more difficult to separate the value of landings from estuaries and waters very close inshore and those from more distant waters. The situation is complicated by the fact that statistics from harbours with a fish market make no distinction between the various sources of fish, while at towns which do not have a fish market on the quay, estimates of landings and of values are usually the result of informed guesswork. Nevertheless, in the state of Oregon, in 1965, the commercial fishery landed 68 095 514 lb of fish and shellfish, to the total value of $9 341 346 (£3 860 000), of which shellfish were valued at $1 449 759 (£600 000)[50]. In the State of Washington, commercial landings of fish and shellfish to the value of $18 930 000 (£7 820 000) and $21 170 000 (£8 750 000) were made in 1965 and 1966 respectively: in these years, commercial shellfish landings were valued > $3 310 067 (£1 370 000) and > $3 813 053 (£1 575 000)[51].

In Britain, in 1969, 560 750 tons of fish and shellfish to the value of £42 849 000 were landed in England and Wales, and 369 450 tons to the value of £22 559 000 were landed in Scotland[52]. In 1970, the total landings of fish and shellfish in Scotland amounted to 412 839 tons which had a value of £27 200 000. Of this total catch 250 000 tons (value £18·9 million), 144 000 tons (value £4·3 million) and 2200 tons (value £3·9 million) were demersal fish, pelagic fish and shellfish respectively. Most of the pelagic fish and shellfish were taken in Scottish coastal waters, and about 60% of the demersal fish were derived from this source[52a]. According to unpublished data from the D.A.F.S. Marine Laboratory, Aberdeen, it may be taken that 90% and 100% of shellfish are taken within the 3 mile limit and within 12 miles respectively; 80% and 90% of herring landed are taken within 3 and 12 miles respectively; 100% of sprat are taken within 3 miles and 10–15% and 45% of demersal fish are taken within 3 and 12 miles respectively.

Invertebrate Fisheries

Although invertebrate fisheries are concerned with shellfish for the most part, they are not confined to shellfish, i.e. molluscs and crustaceans, alone. The holothurian sea-cucumber, or bêche-de-mer, or trepang, has long been a delicacy and landings of 1000 metric tons per annum are recorded[53]. In addition, some species may figure more or less briefly in the "fashion trade": an example of this trade was the exploitation of the hydroids *Sertularia argentea*, Sea Moss or Sea Fern, and *Hydrallmania falcata* in the late 1950s. These hydroids were raked from the seabed in areas like the mouth of the Thames estuary, dyed and used for decorative purposes[54, 55]. Currently, sea urchin, *Echinus*, tests are being used in lamp stands and the like; many mollusc shells are used in the preparation of trinkets for the tourist trade. However, although these fashionable items may be lucrative to the individual, they cannot compare in size, scale and value with the true fisheries which are based upon crustaceans and molluscs.

Crustacean Shell Fisheries

In sandy areas, like the Solway Firth, Morecambe Bay, the Thames estuary, and the Dutch coast, the fisheries for brown shrimp, *Crangon*, and pink shrimp, *Pandalus*, and prawn, *Palaemon*, are valuable. In North America, *Palaemon*, *Crago* and *Penaeus* are taken in such fisheries[22, 56–61]. These fisheries may be seasonal, that is they are dependent upon the migrations which these animals make to and from the shallow, estuarine areas, which depend upon the physiological response of these animals to the estuarine environment (p. 164). In the waters of Texas and Florida catches of *Penaeus setiferus* and *P. duorarum* respectively are positively correlated with the rainfall of the previous year[61]. In the Solway Firth, fishermen claim that catches of *Crangon crangon* are poor in summers with a low rainfall. Long term sediment movements may affect the site of these fisheries[22, 62].

Fishing for these crustaceans may be by means of trawl, either simple beam[22, 57], or a Dutch "double-decker" shrimp trawl[60]; fixed engines, i.e. nets or basket filters, secured to stakes driven into the substratum of the foreshore: in Britain, this method is apparently peculiar to the Severn estuary[56]; and small scale fishing may be conducted by means of push nets[57].

In some estuaries, e.g. the Welsh Dee, the shrimp *Crangon crangon* may be the main fishing industry[57]. In others, e.g. San Francisco Bay, agricultural and other development has reduced once thriving shrimp fisheries to negligible proportions: here in 1962, the total catch of *Crago* spp. was 1075 lb which had a value of $140 (*ca.* £58)[58].

Those areas which are good for the shrimp *Crangon crangon* are often good nursery grounds for flatfish. In 1889, it was realized, that shrimp trawling was bad for young fish[63]; at this time, it was recommended that shrimp-trawls should be replaced by machines and pots. The essential problem, apart from mechanical damage, is that the fine mesh required to catch shrimp also catches very large numbers of young flatfish, which, even if they are returned promptly to the sea, are eaten in large numbers by gulls[22].

Shrimp grounds, such as the Solway Firth, are characteristically inhabited by Britain's only venomous fish, the weevers, *Trachinus draco* and *T. vipera*, of which the more abundant, *T. vipera*, is taken frequently in shrimp nets. The pain induced by the venom, though of relatively short duration, is such that it has induced many fisherman to abandon shrimping as a livelihood. The larger *T. draco* is much less abundant than *T. vipera* and occurs nearer to the mouth of the estuary[22, 23, 46, 64, X].

The larger Crustacea, viz. crabs and lobsters, are normally associated with rocky areas, although some may be trawled from coarser sediments.

Because of the nature of the substratum, these shellfish are taken by baited pots and creels; in some areas the ancient hoop net is still used (Fig. 13.1). The form of trap, and method of use, are characteristic of the individual area[65–72]; indeed the "ideal" trap may differ from area to area and season to season[71]. Unlike the oysters and mussels discussed below, the crabs and lobsters have been captured from the sea, but have in no sense been farmed. However, an experiment, with lobsters, widely reported in the daily, but not the scientific press, is currently in progress on the west coast of Scotland.

Both lobsters and crabs have planktonic larvae to which must apply the factors controlling the survival of teleost fish larvae (p. 401). In British waters, the number of lobsters, *Homarus vulgaris*, landed varies seasonally, both in total number and in the catch per unit effort. The seasonal variations are related to the biology of the lobster, viz. recruitment, the moult cycle, and to differences in fishing habits from place to place; the last may mask differences due to recruitment and the moult cycle[73, 74]. In Maine, the fluctuations in the abundance of the American lobster, *Homarus americanus* and sea water temperature are correlated[75].

Unlike the European *Cancer pagurus*, the North American *Cancer magister*, Dungeness crab, lives upon sandy bottoms, and is fished, by nets

(a)

Fig. 13.1. Methods of lobster and crab fishing (after Thomas[69]). (a) By use of a crook.

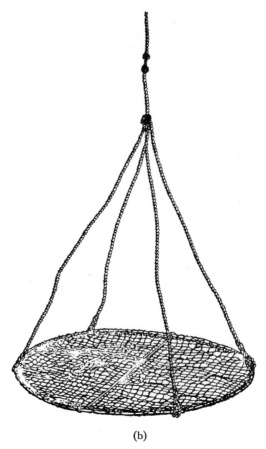

(b)

Fig. 13.1. Methods of lobster and crab fishing (after Thomas[69]). (b) Hoop net.

and traps, from the Aleutian Islands to Mexico[76, 77]. On the east coast of North America, the blue crab, *Callinectes sapidus*, is a commercially important inhabitant of estuaries[78]. In the Navesink estuary, New Jersey, this species was abundant formerly, but conditions induced by pollution have brought about its loss.

Molluscan Shellfisheries

Molluscan shellfisheries readily fall into two categories, viz. those which are subjected to managed exploitation, viz. winkles, whelks, cockles, scallops, and clams generally, and those which are farmed, viz. oysters and mussels.

The most important molluscan fisheries are based on bivalves. However, especially in south-east England, the winkle (or periwinkle) *Littorina littorea*, is collected by hand from between the tidemarks on suitable shores: the best are of bound shingle. The whelk, *Buccinum undatum*, is normally found below the low-water mark and is taken by means of suitable baited pots[79, 80]: this species may be found upon a variety of sea-bottoms, but it normally occurs upon mud mixed with sand and shells. It inhabits the outer areas of estuaries, but the best commercial whelks are taken in the open sea[79].

Scallops occur on sandy-gravel sea-bottoms in outer estuarine areas and in the open sea. Formerly, only the escallop, *Pecten maximus*, was taken, but, in recent years, the queen *Chlamys opercularis* has been fished also; indeed, a valuable export trade, based upon the latter, has developed in the United Kingdom[81-87]. Both species are taken by means of a heavy, toothed dredge; by this means, exploitation has been ruthless, and serious doubts are felt concerning the future of this fishery, and the grounds exposed to this heavy gear.

Cockles, *Cerastoderma (Cardium) edule*, are taken intertidally, although they do occur below the low-water mark. They burrow in coarse sand, sandy mud or mud and are gathered either by a rake, similar to a common garden rake, or the soil is first agitated by means of a "jumbo" and then raked. A jumbo is a heavy, wooden board some $4\frac{1}{2}$ ft long $\times$ 14 in. wide, with two transverse handles at waist height; this device is rocked vigorously on the surface of the sand and causes the cockles to rise to the surface. The same effect can be produced by stamping the feet[88-90]. In recent times, a type of suction dredge has been developed; this is highly efficient in terms of catch per unit effort, but its long term merit has to be demonstrated.

The clams of North America are justly world famous. Of these, *Mercenaria mercenaria*, the hard-shelled clam, has been introduced into British estuaries with some success. This species is taken by raking in shallow areas of fairly firm bottom, of which sand is an important constituent; this method of fishing results in certain sandy areas being worked to the exclusion of others[91]. The soft-shelled clam, *Mya arenaria*, occurs on both sides of the Atlantic, but is commercially important only in North America. This species inhabits grounds which range from fine sand to bound shingle and boulder clay from which it is dug by fork or spade. On the Pacific coast of the U.S.A. a variety of clams are sought eagerly, these include *Schizothaerus nuttalli*, gaper clam; *Clinocardium nuttalli*, the cockle; *Mya arenaria, Saxidomus giganteus*, butter clam; *Protothaca staminea*, littleneck clam; *Macoma nasuta*, bent-nose clam; *M. secta*, sand clam; *Panopea generosa*, the Geoduck; *Tellina bodegensis*, the Bodega tellin and *Siliqua patula* the Pacific razor clam. These clams are taken by digging with fork or shovel, pipe or tube, and mechanical clam diggers; of these, digging by fork or shovel is the most important in the commercial fishery[77, 92, 93]. No licence is required for personal use, but personal and commercial fishing is regulated[94, 95].

In effect, although rarely stated formally, oysters, have been farmed since Roman times at least. In Europe, oyster culture has been concerned primarily with two species, viz. *Ostrea edulis*, the much valued European oyster and *Crassostrea angulata*, the Portuguese oyster, which spread from the Mediterranean to the Atlantic coast of France during the mid-nineteenth century[96]. In North America *Crassostrea virginica* is commercially the most important species on the Atlantic and Gulf of Mexico coasts, although introductions of the Pacific oyster, *Crassostrea gigas*, and the Olympia oyster, *Ostrea lurida*, have been made[78]. Like all the bivalves which are considered here, the oyster has a planktonic larva, which, after a period in the plankton, settles and undergoes metamorphosis. Initially, the number of larvae produced depend upon the fertility of the adults[97], but larval survival depends upon the environmental conditions and the amount and quality of food available[78, 88]. The distribution of the larvae is a function of the hydrography of the individual estuary (Chapter 4), and spat falls may be heaviest where eddies form at the junction of a side stream with a main one, or in other regions of relatively still water[98].

It is an essential feature of ostreiculture that, at the time when eyed-larvae are present in the water and about to settle, materials are provided so that they may settle without undue interference from other organisms. These materials, or *cultch*, may range from the spreading of clean fresh shell, e.g. cockle shell, in suitable areas, to the provision of cultch racks, ropes, stakes, baskets[99], oyster shell retained in chicken wire bags and suspended in the water[100] and tiles. Whatever type of cultch is used, the objective is to produce finally a well-shaped adult oyster of high commercial value. Where oysters settle upon rocks or in marked aggregations, they are separated and may have to be hacked apart, a process fatal to many. The British Columbia Research Council in conjunction with the Nanaimo laboratory of the Fisheries Research Board of Canada have developed an artificial cultch which may hang in large numbers upon ropes suspended in the water. This artificial cultch has been formulated so that it is hard and light, is acceptable to the oyster larvae and does not slime. After about one year, it swells, up, breaks down and so releases the oyster.

Once a successful spatfall has been obtained, the process of cultivation begins, for the larvae may not settle in an area which is most suited to the production of the fat, high condition oysters required by the gourmet (Chapter 4). Consequently the young, or brood, oysters are transferred to the fattening grounds. These grounds are characterized by a high primary productivity, but are not precisely of the same type in all parts of the world.

In Essex, England, these grounds are characteristically sub-littoral and in creeks, e.g. Pyefleet Channel and Strood Channel, which receive little fresh water run out. On the Pacific Coast of North America the culture is generally carried out on between the $+3$ and -1 foot tide levels on the extensive tidal flats[101]. Where the bottom is too muddy, although the other conditions, i.e. high productivity and sheltered situation, are suitable, the oysters may be grown in trays placed on racks near the low-water mark as in the Yaquina estuary, Oregon[X]. In Norway, oysters are suspended beneath the low salinity surface layers of the fiordic oyster polls.

In both Britain and North America, seed oysters may be planted first in specially protected areas, and then transferred to the growing or fattening grounds. These grounds may be harrowed with an ordinary farm harrow which has two functions, viz. to improve the quality of the oyster and to improve the quality of the substratum. It is of interest that while excessive use of heavy gear can be damaging to fishing grounds, it is considered by fishermen generally that the working of a fishing ground improves the quality of the fish taken off it. Although its advantages are recognized for ostreiculture, this claim has not been substantiated with respect to other fisheries. However, the influence of agitation of the seabed upon the release of phosphates and other nutrients to estuarine waters (p. 32) suggests that this fishermen's tale may be correct.

Where the oysters are readily accessible, they may be harvested by hand; however, where culture in the sub-littoral is practised, dredges of various kinds are used. Such dredges may be simple, or of a more complex type which can be adjusted according to the type of ground which is being worked[102, 103].

A considerable world trade takes advantage of the fact that oysters are amenable to cultivation and may be transported over great distances without harm. For example, until pollution of the Tagus ruined the trade, Portuguese oysters from this estuary were shipped to England every year. Here, they were cultivated and sold in the summer period when the natives were not suitable for sale. The advantage of these Portuguese oysters was that in the lower temperatures of the British estuaries they grew well, but did not reproduce easily; therefore, no problems related to the reproductive cycle arose. When the English native oyster, *Ostrea edulis*, suffered a catastrophic decline from disease in the 1920s, stocks were built up once more by imports from Brittany. Attempts have been made to introduce the European oyster into American waters, similarly the American oyster, *Crassostrea virginica*, was introduced to Britain in the 1880s; however, the latter introduction was especially unfortunate since these oysters were accompanied by *Crepidula fornicata*, slipper limpet, and *Urosalpinx cinerea* the oyster drill[96]. At the present time, there is a considerable and growing annual importation of oyster seed from Japan, Korea and Taiwan to the producing areas of the States of Oregon and Washington in the U.S.A.[99, 104].

Mussels, *Mytilus*, are cultivated in a number of European countries, from Italy to Scandinavia[105, 107]. This cultivation is performed by three basic methods, viz. bed cultivation, the "buchot" system and rope cultivation[106, 108].

Bed cultivation is practised in N.W. Europe and is effectively the only system practised in the U.K. Seed or brood mussels are transplanted from areas of overcrowding to less crowded areas, normally about the time of an autumn or winter slack water, when loss by washing away of the brood is minimized[106]. Young brood, especially those which have settled around and below the L.W.M., are particularly vulnerable to predation by crabs and flatfish[22, 109]. However, once they have grown to a size which renders them invulnerable to such attack, they may be transferred to beds, below the L.W.M., where growth is better than in the intertidal zone. This method is used in N.W. Europe.

The "buchot" system is carried on in France: a pine stake is driven into the substratum and spat collect and grow upon these "buchots". Periodically, the largest spat are removed to "rearing buchots" of willow or chestnut interlaced with other stakes: initially, the spat are attached by bags of netting which rot and fall away, the spat having become attached by their byssus in the meantime. These mussels grow rapidly, and require frequent thinning to prevent overcrowding: they are marketed after one year[106]. Attempts to introduce the buchot method in Lancashire proved unsuccessful largely due to the constant, heavy settlements which occurred[110].

Rope cultivation of mussels is carried on largely in the rias, i.e. fiords, of the west coast of Spain. Large numbers of collecting ropes are suspended from rafts and hulks, and spat settle chiefly in the period April–June. By October, they have grown to as much as 1½ in. in length, and are then thinned out. These seed mussels, together with those collected from the shore, are attached to "growing" ropes by thin cotton or nylon net. This net permits the mussel to put down a byssus without falling from the rope, but, once this is done, it does not impeded growth. Growth continues and, in the following spring, the mussels are transferred to "commercial ropes". By the succeeding summer the mussels have reached a length of 3 in.: in these waters the temperatures are such that growth does not cease in winter as it does in British waters[106].

In Britain, relatively few places are suitable for rope culture. However, some of the sea lochs on the west coast of Scotland are potentially suitable sites. Indeed, work carried on at Linne Mhuirich, an arm of Loch Sween, has given very promising results, and, in this area at least, this form of cultivation on a commercial scale is possible[111, 112]. The method of rope cultivation has several advantages over the other methods. First, it allows mussels to take advantage of the phytoplankton food available at all depths; it protects the mussels from predators, e.g. starfish and shore crabs, which live on the seabed; it keeps the mussels clear of the most heavily silt laden waters; and finally, in areas where *Mytilicola* is present, it reduces the risk of infestation[106].

In Britain, shellfish have formed the basis of a "small man's" industry for many years[113], and although this is still true, there has been a tendency in recent times for this industry to pass into the hands of the big company. Clearly, when a large company introduces its expertise into such a field, gains in efficiency are possible, particularly in relation to marketing. Exploitation of resources beyond the means of the "small man" become possible also; for example, on the south west coast of Scotland, crabs are abundant and of good quality, however they are not marketed simply because the cost of transport to the market renders the return unacceptable to the small man. Again a succession of adverse conditions, viz. disease in the 1920s, a severe winter in 1940, and the extremely severe winter of 1962–63, have created great difficulties in the oyster industry of south-east England and most small men do not have the resources to meet such a situation.

Nevertheless the shellfisheries are of considerable value. In 1954, the annual value of lobsters landed in Scotland was £180 000, and this rose to £580 000 by 1964[114]. On the average, some £2 000 000 worth of canned crab meat (*Cancer pagurus*) are imported annually into Britain, but the indigenous fisheries for this animal are underexploited[72]. Shrimps and prawns too are valuable, for example; the average annual landing at Annan, in the Solway Firth, in the period 1963–66 was valued at £12 300[60] and at this time this port was on the decline[22, 62].

Although the individual European native oyster is of great commercial value, this industry as a whole has undergone a marked decline since the nineteenth century. Loss has occurred due to disease and severe winters, poor husbandry and reckless overfishing[115], pollution, loss of grounds as in the Schelde and competition and destruction from introduced pests. Indeed, the situation is a far cry from the days when Charles Dickens could portray David Copperfield's empty-headed wife Dora purchasing a barrel of oysters for their consumption: today even the most flighty individual would be staggered by the cost even if the purchase could be made readily. Another measure of the decline of this industry can be seen in the River Blackwater where it was customary to pay a bounty for *Urosalpinx cinerea*; the slipper limpet, *Crepidula*, was dredged and thrown above the high-water mark to die, and the estuary was patrolled by water-borne river police. All this is gone, and the sole witness of this former greatness is the large area of overgrown oyster pits in the salt marshes, best seen from the air. Surely, before considering the farming of bony fish, a real effort to restore former oyster beds to something like their former productivity might be justified. Oyster hatcheries based on the culture of larvae of *Ostrea edulis*[116-118] and the development of more productive varieties may help this process, but perhaps a less ambivalent attitude to such potentially valuable resources is required. Not least is this necessary with respect to pollution, for, if an improvement in water quality sufficient to improve existing fisheries cannot be achieved, what hope is there for mariculture. In Scottish waters, where these prospects are perhaps brightest, oysters were once abundant almost everywhere, especially on the west coast. Detailed investiagtion of these former sites[119] led to the initiation of a pilot scheme for oyster cultivation in Loch Ryan; this scheme has achieved a modest success which it is to be hoped is an augury for the future.

In 1965, the commercial production of shellfish, excluding several small fisheries for crawfish, ghost shrimp, blue mussels and a miscellaneous collection of tide pool animals, in the State of Oregon was valued at $1 449 759 (£600 000). In the same period, the personal use harvest of razor clams totalled 1.2×10^6 clams[50]. In Washington State, oyster landings were valued at $1 576 957 (£652 000) and $1 830 675 (£760 000) to growers in 1965 and 1966 respectively, and the wholesale values were $5 836 621 (£2 410 000) and $5 684 071 (£2 350 000) respectively; in the same periods clam landings were valued at $296 525 (£122 500) and $264 186 (£109 170) to diggers and to wholesalers at $520 000 (£214 880) and $475 293 (£196 400) respectively[51].

Like the British fisheries, these resources have not always been well managed, have been catastrophically affected by diseases and in places abolished by pollution[120].

Diseases and Parasites

The parasitic habit has been widely adopted by living organisms. However, these organisms have only rarely been shown to have a deleterious effect upon those animals which constitute a fishery; only such positively harmful species will be referred to below. With regard to disease, relatively little is known, especially with respect to the invertebrate species.

Fish

Fish may be parasitized by a number of species of parasitic copepod:

indeed *Lernaea* may be more abundant in some tropical fish farms than it is in natural conditions[121]. Whiting and other gadoids may be reduced almost to skin and bone, by the one or more large *Lernaea* hanging from the gills. Moreover, large infestations by *Lepeophtheirus pectoralis* may produce irritation and destruction of the delicate fin margins of the flounder, *Platichthys flesus*[122].

Of all fish diseases, the incidence of tumours is a cause of great concern at the present time; an increase of this form of disease has been associated with sources of pollution (p. 480) and the implication for humanity is evident. Such diseases have long been known, but have been relatively little studied[123–128]. Tumours and tumour-like conditions have been reported from protozoans, coelenterates, platyhelminths, annelids, sipunuclids, arthropods and molluscs[129, 130].

Crustacea

The gills of lobsters may be infested by the choniostomatid copepod *Nicothoë astaci*, however, at the usual level of infestation, i.e. 100 per lobster, this parasite does little harm[131].

Manx fishermen know as "granny" crabs those individuals of *Cancer* which are typified by dirty, discoloured shells, broken claws and a characteristic pitting of the carapace; when eaten, these crabs have a bitter taste and a powerful purgative effect. Evidently, this black necrosis is of commercial importance. Two species of chitinoclastic bacteria can be isolated from such crabs, but attempts at reinoculation have not proved successful. A black necrosis which occurs on *Carcinus* may result in the loss of legs and mouth parts[109]; the death of a *Carcinus*, maintained in the laboratory for a long period, was apparently due to an extensive infection of black necrosis. The shrimp, *Crangon crangon*, has been found to be infected with black necrosis[X].

Mollusca

Oysters and mussels may be infested by the pea crabs, *Pinnotheres*. In North America, *P. ostreum* is found in the oyster, and in Europe, *P. pisum* is found in the mussel, *Mytilus*. The adults are found upon the lamellibranch gills facing the inhalent current. They deprive the mollusc of food, but in addition they damage its gills which may occasionally serve as a source of food. Although no lamellibranch death has been traced to *Pinnotheres* infection alone, infested molluscs are in poorer condition than those which are not infested[78, 132]. Larger individuals of *P. pisum* are found only in large mussels which are found most frequently at low shore levels[132].

Copepod parasites of the genus *Mytilicola* occur in European and North American lamelli branch molluscs. *M. intestinalis* lives in the recurrent intestine, or rectum, of the mussel, *Mytilus edulis*, in which it caused a considerable mortality among British populations in the late 1940s and 1950s[133, 134], when it was a comparatively recent immigrant. High densities of infestation were found in harbours and estuaries, however it also occurred in exposed and fully marine situations[133, 135]. Mussels which are raised above the seabed are colonized relatively slowly by this parasite[112, 133]. *M. intestinalis* may occur in oysters, *Ostrea edulis*, and in *Venerupis decussata* and *V. aurea*. It has no marked epidemic effect upon the oyster, but these animals may serve to transport the parasite to areas formerly unaffected by *Mytilicola*[136].

In North America, *M. orientalis* originally described in *Crassostrea gigas* and *Mytilus crassitesta* from Japan, has been recorded from the oysters *C. gigas* and *Ostrea lurida*; mussels, *Mytilus edulis* and *M. californianus*; clams, *Venerupis staminea*; and the slipper limpet, *Crepidula fornicata*[137]. *M. orientalis* causes loss of condition in *Ostrea lurida*[136]. *M. intestinalis* affects the biochemical composition of *Mytilus*, especially at the times of year when the greatest change occurs[138].

It is of interest that at Whitstable, Kent, mussels, *Mytilus edulis*, which are infested by the boring polychaete, *Polydora ciliata*, have significantly smaller infestations of *Mytilicola intestinalis*[139].

The Pyramidellid molluscs are parasitic upon a number of lamellibranch molluscs including mussels and oysters. Full grown oysters may be weakened and killed, and young oysters may have their growth retarded by *Odostomia*[79, 140].

A number of protozoan and fungal parasites are known to cause serious oyster mortality, others e.g. *Hexamita*, *Nematopsis ostrearum* and *Crististira* have been implicated in oyster mortality, but the case against them is not proven[79, 89]; however, the fungus *Sirolpidium zoophthorum* may be responsible for the death of oyster larvae in cultures[78]. Another fungus *Dermocystidium marinum* is highly pathogenic to oysters, especially in warmer waters[78, 141]. The protozoan *Minchinia nelsoni* and another probably belonging to the genus *Haplosporidium* have caused a heavy mortality in oysters[78, 142].

The margin between parasitism and predation may be a little imprecise. The trematode infections of the hepatopancrease of the winkle *Littorina saxatilis* are clearly parasitic, may bring about parasitic castration, but do not apparently cause death[143, 144]. The trematode worm *Bucephalus cuculus* inhabits the gonad of oysters and cockles in which it may cause parasitic castration and death[78, 89]. The rhabdocoel turbellarian, *Paravortex cardii*, may be common in the gut of cockles, but does not apparently cause their early death[89]. On the other hand, the flatworms *Stylochus ellipticus* and *S. inimicus* which are of the order of $\frac{1}{2}$ to $\frac{3}{4}$ in. in length, crawl inside an oyster shell and feed on its meat until death results[78, 145]: these species are regarded as predators[145] although they have the attributes of ectoparasites. An active predatory flatworm *Pseudostylochus ostreaophagus* preys on oysters from Japanese waters[99].

Pests of Shellfisheries

Predators

The pests of a shellfishery may range from fouling organisms, which may kill shellfish or be a nuisance to the fishermen[78, 146, 147], and the dinoflagellate, *Ceratium fusus*, which may cause death of oysters[148], to predators and competitors. Because many of the worst pests, both predator and competitor, are immigrants, considerable care is now taken to avoid such introductions and a careful inspection of seed molluscs is undertaken[99]. In Britain, movement from infested areas may be prohibited.

In British waters, shellfish are attacked by a number of predators. The gastropod whelk tingle, *Urosalpinx cinerea*, was introduced from the eastern seaboard of North America[96, 149, 150] and is found principally below the L.W.M. in estuaries like the Blackwater, Essex. Shellfish living below the L.W.M. are especially subject to predation by the starfish *Asterias rubens*[151-154]. Mussels, *Mytilus*, may be preyed upon by the crabs *Cancer pagurus*, *Xantho incisus*, *Carcinus maenas* and *Portunus puber* in addition to the starfish *Asterias rubens*, *Marthasterias glacialis* and *Asterina gibbosa*. The gastropod *Thais* (*Nucella*) *lapillus* may be abundant upon mussel beds and is itself preyed upon by *Carcinus maenas*[155-160].

Carcinus is a curious instance of the manner in which an abundant animal of considerable economic importance may pass almost unrecognized as such for many years. Indeed Orton[161], in 1926, recognized that *Carcinus* was an important predator of small cockles, but this significant observation largely escaped notice. However, in North America, in the late 1940s and early 1950s, it was shown to have a marked impact upon shell fisheries, particularly upon the soft-shelled clam *Mya arenaria*[162, 163]. This work too largely escaped notice in Britain where attempts to organize experimental food chains and fish farms finally indicated its importance. Now it is appreciated that not only *Carcinus*, but other and larger crabs, viz. *Portunus puber*, *Cancer pagurus*, *Cancer irroratus*, *Callinectes sapidus*, *Neopanopeus* spp. and *Menippe mercenaria*, are all important predators of shellfish[78, 155, 156, 159, 164-167].

In North America, shellfish are subject to attack by rather more predaceous gastropods than are British species. These predators include

14*

Urosalpinx cinerea which was transplanted from the eastern seaboard to the Pacific coast, as well as to England; the Pacific coast also received the Japanese drill, *Tritonalia japonica*. Both species multiplied rapidly to cause heavy losses to these oyster fisheries. On the east coast, the gastropod predators range in size from that of *Urosalpinx cinerea* and *Eupleura caudata* (maximum length *ca.* 1½ in.), to that of *Thais haemostoma* (maximum length *ca.* 3 in.), which affects oysters in the Gulf of Mexico, especially those of Alabama, to that of *Busycon canaliculatum*, channel conch (maximum size *ca.* 6 in.); other gastropod predators are the apple murex, *Murex pomum*, the lightning whelk, *Busycon contrarium*: the crown conch *Melongena corona* is not considered to be a serious pest of oysters[78].

Korean and Japanese oysters are subject to predation by the drills *Thais tumulosa*, *T. bronni*, *Tritonalia japonica*, *Purpura burnetti* and *Muriciformis thomasiana*. As we have seen, *T. japonica* was inadvertently transferred to the Pacific coast of N.America where it caused serious trouble. Of two drills which live in the waters of Taiwan, viz. *Cymatium pilare* and *Thais tumulosa*, the latter is the more troublesome in its native habitat and is the more likely to be transferred successfully with oysters to the American grounds[99].

As in Europe, starfish are important predators of the North American shellfisheries. The settlement of *Asterias forbesi* is apparently subject to long term fluctuations related to a similar periodicity in adult abundance, viz. a 14 year cycle in the period 1852–1944[168]. Nevertheless, in Long Island Sound no definite relationship was found between the intensity of starfish and oyster settlement[169].

Birds may be important predators of shellfish; gulls dropping mussels to open them are a common sight where shores are rocky or stony. However, in sandy areas, where cockle fisheries are important, the oystercatcher, *Haematopus ostralegus*, may be a powerful adverse influence upon the cockle fishery[170, 171] and they may also affect mussel beds adversely[171, X]. In times of extreme weather conditions, when food sources are reduced or scarce, the pressure of bird predation upon intertidal shellfisheries, e.g. cockle, may be very great[172].

Control of predators

Predators may be controlled to some degree by various means which depend upon the species of predator, area and the attitude of the would be controller. Starfish may be entangled in mops, which are dragged over the seabed, and then killed by dipping in vats of hot water. Saturated salt solutions have been used to control the drills *Urosalpinx cinerea* and *Eupleura caudata* by killing their eggs and larvae: an immersion of eggs for five minutes causes considerable, sometimes 100% mortality, even if the embryos are returned to sea water immediately. This method is also valuable in the destruction of starfish and the flatworms *Pseudostylochus* and *Stylochus ellipticus*; indeed, for starfish, this method is superior to sprinkling lime upon the shellfish bed, a one minute exposure will guarantee 100% mortality[78, 173, 174]. Sprinkling the shellfish bed with quickline (CaO) causes surface lesions on those starfish upon which it falls; however, although many are killed, starfish which are lightly affected recover[78, 175]. Starfish and the adult whelk tingles, *Urosalpinx cinerea*, may be controlled by burying them under a few centimetres of bottom mud[78, 174, 176]: burial may be accomplished by agricultural ploughs, and oyster and clam dredges. The mortality depends upon the depth of burial, viz. 40%, 75% and 92% died when buried at 3, 4 and 6 cm respectively; when buried at 6 cm depth, the time to 100% mortality was 5, 7, 12, 19 and 52 days at temperatures of 25, 20, 15, 10 and 0–5°C respectively[176].

Trapping may be used for drills and for crabs[78, 177], although it is said to be inefficient for the former[177]. However, the crab *Carcinus* was brought under control in the Ardtoe fish farm, Argyll, by means of trapping, and this raises another issue. *Carcinus* was used as food by the poor of London in the 1840s[178] and is still eaten in Europe[179, 180]. Although this animal may not be highly esteemed as food, the world is short of protein and its use in a low grade fishery is surely preferable to destruction. Indeed, bearing in mind man's known rapacity with respect to fisheries, this is likely to be much the most efficient method of bringing *Carcinus* under control. *Carcinus* is so tough that, in the fish farming situation, the normal processes of drainage and exposure of the farm bed are not likely to control this animal[109].

In North America, attempts to control drills and crabs have been made using insecticidal chemicals. Predatory snails have been controlled by use of chlorinated hydrocarbons such as orthodichlorobenzene (1, 2-dichloro-benzene), or a mixture of several such hydrocarbons known as Polystream. These chemicals are mixed with dry sand, broken oyster shells and other inert materials which carry them to the bottom and keep them there. Normally, these materials are applied in belts or barriers to stop, repel, or kill these predators; their effectiveness may be improved by the addition of Sevin (1-naphthyl N-methyl carbamate). An essential feature of all of these substances is a low solubility in water. Generally, a barrier 8–12 in. wide and 1 in. deep may be effective for 6–14 months against most drills. Against the clam-killing snail *Polinices*, which characteristically moves at some depth within the soil, such barriers are likely to be ineffective, but the injection of the chemicals or plugs of treated sand into the bottom may be used to create a deeper barrier[174, 177, 181]. Movement of starfish, *Asterias forbesi*, into protected areas, may be prevented by 3 ft wide barriers of sand mixed with heavy oils and such compounds as 2-chloro-l-nitropropane and Sevin, or orthodichlorobenzene and 2-chloro-l-nitropropane[181].

In the New England fishery for *Mya arenaria*, barriers of baits soaked in lindane (the gamma isomer of benzene hexachloride—B.H.C.), the most toxic of the B.H.C. isomers and one which is used widely in agriculture, have been used to control the green shore crab, *Carcinus*. Such barriers not only prevented the movement of crabs into the protected areas, but, in a period of 9 weeks, reduced the resident crab population by 76%[162]. However, there are dangers inherent in the use of such methods, for commercial species may be tempted to live in these areas, and be killed by such practices; ecological problems may arise also.

Competitors

There are a number of animals which can be broadly classed as competitors with shellfish, although this is rather a poor description overall.

The boring sponge *Cliona* and the polychaete *Polydora* spp. both bore into the shells of molluscs, and are particularly important with respect to the oyster. These infestations may cause debilitation of the mollusc, but, to the fisherman, the principal difficulty arises from the friable shell which they leave and which is not a sound product for the market. All the fouling organisms, e.g. barnacles and sea squirts, compete with the oyster for food.

The most serious competitor to the oyster is probably *Crepidula fornicata*, the slipper limpet, a gastropod mollusc which is a filter feeder and competes directly with the oyster for food. Introduced into British and U.S. Pacific coast waters, it has become a very serious pest of oyster fisheries[78, 96, 182]. Its competition for food has relatively little effect upon the oyster, but the larvae of both species settle about the same time, and the slower growing oyster may be covered by *Crepidula*, suffocate and die[78, 182]. *Crepidula* also produces copious amounts of pseudo faeces which, when its chains are numerous, may be sufficient to smother both *Crepidula* and its coinhabitants alike. Finally, cockle and mussel shell breaks down relatively rapidly once the animal is dead, hence their particular advantage as oyster cultch; on the other hand, the very hard shell of *Crepidula* breaks down relatively slowly. Consequently, oyster spat which settle on these shells grow into malformed or "crippled" adults and have a rather poor market value.

Still other animals such as the ghost shrimps, *Upogebia* and *Callianassa*, burrow into the substratum and, where they are abundant, do much to render the grounds unsuitable for oyster culture[78].

The boring sponges and polychaetes, sea squirts and other suitably vulnerable organisms may be controlled by dipping the infested oysters in brine solutions for varying periods depending upon whether the oyster is to be left exposed to the air or returned to sea water immediately.*Crepidula* may be controlled by dipping in solutions of corrosive sublimate, or more preferably by dipping in a saturated salt solution: slipper limpets exposed to brine for 3–5 min, followed by a 30–60 min retention on the ship's deck, all die within 5 days of their return to sea water. Various other chemical dips may also be used against these animals[78, 96]. The ghost shrimp may be controlled by coating affected areas with dry sand, or other inactive carrier, mixed with chlorinated benzenes, usually lindane or Polystream, and an insecticide such as Sevin[78, 96, 183]. However, knowledge of biological concentration processes and sub-lethal toxicity (Chapter 15) suggests that these methods must be viewed with caution.

Productivity

In 1948, the world's fisheries produced 20 million tons of all types of fish and shellfish. By 1967, annual production had reached 60 million tons[184], and hopes are expressed that this will reach 100 million tons per annum in the foreseeable future: a view considered to be optimistic by many concerned with the growing problems of pollution and of the effects of fishing pressure itself. However, the crustaceans and molluscs represent about 10% of the landings in the world fisheries as a whole, but in some countries, like Australia, may make up to 50% of the total catch. In some instances, an area of 40 sq. miles will produce 80 000 000 oysters per year[53]. Pleasant Bay, Massachusetts, with high and low water areas of 11·4 and 8·4 sq. miles respectively, produced fish and shellfish to the value of $211 601 (£87 450) in 1965; shellfish accounted for 74%, i.e. $156 785 (£64 800) of this value[185].

Where shellfish are grown on the seabed, the average annual production for large estuarine regions is 150 kg meat per hectare per year; the best techniques of bottom culture in the U.S.A. and Europe may raise this to the order of 5000 kg per hectare per year. The hanging culture method, i.e. from rafts, may yield 50 000 kg of Japanese oysters and 500 000 kg of Spanish mussels per hectare per year[186].

As one might deduce from a knowledge of the nutrient regime (Chapter 2), estuarine waters are more productive than those of the open sea. Indeed the average estuary is capable of producing organic matter at an annual rate of about three metric tons (dry weight) per hectare which is 10–100 times greater than that of the open sea[186]. In Australian estuaries, an experimental addition of equal parts of superphosphate and linseed oil meal was made to the surface layers of mud flats; this enrichment apparently favoured oyster production[187]. Conversely, the River Don has been extensively diverted since 1951, and the flow reduced from 26·2 km³/yr to 19·4 km³/yr; in the sea of Azov, the resulting loss of nutrients caused a fall in fish production to a level 2·5 times less than that of earlier years[58].

If these are the figures for production based on the waters of an estuary, what of the tidal flats and the soils below the low-water mark? Clearly, the production is variable since it is dependent upon a population which is essentially discontinuous in distribution. However, in the mouth of the Exe estuary, Devon, the biomass of the macrofauna ranged from 4·6 g/m² at the poorest stations, to 36·9 g/m² where the polychaete *Ophelia* was abundant, to 45·4, 69·3 and 101·7 g/m² where *Arenicola, Scrobicularia, Cardium* and *Ampharete* were present[188]. Similarly, the biomass of the macrofauna of a sand bank in the mouth of the Welsh Dee ranged from 36 g/m² to 340 g/m²[189], and like the Exe, mud swallowers formed the bulk of this population. In Biscayne Bay, Florida, the total animal biomass had an estimated density of the order of 30 g/m² dry weight, of which 96% was contributed by molluscs[190]. At Firemore Bay, Loch Ewe, the infauna had a mean density of 755 individuals and a biomass of 1·3 g dry weight/m² on the shore and 3055 individuals and 3·7 g dry weight/m² in the sub-littoral[191]; along the west and north coasts of Scotland the biomass of the macrofauna of intertidal sand ranged from 0·3 to > 22 g/m² dry weight[192]. In Long Island Sound, the sub-littoral epifauna had a mean density of 77·00/m² and a mean dry weight of 1·098 g/m²: numbers ranged from 29·77 to 163·83/m² (standard deviation, $\sigma = 48·12$) and the biomass from 0·272 to 2·582 g/m² ($\sigma = 0·716$)[193].

Of course, the biomasses quoted are *standing stock* and are not a measure of production. At Loch Ewe, assuming that, for the type of animals involved, their production is about twice the standing stock, viz. 1·26 g C/m² for the whole bay, the production of this stock may be estimated as 2·5 g C/m²/yr. Taking the *ecological efficiency* to be 10%, then the annual requirement of this population is 25 g C/m² [191]. The term ecological efficiency is concerned with the effective passage of production to the next trophic level. At each stage of the food chain about 90% is lost, so that 100 g of autotrophic diatoms yield 10 g of herbivorous animal, 1 g of carnivorous animal and 0·1 g of fish if it feeds directly at this stage, or 0·01 g at one stage further removed; similarly 0·01 or 0·001 g of man depending upon the stage at which he took the fish[186, 194].

In Firemore Bay, the primary production due to littoral diatoms was 4–9 g C/m²/yr, and is evidently inadequate to meet the needs of the animal population[191, 195]; the production of the water column here is 95 g C/m²/yr, but only a portion is likely to be available to the benthos[191, 195]. However, this takes no account of bacterial sources of production, which in mud, at least, may produce *ca.* 10 g/day/ft³ [196].

Other efficiencies may be defined and are concerned with assimilation and growth in the individual. The terms efficiency of energy conversion, gross efficiency, growth efficiency and feeding efficiency are all terms concerned with the amount of food converted into new tissue as a percentage of the total food captured. However, they are not necessarily strictly synonymous since the units of measurement may be different[197]. The individual *gross growth efficiency* may be defined as:

$$\frac{\text{calorific content of non-respired assimilation}}{\text{calorific content of food intake}} \times 100$$

and the *net growth efficiency* can be defined as:

$$\frac{\text{calorific content of biomass of growth}}{\text{calorific content of total assimilation}} \times 100.$$

Reports of individual gross growth efficiencies range from 3·5 to 40%, and for individual net growth efficiencies from 5 to 65%[198]. Another expression used is the *conversion ratio* or *conversion factor* which is the quotient derived from the weight of food ingested divided by the growth resulting. The efficiency of conversion of some estuarine fishes is summarized in Table 13.4; it may be expected to differ from year to year, and is influenced by the size of the fish and the temperature of the water[205]. The known efficiency of conversion of aquatic invertebrates ranges from 25 to 90%; the very high values are associated with larvae, and this falls markedly once the animal has metamorphosed[206]. Although the efficiency of conversion in the individual may be very high, the ecological efficiency may be very low. For example, the average energy flow in a population of *Scrobicularia* may be 336 k cal/m²/yr, and the average net growth efficiency reach 21·0%. However, only a small proportion of this production passes on to the next trophic level, viz. the oystercatcher, *Haematopus*, when the ecological efficiency only attains 2·4%[206].

Clearly, these considerations are a simplified expression of a complex situation. Energy does not always flow from the lowest to the highest trophic levels, level by level; sources of food are extremely varied, and animals which are potential competitors do not compete because they take advantage of different sources of food. In other cases, as in the meiobenthos, for example, it appears as if little energy passes to higher trophic levels and most is expended within the meiobenthic system[20, 22, 207–233]. The complexities of estuarine food webs are summarized in Figs. 13.2, 13.3, 13.4.

Mariculture

Fish culture is a very ancient practice[234], and one which is performed successfully in fresh and brackish water ponds, particularly in warmer

Table 13.4. The efficiency of conversion of some species of flatfish (after Colman[205])

Species	Weight of fish (g)	Temperature °C	Food	Conversion ratio	Author
Platichthys flesus (Flounder)	0·01–0·06	19–21	Paranais litoralis	3–7	Bregnballe[203]
	0·124–1·278	Environmental conditions and feeding		4·2–6·0	Müller[204]
Pleuronectes platessa (Plaice)	17–205	12–20	Mytilus	7·4–14·4	Dawes[199]
	10–80	15–19	Mytilus	5·8–8·5	Buckman[200]
	0·085–0·421	Environmental conditions and feeding		5·6	Müller[204]
	0·3–5·1	13–16	Mytilus	4·0–6·2	Colman[205]
Scophthalmus maximus (Turbot)	256–2676	Environmental conditions and feeding		4·7	Müller[204]
Limanda yokohamae	74–107	7·7–9·5	Tylorrhynchus heterochaetus (annelids)	6·2	Hatanaka et al.[201]
	42–215	17–21	Tylorrhynchus heterochaetus (annelids)	5·0–7·3	Hatanaka et al.[201]
	68–279	11–15	Tylorrhynchus heterochaetus (annelids)	5·5–9·6	Hatanaka et al.[201]
Kareius bicoloratus	110–222	11–15	Venerupis japonica	7·4	Hatanaka et al.[201]
	102–175	7–13	V. japonica	6·6	Hatanaka et al.[202]
	120–212	14–18	V. japonica	5·2	Hatanaka et al.[202]
	72–275	18–19·5	V. japonica	5·6	Hatanaka et al.[202]
	87–234	11–18	T. heterochaetus	5·2	Hatanaka et al.[202]

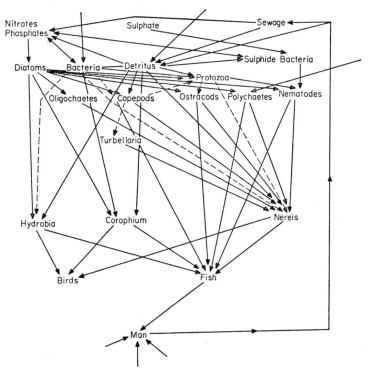

Fig. 13.2. The food web relationships of the microbenthos of a sandy flat at Whitstable, Kent (after Perkins[215]).

countries, where carp and other herbivorous fish are grown[194, 235]. In recent times, it has become customary to view the sea as the major untapped source of food which will solve the problems created by mankind's soaring populations. Superficially, this is a marvellous idea, but oceanic productivity is finite, viz. *ca.* $1.2-1.5 \times 10^{10}$ tons of carbon per year, i.e. *ca.* 30×10^9 tons dry weight or 5×10^{11} tons wet weight of organic matter[194]. Clearly with an ecological efficiency of 10% at every trophic level between algae and man, very much less food is available for the latter than was produced at the primary trophic level, but, even if it were all available, it is still a finite source of food. In addition, not all of the world's marine waters are equally productive; in general, the estuaries, waters close inshore and areas of upwelling are responsible for a major proportion of marine primary production. It is in the first two of these areas that fish farming is most likely to be a success and equally it is these areas which are most at risk from man's disposal of waste products and his rapacity in exploiting the world's resources, viz. oil and minerals. Further, it is these areas, so

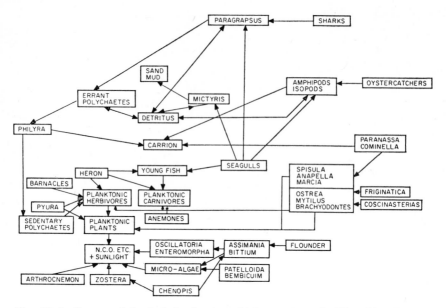

Fig. 13.3. Suggested food-chain for intertidal organisms in Pipe Clay Lagoon (after Guiler[210]).

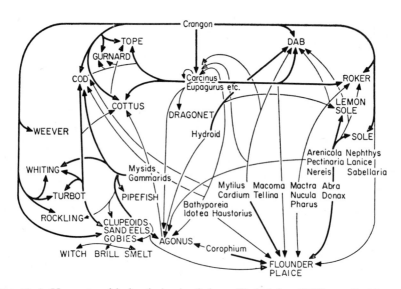

Fig. 13.4. Upper trophic levels in the Solway Firth (after Williams, Perkins and Hinde[22]).

important to marine productivity, that are the sites for colossal new harbour developments, and for increased leisure activities with all that this implies. In all, these areas comprise only some 3% of the earth's surface. On this basis alone, the prospect of mariculture providing much more than a very good living for some individuals does not appear favourable, particularly for highly industrialized nations such as Britain.

In Britain, attempts to improve fisheries and repair the damage of overfishing, date from the last century. At this time, fish hatcheries, constructed in Scotland and the Isle of Man, had the object of raising young fish and releasing them to the sea (see Reports of the Fishery Board for Scotland from 1892 onwards). Later, in the 1940s, attempts were made to increase fish production by the addition of fertilizers to an enclosed arm of a Scottish west coast sea-loch. Productivity was raised by the application of fertilizer, but because the fish produced were not of real interest to the housewife, the experiment as a whole was not a success[194, 236-244]

From the mid 1940s until the early 1960s, it became a dormant topic in British marine biology; however, under the stimulus of new techniques[245] it was reopened once more in the 1960s. As before, the attempt was centred upon flatfish, plaice at first and later sole, but on this occasion much more attention was paid to detailed engineering in relation to ponds for rearing fish which were also provided with supplementary food supplies: young fish were also released to the environment[245-250].

Advantages can be gained by culturing these animals in the warm effluent from power stations[251-253], although in this situation, chlorine could be, but apparently is not, harmful to the animal under culture[253-255]. However, despite these successes, the fish cultivated barely come into the luxury bracket and as such are not especially prized by the housewife or gourmet; moreover, catches of these species made by conventional means seem to satisfy the market at this time.

Although shellfish culture does not seem a particularly attractive prospect to some authors[186], an examination of Table 13.5[256] shows that not only are they as appetizing as fin fish, but the calorific value of the two groups is very similar. Moreover, most of the fin fish, except the mullet, live at a higher trophic level than the bivalve molluscan shellfish. Clearly, use of the latter animals as supplementary foods for fish in fish farms is ecologically very inefficient. Might it not be wiser to ensure that our inshore waters are of such a quality that their potential can be fully realized.

In Japan, the valuable marine fish, the yellowtail, *Seriola quinqueradiata*, and the prawn, *Penaeus japonicus*, are farmed on a national scale[257, 258]. The captive stocks of *Seriola* are fed on a diet of trash fish, fish meal and mollusc meat. In these conditions, a 40 g fry can reach a weight of 1·5 kg in nine months (contrast the plaice and sole which grow much more slowly),

Table 13.5. Chemical composition of fish[1] and shellfish flesh (after Murray and Burt[256])

Species		Water %	Fat %	Protein[2] %	Energy value (Cal/lb)
Scophthalmus rhombus	Brill	76	2·5	19·8	470
Gadus morhua	Cod	78–83	0·1–0·9	15·0–19·0	310–360
Mugil sp.	Grey mullet	76	3·9	19·5	530
Gadus aeglefinus	Haddock	79–84	0·1–0·6	14·6–20·3	280–380
Merluccius merluccius	Hake	80	0·4–1·0	17·8–18·6	320–380
Hippoglossus hippoglossus	Halibut	75–79	0·5–9·6	18·0–18·8	440–740
Microstomus kitt	Lemon sole	79	0·5–3·8	16·4–18·4	330–370
Pleuronectes platessa	Plaice	81	1·1–3·6	15·7–17·8	360–490
Solea solea	Sole	78	1·3	18·8	430
Thunnus sp.	Tuna	71	4·1	25·2	630
Scophthalmus maximus	Turbot	78	2·1–3·9	16·8–20·6	390–420
Cancer pagurus	Crab White meat	74	0·2	22·4	430
	Brown meat	70	7·5	13·1	560
Nephrops norvegica	Scampi White meat	77	0·6–2·0	19·5	400
Palaemon serratus	Prawn	71	1·3	22·8	480
Crangon crangon	Shrimp	68–70	0·9	10·5–23·2	450–500
Mytilus edulis[3]	Mussel	80–84	0·8–2·3	8·9–11·7	270–300
Ostrea edulis[3]	Oyster	77–83	1·1–2·5	8·6–12·6	320–460
Pecten maximus[3]	Escallop	73–79	0·5–1·0	19·5	350

1. The values in the table are for raw flesh or organs. Where a range of values are given a number of analyses were made, but when only one value is given this is the result of one analysis only and not of an invariable result.

2. Protein values are derived by multiplying total nitrogen content by 6·25; since 10–15% of the total nitrogen content of the fish is not in protein form, the values given are higher than the true protein content available for nutritional purposes.

3. Values are for total content of shell.

and the total farm output in Japan for 1964 reached 15 000–20 000 metric tons. The penaeid prawns are tank reared from the fry released by wild, ripe females. They are fattened to a market size, on shredded mollusc flesh, in a variety of shallow, sandy-bottomed ponds or saltings generally irrigated by pumped sea water[258], a method which might be feasible in Britain, where the winds, which often blow, could provide a cheap source of power[259]. In 1964, the Japanese production of farmed prawns was 100 metric tons[258], but this farm production, both of prawns and of yellowtail, is trivial compared with production by conventional methods[50, 51].

In some areas, it has been sugggested that the production of fish might be raised by the enrichment of waters with nutrients and sewage[259–263], similar to the usage in fresh water systems[235, 264–266]. However, there is the inherent danger that the micro-organisms which would result from such treatment may be less suitable for the intended purpose or may even be undesirable[186]. Indeed, the red tides observed in the North Sea in recent times may be an example of just such an effect (Chapter 1).

Mullet are herbivorous and must be considered suitable candidates for exploitation in fish culture. Such use has been made of this fish in Russia[267], in France[268], and Egypt[269]. In France, 1437 tons of mullet are harvested annually from 250 ponds of surface area 375 acres. Wild fry are induced to enter the ponds by a current from the sluice, the contra-current response taking them into the pond. Once in the pond, the fish grow naturally without supplementary feeding and are harvested annually. The sea water, in the ponds, is replenished regularly. Some loss of fish occurs if deoxygenation develops in the ponds and there is some loss to seabird predators, e.g. cormorants, herons, gulls, and to otters. The fry of another fish, the bar, which predates upon mullet enters the ponds at the same time as the mullet fry; however, this fish commands a higher price than mullet and is allowed to remain. At the time of harvest, every 14 tonnes of fish is composed of 11 tonnes of mullet, white, black and golden, and 3 tonnes of bar: for every 10 tonnes of white fish, 5 tonnes of eel are harvested[268].

In Egypt, advantage has been taken of the salinity of Lake Quarun which lies 83 km. S.S.W. of Cairo. The lake is stocked with mullet, *Mugil cephalus*, and *M. capito* and *M. saliens*, but only *M. saliens* can reproduce in the lake. The lake is also stocked with *Solea solea*, which spawns in February and March. Species available as fish food include *Abra ovata*, *Cardium edule* and *Gammarus vulgaris*[269]. Evidently, the lake will eventually become too saline even for these animals, but it does represent a novel use of an otherwise difficult resource.

References

1. Johnston, C. S. (1968). *Process Biochemistry* **3**, 11–14.
2. Keary, R. (1968). *Hydrospace* **1**, 42.
3. Fowler, G. H. and Allen, E. J. (1928). "Science of the Sea", Clarendon, Oxford.
4. Pocock, M. A. (1956). *Proc. Linn. Soc., Lond.* **167**, 11–41.
5. Boney, A. D. and White, E. B. (1967). *J. mar. biol. Ass. U.K.* **47**, 223–232.
6. Blake, B. (1959). "The Solway Firth", Hale, London.
7. Perkins, E. J. and Williams, B. R. H. (1963). H.M.S.O., U.K.A.E.A., P.G. Report 500 (CC).
8. Haw, F., Wendler, H. O., Deschamps, G. (1967). *State of Washington, Dept. of Fisheries. Res. Bull.* **7**, 1–192.
9. Washington State Department of Fisheries (1967). "Fisheries Resources of Washington", 17 pp. Olympia, Washington.
10. Washington State Department of Fisheries (1967). "Pacific Northwest Marine Fishes", Olympia, Washington.
11. Wright, S. C. (1968). State of Washington, Dept. of Fisheries, Information Booklet No. 1: 25 pp.
12. Burgner, R. L., Di Constanzo, C. J., Ellis, R. J., Harry, G. Y., Hartman, W. L., Kerns, O. E., Mathiesen, O. A. and Royce, W. F. (1969). *U.S. Fish and Wildlife Service, Fishery Bulletin.* **67**, 405–459.
13. Banks, J. (1969). *J. Fish. Biol.* **1**, 85–136.
14. Finn, E. L. and Phinney, L. A. (1968). "Controlled Natural Rearing Experiment 1961–1965", State of Washington, Department of Fisheries, Puget Sound Stream Studies, Progress Report.
14a. Washington State Department of Fisheries (1969). "Salmon Hatcheries", 20 pp. Olympia, Washington.
15. LeMier, E. H. and Senn, H. G. (1968). "Appraisal of Project Results (July 1, 1967 through June, 30, 1968)", Submitted to the U.S. Fish and Wildlife Service, Bureau of Commercial Fisheries under the Columbia River Fishery Development Plan. Contract No. 14-17-0001-1676: 13 pp.
16. Bergman, P. and Hager, R. (1969). "The Effects of Implanted Wire Tags and Fin Excision on the Growth and Survival of Coho Salmon (*Oncorhynchus kisutch*, Walbaum", State of Washington, Dept. of Fisheries, Marine Fisheries Investigations, Progress Report: 120 pp.
17. Fisheries Association of B.C. *et al.* "Salmon, the Living Resource", 23 pp. Agency Press Ltd, Canada.
18. Senn, H. G. and Noble, R. E. (1968). *Washington Department of Fisheries, Fish. Res. Pap.* **3**, 51–62.
19. Pearcy, W. G. and Richard, S. W. (1962). *Ecology* **43**, 248–259.
20. Haertel, L. and Osterberg, C. (1967). *Ecology* **48**, 459–472.
21. Rae, B. B. (1955). *Trans. J. Proc. Dumfries. Galloway Nat. hist. Antiq. Soc.*, Ser. 3. **32**, 93–109.
22. Williams, B. R. H., Perkins, E. J. and Hinde, A. (1965). H.M.S.O., U.K.A.E.A., P.G. Report 611 (CC).
23. Perkins, E. J. (1968–1971). *Trans. J. Proc. Dumfries. Galloway Nat. hist. Antiq. Soc.*, Ser. 3. **45**, 15–43, 1968; **46**, 1–26, 1969; **48**, 12–68. 1971.
24. Renfro, W. C. (1960). *Tulane Stud. Zool.* **8**, 83–91.
25. Pearcey, W. G. (1963). *J. Cons. perm. int. Explor. Mer* **27**, 232–235.
26. Alderdice, D. F. and Forrester, C. R. (1967). *J. Fish. Res. Bd. Can.* **25**, 495–215.

27. Forrester, C. R. and Alderdice, D. F. (1965). *J. Fish. Res. Bd. Can.* **23**, 319–340.
28. Hartley, P. H. T. (1940). *J. mar. biol. Ass. U.K.* **24**, 1–68.
29. Hartley, P. H. T. (1947). *J. mar. biol. Ass. U.K.* **27**, 53–64.
30. Shelbourne, J. E. (1957). *J. mar. biol. Ass. U.K.* **36**, 539–552.
31. Shelbourne, J. E. (1962). *J. mar. biol. Ass. U.K.* **42**, 243–252.
32. Cushing, D. H. (1961). *J. mar. biol. Ass. U.K.* **41**, 799–816.
33. Ryland, J. S. (1964). *J. mar. biol. Ass. U.K.* **44**, 343–364.
34. Katz, M. (1945). *Trans. Am. Fish. Soc.* **75**, 72–76.
35. Katz, M. and Erikson, D. W. (1950). *Copeia* 1950, No. 3, 176–181.
36. Bagenal, T. B. (1955). *J. mar. biol. Ass. U.K.* **34**, 297–311.
37. Bagenal, T. B. (1957). *J. mar. biol. Ass. U.K.* **36**, 339–375.
38. Bagenal, T. B. (1963). *J. mar. biol. Ass. U.K.* **43**, 391–399.
39. Bagenal, T. B. (1963). *J. mar. biol. Ass. U.K.* **43**, 401–407.
40. Bagenal, T. B. (1965). *J. mar. biol. Ass. U.K.* **45**, 599–606.
41. Bagenal, T. B. (1966), *J. mar. biol. Ass. U.K.* **46**, 161–186.
42. Russell, E. S. (1942). "The Overfishing Problem", University Press, Cambridge.
43. Graham, M. (1943). "The Fish Gate", Faber and Faber, London.
44. Graham, M. (1956). "Sea Fisheries: Their Investigation in the United Kingdom", Arnold, London.
45. Beverton, R. J. H. and Holt, S. J. (1957). *Fishery Invest.*, London. Ser. 2. **10**, 1–533.
46. Fulton, T. W. (1903). *21st Annual Rept. Fishery Board for Scotland.* Part III, 15–108.
47. Murie, J. (1903). "Report on the Sea Fisheries and Fishing Industries on the Thames Estuary", Pt. I: 1–250, Kent and Essex Sea Fisheries Committee, London.
48. Coward, T. A. (1910). "The vertebrate fauna of Cheshire and Liverpool Bay", Vol. 2. Reptiles, Amphibians and Fishes, London.
49. Huntsman, A. G. (1952). *Fundy Fisherman* Nov. 26, 1952, 8 pp.
50. Fish Commission of Oregon (1967). "Oregon Commercial Fisheries Landings and Value at Fishermen's Level in 1965", 4 pp., Fish Commission of Oregon.
51. Crutchfield, J. A. and Macfarlane, D. (1968). *State of Washington, Dept. of Fisheries, Res. Bull.* **8**, 57 pp.
52. H.M.S.O. (1970). Sea Fisheries Statistical Tables 1969. Scottish Sea Fisheries Statistical Tables 1969. H.M.S.O., London.
52a. Marine Laboratory, Aberdeen (1971). *In* "Institute of Food Science and Technology, Scotland Symposium on Food Industries in Rural Scotland, 23rd April 1971, Aviemore."
53. Richardson, L. R. (1962). *Proc. N.Z. ecol. Soc.* **9**, 47–50.
54. Hancock, D. A. (1955). M.A.F.F. Laboratory Leaflet No. 8, 6 pp.
55. Hancock, D. A., Drinnan, R. E. and Harris, W. N. (1956). *J. mar. biol. Ass. U.K.* **35**, 307–325.
56. Lloyd, A. J. and Yonge, C. M. (1947). *J. mar. biol. Ass. U.K.* **26**, 626–661.
57. Stopford Meredith, S. (1950–52). *Proc. Trans. Lpool biol. Soc.* **48**, 75–109.
58. Hedgpeth, J. W. (1966). *Am. Fish. Soc., Spec. Publ.* **3**, 3–11.
59. Warren, P. J. and Sheldon, R. W. (1967). *J. Fish. Res. Bd. Can.* **24**, 569–580.
60. Mason, J. (1968). "The Scottish Fishery for Brown Shrimps, *Crangon crangon* (L.), in the Solway Firth", I.C.E.S., C.M. 1968/K: 22–12 pp.
61. Hughes, D. A. (1969). *Biol. Bull. mar. biol. Lab., Woods Hole.* **136**, 43–53.

62. Perkins, E. J. and Williams, B. R. H. (1966). H.M.S.O., U.K.A.E.A., P.G. Report 587 (CC).
63. Giard, M. and Roussin, M. (1889). *7th Annual Rept. Fishery Research Board for Scotland* 1889.
64. Carlisle, D. B. (1962). *J. mar. biol. Ass. U.K.* **42**, 155–162.
65. Thomas, H. J. (1953). *J. Cons. perm. int. Explor. Mer* **18**, 333–350.
66. Thomas, H. J. (1954). *J. Cons. perm. int. Explor. Mer* **20**, 87–91.
67. Thomas, H. J. (1958). *Mar. Res.* No. 8, 107 pp.
68. Thomas, H. J. (1959). *J. Cons. perm. int. Explor. Mer* **24**, 342–348.
69. Thomas, H. J. (1966). *D.A.F.S. Scottish Fisheries Information Pamphlet* 2nd edn. 1966, 19 pp.
70. Mason, J. (1965). *Cons. perm. int. Explor. Mer, Rapp. et Proc.-Verb.* **156**, 95–97
71. Mason, J. (1965). *Cons. perm. int. Explor. Mer, Rapp. et Proc.-Verb.* **156**, 98–99.
72. Edwards, E. and Early, J. C. *Ministry of Technology, Torry Advisory Note* No. 26 (revised): 18 pp.
73. Thomas, H. J. (1955). *J. Cons. perm. int. Explor. Mer.* **20**, 295–305.
74. Thomas, H. J. (1958). *J. Cons. perm. int. Explor. Mer.* **24**, 147–154.
75. Dow, R. L. (1969). *Science, N.Y.* **164**, 1060–1063.
76. Johnson, M. E. and Snook, H. J. (1967). "Seashore Animals of the Pacific Coast", Dover, New York.
77. Washington State Department of Fisheries (1969). "Washington State Shellfish", 16 pp. Olympia, Washington.
78. Loosanoff, V. L. (1965). "The American or Eastern Oyster", U.S. Dept. of the Interior, Fish and Wildlife Service, Circular 205: 36 pp.
79. M.A.F.F. (1963). "Whelks". *Fishery Laboratory, Burnham-on-Crouch, Laboratory leaflet*: 11 pp.
80. Hancock, D. A. (1963). *I.C.N.A.F. Spec. Publ.* **4**, 1963, 176–187.
81. Baird, R. H. and Gibson, F. A. (1956). *J. mar. biol. Ass. U.K.* **35**, 555–562.
82. Mason, J. (1957). *J. mar. biol. Ass. U.K.* **36**, 473–492.
83. Mason, J. (1957). *Wld Fish.* **6**, 35–37.
84. Mason, J. (1958). *J. mar. biol. Ass. U.K.* **37**, 653–671.
85. Baird, R. H. (1966). *J. mar. biol. Ass. U.K.* **46**, 33–47.
86. Mason, J. and McIntyre, A. D. "The Scottish Scallop Fishery", D.A.F.S. Scottish Fisheries Information Pamphlets: 11 pp.
87. Hardy, R. and Smith, J. G. M. (1970). *Ministry of Technology, Torry Advisory Note* No. 46, 10 pp.
88. Waugh, G. D. (1960). *Proc. malac. Soc., Lond.* **34**, 113–122.
89. Hancock, D. A. and Urquhart, A. E. (1965). *Fishery Invest., Lond.* Ser 2. **24**, 40 pp.
90. Mason, J. "Cockles in Scotland", D.A.F.S., Scottish Fisheries Information Pamphlets: 7 pp.
91. Wells, H. W. (1957). *Ecology* **38**, 123–128.
92. Marriage, L. D. (1958). "The Bay Clams of Oregon", Fish Commission of Oregon, Educational Bulletin No. 2: 29 pp. Portland, Oregon.
93. Oregon Fish Commission (1963). "Razor clams" Educational Bulletin No. 4: 13 pp. Portland, Oregon.
94. Washington State Department of Fisheries (1969). "Washington Razor Clams. 1969 Regulations", 8 pp. Olympia, Washington.
95. Oregon Fish Commission (1969). "Personal Use Harvest of Marine Food Fish Shellfish Intertidal Animals. Synopsis of Regulations", Portland, Oregon.

96. Loosanoff, V. L. (1955). *Science, N.Y.* **121**, 119–121.
97. Walne, P. R. (1964). *J. mar. biol. Ass. U.K.* **44**, 293–310.
98. Orton, J. H. (1937). *Nature, Lond.*, **140**, 505–506.
99. Lindsay, C. E. (1967). "Potential Sources of Pacific Oyster Seed in Korea and Taiwan", State of Washington, Department of Fisheries, Report on Contract No.: 1-24-D: 52 pp.
100. Marshall, E. (1959). *Ecology* **40**, 298.
101. Lindsay, C. E. "Oyster Culture in the State of Washington", State of Washington, Department of Fisheries, Shellfish Laboratory, Brinnon, Information Leaflet: 6 pp. Mimeo.
102. Baird, R. H. (1959). "Factors Affecting the Efficiency of Dredges. Modern Fishing Gear of the World", Fishing News (Books) Ltd., London.
103. Baird, R. H. (1959). *I.C.E.S., Shellfish Committee, C. M.* 1959 Paper No. 105, 3 pp.
104. Editor (1968). "An oyster Seed Industry for Oregon", Oregon's Agricultural Progress, 15(2), Agricultural Experimental Station, Oregon State University, Corvallis.
105. Hancock, D. A. (1965). *Ophelia* **2**, 253–267.
106. Mason, J. (1965). "Mussels in Scotland", D.A.F.S., Scottish Fisheries Information Pamphlet: 8 pp.
107. Fauretto, L. (1968). *Boll. Soc. adriat. Sci. nat.* **56**, 243–261.
108. M.A.A.F. (1927). "Mussel cultivation", *Fish. Nat. Lond.* No. 13, 4 pp.
109. Perkins, E. J. (1967). *Trans. J. Proc. Dumfries. Galloway Nat. hist. Antiq. Soc.*, Ser. 3, **44**, 47–56.
110. Lancashire and Western Sea Fisheries Joint Committee (1968). "Survey of Mussel Beds at Heysham", June 1968: 6 pp. Mimeo.
111. Mason, J. (1967). *Scott. Fish. Bull.* **27**, 26–27.
111a. Mason, J. (1971). *Underwat. J.* **3**, 52–59.
112. Mason, J. (1968). I.C.E.S., Fisheries Improvement Committee Ref. K. (Shellfish and Benthos C.) C.M. 1968/E : 4. 9 pp. Mimeo.
113. Thomas, H. J. (1956). "Shellfish—A 'Small Man's' Industry", Survey of the Fishing Industry, The Scotsman, Friday 23rd Nov., 1956.
114. Thomas, H. J. (1966). "Handling Lobsters and Crabs", D.A.F.S. Scottish Fisheries Information Pamphlet, 2nd edn: 14 pp.
115. Fulton, T. W. (1896). "Oyster Beds of the Firth of Forth", H.M.S.O., Report of the Fishery Board for Scotland.
116. Loosanoff, V. L. and Davis, H. C. (1963). *Comm. Fish. Rev.* **25**, 1–11.
117. Walne, P. R. (1966). *Fishery Invest., Lond.* Ser. 2. **25**, 53 pp.
118. Walne, P. R. (1970). *Fishery Invest., Lond.* Ser. II, **26**(5): 62 pp.
119. Millar, R. H. (1961). *Mar. Res.* 1961. 3: 76 pp.
120. Woelke, C. E. (1969). "A History and Economic Evaluation of Washington State Oyster reserves", State of Washington, Department of Fisheries, April 1969: 16 pp.
121. Harding, J. P. and Wheeler, A. C. (1958). *Nature, Lond.* **182**, 542–543.
122. Scott, T. (1909). *26th Annual Rept. of the Fishery Board for Scotland.* III 73–92.
123. Williamson, H. C. (1913). *Fisheries, Scotland, Sci. Invest.*, 1911. 2, 39 pp.
124. Schlumberger, H. G. and Katz, M. (1956). *Cancer Res.* **16**, 369–370.
125. Wellings, S. R., Chuinard, R. G., Gourley, R. T. and Cooper, R. A. (1964). *J. natn. Cancer Inst.* **33**, 991–1004.

126. Wellings, S. R., Chuinard, R. G. and Bens, M. (1965). *Ann N.Y. Acad. Sci.* **126**, 479–501.
127. Wellings, S. R., Cooper, R. A. and Chuinard, R. G. (1966). *Bull. Wildlife Dis. Assoc.* **2**, 68.
128. McArn, G. E., Chuinard, R. G., Miller, B. S., Brooks, R. E. and Wellings, S. R. (1968). *J. natn. Cancer Inst.* **41**, 229–242.
129. Sparks, A. K., Paule, G. B. and Chew, K. K. (1968). *Proc. natn. Shellfish Ass.* **59**, 35–39.
130. Sparks, A. K. (1969). *Nat. Cancer Inst., Monogr.* No. 31, 671–682.
131. Mason, J. (1958). *Mar. Res.* 1958. 8, 8 pp.
132. Houghton, D. R. (1963). *J. Anim. Ecol.* **32**, 253–257.
133. Hockley, A. R. (1951). *J. mar. biol. Ass. U.K.* **30**, 223–232.
134. Waugh, G. D. (1954). *J. Anim. Ecol.* **23**, 364–367.
135. Monteiro, M. and Figuiredo, J. (1959). *I.C.E.S. Shellfish Committee*, 1959, No. 113: 7 pp.
136. Hepper, B. T. (1956). *J. Anim. Ecol.* **25**, 144–147.
137. Johnson, E. A. and Chew, K. K. (1969). *J. Fish. Res. Bd. Can.* **26**, 2245–2246.
138. Williams, C. S. (1969). *J. mar. biol. Ass. U.K.* **49**, 161–173.
139. Williams, C. S. (1968). *J. Anim. Ecol.* **37**, 709–712.
140. Cole, H. A. and Hancock, D. A. (1955). *J. mar. biol. Ass. U.K.* **34**, 25–31.
141. Andrews, J. D. and Hewatt, W. G. (1957). *Ecol. Monogr.* **27**, 1–26.
142. Sindermann, C. J. (1968). *U.S. Dept. of the Interior, Fish and Wildlife Service, Spec. Sci. Rept.—Fisheries* No. 569: 10 pp.
143. Berry, A. J. (1962). *Parasitology* **52**, 237–240.
144. Logan, J. (1971). Personal communication.
145. Loosanoff, V. L. (1965). *Science, N.Y.* **123**, 1119–1120.
146. Dexter, R. W. (1955). *Ecology* **36**, 159–160.
147. Editor (1967). *Scott. Fish. Bull.* No. 27, 29.
148. Scholz, A. J., Westley, R. E. and Tarr, M. A. (1968). "Pacific Oyster Mass Mortality Studies". Seasonal summary report No. 2, May 1966 to April 1967, 36 pp. State of Washington, Department of Fisheries.
149. Cole, H. A. (1942). *J. mar. biol. Ass. U.K.* **25**, 477–508.
150. Hancock, D. A. (1959). *Fishery Invest., London.* **22**, 1–66.
151. Vevers, H. G. (1949). *J. mar. biol. Ass. U.K.* **28**, 165–188.
152. Barnes, H. and Powell, H. T. (1951). *J. mar. biol. Ass. U.K.* **30**, 381–385.
153. Hancock, D. A. (1955). *J. mar. biol. Ass. U.K.* **34**, 313–331.
154. Hancock, D. A. (1958). *J. mar. biol. Ass. U.K.* **37**, 565–589.
155. Kitching, J. A., Sloane, J. F. and Ebling, F. J. (1959). *J. Anim. Ecol.* **28**, 331–341.
156. Ebling, F. J., Kitching, J. A., Muntz, L. and Taylor, C. M. (1964). *J. Anim. Ecol.* **33**, 73–95.
157. Ebling, F. J., Hawkins, A. D., Kitching, J. A., Muntz, L. and Pratt, V. M. (1966). *J. Anim. Ecol.* **35**, 559–566.
158. Kitching, J. A., Muntz, L. and Ebling, F. J. (1966). *J. Anim. Ecol.* **35**, 113–126.
159. Muntz, L., Ebling, F. J. and Kitching, J. A. (1965). *J. Anim. Ecol.* **34**, 315–329.
160. Largen, M. J. (1967). *J. Anim. Ecol.* **36**, 207–214.
161. Orton, J. H. (1926). *J. mar. biol. Ass. U.K.* **14**, 239–279.
162. Hanks, R. W. (1961). *Proc. natn. Shellfish. Ass.* **52**, 75–86.
163. Ropes, J. W. (1969). *Fish. Bull., U.S. Fish. Wildlife Serv.* **67**, 183–203.

164. Perkins, E. J., Williams, B. R. H. and Hinde, A. (1964). H.M.S.O., U.K.A.E.A., P.G. Report 578 (CC).
165. Perkins, E. J., Williams, B. R. H. and Gorman, J. (1965). U.K.A.E.A., Chapelcross Health Physics Dept., Note No. 2.
166. Perkins, E. J. and Penfound, J. M. (1969). *Proc. Challenger Soc.* 4, August 1969.
167. Perkins, E. J. and Penfound, J. M. (1971). *Spectrum. British Science News* 1971. No. 84: 708.
168. Burkenroad, M. D. (1957). *Ecology* 38, 164–165.
169. Loosanoff, V. L. (1964). *Biol. Bull. mar. biol. Lab. Woods Hole,* 126, 423–439.
170. Drinnan, R. E. (1957). *J. Anim. Ecol.* 26, 441–469.
171. Drinnan, R. E. and Cole, H. A. (1957). *Nature Wales.* 3, 499–503.
172. Williams, B. R. H., Perkins, E. J. and Bailey, M. (1964). *Trans. J. Proc. Dumfries. Galloway Nat. hist. Antiq. Soc.* Ser. 3. 41, 30–44.
173. Loosanoff, V. L. (1958). *Comm. Fish. Rev.* 20, 45–47.
174. Loosanoff, V. L. (1960). *Gulf and Caribbean Fish. Inst.,* 13th Annual Session, Nov. 1960. 113–127.
175. Loosanoff, V. L. (1961). *U.S. Dept. of the Interior, Fish and Wildlife Service, Fishery Leaflet.* 520, 11 pp.
176. Loosanoff, V. L. and Nomejko, C. A. (1958). *Proc. natn. Shellfish Ass.* 48, 83–89.
177. Haydock, C. I. (1964). *Calif. Fish. Game.* 50, 11–28.
178. Maunder, S. (1948). "The Treasury of Natural History", Longman, Brown, Green and Longmans, London.
179. Abattucci, L. (1966). Private communication.
180. Frimodt, J. (1966). Private communication.
181. Loosanoff, V. L., Mackenzie, C. L. and Shearer, L. W. (1960). *Science, N.Y.* 131, 1522–1523.
182. Chipperfield, P. N. J. (1951). *J. mar. biol. Ass. U.K.* 30, 49–71.
183. Lindsay, C. E. (1961). *Proc. natn. Shellfish Ass.* 52, 87–98.
184. F.A.O. (1968). "Yearbook of Fishery Statistics for 1967", Food and Agriculture Organisation of the United Nations, Rome, 1968.
185. Fiske, J. D., Watson, C. E. and Coates, P. G. (1967). *Division of Marine Fisheries, Dept. of Natural Resources, Commonwealth of Massachusetts. Monograph Series* No. 5, 56 pp.
186. Ryther, J. H. (1968). *Proc. natn. Shellfish Ass.* 59, 18–22.
187. Rochford, D. J. (1951). *Aust. J. mar. freshwat. Res.* 2, 116.
188. Holme, N. A. (1949). *J. mar. biol. Ass. U.K.* 28, 189–237.
189. Perkins, E. J. (1965). *Ann. Mag. nat. Hist.,* Ser. 12. 9, 112–128.
190. Moore, H. B., Davies, L. T., Fraser, T. H., Gore, R. M. and López, N. R. (1968). *Bull. mar. Sci. Gulf Caribb.* 18, 261–279.
191. McIntyre, A. D. and Eleftheriou, A. (1968). *J. mar. biol. Ass. U.K.* 48, 113–142.
192. McIntyre, A. D. (1970). *J. mar. biol. Ass. U.K.* 50, 561–575.
193. Richards, S. W. and Riley, G. A. (1967). *Bull. Bingham oceanogr. Coll.* 19, 89–135.
194. Fraser, J. H. (1958). *Proc. Nutr. Soc.* 17, 127–132.
195. Steele, J. H. and Baird, I. E. (1968). *Limnol. Oceanogr.* 13, 14–25.
196. ZoBell, C. E. and Feltham, C. B. (1942). *Ecology* 23, 69–77.
197. Reeve, M. R. (1969). *J. mar. biol. Ass. U.K.* 49, 77–96.

198. Hunter, W. Russell (1970). "Aquatic Productivity", Macmillan, London.
199. Dawes, B. (1930). *J. mar. biol. Ass. U.K.* **17**, 103–174.
200. Buckmann, A. (1952). *Kurze Mitt. fischbiol. Abt. Max-Planck-Inst. Meeresbiol. Wilhelmsh.* **1**, 8–20.
201. Hatanka, M. (1956). *Tohoku J. agric. Res.* **7**, 151–162.
202. Hatanka, M. (1956). *Tohoku J. agric. Res.* **7**, 163–174.
203. Bregnballe, F. (1961). *Meddr kommn Danm. Fisk. -og. Havunders.* **3**, 133–182.
204. Muller, A. (1969). *Ber. dt. wiss. Kommn Meeresforsch.* **20**(H2), 112–128.
205. Colman, J. A. (1970). *J. mar. biol. Ass. U.K.* **50**, 113–120.
206. Hughes, R. N. (1970). *J. Anim. Ecol.* **39**, 357–379.
207. ZoBell, C. E. and Landon, W. A. (1937). *Proc. Soc. exp. Biol. Med.* **36**, 607–609.
208. Ritchie, A. (1938). *Cons. perm. int. Explor. Mer, Rapp et Proc.-Verb.* **107**, 49–56.
209. Enequist, P. (1949). *Zool. Bidr. Upps.* **28**, 299–492.
210. Guiler, E. R. (1951). *Pap. Proc. R. Soc. Tasm.* 1950, 29–52.
211. Smidt, E. L. B. (1951). *Meddr Kommn Danm. Fisk. -og Havunders.* **11**, 1–151.
212. Longhurst, A. R. (1957). *J. Anim. Ecol.* **26**, 369–387.
213. Qasim, S. Z. (1957). *J. Anim. Ecol.* **26**, 389–401.
214. Gaevskaya, N. S. (1958 (1959)). *Tr. Mosk. Tekhnol. Inst. Rybn. Promyshleunosti i Khoz.* **9**, 49–62, 1958. *Referat. Zh. Biol.*, 1959, No. 39511 (Translation).
215. Perkins, E. J. (1958). *Ann. Mag. nat. Hist.* Ser. 13. **1**, 64–77.
216. Drummond, D. C. (1960). *J. Anim. Ecol.* **29**, 341–347.
217. Noble, A. (1962). *Indian J. Fish.* **9A**, 701–713.
218. Chepurnov, V. S. et al. (1962 (1963)). *Uchen. Zap. kishinev. Univ.* **62**(1): 73–1962. *Referat. Zh. Biol.*, 1963. (Translation).
219. Segerstråle, S. G. (1962). *Commentat. biol.* **24**(7), 26 pp.
220. Paine, R. T. (1963). *Ecology* **44**, 63–74.
221. Paine, R. T. (1963). *Veliger* **6**, 1–9.
222. Paine, R. T. (1963). *Ecology* **44**, 402–403.
223. Wheeler, A. C. (1963). *Rep. Challenger Soc.* **3**(15), 1963.
224. Newell, R. C. (1965). *Proc. zool. Soc., Lond.* **144**, 25–45.
225. Segerstr°le, S. G. (1965). *Bot. Goth.* **3**, 195–204.
226. Macer, C. T. (1967). *Helgoländer wiss. Meeresunters* **15**, 560–573.
227. Rae, B. B. (1967). *Mar. Res.* 1967. **1**, 68 pp.
228. Edwards, R. and Steele, J. H. (1968). *J. exp. mar. biol. Ecol.* **2**, 215–238.
229. Green, J. (1968). "The Biology of Estuarine Animals", Sidgwick and Jackson, London.
230. Muller, A. (1968). *Kieler Meeresforsch.* **24**, 124–143.
231. Hughes, R. N. (1969). *J. mar. biol. Ass. U.K.* **49**, 805–823.
232. Nilsson, L. (1969). *Oikos* **20**, 128–135.
233. Rae, B. B. (1969). *Mar. Res.* 1969. **2**, 23 pp.
234. Prowse, G. A. (1963). *I.P.F.C. Current Affairs Bulletin.* **36**, 10 pp.
235. Hickling, C. F. (1962). "Fish Culture", Faber and Faber, London.
236. Gross, F. (1941). *Nature, Lond.* **148**, 71–74.
237. Gross, F., Raymont, J. E. G., Marshall, S. M. and Orr, A. P. (1944). *Nature, Lond.* **153**, 483–485.
238. Orr, A. P. (1947). *Proc. R. Soc. Edinb.*, B. **43**, 3–20.
239. Gross, F. (1947). *Proc. R. Soc. Edinb.*, B. **43**, 56–95.

240. Marshall, S. M. (1947). *Proc. R. Soc. Edinb.*, B. **43**, 21–33.
241. Marshall, S. M. and Orr, A. P. (1948). *J. mar. biol. Ass. U.K.* **27**, 360–379.
242. Gross, F. (1949). *J. mar. biol. Ass. U.K.* **28**, 1–8.
243. Raymont, J. E. G. (1949). *J. mar. biol. Ass. U.K.* **28**, 9–19.
244. Nutman, S. R. (1950). *Proc. R. Soc. Edinb.*, B. **64**, 5–34.
245. Shelbourne, J. E. (1964). *Adv. mar. biol.* **2**, 1–83.
246. Riley, J. D. and Corlett, J. (1965). *Rep. mar. biol. Stn. Port Erin* **78**, 51–56.
247. Corlett, J. (1967). *Ir. Nat. J.* **15**, 249–254.
248. Milne, P. H. (1969). *Proc. Challenger Soc.* **4**, 20.
249. Milne, P. H. (1970). "Fish Farm Enclosures", *In* "World Fishing, Dec. 1969–July 1970" (H. S. Noel, ed.), 20 pp, Commercial Exhibitions and Publications, London.
250. Milne, P. H. (1970). *Mar. Res.* 1970. 1, 31 pp.
251. Iles, R. B. (1963). *New Scient.* No. 324, 31 Jan. 1963, 227–229.
252. Nash, C. E. (1970). *Mar. Pollut. Bull.* **1**, 5.
253. Nash, C. E. (1970). *Mar. Pollut. Bull.* **1**, 28–30.
254. Waugh, G. D. (1964). *Ann. appl. Biol.* **54**, 423–440.
255. Alderson, R. (1970). *In* "F.A.O. Technical Conference on Marine Pollution and its Effects on Living Resources and Fishing", F.I.R.: MP/70/E-3: 8 pp. Rome, 9–18 December 1970.
256. Murray, J. and Burt, J. R. (1969). *Ministry of Technology, Torry Advisory Note* No. 38: 14 pp.
257. Allen, J. H. (1969). *Proc. Challenger Soc.* **4**, 19–20.
258. Shelbourne, J. E. (1969). *Proc. Challenger Soc.* **4**, 18–19.
259. Perkins, E. J. (1960). "Estuarine Biology—The Next Ten Years?", Symposium, Association of British Zoologists, held at the Zoological Society of London, Regents Park, London, 9 January 1960.
260. Allen, G. H. (1970). *In* "F.A.O. Technical Conference on Marine Pollution and its Effects on Living Resources and Fishing", F.I.R.: MP/70/R-13: 26 pp. Rome, 9–18 December 1970.
261. Chan, G. (1970). *In* "F.A.O. Technical Conference on Marine Pollution and its Effects on Living Resources and Fishing", F.I.R.: MP/70/E-10: 4 pp. Rome, 9–18 December 1970.
262. Parsons, T. R. *et al.* (1970). *In* "F.A.O. Technical Conference on Marine Pollution and its Effects on Living Resources and Fishing" F.I.R.: MP/70/E-58: 14 pp. Rome, 9–18 December 1970.
263. Stirn, J. (1970). *In* "F.A.O. Technical Conference on Marine Pollution and its Effects on Living Resources and Fishing", F.I.R.: MP/70/E-105: 7 pp. Rome, 9–18 December 1970.
264. Prowse, G. A. (1961). *I.P.F.C. 9th Proceedings.* Section II, 73–75.
265. Gopalakrishnan, V. and Srinath, E. G. (1963). *Proc. Indian Acad. Sci.* **57**, 379–388.
266. Prowse, G. A. (1964). *Verh. int. Verein. theor. angew. Limnol.* **15**, 480–484.
267. Chepurnov, V. S. and Dimitrev, Ya. I. (1962 (1963)). *Uchen. Zap. kishinev Univ.* **62**(1), 53–62, 1962. *Referat. Zh. Biol.* 1963, No. 18 I 36 (Translation).
268. Milne, P. H. (1965). "Visit to Fish Farms in the Gironde", University of Strathclyde, Department of Civil Engineering, Fish Farming Investigations for the White Fish Authority. HO-65-12: 24 pp.
269. El-Zarka, Salah El-Din (1963). *J. Cons. perm. int. Explor. Mer* **28**, 126–136.

14

Pollution

Pollution of the marine environment may be manifest in a number of ways, none of them pleasant. Examples which come readily to mind include the outbreak of Minimata disease which resulted from mercury poisoning, the closure of valuable shell fisheries contaminated by infective hepatitis and the spoiling of fine estuaries by an excessive input of sewage and industrial wastes. However, quotation of such examples of the distortion of natural systems neither enables us to understand why pollution occurs nor do they help us to appreciate that any one example of the pollution phenomenon is a part of a multiphase system which may have components at opposite ends of the earth. It is this strategic concept of the pollution phenomenon that is considered below.

The word pollution is often used in a social context with the implication that it is something unpleasant to man alone, but this too does little to further an understanding of the problem. To use the simple definition that pollution results when undesirable substances are present in a situation is to adopt far too naive a view. Similarly, to say that a state of pollution exists when there are attributes of the environment which adversely affect the balance of nature and man's use or enjoyment is far too narrow and self-centred a view: particularly since it places a severe winter or a drought on a par with a stinking sewer and, since it fails to recognize that much of the worst pollution arises from the influence of essential growth factors, contributes nothing to an understanding of why pollution occurs. Indeed such views lead some engineers to consider an estuary polluted because salt water intrudes into the freshwater run-out, but while there are certain characteristics which estuaries have in common with the influence of introduced toxicants, to regard an estuary as polluted when it is little influenced by man and has nutrient levels not outstandingly influenced by the wildlife which live around its shores and receives only a fresh water input from the land is patently nonsense. Nevertheless, when the activities of man are such that unpleasant symptoms become overt and result either in a widespread mortality, or a marked decline in diversity and depressed production, no one is in any doubt that the system is polluted; unfortunately, however, in the sub-lethal region the situation is not one in which positive judgements are readily made.

A consideration of the effect of necessary growth factors, such as sewage and trace metals, upon biological systems yields both a more lucid view of what is meant by pollution and gives a clear indication of the essentially multiphase nature of the system of which the polluted situation forms a part. The response of biological systems to increasing concentrations of materials which are growth factors of whatever kind, is an increase in the rate of growth or productivity up to an optimum, after which any increase in the concentration of the added factor merely produces a decline in the individual species or population (Fig. 14.1). Clearly, any addition up to the

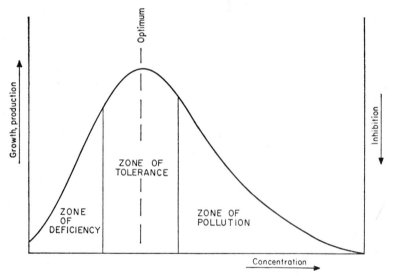

Fig. 14.1. The relationship between the concentration of growth factors optimal conditions and the development of deficient and polluted environments.

optimal concentration is advantageous to the individual species or population, whereas concentrations which exceed the optimum are disadvantageous. Equally clearly, few living organisms are living in optimum conditions with respect to one growth factor, much less to all, that it requires and most organisms are tolerant to, and thrive in, a wide range of sub- or supra-optimal conditions, i.e. about the optimum condition there is a zone of tolerance. When the concentrations of the essential growth factor falls below the minimum level necessary for the maintenance of a healthy organism, or a healthy population, or a healthy community, then the system enters the zone of deficiency. Deficiency may have many manifestations, in man, for example, vitamin deficiency may cause rickets, beri-beri and anaemia.

Certain of the grasslands of Australia and New Zealand are deficient in cobalt. Such grass is lush and consumed readily by sheep, but since the microflora of their gut is unable to synthesize cobalmines, the sheep lose their appetite and die when the cobalt content of this grass falls below 0·2 p.p.m.[1, 2]. Likewise a deficiency of copper caused crop failure in the reclaimed lands of Denmark and Holland and may produce generally poor health in farm animals[2, 3]. Conversely, when the concentration of the growth factor exceeds that of the maximum limit for the zone of tolerance then the system is progressively poisoned. Such poisoning may be typified by the haemolysis and death which results when excessive amounts of copper are stored in the liver of sheep[4]. In man, fatalities have been brought about by a large excess of vitamin A, and the calcification of arteries and kidneys has resulted from an excessive intake of vitamin D; children have been killed by small doses of ferrous sulphate[2].

In general, the response to environmental factors is one to combinations rather than to individual factors alone[5] and is described by response surfaces rather than the simple two dimensional curve of Fig. 14.1[6, 7]. However, it is more convenient here to consider these phenomena in relation to the oligodynamic response curve of Fig. 14.1, since its characteristics are more evident, more is known of its behaviour and it does seem to be a peculiarly powerful expression of observed phenomena. To summarize then, the response of biological systems to environmental parameters including growth factors is typified by the oligodynamic response curve which may be divided into three phases: (1) the sub-optimal zone when the system is starved of a growth factor or is held at a level below the minimum for a tolerable existence; (2) a zone of tolerance around the optimum; and (3) a supra-optimal zone when the system either receives an excess of a growth factor and is unable to deal with it or is held at a level above the maximum for a tolerable existence. By observation and definition, both (1) and (3) lead to a decline of the individual species, population or community.

Consider first what would happen if we investigated a transect drawn from the central water masses of the ocean to the head of an estuary (Fig 14.2). The impoverished central water masses of the oceans are characterized by low density populations of a great number of species, the individuals of which are much reduced in size (p. 3): these areas make no effective contribution to the world production of fish. The continental shelf areas of the world are supplied with much greater quantities of nutrients and trace elements than the central water masses and are characterized by a great number of species which occur at a very much higher density of population, the individuals of which are varied in size, but larger species abound: these areas, some 3% of the earth's surface, are the

15

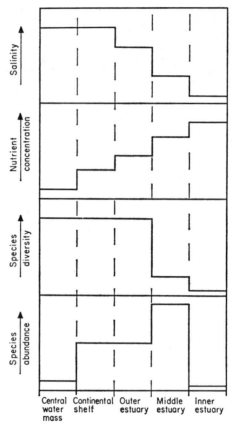

Fig. 14.2. Hypothetical transect drawn from the central water masses to the head of an estuary, and showing the changes in species diversity and abundance in relation to the salinity and nutrient status of the water.

site of the world's major commercial fisheries. In an estuary, the general species composition in the mouth is much the same as that of the open sea, but with increasing distance moved upstream the number of species present decline steadily, and finally the marine and estuarine species disappear and are replaced by those typical of fresh water (p. 173). Concurrent with the changes in the number of species the abundance of the individual species, e.g. *Corophium volutator*, *Macoma balthica*, and *Arenicola marina*, in the middle reaches of the estuary increases significantly: here, compared with the open coast, the nutrient and trace element levels are much higher than the open sea, but the salinity is reduced (Chapter 2). With a greater reduction in the salinity, these resistant estuarine forms decline in abundance, until they disappear completely under the pressure of a much reduced salinity: such changes are typical both of the plankton (Chapter 4) and of the benthos (Chapter 7). Evidently, this progression of change, which can be traced from the central water masses across the continental shelf to the head of an estuary, closely parallels, i.e. "mimics", the influence of gradually increasing concentrations of some kinds of pollutant. However, this parallel may be drawn further in that marine animals subjected to decreasing salinity, and freshwater animals subjected to an increasing salinity evince a normal toxicity response curve (p. 172). Moreover, it has been shown that the survival time of freshwater animals exposed to a low salinity at moderate temperatures, is increased 2–3 fold by a decrease in temperature of $10°C$[8]: an effect which parallels the relationship between temperature and the toxicity of solutions to aquatic organisms.

The Solway Firth, famous for its shifting sand banks and channels, has provided an interesting study of the influence of sediment movement upon the infauna. Some areas, e.g. the Nith estuary, are so mobile that frequent shifts of channel accompanied by the translocation of millions of tons of sediment are known to occur: here the infauna, e.g. *Macoma balthica*, *Mya arenaria*, are washed out into the bed of the channel and transported away[9]; similarly, the central banks may be very impoverished[10]. Elsewhere, and in more sheltered areas less susceptible to movement, populations of infauna are abundant and successful larval settlements occur frequently. However, by far the greater proportion of the sediments in this estuary occupy an intermediate position, i.e. one in which the sediment movements are continual, but do not reach the obviously vast proportions of the movements observed in the Nith estuary. One such example is the shore at Beckfoot on the coast of Cumberland.

The unconsolidated sands of the Solway rest upon a bed of bound shingle. On the shore, the amount of bound shingle exposed can be used as a measure, although imperfect, of sediment movement. At Beckfoot, the shore was relatively stable early in the period 1963–71. However, movements on this shore became pronounced from 1968 onwards (Fig. 14.3). The length of shore occupied by the bivalve molluscs *Macoma balthica* and *Tellina tenuis* decreased significantly and even the large spatfall of 1967–68 had little influence on this trend. Concomitant with these changes the abundance of both species declined: in 1963 *Macoma* was about three times more abundant than *T. tenuis*, but in 1966 the abundance ratio reached equality and in 1967–69 the abundance ratio of *Macoma* to *T. tenuis* fell below 1. In 1970–71 with a marked reduction in number the abundance ratio changed once more, but this probably had little meaning. Clearly, the movement of sediment had a marked effect upon these populations[11]. Where information can be compared, the

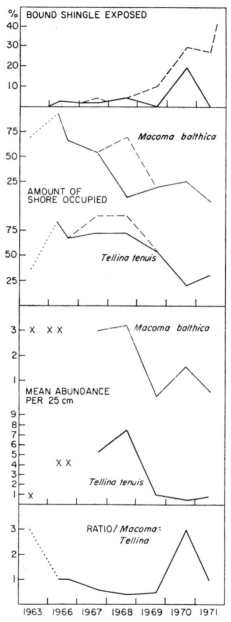

Fig. 14.3. Changes in the distribution and abundance of *Macoma balthica* and *Tellina tenuis*, in relation to movement of the substrata as indicated by exposure of the bound shingle at Beckfoot, Cumberland (after Perkins[11]).

spatfalls of *Macoma* were much less successful than at the more stable Powfoot, where total populations reached 1600–3200 per m². However, these changes at Beckfoot do not indicate that *Macoma* and *T. tenuis* were killed by the sediment movement, but since these molluscs must behave in part as sedimentary particles it is likely that they were transported away. Nonetheless, the changes noted bear a superficial resemblance to, and could be confused with, the effect of an introduced toxin.

The influence of oxygen is perhaps the neatest example to illustrate the effect of changes in necessary factors upon animal life, at low concentrations death or distress may be evident whereas at supra-optimal levels, oxygen is poisonous[12].

Consider now what is known of the polluted situation. Here, marine scientists have much to learn from the experience of limnologists (i.e. freshwater scientists)[13–19]. The effects of pollutants in rivers may be illustrated by reference to the following examples:

1. In the River Rheidol, Wales, the effects of pollution, due to lead mining, could be recognized in four stages[19, 80] thus:

   (a) No life at all.
   (b) A sparse fauna composed almost entirely of insect larvae, among which those of the Trichoptera were notable absentees; a sparse flora with such algae as *Lemanea* and *Batrachospermum* and a few mosses and liverworts.
   (c) A richer fauna, including oligochaetes, turbellarians and Trichoptera; flora including Chlorophyceae.
   (d) Fauna completed by the appearance of molluscs and fishes; flora by the appearance of higher plants.

2. In the rivers Churnet and Dove low concentrations of Cu acted as a growth factor to the plants; at an increased, but still low, concentration the abundance, though reduced, represented an increase over that above the outfall, but at higher concentrations still, i.e. *ca.* 1·0 p.p.m. Cu the abundance was reduced to very low levels. On the other hand, the animals showed a decrease in the number of species and of abundance at all concentrations[14].

3. After 1913, the addition of organic pollution to the Illinois River led to a decrease in the number of clean water species present, viz. from 49 to 3, and indeed to a general decrease in the abundance of all species, viz., from 91 to 15, except chironomids, by 1920; some improvement was shown by 1925[14].

In the sea, however, such response series are rarely so well defined, if only because, unlike rivers, the motion of the water is not unidirectional.

However, at Marseilles, for example, a series of facies which occupied concentric circles of population around a sewage outfall have been traced thus:

1. an azoic zone of maximum pollution which also lacks macrophytes;
2. a polluted zone characterized by *Capitella capitata* and *Scolecolepis fuliginosa*;
3. a sub-normal zone characterized by the molluscs *Corbula gibba* and *Thyasira flexuosa*; and
4. a zone of pure water lying beyond a more or less well-defined transitional margin[20].

In summary, a pollutant may have one of two main effects. An increase in the concentration of the trace metals and sewage which are vital to the well being of the individual and system, first induces an increased biological production, which may be accompanied by a decrease in diversity, and then leads to a decreased abundance and a further decrease in species diversity, followed by an elimination of all the former inhabitants as the concentration gradually increases still further. Evidently the ecological effect, observed in the field, belongs to the family of phenomena described by the oligodynamic response curve and observed both in the field and in the laboratory where a pollutional situation does not exist. This conclusion is perhaps rather less surprising if we accept the contention that the study of ecology may be regarded as that of physiology "on a grand scale" and the fact that ecological studies can be interpreted more satisfactorily if they are supported by well-found studies in physiology.

On the other hand, the addition of a non-utilizable *biocide*, e.g. Pb, or of radioactivity does not lead to enhanced production, and once the concentration exceeds a level which is tolerable to the community increasing impoverishment of the faunas and floras results. Moreover, both this response and the sigmoid dose response curve to these substances is not markedly different from the optimal and supra-optimal portion of the oligodynamic response curve.

Pollution then represents a physiological condition which is only one facet of the response to supra-optimal conditions evinced by living systems. *By definition, once the system has reached this condition pollution in contra distinction to waste disposal as such or the recycling of essential nutrient materials, is necessarily bad and failing the application of remedial measures can only become worse. Conversely, waste disposal and the recycling of nutrient materials is not pollution and does not necessarily lead to pollution.*

At first sight, these conclusions may seem to obscure rather than clarify the issue. However, the appreciation that pollution arises once a particular threshold has been crossed, and may be corrected readily given an adequate physiological understanding, is a more valuable approach to the problem than the almost charismatic and defeatist view that it is something special and merely an attribute of population growth. Clearly, it is a function of an excessive input of a given attribute of the environment which may be induced by a human population of any size. Because the magnitude of the departure from optimal conditions depends upon the scale of the distortion applied to the equilibrium which existed before the situation became polluted, it follows that a consideration of the behaviour of the appropriate oligodynamic curve can indicate both the quality and magnitude of the remedy necessary to bring about a return to tolerance zone conditions. Indeed, failure to appreciate the limits of optimal or tolerance zone growth could lead in relation to a river, for example, to such over treatment of effluents that the input of excessively pure water could carry the river as a whole into the zone of deficiency which would be damaging though, unlike pollution, not overtly unpleasant. This in essence is a statement that to maintain the environment in a healthy condition it must be managed adequately, i.e. the principles of conservation must be applied for conservation means management of the earth's resources and not just the protection of a few failing species. Two implications follow from these conclusions: (1) workers in the "pollution" field may expect to find the answers to many of their problems in the past and present literature and furthermore, the purest scientist may find valuable information in the literature of "pollution"; and (2) since widespread and acute pollution is observable on a global scale man is not managing the earth's resources adequately.

On the first of these implications, "radioactive pollution" provides an excellent example. With the discovery of radioactivity and atomic radiation, mankind, without first considering or investigating the physiological consequences of their use, utilized radioactive materials in a number of ways which in the light of later knowledge were disastrous, viz. to watch dial painters and to those who drank radium solutions in order to alleviate mental, rheumatic and other afflictions[21]. The consequences of this misuse (paralleled perhaps by the contemporary use of exotic substances) were horrifying and to the general public today the word radioactivity is powerfully emotive. And yet properly used radioactive isotopes are employed widely with beneficial effect[22].

In marine science, especially before atomic absorption spectrophotometry was developed, radioactive nuclides were measured more readily than the stable isotope of the same element and made an invaluable and otherwise unobtainable contribution to our understanding of ion uptake, transport, exchange, balance and deposition in marine organisms[23–40]. Similarly, these isotopes contributed to our understanding of the processes of adsorption of ions by sediments (p. 122) and by use of radioactive tags, both from specific release experiments, e.g. ^{46}Sc, or by release in an effluent or fallout, e.g. ^{106}Ru, ^{95}Zr/^{95}Nb, of the processes of sediment transport and salt marsh accretion (Chapter 6).

If we exclude military misuse of nuclear energy, it is probably true to say that if the physiological effect of all other pollutants were as well understood and if all other pollutants were as well controlled as radioactivity, the world would be a cleaner, pleasanter place in which to live and mankind would not be undergoing its present traumatic awakening. However, this example merely serves to confirm man's poor record in the management of his resources.

With regard to man's mismanagement of the earth's resources, the reasons are complex and deep-rooted, moreover it is essential to distinguish between mismanagement in the pollutional sense and that which arises for other reasons, e.g. excessive use or misuse. In recent years, it has become customary to refer to man himself as a pollutant, and with the prospect of an increase in the world population from 3.4×10^9 in 1966 to 5.0×10^9 in 1985[41], this statement looks reasonable. Moreover, although it is fashionable to deride all the conclusions of Malthus[42], the expedients to which man has resorted have not solved the essential problem which Malthus diagnosed even though he was a political pamphleteer. Unfortunately, man is a large mammal for the world to support in great numbers, and his record of management, characterized by sloth, greed, insanitary habits, the pursuit of short term gain and an inability to profit by experience, has always been very poor.

Man's essentially slovenly nature is characterized by an aptitude for disposing of waste products by either throwing them from the door of his hut on to a midden, or of moving on when his residence has become intolerably foul. This characteristic has been advantageous to the archaeologist, but is hardly admirable, and is still evident in the behaviour of modern man. Again archaeology indicates that, from the earliest times, man has been able to live beside filth without seeing it, and this is, perhaps, one of the more insidious aspects of the pollution problem. Of course when the world population of mankind was relatively small, the environmental impact of such an attitude was equally small.

These attitudes have been paralleled by man's attitude to the sea which, because it covers some 70% of the earth's surface, has seemed to many to be a vast carpet beneath which filth can be swept and forgotten. Indeed, even as late as 1959, a reputable scientific journal congratulated a youth organization who cleaned the Isle of May of an army's debris by throwing it into the Firth of Forth: an attitude which is little different from that which makes every effort towards cleanliness, economy and good housekeeping within a factory, but sees no such responsibility at the end of the effluent line.

However, serious as this consideration may be, it is man's evident inability to profit from experience, and his capacity as a desert maker, which seems likely to lead him into the most serious trouble. There are many instances in which man has vitiated the earth's green mantle[43], most of them performed at a time when his capacity for mischief was infinitesimal compared to that which he has in the twentieth century. Yet despite all the experience accumulated, the 1960s saw a dust bowl incipient in the eastern counties of England: a country not lacking in rainfall in the eyes of the rest of the world. All this and more has been perpetrated upon *ca.* 30% of the earth's surface, which constitutes the terrestrial environment. What of the remainder of *ca.* 70%, i.e. the aquatic environment? Here too, the same path has been followed with large bodies of fresh and salt water. Both the popular and scientific press contain reports referring to the marked deterioration of Lakes Erie and Ontario, the Caspian Sea, the Baltic Sea[44] and the Mediterranean Sea[45]. In each case, the awakening has been belated and somehow viewed with surprise.

Perhaps the essential problem is man's ability to pursue fantasies which obscure reality and create yet greater problems[46]. This attribute could explain both the repetitive nature of certain difficulties and the lack of ready recollection that the three major environments, viz. terrestrial, aquatic and atmospheric, are very closely related. So closely related, indeed, that a major event in one may affect the others, and that such an event in one part of the Earth ultimately affects the whole. This failure seems to be unjustifiable since, in general, the manifestations are well known. However, it is worth reviewing the more important of these events which have had so little lasting impact. In 1883, Rakata, a volcanic cone on Krakatoa, one of a group of islands in the Straits of Sunda, erupted and exploded violently, and some 2×10^{11} cubic feet of island were lost in the process. Seven major surges in atmospheric pressure were generated, and simultaneously giant tidal waves or *tsunami* (as they are more properly called since they are not tidal in origin), were generated, and spread round the world, doing immense damage in the process. For some years afterwards, the dust which this explosion threw up into the atmosphere, gave rise to blue suns and green suns, and strange twilights and sunsets were seen all over the world[47, 48]. Tsunamis are a particular danger to peoples who live on the margins of the Pacific; these waves, generated by submarine earthquakes, result in a substantial loss of life and property, often at points far removed from the epicentre. Indeed these waves are so important that in Japan, at least, a warning service has been created.

During the 1939–45 war, the great tank battles which took place in the

15*

Libyan Desert raised great clouds of dust, some of which settled on islands in the Caribbean, some 6000 miles away.

The Antarctic region is considered to influence the weather of the whole world; a contention which is not difficult to believe in the light of recent work concerning the effect of sea surface temperature anomalies upon terrestrial climate. In temperate latitudes, depressions undergo abrupt changes of track when they approach pools of anomalous warm sea surface: indeed a prolonged drought in the north eastern United States which occurred during the period 1962–65 has been ascribed to this cause. While in 1957, anomalous sea temperatures over the western and central equatorial Pacific, shifted the N. Pacific low to the Alaskan side of the ocean, and the Icelandic low to the eastwards[49]. Perhaps the finest example of all derives from the nuclear H-bomb tests, which were conducted in a relatively few localities: principally some islands in the Pacific e.g. Bikini or Christmas Island, and in the Barents Sea at Novaya Zemlya. Despite the essentially local nature of these tests, the radioactive nuclides which resulted were thrown into the upper atmosphere from which they returned as "fall-out" upon the earth and sea alike over the entire Earth's surface. Because radioactivity is measured relatively easily, the deposition and transport of these materials in the terrestrial and aquatic environments and their subsequent accumulation by living organisms is relatively well known.

Bearing in mind the scale of these phenomena and the impact which some of them have had upon the consciousness of man, it seems a little odd that mankind should be surprised yet again when it found that D.D.T., and similar pesticides, and plasticisers of the polychlorinated biphenyl (P.C.B.) type should occur throughout the marine environment and in some places at least, give rise to severe biological problems[50]. Curiously, it is the P.C.B.'s which have furnished a recent example of the influence of man's terrestrial activity upon pollution in the sea. During the autumn of 1969, the death of thousands of sea birds in the Firth of Clyde and Irish Sea caused widespread concern. Investigation implicated the P.C.B.s[51], although this view was somewhat modified subsequently[52]. However, and as a result of this sea bird wreck, these substances came under much greater scrutiny than hitherto. It was found that a major source of the P.C.B.s introduced into the Irish Sea was sewage sludge dumped by the Corporation of Glasgow off the Garroch Head, Isle of Bute, Firth of Clyde. Despite a voluntary restriction on the sales of P.C.B.s by the Monsanto Company Limited, after the sea bird wreck, it was found that one sewage works, viz. that of the Burgh of Hamilton, which contributed to this sludge did not reflect the general downward trend which followed Monsanto's restriction. Careful work by scientists of the

Clyde River Purification Board showed the source to be the British Rail maintenance depot at Hamilton which is the largest of its kind in Scotland and where diesel multiple units are serviced. Once it was appreciated that an important source of marine pollution had arisen in this way remedial measures were undertaken and an alternative material substituted in these engines[53].

That problems should arise from the widespread use of toxic metals such as lead and mercury is surely to be expected rather than to cause surprise. Of course, a failure to anticipate events of this sort must lead inevitably to scares such as that witnessed during December 1970–January 1971, when mercury was found in canned tuna from Japan and many other fish also; in some, at least, of the material tested, the mercury found resulted from the natural biological concentration of this element which is naturally present in sea water at a concentration of 0·00003 mg/litre (see Table 1.2). Unnecessary though this particular scare was there seems little doubt that there will be others in the future. Furthermore, there will inevitably be those which are well-founded, perhaps of a more localized nature, but similar to those which mankind has already experienced with mercury and cadmium for example. Lacking the information necessary, it is not possible to forecast where the next trouble will arise, but given a sufficient knowledge of the production of exotic chemicals, together with details of their use and misuse, and information upon their sub-lethal physiology, a forecast could be made with a fair degree of accuracy.

Before moving on from this general appreciation of the pollution problem to that specifically concerned with estuaries and coastal waters, there are a number of general effects observable in these areas generally, not apparently a part of the seaboard pollution problem, but which have the same net result. The first of these arises from the rather general lack of respect which man has for the *wetlands* of the world. Until recently, these areas have been regarded as wastelands, either to be reclaimed as soon as possible or, since they have *no worth*, to be used as sites for spoil heaps and the like. That they have an important influence upon the regulation of surface water in river basins and upon subsoil water levels, that they are areas of high biological productivity, and that they are important nursery grounds for commercially important fish and crustacea which are captured as adults in the offshore area, has been ignored. Where drainage schemes have been carried out, the returns have often been poor and made little contribution to the need for increased food production[54]. Moreover, although the influence of offshore areas upon those inshore has become appreciated widely in recent times (Chapters 2, 6), the influence of the marshlands of estuarine areas upon the waters of the continental shelf is

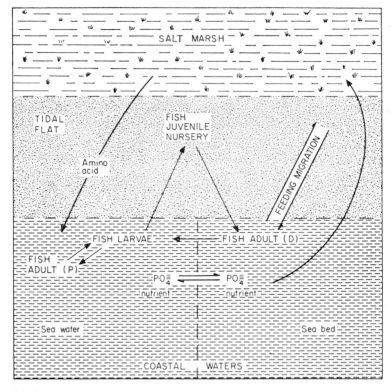

Fig. 14.4. An aspect of the relationship between salt marshes, tidal flats, coastal waters and their inhabitants. Fish Adult (P) = Pelagic Fish, Fish Adult (D) = Demersal Fish (after Perkins [55]).

largely unknown, except in the fish nursery relationship; nevertheless, there seems no reason to doubt that its importance will be demonstrated eventually[55].

A possible mechanism can be considered now. We have already seen (p. 13) how the external metabolites present in a body of sea water affect its productivity and the success of its inhabitants. Inshore waters are particularly rich in amino acids, many of which are produced by the blue-green algae which live in the salt marsh fringes (p. 000). Similarly, vitamin B_{12} production in estuaries is associated with soils of high silt content (p. 124) (Fig. 14.4). Bearing in mind the powerful adsorbtive capacity of these soils (Chapter 5), it is not difficult to visualize the transport of contaminated sediment to, and deposition in, these important systems which may thus become poisoned. Once poisoned, it seems likely that production of the vital external metabolites must fall, and be accompanied by a similar fall in the productivity and quality of life in the adjacent waters[55].

In North America, particularly the United States, a very substantial proportion of the very fine and productive marshlands has been lost either by dredging of channels to permit the drilling of oil wells, as in the Mississippi area, or by land-fill reclamation schemes, for the construction of towns, chemical factories, oil refineries and the like. For example, in the past 20 yr, more than 81 000 ha of shallow, coastal bays have been altered in the Gulf of Mexico and South Atlantic areas alone. In Texas, the development and deepening of 700 miles of Federal navigation channels has altered 5265 ha of bay bottom and has destroyed 2380 ha of salt marsh. The spoil dredged from these areas has filled 2025 ha of shallow bay area and has covered 9315 ha of salt marsh[56]. Elsewhere, as along vast areas of the New Jersey coasts, these marshlands have been used as dumping grounds for toxic spoil heaps with visible adverse results. Although this issue is now causing great concern, especially in relation to fish nursery grounds, the results, in many cases, must be considered irreversible.

The effects of spoil heaps are not confined to marshlands alone, nor are they due solely to the toxic materials which leach out of the heaps. The coast of Cumberland was an important centre of the Industrial Revolution due largely to the close proximity of coal and a high quality haematite iron ore which was particularly suitable for use in the Bessemer process. The spoil heaps, or bings, derived from both processes both dominated and despoiled the coast, and both were dumped on the shore to M.T.L. and below. Formerly, very hot blast furnace wastes were poured upon the shore where they set to form an artificial rock known to the industry as *slagcrete*. At Siddick and Flimby, coal mine spoil became consolidated upon contact with sea water and formed a reef. Both the slagcrete and consolidated coal bing, having replaced the original shore substrata of sand or bound shingle, provided a poor substitute for the more usual surfaces inhabited by rock-dwelling plants and animals: these artificial surfaces may be friable, give a poor key for attachment of sessile organisms and in some cases are evidently toxic. Consequently, the biota which do settle upon them may be confined to a relatively few species, often stunted in growth. On the whole, these *rocks* erode readily and in some places a return to pre-Industrial Revolution condition is becoming apparent, thus adding further confusion to a situation intrinsically difficult to analyse because of the mobility of the sediments.

The influence upon biological productivity may not be confined to those shores which were the immediate sites of coal and iron works spoil heaps. Erosion contributes a supply of such materials to that of a relatively large size carried by longshore drift at the high-water mark; at lower shore levels, the size of these materials mixed in with the sands is finer, but these sediments contrast markedly with those on the north shore where coal, shale and iron works waste are conspicuously absent. Although the sediments of the south shore are mobile and such movement can, as we have seen, influence adversely those species important as the food of fish, the productivity of this shore, as a whole, is much poorer than one would otherwise expect, particularly since rich diverse faunas do occur at low water levels which tend to lie outside the ambit of these materials, whose influence may arise mechanically or through chemical leachates though which has yet to be determined[55].

Currently, considerable emphasis is placed on the influence of chemical entities upon living systems. However, in general, though by no means always, relief of such a pressure upon a living system will, as we can see by inspection of oligodynamic response curve, lead within a foreseeable period to a restoration of zone of tolerance conditions. On the other hand, some physical act, such as the dumping of an apparently inert solid waste, may have more long lasting and more potent effects upon the environment. Moreover, once such physical alterations are set in train they may be very costly, if not impossible to reverse. It is worth reflecting that the fine archaeological relics, e.g. chambered cairns and stone circles left by our Mesolithic forefathers were erected when the population of these islands was a few tens of thousands. How long may the physical changes which 50×10^6 people have brought about be expected to persist? The artificial shores in Cumberland are at least 100 yr old, and will persist for some time yet despite erosion. This seems to exemplify one of the problems which a rash exploitation of the environment presents, viz. that even if all the measures necessary to ensure its proper management are initiated now a considerable time must elapse before the decline in environmental quality is halted and that this subsequent equilibrium may be at a quality level lower or much lower than that which presently prevails. The return to present conditions may be difficult, and indeed may only occur on a geological time scale.

Other examples of the misuse of the earth arise from the increasing amounts of leisure, coupled with the near universal possession of the motor car enjoyed by the western world. An increased tourism and the development of holiday shanty-towns is having a fatal impact upon the fragile supra-littoral in many parts of the western world. Here even a limited pressure leads to changes in species composition, and by reducing the number of species leads to physical and biological instability, which may not detract from its amenity value[57]. However, the change may be both rapid and adverse: even as late as 1964, the fragile supra-littoral grass and dune lands of the Solway Firth were virtually unspoiled. With increased tourist pressure changes became evident, and from 1967 onwards were pronounced in many areas: at Allonby the fine grass sward became replaced by a more open community typified by the yarrow, *Achillea millefolium*[58]. By 1971, active protective measures were necessary at Allonby and elsewhere, i.e. within 4 yr of the pressure becoming marked. Since loss of dune lands may result from such hyper-activity, and since the sandy shore represents an equilibrium, influenced by gains and losses of sand at the high and low-water marks, it too cannot remain unaffected by such changes. Likewise, another attribute of increased leisure is the vast increase in sea angling as a recreation. On this scale, the large quantities of bait required can be provided only if a corresponding amount of shore is disturbed. In general little is known of the ecological impact of this activity, although on the coast of Cumberland it led, in the south, to the loss of a potentially valuable mussel fishery[55], and, elsewhere, it is now a common complaint among sea-anglers that those areas which were formerly good for bait production and thus dug over regularly no longer produce the bait required[X]. Excessive bait digging may cause changes in shore profile and death through damage and exposure of unwanted invertebrates: on the coast of Dorset such digging was considered to have caused a scarcity of *Nereis virens*, *N. diversicolor*, *Arenicola marina*, *A. ecaudata*, *Nephtys caeca* and *N. hombergi*[59]. Deep burial may also have adverse effects. Both the over use of the dune-lands and the excessive digging of shores for bait represent the results of human activity in producing extra-optimal conditions leading to impoverishment of the flora and fauna, and while an oligodynamic response curve may describe this effect, it is nevertheless, like poor land use, not an example of pollution.

Further problems arise from our affluent, throw-way society. Apart from the poor management of our natural resources, implicit in such an attitude, the throw-away plastic cup and the non-returnable bottle seem to be the contemporary equivalent of the Mesolithic oyster shell. However, polythene, being nearly indestructible, is causing massive problems in inshore areas[60, 61], although the effect is not limited to these areas. In the Wash, plastic cups are trawled by the bushel and consequently interfere with normal fishing activity[62]. Polythene sheeting fouls trawls, ships propellers and cooling water intakes[60]; Portuguese fishermen have found bonito and tunny killed by internal obstruction due to eating expendable P.V.C. drinking cups[63]. Yet a further impact upon marine life results from synthetic fishing gear which, when lost, continues to fish[60].

Of all the parts of the sea likely to be subjected to man's activity, estuaries and coastal waters are most at risk. This situation arises for three principal reasons:

1. Pollutants which reach the rivers directly either from effluent input or wash out of aerial pollutants, or indirectly by drainage from the land, are transported inevitably to an estuary.
2. Materials released to an offshore area may end up in an estuary, particularly where the release takes place upon the Continental Shelf (see Chapter 6). These materials include those released by open coast effluents, radioactive fall-out, wash out of atmospheric pollution and solid wastes dumped offshore. The problem of the dumping of solid toxic wastes (from plastic production for example) offshore has been exacerbated, in recent years, by the application of increasingly stringent standards to effluents released to rivers. Disposal may occur with[64] or without containers[65]. Such dumping is a hazard to fishermen as well as to marine organisms[64]: it has been alleged to cause the death of planktonic animals, and O-group fish living near the surface[65], but the most insidious effects may result from the transport of these materials to the fertile fish nurseries of estuaries.
3. Estuaries themselves receive a wide variety of pollutants directly, either as effluents or as wash out of atmospheric pollution. Both of these parts of the problem are bound to become worse if great care is not exercised in the future. This situation arises because industry will use increasingly large amounts of water for cooling purposes and, in the absence of better methods, for effluent dilution; tourism and recreational use is increasing, and in general the "needs" of our high energy society "dictate" that estuaries will be used increasingly for a multitude of purposes[66]. Evidently, the conflict between economics and bioeconomics will worsen progressively.

It is of great interest that the process whereby rain may wash atmospheric pollutants out of the air into the sea is only one aspect of an important phenomenon. In those estuarine and inshore areas which are adjacent to large industrial installations, e.g. oil refineries or power stations, or towns which emit large amounts of SO_2, industrial hazes are often present, e.g. Teeside. These hazes result when ammonia which passes from the sea surface to the atmosphere reacts with water vapour and the SO_2 to form ammonium sulphate[67]. In turn, such hazes reduce the quality of the incoming solar radiation and adversely influence the photosynthetic activity of plant life in the adjacent areas of land and sea; moreover the quality of human life is directly lowered also.

Evidently, the foregoing discussion has been concerned with a number of phenomena which arise from the development of supra-optimal conditions. Some occur when for some intrinsic reason these conditions are induced by the normal attributes of the environment, but others arise from the activities of man. All can be discussed in relation to an oligodynamic response curve but not all are pollution, although since an oligodynamic response curve is involved they resemble it. What then distinguishes pollution from the associated family of supranormal conditions?

Industrial or sewage pollution arises because the supply of these materials exceeds the ability of a given part of the environment to utilize them, that is a concentration of input has occurred. Very clearly any tendency to the development of towns or massive industrial complexes exacerbates this situation which normally arises for apparently sound economic reasons. Therefore, polluted situations have developed at their worst because of the concentration of industrial and commercial activity which arose with the Industrial Revolution and continues today. It is important to note here that an Industrial Revolution cannot take place until a population reaches a critical size, but that once it has occurred people become increasingly less necessary, though environmental problems become more acute. In the wheat producing states of the prairie, an enlarged labour force was vital to the early stages of the mechanization that subsequently produced large scale unemployment[68]. This highly economic process now requires that some 5 g cal of energy shall be expended to harvest 1 g cal of energy as wheat[69]: both illustrate yet another example of an oligodynamic response curve. Thus, although economics can be used as a measure of energy flow and utilization, what is economically sound is not necessarily balanced in environmental terms and another yardstick by which man's activities can be measured can be termed *bioeconomics*. In essence a situation which is bioeconomically sound is one in which the metabolites necessary to life remain in a balanced circulation, each part of the system being approximately in equilibrium with every other part and in which energy production and consumption are balanced. The cottage industries and agriculture of more primitive communities are essentially in this state, while the more highly civilized communities tend to move further away from this stable condition.

The problems associated with sewage disposal form an excellent illustration of the events which take place when departures from a simple bioeconomic system occur. Primitive pastoral or agricultural communities, such as existed even in Britain until the late 1950s, feed off the adjacent land. Their excremental products are first stored in primitive lavatories and then returned to the land periodically. In these situations nothing which we now understand as a polluted situation arose. The whole system was in equilibrium and it was bioeconomically sound, although as anyone who has lived in such a community must be aware human faeces is not an ideal manure and in some situations at least periodic additions of animal manure are essential. Not only are bioeconomical principles met, but the

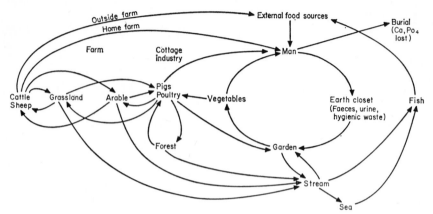

Fig. 14.5. Nutrient cycles in relation to the life of rural man. *Note*: lines of equal shading indicate a system in approximate equilibrium.

opportunities for the water-borne carriage of disease are limited (Fig. 14.5).

Once man begins to accumulate large quantities of sewage and to release it to relatively restricted waterways then pollution results. In this case he is no longer working within a cyclical system which is in equilibrium and the system is no longer bioeconomic; although to the individual it may be greatly saving in time and effort and thus viewed as economic. The classic case of this breakdown occurred in relation to London with the introduction of the water closet in the 1850s: prior to this date the domestic sewage as "night soil" was dumped on fields, and fish including salmon were taken in the metropolitan Thames. However, from about 1850 onwards this part of the river was virtually lifeless[70], i.e. with respect to aerobic species.

However, the problem is not solely one of a large input of sewage leading to the development of anaerobic conditions in the aquatic system and increased opportunities for the spread of disease, but by definition such an imbalance is produced by a large population which does not, and indeed cannot, feed off the adjacent land, and in the case of Britain as a whole, which is both overpopulated and not self-sufficient, a massive movement of food stuffs inwards is necessary. Ancient Rome drained North Africa of nutrients, an effect so pronounced that an unusually high phosphate content is still found in the soils around modern Rome[71]. To return now to the case of Britain, and in particular that of London. For its sustenance London depends on the outside world and having consumed this food, the products of sewage are transferred via the Thames to the North Sea where, it is true, some residual benefit from improved fisheries results. However, the wheat used by London is grown on the prairies of North America or the plains of Australia, which areas, since they do not as

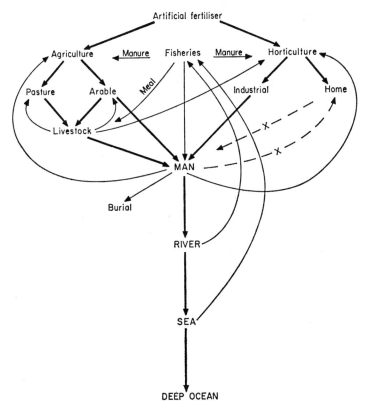

Fig. 14.6. Nutrient transfer in relation to urban man. *Note*: heavy lines indicate paths by which the equilibrium situation of Fig. 14.5 is disrupted and massive transfer to the sea takes place.

they would in a truly bioeconomic system receive return cargoes of human ordure to maintain the content of humus, nutrients and trace metals, become progressively stripped of these materials. Therefore a situation of pollution in Britain is balanced by a situation of deficiency in North America or Australia. Once more it is not just a very large human population which has created this situation, but that a process of management which ignores the essential requirements of biological systems has been developed.

By the same process it can be shown that most polluted situations result when man has brought about a massive translocation of materials as a result of the Industrial Revolution. In each case, the process which has caused pollution at one site may have had the same effect at more than one other site which forms an essential stage in the same process, and impoverishment at a source site is always a concommitant of pollution in some other

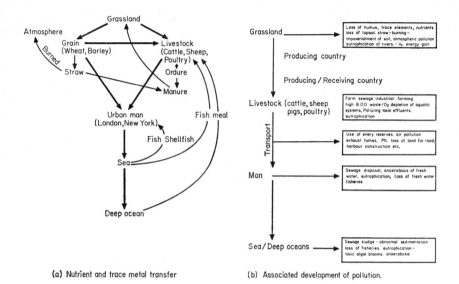

(a) Nutrient and trace metal transfer   (b) Associated development of pollution.

Fig. 14.7. Process by which a massive transfer of nutrients from the grasslands to the metropolitan areas of the world (e.g. the prairie to London). Heavy lines indicate the means by which the equilibrium is disrupted.

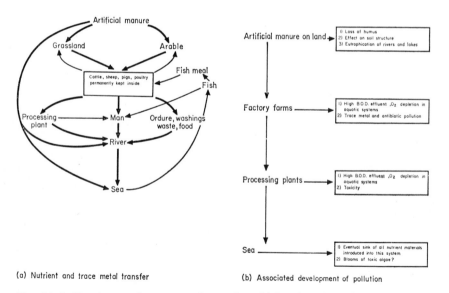

(a) Nutrient and trace metal transfer   (b) Associated development of pollution

Fig. 14.8. Nutrient and trace metal transfer which arises from factory farming. Heavy lines indicate the means by which the equilibrium is disrupted.

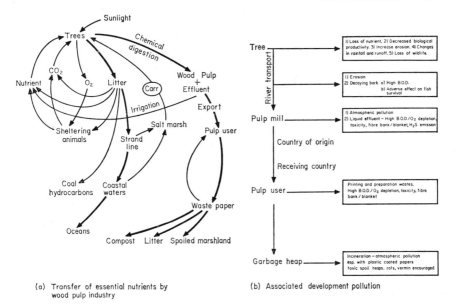

(a) Transfer of essential nutrients by wood pulp industry

(b) Associated development pollution

Fig. 14.9. Removal of nutrients from woodlands to the sea by means of the wood pulp industry. Paths contributing most strongly to the lack of balance marked by heavier lines. Associated development of impoverishment and pollution indicated in part (b).

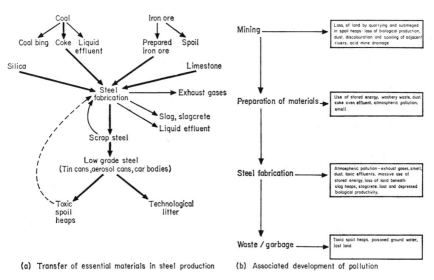

(a) Transfer of essential materials in steel production

(b) Associated development of pollution

Fig. 14.10. Transfer of energy and raw materials with the associated development of pollution by the steel industry.

part of the world (Figs. 14.6–14.10): therefore, man-induced pollution is a multiphase process.

To sum up, man-made pollution arises when by a concentration of activity an environmental situation formerly in a state of dynamic equilibrium is carried into a supra-optimal condition. This situation is brought about by a replacement of the normal modes of energy transfer and of the cycling nutrient and other materials by systems of massive linear transfer. Each such polluted situation may be accompanied by one or more other polluted situations along the course of the linear transfer system, while at the point of initiation impoverishment of the environment frequently occurs. Evidently, the means by which the cyclical systems have been transformed into linear transfer systems is technology. Given this conclusion, it is doubtful if, in the strictest terms of adequate management of the environment, technology can ever be regarded as beneficial. However, the use of technology is clearly in man's interest, it follows, therefore, that this technology should be managed to the highest possible standards if the world is to remain a tolerable place in which to live.

Evidently, a global strategy is required for if a correction is applied to a particular situation as a result of a consideration of the oligodynamic response curve for that point or area, such a correction may not, in itself, in any way contribute to a restoration of the global equilibrium. Further, it is questionable if those supra-optimal situations which occur naturally, e.g. a freshwater pool which may become anaerobic due to the breakdown of leaves introduced each fall, or a backwater of a river which has become fouled by cattle seeking both to keep cool and to avoid the warble fly, should ever be defined as polluted, since the characteristics of pollution in the wider sense are absent.

What then are the substances with which man is polluting his environment. The bewildering array of materials which reach the sea either directly, or indirectly, are summarized briefly in Fig. 14.11[72], which, despite its seemingly exhaustive nature, must be far from complete[44, 45, 73].

In general, it is only in severe cases that the effects of pollution, upon production in the sea, can be recognized readily[61]. However, as we have seen, such major bodies of water as the Caspian, the Baltic and the Mediterranean are already at risk. In the North Sea, to which vast amounts of industrial wastes and sewage are released annually, there are no definite signs of pollution. Nevertheless, despite a lack of direct evidence, there may be some connection between the build up of nutrients from sewage, and the occurrence of toxic blooms of dinoflagellates off the north-east coast of Britain and the west coast of Denmark since 1968[61].

In the North Sea, man has a great deal to lose if the fisheries fail due to poor management. Excluding shellfish, the catches of fish have risen from

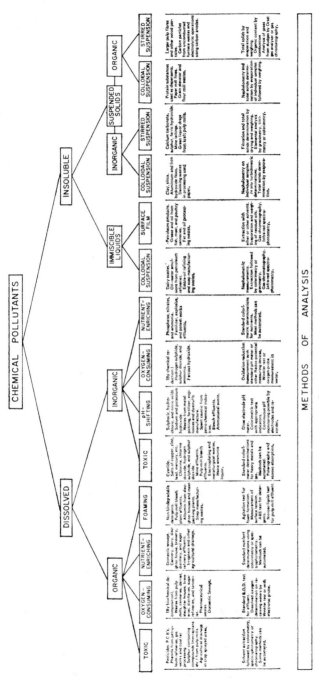

Fig. 14.11. Sources of chemical pollution and methods of analysis (after Waldichuk[51]).

Table 14.1. Average number of salmonid fish caught in the River Tees (after Macan and Worthington[16], Alexander *et al.*[74], Turing[75])

Period	Annual catch Salmon In nets at mouth of estuary	By rod in the River Tees	Sea trout in nets
1904–08	6411	—	5229
1909–13	8067	*ca.* 400	5816
1914–18	5122	*ca.* 190	1825
1919–23	5730	*ca.* 190	4423
1924–28	2850	*ca.* 70	1274
1929–33	1826	*ca.* 3	816
1937	23	0	—
1947	0	0	—

1 529 745 metric tons in 1950, to 1 553 106 in 1959, to 3 237 226 metric tons in 1969: this increase has been due in part to exceptionally rich year classes of haddock and cod following the severe winter of 1962–63; in recent years, the mackerel has undergone a great increase of exploitation. Landings of other species, viz. herring, Norway pout, sandeels and sprats, have been highly variable due to year class differences, but turbot has declined[61].

Formerly the River Tees estuary, N.E. England, supported a flourishing salmon and sea trout fishery. However, with the marked increase in the pollution of the estuary, especially during the late 1920s and early 1930s, it declined dramatically and by 1947 it was non-existent[16, 74 75] (Table 14.1). Even in 1930, 2246 dead bodies were picked up on the banks of the estuary during the smolt migration[16].

Lest there should be doubt regarding the financial loss which can be sustained by fishery interests from the effects of pollution, reference can be made to the experience in Raritan Bay; an area which forms a triangle between Staten Island, New York State and New Jersey. In the early years of the nineteenth century, these waters supported a highly productive commercial fin and shellfish industry. However, overfishing and the release of noxious wastes, both sewage and industrial, led first to a decline in the fisheries, e.g. the oyster industry became virtually extinct due to the effect of metallic copper, only a few years after this was predicted in 1916. Then an epidemic of infective hepatitis, traceable to the consumption of raw clams, *Mya arenaria*, led to the closure of most of this fishery. Studies to ameliorate this position were initiated in 1961. At present the standing crop of hard clams, *Mercenaria mercenaria*, has a value of 35×10^6 and could produce an annual crop worth 3.85×10^6 if the water quality were suitable. Similarly, commercial fin-fish have an annual value of $200 000, but an improvement both of water quality and of fishing methods could probably lead to a 100% increase in this figure. Most startling, however, is the fact that currently *Mya arenaria* yields no significant harvest, but, if the water quality were improved to an appropriate level, this fishery has a potential commercial value of 18×10^6 p.a.[76].

It is worth remembering that with good management, i.e. taking care neither to over-fish nor to produce a poor water quality, this resource at this value is there for man's taking forever. On the other hand, most industrial plants require expensive maintenance, are written off in at most 25 yr, and to last this time require good management.

It is worth remembering also that many industrial wastes have a high economic value when properly treated. Yet bad management permits these materials to be thrown into the sea to the detriment of all living organisms, including man[77, 78]. Nevertheless, even where management of waste waters is good, it is not in every case that the recovery of valuable products is possible[79], and release of dilute plant washing to the sea may be unavoidable.

To conclude, man often concentrates his activities for what appear to be good, sound economic reasons. Where such a concentration of activity occurs, the basic equilibrium of a highly bioeconomical system is disrupted, and in the long run, this is rarely to man's advantage. In the development of new schemes in which it is customary to receive the assurance that it is for the good of man and excellent economic value, it would be well if the project could be viewed by bioeconomic principles to ensure that it really is good for mankind. Rivers were once the site of bioeconomic breakdown, and are now being reinstated at great cost. Some estuaries have also followed this path, but bioeconomic concepts could help to save many. Fisheries which are neither under- nor over-fished constitute a cyclical, self perpetuating and highly bioeconomic system, moreover they are highly economic in this condition. If once disrupted or lost by bad bioeconomics, i.e. by over-fishing or due to pollution, their reinstatement may be very expensive or even impossible.

References

1. Marston, H., Allen, S. and Smith, R. (1961). *Nature, Lond.* **190**, 1805.
2. Albert, A. (1968). "Selective Toxicity". Methuen, London.
3. Bennetts, H. and Chapman, F. (1937). *Aust. vet. J.* **13**, 138.
4. Albiston, H., Bull, L., Dick, A. and Keast, J. (1940). *Aust. vet. J.* **16**, 233.
5. Mills, C. F. and Mitchell, R. L. (1971). *Br. J. Nutr.* **26**, 117–121.
6. Sprague, J. B. (1970). *Water Res.* **4**, 3–32.

7. Alderdice, D. F. (1972). *In* "Marine Ecology" (O. Kinne, ed.), Vol. 1, Part 3, 1659–1722. Wiley-Interscience, London, New York, Sydney, Toronto.
8. Kinne, O. (1964). *Oceanogr. Mar. Biol., Ann. Rev.* **2**, 281–339.
9. Wilson, J. B. (1967). *Scott. J. Geol.* **3**, 329–371.
10. Perkins, E. J. and Williams, B. R. H. (1964). H.M.S.O., U.K.A.E.A., P.G. Report 587 (CC).
11. Perkins, E. J. (1972). *Spectrum, British Science News.* No. 97, 12–13.
12. Fox, H. Munro and Taylor, A. E. R. (1954). *Nature, Lond.* **174**, 312.
13. Klein, L. (1959). "River Pollution. **I.** Chemical Analysis", Butterworths, London.
14. Klein, L. (1962). "River Pollution. **II.** Causes and Effects", Butterworths, London.
15. Klein, L. (1966). "River Pollution. **III.** Control", Butterworths, London.
16. Macan, T. T. and Worthington, E. B. (1951). "Life in Lakes and Rivers", Collins, London. New Naturalist Series No. 15.
17. Jones, J. R. E. (1964). "Fish and River Pollution", Butterworths, London.
18. Hynes, H. B. N. (1966). "The Biology of Polluted Waters", University Press, Liverpool.
19. Carpenter, K. E. (1926). *Ann. appl. Biol.* **13**, 395–401.
20. Bellan, G. (1970). *Mar. Pollut. Bull.* **1**, 59–60.
21. Alexander, P. (1959). "Atomic Radiation and Life", Penguin (Pelican), Harmondsworth, Middlesex.
22. Putman, J. L. (1960). "Isotopes", Penguin (Pelican), Harmondsworth, Middlesex.
23. Boroughs, H., Townsley, S. J. and Hiatt, R. W. (1956). *Biol. Bull. mar. biol. Lab. Woods Hole.* **3**, 336–351.
24. Boroughs, H., Townsley, S. J. and Hiatt, R. W. (1956). *Biol. Bull. mar. biol. Lab. Woods Hole,* **3**, 352–257.
25. Bryan, G. W. (1961). *J. mar. biol. Ass. U.K.* **41**, 551–575.
26. Bryan, G. W. (1963). *J. mar. biol. Ass. U.K.* **43**, 519–539.
27. Bryan, G. W. (1963). *J. mar. biol. Ass. U.K.* **43**, 541–565.
28. Bryan, G. W. (1965). *J. mar. biol. Ass. U.K.* **45**, 97–113.
29. Bryan, G. W. and Ward, E. (1962). *J. mar. biol. Ass. U.K.* **42**, 199–241.
30. Bryan, G. W. and Ward, E. (1965). *J. mar. biol. Ass. U.K.* **45**, 65–95.
31. Fretter, V. (1953). *J. mar. biol. Ass. U.K.* **32**, 367–384.
32. Hiatt, R. W., Boroughs, H., Townsley, S. J. and Kau, G. (1955). "Radioisotope Uptake in Marine Organisms with Special Reference to the Passage of Such Isotopes as are Liberated from Atomic Weapons Through Food Chains Leading to Organisms Utilized as Food by Man". U.S.A.E.C., A.E.C.U.—3079; U.S.A.E.C. contract AT (04-3)-56 Annual report 1954–55: 46 pp. Unclassified.
33. Polikarpov, G. G. (1966). "Radioecology of Aquatic Organisms", Reinhold, New York.
34. Spooner, G. M. (1949). *J. mar. biol. Ass. U.K.* **28**, 587–625.
35. Townsley, S. J. (1963). *In* "Radioecology. Proc. 1st National Symposium on Radioecology". (V. Schulz and A. Klement Jr., eds.), Reinhold, New York.
36. Mauchline, J. (1961). H.M.S.O., U.K.A.E.A., P.G. Report 248(W): 18 pp.
37. Chipman, W. A., Rice, T. R. and Prince, T. J. (1958). *U.S. Dept. of the Interior. Fish. Bull.* **135**, 279–292.
38. Bryan, G. W. (1964). *J. mar. biol. Ass. U.K.* **44**, 549–563.

39. Morgan, F. (1964). *J. mar. biol. Ass. U.K.* **44**, 259–271.
40. Battaglia, B. and Bryan, G. W. (1964). *J. mar. biol. Ass. U.K.* **44**, 17–31.
41. Editorial (1966). Sunday Times, 16th October, 1966.
42. Malthus, T. (1960). *In* "On Population. Three essays", New American Library, New York.
43. Worthington, E. B. (1970). "The Environmental Crisis", Hodder and Stoughton, London.
44. Dybern, B. I. (1970). *In* "F.A.O. Technical Conference on Marine Pollution and its Effects on Living Resources and Fishing", F.I.R.: MP/70/R-3: 17 pp. Rome, 9–18 December 1970.
45. Le Group d'Experts C.G.P.M. (1970). *In* "F.A.O. Technical Conference on Marine Pollution and its Effects on Living Resources and Fishing", F.I.R.: MP/70/R-24: 17 pp. Rome, 9–18 December 1970.
46. Booker, C. (1969–70). "The Neophiliacs", Collins and Fontana, London and Glasgow.
47. Ball, R. S. (1892). "Time and Tide", S.P.C.K., London.
48. Geelan, M. (1938). "The Krakatoa Eruption", *In* "Fifty Great Disasters and Tragedies that Shook the World", Odhams, London.
49. Hay, R. F. M. (1967). *Rep. Challenger Soc.* 3(29), 51.
50. Haderlie, E. C. (1970). *Proc. Challenger Soc.* **4**(2), 81–82.
51. Bourne, A. G. (1969). *New Scient.* **44**, 292–293.
52. Holdgate, M. V. (ed.) (1971). "The Sea Bird Wreck in the Irish Sea Autumn 1969", The Natural Environment Research Council, Publ. Ser. C, No. 4.
53. Waddington, J. I., Best, G. A., Dawson, J. P. and Lithgow, T. (1973). *Mar. Pollut. Bull.* **4**, 26–28.
54. Hoffman, L. (1970). *New Scient.* **48**, 120–122.
55. Perkins, E. J. (1972). *The Environment this Month.* **1**(2), 47–54.
56. Trent, W. L., Pullen, E. J. and Moore, D. (1970). *In* "F.A.O. Conference on Marine Pollution and its Effects on Living Resources and Fishing", F.I.R.: MP/70/E-60: 10 pp. Rome, 9–18 December 1970.
57. Goldsmith, F. B., Munton, R. J. C. and Warren, A. (1970). *J. Linn. Soc. (Biol.)* **2**, 287–306.
58. Perkins, E. J. (1968–71). *Trans. J. Proc. Dumfries. Galloway Nat. hist. Antiq. Soc.* Ser. 3., **45**, 15–43; **46**, 1–26; **48**, 12–68.
59. Natural Environment Research Council. (1973). "Marine Wildlife Conservation", N.E.R.C. Publ. Ser. B. No. 5, 1–39.
60. O'Sullivan, A. J. (1969). *Mar. Pollut. Bull.* No. 13: 14 July 1969.
61. Cole, H. A. (1970). *In* "F.A.O. Technical Conference on Marine Pollution and its Effects on Living Resources and Fishing", F.I.R.: MP/70/R-20: 12 pp. Rome: 9–18 December 1970.
62. French, P. (1970). Personal communication.
63. Hills, L. D. (1970). *Ecologist* **1**, 14
64. Berge, G., Ljøen, R. and Palmork, K. H. (1970). *In* "F.A.O. Technical Conference on Marine Pollution and its Effects on Living Resources and Fishing", F.I.R.: MP/70/E-73 Rome: 9-18 December 1970.
65. Jensen, S., Lange, R., Jernelov, A. and Palmork, K. H. (1970). *In* "F.A.O. Technical Conference on Marine Pollution and its Effects on Living Resources and Fishing", F.I.R.: MP/70/E-88. Rome: 9–18 December 1970.
66. Drew, R. L. (1967). *Philos. J.* **4**, 51–64.

67. Eggleton, A. E. J. (1969). *Atmosph. Environ.* **3**, 355–372.
68. Allsop, K. (1972). "Hard Travelling'", Penguin, Harmondsworth, Middlesex.
69. Cole, L. C. (1970). *In* "The Environmental Crisis". (H. W. Helfrich, ed.), Yale University Press.
70. Wheeler, A. (1970). *Science Journal.* **6**(11), 28-35.
71. Cantlon, J. E. (1969). *Centenn. Rev. Arts. Sci. Mich. St. Univ.* **13**, 123–137.
72. Waldichuk, M. (1967). *Can. Fish. Rept.* **9**, 24–32.
73. Wastler, T. A. and De Guerrero, L. C. (1970). *In* "F.A.O. Technical Conference on Marine Pollution and its Effects on Living Resources and Fishing", F.I.R.: MP/70/R-11. Rome. 9–18 December 1970.
74. Alexander, W. B., Southgate, B. A. and Bassindale, R. (1935). "Survey of the River Tees. II. The Estuary—Chemical and Biological Data", Tech. Pap. Wat. Pollut. Res. Lond., 5: i-xiv, 1–171.
75. Turing, H. D. (1947). *Pollution* **1**, 1–48; **2**, 140. British Field Sports Society, London.
76. Dewling, R. T., Walker, K. H. and Brezebski, F. T. (1970). *In* "F.A.O. Technical Conference on Marine Pollution and its Effects on Living Resources and Fishing", F.I.R.: MP/70/E-78: 14 pp. Rome: 9–18 December 1970.
77. Halstead, B. W. (1970). *In* "F.A.O. Technical Conference on Marine Pollution and its Effects on Living Resources and Fishing", F.I.R.: MP/70/R-6. Rome: 9–18 December 1970.
78. Clutterbuck, D. (1971). *New Scient.* **49**, 58–60.
79. Bock, K. J. (1970). *In.* "F.A.O. Technical Conference on Marine Pollution and its Effects on Living Resources and Fishing", F.I.R.: MP/70/E-8: 4 pp. Rome: 9–18 December 1970.
80. Carpenter, K. E. (1928). "Life in Inland Waters", Sidgwick and Jackson, London.

15

The Parameters of Waste Disposal

Introduction

Until very recently, it has been customary to release, to the sea, divers wastes with little regard for their behaviour and fate other than dilution. Often sewers have debouched upon the shore merely to spew forth their contents at the E.H.W.M.S.T. All shore life has been killed as the effluent crossed the shore and the health of man himself has been endangered. Apart from the infections likely to result from sewage wastes, some acid and toxic wastes are inherently hazardous to persons with whom they come into contact.

The Rivers Pollution Act, 1876, and its successors indicate a long standing concern for the state of British rivers. However, estuarine and coastal waters were disregarded except in those instances where action was taken on sanitary grounds subsequent to a local inquiry. This was partly remedied by the Clean Rivers (Estuaries and Tidal Waters) Act 1960[1]. By this act some estuaries and tidal waters came under the control of statutory bodies defined as River Purification Boards. In general, these regulatory bodies do not cover a wide enough area and, since the Jeger Committee of 1970[2], their function, and that of the District Sea Fisheries Committees of England, has been under review. These latter bodies have a measure of control over effluents released in estuaries and open coastal areas, but unlike the River Purification Boards, they have no control over the disposal of sewage at present. In practice, these bodies confer with, and are advised by, the Ministry of Agriculture, Fisheries and Food, the Department of Agriculture and Fisheries for Scotland, the Ministry of Housing and Local Government and, in special cases, the Water Pollution Research Laboratory and the Commissioners for Crown Lands. Much of the British organization concerned with the use of water is under review and it seems likely that considerable changes will take place in the near future, particularly in relation to the new Regional Water Authorities.

Radioactive wastes are excluded from these arrangements and are controlled by the Atomic Energy Act, 1954 and the Radioactive Substances Act, 1960. Under the latter act, all persons holding radioactive materials on their premises must register these premises with the appropriate Ministry and the disposal of any radioactive waste, by whatever means, is permitted only under a licenced authorization[1]. Nearly all this waste is produced by the United Kingdom Atomic Energy Authority (U.K.A.E.A.) who process all the spent reactor fuel from their own plants and the power generating stations of the Central Electricity Generating Board (C.E.G.B.), the South of Scotland Electricity Board (S.S.E.B.) and those at Latina, Italy and Toko Mara, Japan. There is an absolute liability upon the disposing authority, e.g. U.K.A.E.A., C.E.G.B., to ensure that no individual is harmed by their activities. Furthermore, no discharge is made from such an establishment without the sanction of a Radiochemical Inspectorate advised by two Ministries, viz. the Ministry of Agriculture, Fisheries and Food, and the Ministry of the Environment. The Government Radiochemical Inspectorate has a right of entry to such installations at all times; the details of all effluent releases must be recorded and are submitted to the Inspectorate regularly.

Under ideal conditions, an effluent authorization is arrived at by a detailed investigation based on the following four essentials:

1. A consideration of existing conditions.
2. The characteristics of dilution in the individual area, and the likely characteristics of chemical and physical reconcentration processes.
3. Direct effects upon the biota including toxicology and biological reconcentration processes, especially those along the food chain.
4. Both (2) and (3) are accompanied by surveys which will determine the situations and species likely to be most at risk.
5. The possible effects upon man who either ingests sea food from the area, or is directly exposed to the substance released.

Clearly, the forecasts made in the planning stage should be confirmed in the post-commissioning period, and adjustments made where necessary. However, in Britain, only radioactivity is subjected to this process presumably because its emotive power.

In the U.S.A., it is considered that five considerations are essential in the management of the aquatic environment. They are that:

1. By extensive, intensive and long-term laboratory and field research the water quality requirements for each water use and the various levels of such use shall be determined.
2. Water quality *standards* shall be set and the manner in which they shall be achieved in the various waters shall be specified.
3. By a popular referendum or by legally constituted authority the use or uses of a given water or portion thereof shall be determined.
4. The location and extent of mixing zones shall be determined administratively in accordance with the local situation and conditions.

5. A monitoring and enforcement program shall be established to ensure that the water quality standards are met outside the established zones of mixing[3].

The approach to an authorization for a radioactive effluent is essentially the same in the U.K.[4], the U.S.A.[5] and France[6]; the factors investigated at each site are given in Fig. 15.1[5]. Such discharges are related to

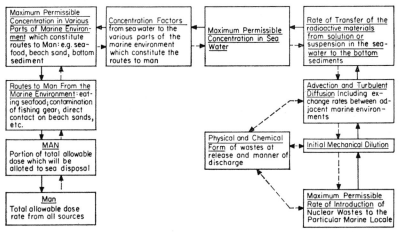

Fig. 15.1. Schematic presentation of the step by step considerations which should be made in evaluating the suitability of any marine locale as a receiver of nuclear wastes (after Pritchard[5]).

the permissible limits of human exposure. They depend upon the make-up of the effluent, the characteristics of the receiving waters and their inhabitants, and the mode of life of those human populations which may be exposed to the waste. The physical, chemical, biological and social factors involved are so diverse that individual assessment of each effluent or discharge is necessary. From these order of magnitude assessments, it is possible to provide estimates of the maximum discharge rates permissible and the foundation for monitoring programmes to provide a logical evaluation of human exposure, once the discharge has commenced[4].

The standards used to control radiation doses and intakes of radioactivity are based upon recommendations by the International Commission on Radiological Protection (I.C.R.P.)[5a]). These standards define the maximum permissible radiation dose, the maximum permitted concentration (M.P.C.) in air or water, a maximum permissible level (M.P.L.) of contamination of surfaces, and a maximum permissible body burden (M.P.B.B.) of a particular nuclide[7]. Evidently, there is a considerable difference in meaning between the concepts of M.P.C. and M.P.B.B. and the total amount of radioactivity, and these concepts should be used with care[8]. Nevertheless, these I.C.R.P. standards have been of great value and upon them working limits for individual situations based upon the factors defined above are derived: these limits are known as the Derived Working Limits (D.W.L.)[7]. In the River Blackwater, the concentration of ^{65}Zn in the oyster was assessed to be the critical path to man in this particular location and the D.W.L. was derived thus[9]:

> Maximum permissible daily intake of zinc 65 (as recommended by I.C.R.P.) for members of the public living in the vicinity of a nuclear installation = 220 000 pCi (1 pCi = one million millionth of a Curie).
> Maximum individual rate of consumption of oyster flesh as found by local survey = 75 g/day (Approx. 2·7 ozs/day).
> ∴D.W.L. for zinc 65 concentration in oyster flesh

$$= \frac{220\,000}{75} \text{ pCi/g, i.e. 2900 } \mu\text{Ci/g.}$$

The maximum individual rate of consumption of oyster flesh is assumed to be consumed every day over the full lifetime of the individual in question.

In the early years of operation, a restriction to one-tenth of the calculated permissible maximum rate of discharge is imposed upon the operators. During this period, the validity of the original estimates is checked by a careful environmental monitoring programme[4].

With non-radioactive effluents the standard of disposal practice rarely approaches that for radioactivity. Commonly it is considered sufficient to perform a dilution study upon which, coupled with toxicological data of doubtful validity, most unscientific projections are made. Scant attention is paid to physico-chemical reconcentration phenomena, of which the most obvious, acute and well-known, is the ability of flood channel systems to return sedimentary materials to the shore. Biological reconcentration is similarly ignored.

In 1959, Carritt et al.[9a] stated that "Nevertheless, several areas stand out as being probably unsuited as disposal sites. They are coastal estuaries, bays and regions immediately seaward of these areas. Shoreward transport along the bottom in these regions would tend to intensify the rate of return of a contaminant to Man". This statement is true both for radioactive materials, to which these authors referred specifically, and, for physico-chemical reasons (Chapters 3 and 5), to waste materials in general. However, countries such as the United Kingdom and the U.S.A. increasingly dispose of waste materials in these waters. Moreover, even with good management techniques which produce high quality effluents, it is difficult to see how this situation can be avoided entirely, or, in the case of sewage materials, whether it should be avoided entirely since the nutrients which we remove as food must be recycled to maintain biological production. Continental countries which criticize these methods without revealing their own, leave little basis for objective discussion. In brief, many countries use little management or science in their waste disposal practices which are characterized by lavish guesswork and much hope that all will be well.

Waste Dispersion

The manner in which a waste disperses from an effluent line depends inherently upon whether it will dissolve in sea water or if it is insoluble. Insoluble substances with a specific gravity lower than that of sea water will float on the surface, whereas a substance of neutral buoyancy, e.g. a colloidal or organic particulate suspension, will float at any level in the water mass and will eventually become adsorbed upon denser organic and inorganic matter, and with it contribute to deposits on the sea floor. A substance of greater density than sea water will sink to the sea bed. The fate of these insoluble substances depends upon these characteristics.

The insoluble substances of low specific gravity will be transported in and on the surface layers and be subject to wind induced water movements particularly. The substances of neutral buoyancy will tend to diffuse throughout the whole estuarine system, and, if they do not become attached to a particle of greater density, will move in a manner characteristic of the hydrography of the individual estuary (Chapter 2). An insoluble particle which is more dense than sea water will tend to move upstream in a manner characteristic of the individual estuary (Chapter 6). Wood-fibres, which are buoyant when first exposed to sea water, but later sink, evince an interesting behaviour which is intermediate between that of the sedimentary and the fully soluble waste[X].

Soluble substances, which are not precipitated upon contact with sea water, will become dispersed by diffusion and transport; initially by an amount which depends upon the tide induced turbulence, i.e. tidal range, and the wave action characteristic of the individual estuary.

In the sea, turbulence is difficult to measure directly, unlike diffusion which may be measured by a number of means. The salinity distribution is itself a natural tracer, but various artificial means have been used also. Currently, fluorescent dyes, e.g. Fluorescein, Amino-G acid and Rhodamine B, are used widely. Rhodamine B is of particular importance since it is non-toxic, light stable and can be detected at concentrations *ca.* 1 part in 10^{11}. Radioactive tracers, e.g. ^{24}Na, have been employed, but because some risk is involved, their use is now avoided. Based upon such information either simple conclusions or complex mathematical models may be raised[10-30], although the pattern of dispersion outside the effluent plume will tend to conform to the general pattern of the particular estuary (Chapter 3).

Individual studies have employed novel and apparently strange techniques well fitted to the particular situation. Many of these studies have used drogue or other trajectories to define the probable path of the waste particle in the 12·5 hr period immediately after release. Other methods include pH determinations related to a plant producing acid wastes[15]; Secchi Disc and Nessleriser determinations of water clarity and colour[31]; sulphite mill effluents may be traced chemically[31a]; measurements of the distribution of coliform bacteria[32]; the dispersal of the distinctly coloured, non-marine bacteria *Serratia indica*[33, 34]; tracking of various target materials by means of aerial photography[35, 36] (Table 15.1); and more recently infra-red photography has been used to detect and investigate water pollution[37]. In the Solway Firth, the dispersal of wood fibres, derived from a mechanical pulping process, i.e. with the lignin still present, has been studied by use of the Phloroglucinol Test performed upon fibres recovered from both water and sediment[X].

Methods which employ the natural characteristics of an effluent may be used to study dispersal over longer periods than the initial 12·5 hr. The wood fibres noted above have this advantage, but the studies based upon radioactive effluents are outstanding[38-45]. Drift cards encased in a polythene envelope have been used also[28], but these have the disadvantage that the wind blows them from wave to wave through the air. The development of sea surface and seabed drifters (p. 65) has been particularly valuable, and these have the particular function that they yield information for periods greater than that of the 12·5 hr for which drogues are normally used. To summarize, dilution studies are performed in three essential stages, thus:

1. By use of dyes, current measurements at fixed stations, salinity measurements and tracers dilution over a short period may be studied.
2. By use of drogues of various kinds the trajectory of the water particles may be traced over a period from H.W. to H.W., L.W. to L.W., or some other period of 12·5 hr, and the influence of the wind, in particular determined.
3. By use of drift envelopes, sea bed and sea surface drifters and its natural attributes, the dispersal of an effluent over periods exceeding 12·5 hr and the factors which influence it may be studied.

Because of the emphasis upon the production of mathematical models and the "need" to produce mean or average pictures for the individual situation, the presence of "cells of water", cores of water derived from

Table 15.1. Characteristics of tracers and targets for aerial photography (after Waldichuk[36])

Tracer or target	Appearance in black and white photograph	Behaviour	Advantages	Limitations
River silt	Cloud-like in appearance; white or grey against dark background of clear water.	Moves on surface of sea water. Drifts with tidal and wind current. Mostly in colloidal suspension in fresh water. Precipitates in water having salinity > 5 ppt.	Naturally available with silt-laden runoff. Clearly visible in aerial photographs. No expense involved in seeding or recovery.	Requires turbid river water. Uncontrolled release. Difficult to identify discrete silt patches for quantitative current measurements.
Paper	White specks against dark background.	Drifts in upper few cm of surface water when lying flat. If buckled, twisted, or torn, may respond to current down to 1 m.	Comparatively inexpensive. Simple to dispense from roll mounted on A-frame.	To be visible from high elevation, large sheets are necessary. Suitable mainly in calm water. Paper must be strong and buoyant when wet. Tends to buckle, twist and tear if sea surface choppy.
Meteorological balloons filled with sea water and/or fresh water.	White or grey specks against black background, if balloons are brightly coloured and proper filter is used in camera.	Drifts in upper layer of water of balloon thickness.	Offers little wind resistance. Represents movement in greater thickness of water than other surface drift techniques. Can be made neutrally buoyant by varying salinity of water used for inflation.	Much effort required in preparation of balloons. Difficult to handle. Can be seen only from low-flying aircraft.
Dye solutions	Light green (fluorescein) or red (rhodamine B) dyes show up as white or grey patches against black background, if suitable camera filter is used.	By adjusting density of dye solution, it can be made to drift at or near surface. Depending on currents and turbulence, dye may disperse rapidly.	Give a measure of horizontal diffusion as well as transport of surface water. Can be sampled for quantitative dilution data.	Because of dispersion, dye is suitable mainly for short time-lapse aerial photography. Difficult to introduce into the sea as a point source.

Table 15.1—*continued*

Tracer or target	Appearance in black and white photograph	Behaviour	Advantages	Limitations
Dye packages (water-soluble fluorescein powder or other light-coloured dye) in paper bag for aircraft release or in cloth sack enclosed in punctured plastic bag for ship release	Shows up as a light patch against a dark background, if proper camera filter is used.	Can be made to burst on impact when released from aircraft or to float on the water surface, providing a controlled rate of leaching. Released dye from burst package disperses rapidly.	Convenient to distribute from boat or aircraft. Attached to a buoy, dye packages can be used to track subsurface drogues from the air.	Dye patch may become considerably extended by wind effect on intact package.
Aluminium powder	Shows as a sleek, white, shiny patch against dark background.	Floats on water surface as film of powder. Responds only to movement of thin surface layer of water.	Comparatively inexpensive. About 0·5 lb (0·23 kg) makes a large target (50 ft or 15 m in diameter). Easily distributed by helicopter or fixed-wing aircraft, in paper bags bursting on impact.	Patches disintegrate rapidly in choppy water; frequent reseeding required. Represents movement of only the thin surface film of water.
Natural foam	Appears as white patch against dark background.	Floats on water surface in clumps. Forms in turbulent waters. Collects in convergence lines.	No effort or expense required in seeding.	Only present where water contains surface-active substances and/or is turbulent enough to create the foam.
Artificial foam (produced with commercially available protein-base liquids)	Appears as white patch against dark background.	Floats in thin, even layer on water and present no significant bearing surface to the wind.	Can be dispensed from small boats with spray pumps. Presents no pollution problem or hazard to navigation. Large targets can be formed to produce distinctive patterns readily seen in aerial photographs.	Represents movement only of uppermost surface layer of water. Has limited duration.

Plywood panels	If painted white, show up as bright patches against a dark background.	Float in upper few cm of water.	Remain intact indefinitely. Size can be increased by tying sheets together. Can be recovered, if marked with staff and flag for ease of spotting from boat.	Large panels needed for small-scale photography. Inconvenient to handle. Costly. Present a hazard to navigation.
Polyethylene sheets	White or bright yellow, fluorescent orange or red will show up as white patches against a dark background, if the required camera filter is used.	Float in upper few cm of water.	Can be recovered if appropriately marked for identification.	Inconvenient to handle from boat. Requires cross pieces as stiffeners to hold sheets flat on water. Presents some hazard to navigation. Moderately costly.
Oil film	Appears as a light-coloured patch because of light reflectivity, against a dark background of water.	Confined to a thin film on water surface.	Small quantity may disperse widely to give a large patch. Easily visible in aerial photograph. Inexpensive. Simple to distribute.	May be objectionable because of pollution potential. Disperses rapidly in turbulent waters.
Sawdust or wood chips	Light-coloured patches against dark background.	Float on water surface, representing movements of thin surface layer.	Inexpensive. Clearly identified in aerial photographs or from surface craft, if present in large quantity.	Potentially a source of pollution on beaches. Large quantity required which is difficult to seed with ordinary equipment. Disperses rapidly in choppy water.
Sewage or industrial effluent	Raw sewage appears as a grey turbid plume or cloud against a dark background in clear sea water. Depending on type of industrial effluent, it may be identified as a grey, turbid cloud, or as some other discolouration in the water.	Because it is usually of about the same density as fresh water, effluent will float on the surface of sea water	No cost. If industrial plant and sewer system in operation, tracer always available for photography.	Uncontrolled release. Dispersion observed only from the point of discharge. Usually suitable only for tracing nearshore currents adjacent to the outfall.

individual creeks (p. 61) and the phenomena which take place at the land/ sea interface are very largely ignored despite their tremendous biological importance. The interface phenomena in particular are of great importance when an effluent is released near the low-water mark. They may have the attributes of wind drift or littoral currents, they may be very narrow (*ca.* 50 yd wide) and occur in a situation where measurement may be very difficult and yet may be traced up to 6 miles offshore as jets in which effluent dilution is minimal[X].

It should be emphasized, then, that when physical processes are discussed, especially those related to turbulent diffusion, they are characteristically statistical in nature, i.e. the equations, laws or rules which describe these situations are only valid on an average of a great number of similar processes. It follows that for the most part single events, e.g. traces of single particles, cannot be predicted precisely[28], but such events may have vital consequences biologically especially in relation to larval settlement and survival.

Waste Reconcentration

Waste reconcentration may occur by two routes, viz. by physical or biological processes.

Physical Reconcentration of Waste

In the period shortly after a waste is released, precipitation and adsorbtion upon the sedimentary particles (especially silt and clay, i.e. the grade < 0·0625 mm) may be extremely important. The parameters of adsorption are poorly known, but it is certain that a wide variety of materials are adsorbed by marine sediments[41, 46–61] (see also p. 121); evidently once attachment to the sediment has occurred these substances may either become available to, or affect, the life processes of the microbenthos, or be transported over great distances.

Some materials may remain firmly attached, e.g. ^{51}Cr, ^{46}Sc, while others are only loosely attached to the sediment, e.g. ^{54}Mn, ^{65}Zn, and a general order of displacement $Cu^{2+} > Co^{2+} > Zn^{2+} > Mn^{2+}$: of some metals 75% may be released from the upper layers of the sediment[52, 53]. Biodegradation may occur, but yields noxious by-products in some cases[62]. Nevertheless, an originally low concentration of a dangerous substance may be concentrated by sedimentary particles and stored for protracted periods. Cu, Cr, Sn, Pb and Hg, for example, are known to be toxic to varying degrees, but the process and effect of the sedimentary uptake of carcinogenic or cumulative poisons is largely unknown[54]. The minor element concentrations in American harbour sediments, dredged wastes, and

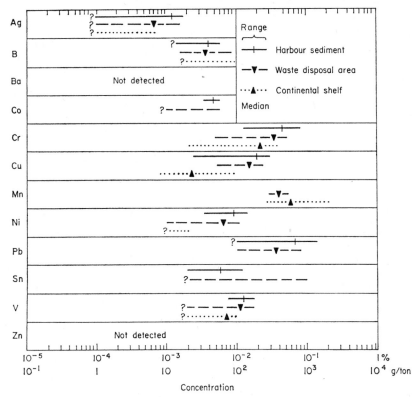

Fig. 15.2. Minor element concentration of harbour sediments and dredged wastes as compared to continental shelf sediment (after Gross[54]).

continental shelf sediments, summarized in Fig. 15.2, are significantly higher than the concentrations normally recorded in sea water (Table 15.2). The chlorinated hydrocarbons, e.g. D.D.T., Dieldrin and P.C.B.s are adsorbed on marine sediments also. The concentration of pesticides and P.C.B.s in sea water may be less than 1×10^{-9} g/ml and 1×10^{-8} g/ml respectively, whereas in the sediment of Long Island Sound, D.D.T. occurred ~ 0.2 p.p.m.; in the Gota Alv, Sweden, P.C.B. occurs in sediments at concentrations from 40 to 160 p.p.b. and decreases with depth, here D.D.T. occurs at a concentration of 14 p.p.b. The rate of decomposition or interchange between the mud and sea is uncertain[55].

The adsorption process will be influenced by the sediment burden of the individual estuary and embayments may facilitate the process[6] (Fig. 15.3). Once attached to the sediment, transport may follow in a pattern consistent with the hydrography of the individual estuary, i.e.

Table 15.2. Elemental composition of some Echinoderms (after Riley and Segar[77]). (Nd = not determined. All concentrations as p.p.m. in dried (60°C) specimens)

| Genus Species | Asteroidea | | | | | Ophiuroidea |
	Asterias rubens	*Solaster* *papposus*	*Solaster* *papposus*	*Porania pulvillus*	*Henricia sanguinolenta*	(unidentified)
Dry wt (%)	25·2	22·5	22·5	28·6	47·9	54·0
Iron (Fe)	37	170	200	16	1 800	890
Manganese (Mn)	6·5	43	31	22	34	82
Cobalt (Co)	<0·30	0·75	1·0	0·66	<0·19	1·5
Nickel (Ni)	1·5	2·3	4·0	<0·28	3·7	3·0
Cadmium (Cd)	3·7	5·3	4·5	9·4	3·5	1·7
Copper (Cu)	4·3	11	6·1	30	8·2	4·1
Lead (Pb)	2·3	5·2	6·8	<0·50	<0·53	<0·67
Zinc (Zn)	190	120	130	100	240	68
Silver (Ag)	Nd	Nd	Nd	Nd	0·60	0·59
Chromium (Cr)	Nd	Nd	Nd	Nd	0·46	<0·58
Aluminium (Al)	<100	240	500	220	90	<90
Sodium (Na)	18 000	18 000	22 000	23 000	13 000	13 000
Potassium (K)	9 000	4 800	5 400	5 000	5 800	3 700
Calcium (Ca)	160 000	140 000	140 000	140 000	170 000	210 000
Magnesium (Mg)	12 000	14 000	14 000	18 000	16 000	18 000
Strontium (Sr)	390	490	260	550	200	490
Phosphorus (P)	4 000	3 400	4 500	1 900	1 900	14 000

* Separate halves of the same animal.

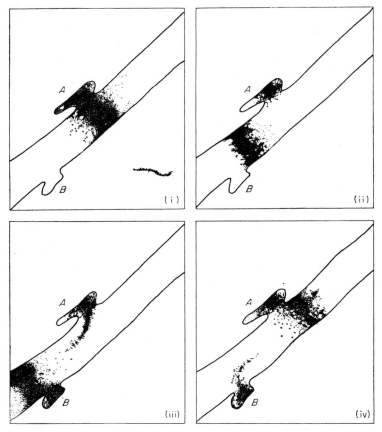

Fig. 15.3. Schematic representation of entrapment by shore features. Concentration is indicated by intensity of shading. As peak concentration of pollution moves downstream on the ebb tide, a portion spreads into shore indenture A, in diagram (i). This portion is entrapped by the indenture as the main volume moves on downstream, as shown in diagram (ii). In diagram (iii) the entrapped pollutant feeds out into the main channel, thus contribution to the longitudinal spread of pollutant, and reducing the concentration. A portion of the main volume is shown spreading into a second shore indenture, B. In diagram (iv) the phenomenon of delayed dispersal following temporary entrapment is repeated for indenture B on the flooding tide (after Pritchard[5]).

depending upon estuary type and the development of the ebb-floop channel system. These processes are known for sediment transport (Chapter 6) and the transport of radioactive waste, e.g. $^{106}Ru^{(56)}$, and organic matter[57].

16*

Biological Reconcentration of Waste

The means by which waste materials may be reconcentrated by biological processes are, on the whole, better understood than the physical processes of reconcentration. Much of this understanding is derived from the need to prevent excessive amounts of radioactive waste reaching mankind. In general, the fate of a radioactive nuclide depends upon its state in sea water, but this depends upon the group of the periodic table from which the nuclide is derived (p. 122). Particulate fission products, e.g. $^{95}Zr/^{95}Nb$, ^{106}Ru, ^{144}Ce, are concentrated only at the second trophic level, i.e. by filter feeding herbivores, but are discriminated against at higher trophic levels. On the other hand, *ionic species*, e.g. ^{90}Sr or ^{65}Zn, may undergo marked concentration at higher trophic levels: a phenomenon both of the open sea and estuaries alike[58, 59]. ^{65}Zn is relatively innocuous to humans, but ^{90}Sr, a bone-seeking isotope, has been a long standing cause for concern[58-73]. Evidently, a radionuclide absorbed by a non-sedentary living organism will be transported by that organism. However in the open ocean, the vertical distribution of radionuclides is influenced by living organisms which may also accelerate the sinking rate of these materials[74-76].

More recently, the development of the atomic absorption spectro-photometer has permitted detailed analysis of the elemental composition of and concentration by marine organisms[77] (Table 15.2 cf. Table 1.2).

Pesticides are concentrated increasingly at the higher trophic levels[78, 79]. Diatoms which contain oils are known to absorb D.D.T., its derivatives and other hard pesticides which may be present in sea water at a concentration of 1×10^{-12} g/ml. In Monterey Bay, *biological magnification* is such that the concentrations of D.D.E. in the livers of fish eating birds have been recorded thus: Brandt's cormorants 107–155 p.p.m.; Western grebes 191–292 p.p.m.; Fork-tailed Petrel 373 p.p.m.; Ashey Petrel 412 p.p.m. and Ring-billed gull 805 p.p.m. The livers of two sick, immature California Sea-lions contained 4 and 89 p.p.m. D.D.E.; 140 p.p.b. was recorded in the sand-dwelling crab *Emerita* [78].

There are many substances susceptible to sedimentary or biological reconcentration that can be considered as possible agents of disease in man and marine animals. Such effects could theoretically result from the build up of acute and cumulative poisons, and carcinogens, e.g. metals and metalloids—Sb, As, Cd, Cr, Co, Pb, Hg, Ni, and Se; petroleum derivatives; petrochemicals and other organics; pesticides, P.C.B.s and 3, 4-benzpyrene (BP). However, there is little evidence from the environment that such a consequence has occurred so far[79-81]. Nevertheless, spoilage of human food by the tainting of fish and shellfish flesh is well known[79, 82], but little is known for certain with regard to P.C.B.s and pesticides[78, 79].

The occurrence of skin tumours and other lesions has been reported in fish species which are neritic and benthic or demersal in habit: most have an inshore dwelling habit, and 9 out of 14 species are flat-fish which live with one surface in permanent close contact with bottom sediments. There is an ecological association between the occurrence of these diseases (in 2·8–18% of fish) and polluted areas in which sewage, petrochemical, and chromium wastes are implicated[79]. The surface of flat-fish most affected by tumours is that most closely in contact with the bottom sediments[79, 83]. Such tumours have been recorded from the Pacific coast of N. America, i.e. from Hecate Strait, British Columbia; Puget Sound, Washington State; San Francisco Bay, Santa Monica Bay, Long Beach and the Santa Ana river outfall, California; the Baltic and the North Sea coast of Holland and Belgium[79]. In the north east Irish Sea, there is some evidence that the incidence of epidermal ulcerations and tumours of fish, particularly flat-fish, may reach 10%[X].

There is only one authenticated case whereby the concentration of toxic materials by marine organsisms has poisoned human beings, viz. when, in the period 1953–61, more than 100 persons in Japan experienced mercury poisoning. This outbreak of mercury poisoning, or *Minimata disease*, arose from the contamination of algae and invertebrates, the food of fishes, themselves taken as food by man. The active agent was methyl mercury derived from an acetaldehyde and vinyl chloride plant. Minimata disease is symptomized by a wide range of serious neurological disturbances, including blindness, deafness, stupor, coma, loss of emotional control and intellectual impairment. The community most affected were fisherfolk who live at Minimata Bay, Kyushu and the River Agano, Niigata Prefecture, Japan and who ate shellfish, crabs and fishes. Human mortality was 40% and 17% at the former and latter respectively: fish-eating birds and mammals were affected also[79, 82].

The problem of mercury poisoning is not confined to areas related to vinyl chloride and acetaldehyde plants alone. In Sweden and Finland, mercurial contamination of the environment arises from a number of sources, viz. the pulp and paper industry (fungicides and slimicides), agriculture (seed dressings and fungicides), the chlorine-alkali industry, electrical installations, heating of ores and clays, burning of oil and coal containing small amounts of mercury, burning of paper and sweepings containing mercury, and wastes from dental clinics and hospitals. In contaminated areas, methyl mercury concentrations in the flesh of the Swedish marine fishes have been recorded thus: flounder (*Platichthys flesus*) 50–80 ng/g; plaice (*Pleuronectes platessa*) 71–3100 ng/g; and cod (*Gadus morhua*) 245–2700 ng/g. Whereas in non-contaminated areas, the mercury levels in fish ranged from 25 to 155 ng/g (where 1 p.p.m. = 1 mg/kg

=1000 ng/g). Mammals absorb >90% of their methyl mercury intake; in one group of humans who died of Minimata disease the mercury content of their organs was liver 6–71 p.p.m., kidney 13–144 p.p.m., and brain 1–21 p.p.m.[79]. In view of the relatively small concentration factor required to reach these levels from a diet of contaminated flounder, plaice and cod, the concern currently expressed in Sweden is easy to understand. Viewed in this light too, the failure to recognize and accept ecological evidence in the Japanese outbreaks[84, 85] seems particularly tragic. It is, however, a warning that procrastination and disregard for such evidence is a hazardous luxury.

It is important to emphasize that mercury may enter the food chain of man largely as a direct result of industrial and agricultural technology. Apparently, marine micro-organisms, working under aerobic conditions, transform *ca.* 90% of mercury, regardless of source, into the highly toxic methyl mercury[52, 79]; evidently the behaviour of methyl mercury is unlike that of mercuric pharmaceuticals[86].

Before concluding this topic of fish rendered toxic to mankind, it is worth noting that *ciguatera fish poisoning* may be triggered by water contaminants such as copper, and various other metallic compounds, dumping of war material, industrial pollutants and shipwrecks. This association was recognized by Thomas in 1799[87], and many such occurrences have been noted in the past 100 yr. In the West Indies, ciguatera, clupeoid, scombroid and puffer poisoning occurs, although ciguatera is the greatest cause for concern: a situation which occurs on an even larger scale in some of the tropical Pacific Islands. Indeed, in the West Indies, fish poisoning is such that canned fish are increasingly imported to areas in which fresh fish abound. In some instances the island's economy, i.e. balance of payments, is adversely affected[79].

It appears that among medical investigators there is a growing belief that some of the chronic degenerative diseases of man, viz. cancers and neurological disorders, may be caused by environmental intoxicants. They consider that marked qualitative and quantitative changes in the environmental carcinogenic spectrum and the cancer panorama in man may be paralleled by a high incidence of liver cancer in hatchery reared rainbow trout[79, 88]. While such conclusions are far from proven, recent work with P.C.B.s does suggest that a further careful examination of this problem is necessary[88a, 88b].

Environmental Surveys

Current opinion apparently favours the concept that industrial and other developments in coastal areas, likely to induce biological changes, shall be the subject of scientific appraisal. However the motives behind this opinion are rarely analysed objectively, consequently most of these surveys are inadequate.

Estuarine surveys, in general, were given a much needed impetus by the classic studies carried out in Tees[89, 90] and the Mersey[91] during the 1930s. Although these studies influenced both academic and applied surveys subsequently, relatively few[9, 40, 41, 92-95] (and the Department of Oceanography, Oregon State University, Corvallis, Columbia River Survey) have since attempted to see the whole problem; the content has been either entirely biological with the occasional physical and chemical measurement, or entirely hydrographical. All have been performed for too short a period, while the simple parameter surveys of the biota which take no account of water quality and movement or of sediment conditions, especially movement, are to all intents worthless.

The basic difficulty encountered in all environmental surveys is summarized by the following quotation: "It may be said that if one desires to know how abundant animals are, where they occur, and what they are doing, there is only one way to find out, and that is to go out and live with them, not for a week or a month, or even a year, but for years"[96]. This concept applies not just to the living organisms, but to the water and sediment also, indeed it can be argued that a thorough knowledge of the physical and chemical environment is more important to an understanding of the biology of an estuary than the biology itself. In truth, however, an estuary must be looked at as a whole system, not as a self-contained ecosystem, but as a centre of action, in which the most major effects take place within and are generated in close proximity to the estuarine system. Outside the centre of action other events take place which may not be and are probably not influenced by the individual estuary, but these events do influence the estuary. With increasing distance the influence of the individual event upon the individual estuary is diminished, but, if a proper study were made, its influence could probably be demonstrated.

Many examples of these interactions have been quoted throughout the text, but some form of summary is appropriate here, thus:

1. Atmospheric

General climatic effects which arise through normal meteorological and climatic phenomena are important, especially with regard to exposed sand and mud flats and the inner areas of estuaries. By these means such phenomena can have a marked influence in those estuaries in which the sand and mud flats are particularly extensive (Chapter 2).

Oceanic sea surface temperature anomalies at great distances from the individual estuary may produce exceptionally extreme conditions, either warm or cold (Chapter 14), and may bring about the mass mortality of many animals (Chapter 10).

2. Hydrographic

Each individual estuary may be classified as a particular type with a characteristic circulation pattern which influences sediment movements (Chapters 3 and 6); larval survival and settlement and hence adult success (Chapter 4); and the dispersal, i.e. dilution or reconcentration, of added substances such as pollutants. Such a circulation pattern may change with season depending upon the amount of natural run-out from the land, or may be induced by engineering works. In either case, the effects may be far reaching, considerable and costly (Chapter 6).

The outer areas especially (and therefore the whole estuary) may be influenced by water masses set in motion at vast distances away from the individual estuary. The chaetognath, *S. setosa*, of the Scottish west coast sea-lochs is apparently derived from the Irish Sea, and since these animals are indicators of water movement, these lochs have their composition influenced by that of the Irish Sea, and the whole movement, i.e. the north bound residual, is influenced by forces external to the Irish Sea. The whole balance of water movement in this area is influenced by variations in the Gulf Stream Current and Drift, itself a result of forces in the Gulf of Mexico and influences brought to bear as the water crosses the N. Atlantic Ocean.

Similarly, events in the Coral Sea bring about the water movements which result in changes in the composition of the diatom flora of the estuaries of New South Wales, Australia (Chapter 4).

The supply of sediment to an estuary also may be influenced by such events: changes in the balance of current flows at the mouth of an individual estuary may result in an intermittence in the supply of silt and clay grades (i.e. < 0·0625 mm) to that estuary.

3. Sedimentary

The composition of the sediments is of vital importance to the biota of the estuary. The supply and source of these materials and the sites of deposition depend upon the estuarine type, wave action and tidal regime (viz. range, symmetry of the tidal curve, and tidal harmonics of the individual area). Where such phenomena act favourably, the marshland areas are rich, and conversely, where they are unfavourable, the marshland areas may be poor (Chapter 6).

The sediments may be inherently unstable and therefore relatively unfavourable to would be colonizers. A decrease in diversity and abundance occurs in these conditions which are therefore similar to the effects of a pollutant. Such sediment movements are relatively difficult to measure and are often ignored; they should not be confused with a seasonal build up or loss of sediment to or from a shore, or long term accretion, or long shore drift. Rather they are analogous to a slow motion wave action or intermittent current of the soil, which results in a massive redistribution of shore materials in a relatively short time, i.e. a few days or weeks. The relative lack of information in this field seems to be due to the fact that engineers have required information concerned with long term resultant changes and botanists and geographers have been concerned only with net gain to, or loss from, salt marshes, whereas most of these effects occur in sand flats which are vital fish feeding and nursery grounds. There is an urgent need for studies of such phenomena, especially as they affect conclusions and decisions in relation to pollution.

4. Biological

Biological processes can influence the transport of sediment (Chapter 6), its deposition (Chapters 6 and 11) and erosion (Chapter 12). Such processes are important in the reconcentration of trace elements and of pollutants (Chapter 16).

In most biological surveys, it has been considered sufficient to investigate either numbers of genera and species present only, or, additionally, recruitment. Based on such information it has been usual to conclude that adverse effects have occurred only when species diversity and abundance has declined. The situation is then so obvious, that a survey is scarcely necessary to confirm it, and prompt action is almost inevitable[79]. More serious, however, are the insidious, long-term, low-level effects which may not be recognized until catastrophe approaches[79]; indeed, their effect upon production may be difficult to recognize, except in severe conditions[97]. Great caution is needed when thinking in terms of "base-line" studies to establish suitable littoral organisms as indicators of pollution[97]: for example, the acorn barnacles are very resistant to acid effluents[X], but are poorly resistant to oil refinery effluents[98].

However, there are much greater pitfalls than this simple example. The approach which concerns itself solely with plant or animal numbers and their recruitment is not valid. Pollutants may be particularly lethal to larvae. Nevertheless, larvae may settle and reach an adult size in an apparently polluted situation; but do they reproduce? On the coast of Northumberland, the sea-urchin *Echinocardium cordatum* occurs abundantly in offshore sediments which are rich in organic detritus, whereas those which live at the low-water mark in clean sand are less abundant. The abundant offshore dwellers, show poorer growth than those at

L.W.M.; the former *do not reproduce* whereas the latter reproduce regularly[99]. In relation to pollution it can be argued that all is well if animal numbers remain high in a polluted area, even though they do not reproduce: that this contention is fallacious is self evident.

A living organism may settle and grow well in an area suitable for its growth and yet disappear. If this happens, particularly adjacent to a town, the effect may be ascribed to pollution, whereas the survey may be inadequate. At Silloth, on the Solway Firth, a large settlement of mussels, *Mytilus*, occurred, in 1967, on ground thought to be suitable, but upon which in the period, 1961–67, no markedly successful settlement had occurred. This 1967 settlement grew well through the summer and the individuals showed all the attributes of good growth and condition. A survey party visited the site, on 15th December 1967, and found a large flock of oyster catchers, *Haematopus ostralegus*, feeding upon an estimated $< 0.1\%$ of the settlement. The very many mussel shells littered upon this shore showed the fracture characteristically made by *Haematopus* when opening a mussel. At this time then the evidence clearly showed that the mussel mortality was due to pressure of feeding by *Haematopus*. However, in this part of the Solway, the flood current may have a velocity of 6 kt and may transport material such as valves of the mussel, with ease. Suppose the survey team had visited this site 7 or 14 days later, then few or no typically fractured mussel valves would have been present, and, lacking food, the *Haematopus* would have been elsewhere. It would have been an obvious conclusion that someone in Silloth had released a substance particularly toxic to mussels. Lest such an observation be dismissed as trivial, it is well to recall that removal of one species, by whatever means, may have profound effects upon the community as a whole[99a], and the impact of a single storm may be felt for years afterwards.

The simple two parameter survey of a type which may conclude, in the absence of data upon population composition, that the rise of primary production in raised nutrient concentrations is beneficial[100], is unhelpful here. Indeed, in this instance, the whole of the increased primary production could have been due to algae which produce neurotoxins.

At its very best, an ecological survey, alone, can demonstrate only that a change is taking place, but not why it is taking place: the Continuous Planktonic Recorder Survey of the North Atlantic performed by the Oceanographic Laboratory, Edinburgh, has shown that, in the past 22 yr, the duration of the plankton season has decreased by about $4\frac{1}{2}$ weeks. Some of the species exhibit a complicated pattern of variation, but a number of the common diatom species, viz. *Rhizosolenia styliformis*, *Chaetoceros* (especially *Phaeoceros*), *Nitzschia delicatissima*, *Rhizosolenia imbricata* var. *shrubsolei* and *Dactyliosolen mediterraneus* have declined in abundance. However, although the turbidity of the atmosphere has increased and pollutants could have induced the change, it may be due to a natural variation[101] *But* as J. R. Lewis has stated repeatedly, the most serious problem in baseline studies is the assessment of natural fluctuations. Indeed, natural changes which have occurred in the zooplankton off Plymouth between 1930 and 1970 (*Calanus*—rich, 1930; *Calanus*—poor, late 1930s–60s; *Calanus*—rich, 1970[101a]) demonstrated both the long term nature of such fluctuations and the need for such studies.

Because of the inherent incapability of ecological surveys to do other than indicate that some change or changes are occurring, a different approach to the pollution problem may be preferable, viz. "by which the effects of potentially-toxic pollutants on the processes of growth, physiology, behaviour and breeding are established in controlled laboratory experiments and are then looked for in the field. Where gradients in the concentration of pollutants in the sea can be found, as, for example, off the mouths of rivers carrying sizeable quantities of easily recognizable materials such as metallic wastes, it should be possible to establish in sedentary animals and plants a comparable gradation in pollution effects, for example in growth rates, behaviour, maturation of gonads etc. From such a combination of precise laboratory experiments and field studies of polluted situations it should be possible to build up an appreciation of the ecological consequences of pollution and so reach sound judgements regarding the capacity of the marine environment to accept waste without damage to living resources or productivity"[97]. The ability of the environment to "mimic" such events must not be forgotten and is worthy of further investigation.

Experimental Studies

The failure to appreciate that a substance, which is dissolved in water, may be ingested by man without harm, but is lethal to aquatic animals, seems to have led to many aquatic pollution problems. Probably the best known is chlorinated drinking water which may be essential to man's well being, but as every aquarium keeper knows is lethal to fish[101b].

Most studies of the effects of pollutants upon aquatic life have been concerned with the measurement of median tolerance limits of a relatively narrow range of organisms or single organisms; apparently only pulp-mill wastes have been studied thoroughly[102, 103] (Table 15.3). To these studies investigations of (1) food intake, growth and conversion efficiency, (2) respiratory scope (active and standard rates of oxygen uptake) and (3) active capacity (swimming speed, track made good, fatigue time) have been added[103].

Table 15.3. Responses which have been utilized in detecting and measuring the pollutional effects of pulp-mill wastes upon fishes (after Alderdice[103])

Nature of response	Ultimate effect on organism	Laboratory tests
Lethal or acute General and gross symptoms of distress	Death	Gross determinations of biological activity and relative potency of pollutants (i.e. TL$_m$ studies)
Sub-lethal or chronic Behavioural	Preference or avoidance reactions	Behavioural studies using "choice" situations.
Physiological and histopathological	Changes in: —oxygen uptake —enzyme activity	Respirometry Dehydrogenase, transaminase acetylcholine- sterase reactions.
	—haematological	Red cell counts, differential cell counts, Haematocrits, haemoglobin Electrophoretic studies
	—serum proteins —decrease in glycogen, R.N.A., changes in bile duct, portal vein —decrease in R.N.A., change in structure	Histological and cytochemical examination of: —liver —pancreas
	—necrosis of tubules —changes in secretion of mucus, activity of goblet cells, necrosis of epithelium	—kidney —intestine
	—changes in secretion of mucus, necrosis of filaments	—gills

The median tolerance limit, or TL_m, is the dose required to kill 50% of the test animals; in aquatic studies, this is normally expressed as a L.C.$_{50}$, i.e. the concentration of solute, in parts per million (p.p.m.), required to produce a 50% mortality of the test animals. Clearly, the duration of the exposure of the organism to the toxic solution is an important parameter in defining this value. In practice, therefore, the TL_m is defined as, for example, a 24 hr L.C.$_{50}$ or a 96 hr L.C.$_{50}$ which means the concentration of poison required to produce a 50% mortality in organisms subjected to a test solution for 24 and 96 hr respectively, and allowed to recover for a given number of days, normally 5 or 7; the mortality is determined at the end of this period. There are many variants of this procedure, including no-flow and continuous-flow systems, the results in the former indicating a somewhat lower toxicity than the latter. In general, the results from these determinations are used to decide effluent authorizations by use of an *application factor* which may be 1/10, 1/20 or 1/100 of the 24 or 96 hr L.C.$_{50}$, based upon quite arbitrary values.

Before discussing application factors, it is worth stating that few studies appear to pay much attention to the handling and acclimation of test organisms. Death in the pre-test period is important. Salinity acclimation may be protracted, in mussels, *Mytilus edulis*, it may take 4–7 weeks[104], whereas it takes 5–24 hr in the snail *Potamopyrgus jenkinsi*[105]; moreover the gastropod *Hydrobia ulvae* has distinct physiological races which differ in their salinity tolerance[106]. Temperature acclimation may take *ca.* 14 days[107, 107a, 108]. Care is necessary, to ensure that all the test animals are in a similar physiological condition relative to growth and position occupied on the shore[109]. Normally, in setting the limits of the temperature regime of the tests, little attention is paid to such important phenomena as the variations in the structure of water with temperature[110] with which the observed changes in the respiration rate of some animals, e.g. *Patella*, may be related. In the absence of such experimental definitions, arguments concerning the relative merits of a no-flow or continuous-flow system, which may produce results that are not significantly different statistically[111, 112], do not inspire confidence in the technique. Moreover, the universal adopting of a recovery period of $\leqslant 5$ days precludes the acquisition of information upon delayed mortality and inhibited growth post treatment. Such phenomena are well known in the human situation; they have now been demonstrated for oil-emulsifier detergents in polychaete worm, *Sabellaria spinulosa*, larvae[113] and gastropod molluscs[109].

The response and survival of treated animals liberated to the environment may be very different to that evinced in the laboratory: in the field, for example, the effect of predators may be considerable[109]. In general

Table 15.4. A summary of the effect of sulphite waste liquor on oysters as determined from traditional bio-assays (after Woelke[117])

Type of bioassay	Species	Response utilized	P.B.I.* at which adverse effect occurred.
Ecological survey	Olympia oysters	Mortality and ecological conditions	Not given
Acute toxicity	Olympia and Pacific oysters	Pumping rate	670 p.p.m.
Acute toxicity	Olympia oysters	Pumping rate	100 p.p.m.
Acute toxicity	Pacific oysters	Reproduction	3 p.p.m.
Long term toxicity	Olympia oysters	Mortality	13 p.p.m.
Long term toxicity	Olympia oysters	Mortality	16 p.p.m.
Long term toxicity	Olympia oysters	Reproduction	8–16 p.p.m.
Live box	Olympia oysters	Mortality and fatness	7 p.p.m.
Live box (lagoon)	Pacific oysters	Mortality and growth	13 p.p.m.

* Chemical measure of pulp mill wastes, Pearl and Benson (1940).

Table 15.5. Comparison of median tolerance limits and long term chronic exposure of marine invertebrates to toxic solutions. (All salinities $= 30\%_0^{(X)}$)

Species	Substance	24 hr L.C.$_{50}$	Chronic exposure Concentration	Fraction of L.C.$_{50}$	Time to 50% mortality
Eupagurus bernhardus	Na_2SO_4	> 6400 p.p.m.	500 p.p.m.	< 1/13	12 weeks
Eupagurus bernhardus	Board mill effluent	> 50%	1%	< 1/50	43 days
Eupagurus bernhardus	Gas flume	> 10%*	0·5%	< 1/20	17 weeks
Carcinus maenas	Na_2SO_4	> 6400 p.p.m.	275 p.p.m.	< 1/23	> 8 weeks
Buccinum undatum	CH_3COOH	~ 6400 p.p.m.	100 p.p.m.	~ 1/64	12 weeks
Buccinum undatum	CH_3COOH		10 p.p.m.	~ 1/640	> 15 weeks

* 96 hr L.C.$_{50}$

too, such studies are concerned with toxin concentrations at constant temperature and salinity when a multivariate approach both with regard to these parameters and to toxic mixtures may be desirable[114, 115].

If the experimental determination of TL_m is far from satisfactory, what of application factors? Although widely used, recent views consider that those factors which represent a compromise negotiated upon the TL_m have no sound technical basis[116]; a contention supported by Tables 15.4[117] and 15.5[X]. These and other studies show that "many non-lethal concentrations (i.e. relative to the TL_m) are not safe in the functional sense and bioassay techniques and data are badly needed so that standards and management techniques for industrial areas can be developed to insure vigorously functioning aquatic communities instead of ones that merely survive"[118]. Clearly, low-level, long term, chronic toxicity studies, on a wide variety of test organisms must be an important field of investigation in the future[79]. Since workers in the medical field have already appreciated and make use of differential toxic effects[86], it is a little difficult to understand the apparent mental block in the aquatic pollution field. Nevertheless, the relative difficulty of handling aquatic organisms and the low priority given to such studies in the past are in part responsible. Curiously, human self interest, in relation to fish-control chemicals, has produced the most advanced technique in the field of aquatic toxicology[119].

Water Quality Criteria

If the environment is to remain in a healthy condition, it is essential that all effluents, and their components, released to it are controlled by an authorization which will ensure that the healthy state be maintained Probably, the most advanced attitude, in this respect, is embodied in the standards devised by the United States Federal Water Quality Administration of F.W.Q.A. (formerly Federal Water Pollution Control Administration)[120]: moreover, the individual states of the U.S. have their own standards in addition[121-123]. Canadian Federal practice works within the framework that an effluent shall not be harmful to fish life; within this framework, and that of the individual provincial governments, an efficient form of pollution control is being developed. Still other countries are moving towards either more strict pollution control or demanding more detailed information of their scientists upon which to base standards of control which are meaningful[124-129]. In contrast, and with the exceptions of oil and radioactivity, the United Kingdom has rather poor standards[130] and appreciation of estuarine systems.

The standards devised by the United States Federal Water Quality Administration are so thorough that they represent not only the most significant advance, but a source of inspiration, in this field. The report first discusses what is known of each parameter, and then concludes with a specific recommendation. The recommended criteria of water quality with respect to estuaries are worthy of detailed consideration, and they are quoted therefore:

Salinity:

For the protection of estuarine organisms, no changes in channels, in the basin geometry of the area, or in feshwater inflow should be made that would cause permanent changes in isohaline patterns of more than ± 10% of the natural variation.

Currents:

In view of the requirements of estuarine organisms and the nature of marine waters, no change in basin geometry or freshwater inflow should be made in tidal tributaries which will alter current patterns in such a way as to cause adverse effects.

pH:

Materials that extend normal ranges of pH at any location by more than ± 0·1 pH unit should not be introduced into salt water portions of tidal tributaries or coastal waters. At no time should the introduction of foreign materials cause the pH to be less than 6·7 or greater than 8·5.

Temperature:

In view of the requirements for the well-being and production of marine organisms, it is concluded that the discharge of any heated waste into any coastal or estuarine waters should be closely managed. Monthly means of the maximum daily temperatures recorded at the site in question and before the addition of heat of artificial origin should not be raised by more than 4°F during the fall, winter, and spring (September through May), or by more than 1·5°F during the summer (June through August). North of Long Island Sound and in the waters of the Pacific Northwest (north of California), summer limits apply July through September, and fall, winter and spring limits apply October through June. The rate of temperature change should not exceed 1°F per hour except when due to natural phenomena.

Dissolved Oxygen (D.O.):

For the protection of marine resources, it is essential that oxygen levels shall be sufficient for survival, growth, reproduction, the general well being, and the production of a suitable crop. To attain this objective, it is recommended that dissolved oxygen concentration in coastal waters, estuaries and tidal tributaries of the Nation, including Puerto Rico, Alaska and Hawaii, should be as follows:

1. Surface dissolved oxygen concentrations in coastal waters should not be less than 5·0 mg/litre (p.p.m.), except when natural phenomena cause this value to be depressed.
2. Dissolved oxygen concentrations in estuaries and tidal tributaries shall not be less than 4·0 mg/litre at any time or place except in dystrophic waters or where natural phenomena cause this value to be depressed.

The committee would like to stress the fact that, due to a lack of fundamental information on the D.O. requirements of marine and estuarine organisms, these requirements are tentative and should be changed when additional information indicate that they are inadequate.

Crude oil and petroleum products:

Combined effect of oil and sewage pollution; colour of oil film on the surface of water; adsorption of oil by sand, clay, silt and other suspended particles; toxicity of crude oil and petroleum products; carcinogenic substances from oil polluted waters:

Until more information on the chemistry and toxicology of oil in sea water becomes available, the following requirements are recommended for the protection of marine life. No oil or petroleum products should be discharged into estuarine or coastal waters in quantities that (1) can be detected as a visible film or sheen, or by odour, (2) cause tainting of fish and/or edible invertebrates, (3) form an oil-sludge deposit on shores or bottoms of the receiving body of water, or (4) become effective toxicants according to the criteria recommended in the 'Toxicity' section.

Turbidity and Colour:

No effluent which may cause changes in turbidity or colour should be added to, or discharged into, inshore or coastal waters unless it has been shown that it will not be deleterious to aquatic biota.

Settleable and Floating Substances:

Water quality requirements for specifying the permissible limits of settleable solids and floating materials cannot be expressed quantitatively at present. Since it is known that even minor deposits may reduce productivity and alter the benthic environment, it is recommended that no materials containing settleable solids or substances that may precipitate out in quantities that adversely affect the biota should be introduced into estuarine or coastal waters. It is especially urgent that areas serving as a habitat or nursery grounds for commercially important species (scallops, lobsters, oysters, clams, crabs, shrimps, halibut, flounders, demersal fish eggs and larvae and other bottom forms) be protected from any infringement on natural conditions.

Tainting Substances:

To prevent tainting of fish and other marine organisms, substances that produce taste and off-flavours should not be present in concentrations above those shown to be acceptable by means of bioassays and test panels. Experience has shown that test organisms should be exposed to the materials under test for 2 weeks at selected concentrations to determine the maximum concentration that does not produce noticeable off-flavours as determined by organoleptic tests (cooking should be done by baking the material wrapped in aluminium foil).

Plant Nutrients and Nuisance Organisms (i.e including the events leading to eutrophication, excessive growths of attached and other algae including the producers of neurotoxins):

The ecological factors most often associated with nuisance growths are changes in the natural temperature and salinity cycles and increases in nutrients. The change in any of these factors may directly or indirectly affect the response of the organisms to other factors. Increase or decrease in current and, indirectly, its effect on available nutritional materials have also been found to be important.

To maintain a balance among nutrients and a balanced biota most conducive to the production of a desired crop, it is recommended that:

1. No changes should be made in the basin geometry, current structure, salinity or temperature of the estuary without first studying the effects on aquatic life. For example, these studies should be made before dams are erected, water diversion projects are constructed or dredge and fill operations carried out.

2. The artificial enrichment of the marine environment from all sources should not cause any major quantitative or qualitative alteration in the flora. Production of persistent blooms of phytoplankton, whether toxic or not, dense growths of attached algae and higher aquatics or any other sort of nuisance that can be directly attributed to nutrient excess or imbalance should be avoided. Because these nutrients often are derived from drain-

age from land, special attention should be given to correct land management in a river basin and on the shores of a bay to prevent erosion.

3. The naturally occurring atomic ratio of NO_3-N to PO_4-P in a body of water should be maintained. Similarly, the ratio of inorganic phosphorus (orthophosphate) to total phosphorus (the sum of inorganic phosphorus, dissolved organic phosphorus, and particulate phosphorus) should be maintained as it occurs naturally. Imbalances have been shown to bring about a change in the natural diversity of the desirable organisms and to reduce productivity.

Toxic Substances (including pesticides; heavy metals; ammonia and ammonium compounds; cyanides, sulphides; flourides; detergents and surfactants; pathogenic organisms; tar, gas and coke wastes; petroleum refining and petrochemical wastes; pulp and paper manufacturing wastes; waterfront and boating activites; disposal of laboratory wastes):

A. *General recommendation*: in the absence of toxicity data other than the 96-hour TL_m, an arbitrary application factor of 1/100 this amount shall be used as the criterion of permissible levels.

Additional chronic exposure tests will be conducted within a reasonable period to demonstrate that the estimated maximum safe levels as indicated by the 96-hr TL_m and the application factor do not, in fact, cause decreases in productivity of the test species during its life history.

B. *Recommendation with respect to pesticides*: the pesticides are grouped according to their relative toxicity to the shrimp, one of the most sensitive groups of marine organisms. Criteria are based on best estimates in the light of present knowledge and it is expected that acceptable levels of toxic materials may be changes as the result of future research.

(i) Pesticide group A—The following chemicals are acutely toxic at concentration of 5 μg/litre and less. On the assumption that 1/100 of this level represents a reasonable application factor, it is recommended that environmental levels of these substances not be permitted to rise above 50 nanograms/litre. This criterion is so low that these pesticides could not be applied directly in or near the marine habitat without danger of causing damage. The 48-hour TL_m is listed for each chemical in parts per billion (μg/litre).

Organochloride pesticides

Aldrin	0·04	D.D.T.	0·6
B.H.C.	2·0	Dieldrin	0·3
Chlordane	2·0	Endosulfan	0·2
Heptachlor	0·2	Methoxychlor	4·0
Endrin	0·2	Perthane	3·0
Lindane	0·2	T.D.E.	3·0
		Toxaphene	3·0

Organophosphorous pesticides

Coumaphos	2·0	Naled	3·0
Dursban	3·0	Parathion	1·0
Fenthion	0·3	Rounel	5·0

(ii) Pesticides group B—The following types of pesticide compounds are generally not acutely toxic at concentrations of 1 mg/litre or less. It is recommended that an application factor of 1/100 be used and, in the absence of acute toxicity data that environmental levels of not more than 10 μg/litre be permitted.

Arsenicals	2,4,5-T compounds
Botanicals	Phthalic acid compounds
Carbamates	Triazine compounds
2,4-D compounds	Substituted urea compounds

(iii) Other pesticides—acute toxicity data are available for approximately one hundred technical grade pesticides in general use not listed in the above groups. These chemicals either are not likely to reach the marine environment, or, if used as directed by the registered label, probably would not occur at levels toxic to marine biota. It is presumed that criteria established for these chemicals in fresh water will protect adequately the marine habitat. It should be emphasized that no unlisted chemical should be discharged into the estuary or coastal water without preliminary bioassay tests and the establishment of an adequate application factor.

C. *Recommendations for toxicants other than pesticides*:

(i) Allowable concentrations of metals, ammonia, cyanides and sulphides should be determined by use of the 96-hr TL_m values and appropriate application factors. Preferably the TL_m values should be determined by flow-through bioassays in which environmental factors are maintained at levels under which these materials are most toxic. Tests should utilize the most sensitive life stage of a species of ecological or economic importance in the area. Tentatively, it is suggested that application factors should be 1/100 for pesticides and metals, 1/20 for ammonia, 1/10 for cyanide and 1/20 for sulphides.

(ii) There is evidence that fluorides are accumulative in organisms. It is tentatively suggested that allowable levels should not exceed those for drinking water.

(iii) The further dilution of wastes in marine waters suggests that the adoption of criteria established for detergents and surfactants in fresh water (i.e. the concentration of L.A.S. should not exceed 0·2 mg/litre or 1/7 of the 48-hr TL_m concentration, whichever is the lower) also will protect adequately biota in the marine environment.

(iv) Bacteriological criteria of estuarine waters utilized for shellfish cultivation and harvesting should conform with the standards as described in the "National Shellfish Sanitation Program Manual of Operation". These standards provided that:

(a) Examinations shall be conducted in accordance with the American Public Health Association recommended procedures for the examination of sea water and shellfish.
(b) There shall be no direct discharges of untreated sewage.
(c) Samples of water for bacteriological examination to be collected under those conditions of time and tide which produce maximum concentration of bacteria.
(d) The coliform median M.P.N. of the water does not exceed 70/100 ml, and not more than 10% of the samples ordinarily exceed an M.P.N. of 230/100 ml for a 5-tube decimal dilution test (or 330/100 ml where the 3-tube decimal dilution test is used) in those portions of the area most probably exposed to faecal contamination during the most unfavourable hydrographic and pollution conditions.
(e) The reliability of nearby waste treatment plants should be considered in the approval of areas for direct harvesting.

(v) It is also essential to monitor continuously waste from tar, gas, and coke, petroleum refinery, petrochemical, and pulp and paper mill operations. They all produce complex wastes of great variability, not only from facility to facility, but also from day to day. This should be done on an individual basis with bioassays. These tests should be made at frequent intervals to determine TL_m values as described for other wastes. For the more persistent toxicants, an application factor of $1/100$ should be used while for unstable or biodegradable materials an application of $1/20$ is tentatively suggested.

(vi) Concentrations of other materials with noncumulative toxic effects should not exceed $1/10$ of the 96-hr TL_m value. For toxicants with cumulative effects, the concentrations should not exceed $1/20$ and $1/100$ of the above respective values.

When two or more toxic materials that have additive effects are present at the same time in the receiving water, some reduction in the permissible concentrations as derived from bioassays on individual substances is necessary. The amount of reduction required is a function of both the number of toxic materials present and their concentrations with respect to the derived permissible concentration. An appropriate means of assuring that the combined amounts of the several substances do not exceed a permissible combination for the mixture is through the use of the following relationship:

$$\frac{C_a}{L_a} + \frac{C_b}{L_b} + \ldots + \frac{C_n}{L_n} \leqslant 1$$

where C_a, C_b . . . C_n are the measured concentrations of the several toxic materials in the water and L_a, L_b . . . L_n are the respective permissible concentrations (limits) derived for the materials on an individual basis. Should the sum of the several fractions exceed 1, then a local restriction on the concentration of one or more substances is necessary[123].

Although willing, many regulating bodies may not have the means to monitor all the criteria stipulated. Nevertheless where such means are limited a worthwhile monitoring programme can be undertaken with limited resources[131–133], thus:

1. Dissolved oxygen may be measured either by the classic Winkler test, or by more modern electrode systems.
2. Hydrogen-Ion Concentration (pH) may be measured either by the more accurate glass electrode or by the cheaper, but less accurate, indicator methods.
3. Turbidity and Water Colour may be measured either by a photo-cell system, i.e. a hydrophotometer, or by use of a Secchi Disc. In water transparency studies the Secchi Disc is used in the normal way (p. 19); in colour studies, it may be lowered to a standard depth, viz. 1 m, and the water colour matched against a set of standards.

   The sediment and suspended solids load, which influences these parameters, may be determined by filtration of samples returned to the laboratory.
4. Nutrient Salts and Eutrophication. The concentration and balance of nutrients may be determined in the normal way. Routine sampling and analysis of the plankton, especially the phytoplankton, may be used to support these studies.

   Where eutrophication or sedimentation of organic materials, e.g. wood fibres, occurs, the accumulation of organic matter may be investigated by normal chemical analytical techniques. In such situations, bottom samples may be examined for the development of anaerobic conditions near to the surface: moreover, the content of H_2S, NH_3, and O_2, the pH and oxidation-reduction potential of these materials may be measured readily by electronic methods.
5. Coliform counts may be performed readily by the Millipore filter method[134].

Other parameters such as tastes and odours, films of scum, oil, slime or foam may be readily perceptible to the senses, but are difficult to quantify[132, 133, 135]. Where adequate funds are available, all the techniques of modern science may be used, and automated and remote sensing techniques are being considered for this purpose[136, 137].

To ensure that environmental standards are met, it is essential to relate

Table 15.6. General standards of quantity for trade effluents discharge to river (after Ainsworth[139])

Non-fishing streams	
(i) pH value	≮ 5 or ≯ 9
(ii) Suspended solids content (dried at 105 °C)	≯ 40 mg/litre
(iii) Suspended solids content + dissolved metals*	≯ 40 mg/litre
(iv) Permanganate value (4 hr at 27 °C)	≯ 60 mg/litre
(v) Biochemical oxygen demand (five days at 20 °C)	≯ 40 mg/litre
(vi) Sulphide (as S)	≯ 1 mg/litre
(vii) Cyanide (as CN)	≯ 0·2 mg/litre
(viii) Oils and grease	≯ 10 mg/litre
(ix) Formaldehyde	≯ 1 mg/litre
(x) Phenols (as cresols)	≯ 1 mg/litre
(xi) Free chlorine	≯ 1 mg/litre
(xii) Tar	None
(xiii) Arsenic, barium, cadmium, chromium, copper, lead, mercury, nickel, selenium, silver and zinc	Individually or in total ≯ 1 mg/litre
(xiv) Soluble solids (dried at 105 °C)	≯ 7500 mg/litre
(xv) Temperature	≯ 32·5 °C

Fishing streams	
(i) Biochemical O_2 demand (five days at 20 °C)	≯ 20 mg/litre
(ii) Permanganate value (4 hr at 27 °C)	≯ 20 mg/litre
(iii) Suspended solids content (dissolved metals)*	≯ 30 mg/litre
(iv) pH value	≮ 5 or ≯ 9
(v) Sulphides (as S)	≯ 1 mg/litre
(vi) Cyanide (as CN)	≯ 0·1 mg/litre
(vii) Oils and grease	≯ 10 mg/litre
(viii) Temperature	≯ 30 °C
(ix) The effluent shall not contain any other matter which would cause the stream to be poisonous or injurious to fish and other aquatic life	

* Applies to relatively non-toxic metals in solution (e.g. iron, aluminium) which could give precipitates with the natural alkalinity of the river water.

all that is known of the dilution and reconcentration processes likely to influence the individual effluent to a set of effluent standards which are embodied in a document of consent or authorization. These have not been defined for British inshore and estuarine waters in general, but the standards set for rivers in general, Table 15.6, may be quoted as an example of the form which such standards take[1]. In this definition of standards most of the requirements are not in need of elaboration. However, the question of the oxygen demand of an effluent is important especially when sewage is involved.

The test most widely used is that for *B.O.D.* or *Biochemical Oxygen Demand* which was first devised by the Royal Commission on Sewage

Table 15.7   Processes applicable to waste water treatment (after Eckenfelder[141])

Pollutant	Processes
Biodegradable organics (B.O.D.)	Aerobic biological (activated sludge, aerated lagoons, trickling filters, stabilization basins), anaerobic biological (lagoons, anaerobic contact, deep-well disposal).
Suspended solids (S.S.)	Sedimentation, flotation, screening.
Refractory organics (C.O.D., T.O.C.)	Carbon adsorption, deep-well disposal.
Nitrogen (N)	Maturation ponds, ammonia stripping, nitrification and denitrification, ion exchange.
Phosphorus (P)	Lime precipitation; Al or Fe precipitation, biological coprecipitation, ion exchange.
Heavy metals	Ion exchange, chemical precipitation.
Dissolved inorganic solids	Ion exchange, reverse osmosis, electrodialysis.

Disposal in 1913[138]. In essence, the B.O.D., expressed in p.p.m., is the amount of oxygen in milligrams taken up by 1 litre of a sample during a 5 day incubation at 20°C. However, the 5-day B.O.D. test is not a very satisfactory measure of oxygen demand for effluents released to estuaries, since in this situation the breakdown time may exceed 5 days. Moreover the test is influenced by salinity and it does not account for the oxidation of ammonia to nitrate, a reaction which takes place after 5 days and is known to occur in estuaries. Here a better measure is given by the *Ultimate Oxygen Demand*, or U.O.D., a result obtained when a B.O.D. test is incubated for 2–3 months instead of the usual 5 days. However, even this test may have a poor performance in the presence of materials such as wood fibres. The absorption of oxygen by a sample from a dilute sulphuric acid solution of N/80 potassium permanganate, which yields *3 min and 4 hr permanganate values*, i.e. the 4 hr oxygen absorption or O.A. test, is a poor substitute for the 5-day B.O.D. test since it does not work well in relation to the estuarine situation. Perhaps the best measure is the *Chemical Oxygen Demand Test*, or *C.O.D.*, in which the effluent sample is refluxed with potassium dichromate and 50% sulphuric acid for 2 hr and a more complete oxidation is obtained. With some organic wastes the C.O.D. and B.O.D. are approximately equal, but with others the results of the two tests differ significantly. For a detailed appreciation of these problems reference should be made to the classic work on River Pollution by Klein[1, 140a, 140b].

Where effluents fail to reach the required standards treatment may be required; this may range from simple sedimentation to tertiary treatment.

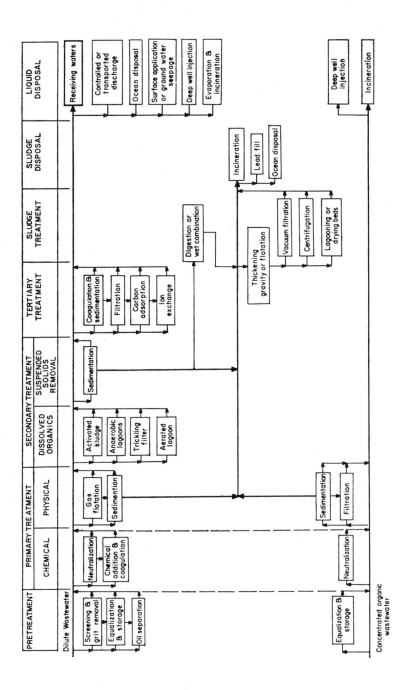

Fig. 15.4. Waste water treatment sequence/processes substitution diagram (after Eckenfelder[141]).

Table 15.8. Maximum effluent quality attainable from the waste treatment processes (after Eckenfelder[141])

Process	B.O.D.	C.O.D.	S.S.	N	P	T.D.S.
Primary treatment						
Sedimentation, % removal	10–30	—	50–90	—	—	—
Flotation[a], % removal	10–50	—	70–95	—	—	—
Secondary treatment						
Activated sludge, mg/l	<25	[b]	<20	[c]	[c]	—
Aerated lagoons, mg/l	<50	—	>50	—	—	—
Anaerobic ponds, mg/l	>100	—	<100	—	—	—
Disposal						
Deep-well disposal	Total disposal of waste					
Tertiary treatment						
Carbon adsorption, mg/l	<2	<10	<1	—	—	—
Ammonia stripping % removal	—	—	—	>95	—	—
Denitrification and nitrification, mg/l	<10	—	—	<5	—	—
Chemical precipitation, mg/l	—	—	<10	—	<1	—
Ion exchange, mg/l	—	—	<1	[d]	[d]	[d]

[a] Higher removals are attained when coagulating chemicals are used.
[b] $C.O.D._{inf}$—(B.O.D. (Removed)/0.9)
[c] N_{inf}—0.12 (excess biological sludge), lb; P_{inf}—0.026 (excess biological sludge), lb.
[d] Depends on resin used, molecular state and efficiency desired.

The processes of waste water treatment are summarized in Fig. 15.4; their application and effectiveness are summarized in Tables 15.7 and 15.8[141]. For a more detailed review of these processes see Klein[1].

Commentary

In essence we are concerned here with the means by which sewage and trade wastes may be released to estuarine and coastal waters without producing a deterioration in that environment. It would be misleading to suggest that high standards of effluent quality alone would solve all waste disposal problems, for this is not true. Indeed, the rigid use of water quality standards without reference to the amounts of effluents released in total may be of little value or may even be damaging, moreover as E. A. Pearson demonstrated very clearly effluent treatment and the long effluent line are not mutually exclusive alternatives to this problem[142].

Clearly, as with radioactivity, each situation must be considered on its merits for only by these means can such difficulties as those which occur in an estuary overloaded by a multiplicity of small discharges be avoided; conversely, there are situations where a series of small outfalls may be less damaging than one big one. However, such considerations imply that a sound knowledge of the biology and hydrography of the receiving waters must be acquired.

In the last analysis, though, there must be a point at which standards are set. All uses of radioactivity are controlled ultimately by the standards for human exposure set by the International Commission on Radiological Protection[5a]; these standards or recommendations are the subject of continuing discussion, but without them the modern use of radioactive materials and of nuclear energy would probably be impossible. Evidently, some equivalent set of standards or recommendations are required to define the acceptable amounts of waste materials, singly and in combination, in terms of human and environmental health. The incidence of phossy-jaw in early match-makers and the fate of the watch dial painters indicates the penalties which await those who have failed to take reasonable precautions. Such a conclusion, implies that we have much more to learn of the acute and chronic toxicity of these substances to all stages of animal and plant life histories. Acute studies are relatively easy to perform, and tell us nothing of the effects of prolonged exposure, but it is important to determine the relative sensitivity of at least the common species, since the loss of one such species by poisoning can have far reaching effects[99a]. Just as chronic exposure may be important, the effect of exposures to very high doses for very short periods is relatively unknown: because of inequalities in the coastline, adverse winds, or the release of abnormally large amounts

17

of toxic substances in emergencies, high concentrations may occur from time to time and may produce demonstrable damage in the field[X]. The definition of effluent standards would help to prevent such events, but they do have another value which is frequently overlooked. In a sense, they may be taken to represent an order of magnitude which the regulating body is prepared to accept for the receipt of an effluent into the waters for which it is responsible. By this means, an industry knows from the outset approximately what measures it will have to undertake in this respect and much time-wasting and costly preliminary discussion is avoided. Unfortunately however, whether the consent for an effluent is arrived at by the use of water quality standards or by investigation of the individual case, the conditions set are invariably regarded by the particular industry as immutable no matter what circumstances may arise subsequently. Such attitudes prevent either an easing of the consent conditions where a suitable post-commissioning investigation has shown it to be possible or a curtailment where undesirable environmental changes indicate the necessity. Inevitably, stricter constraints, than would otherwise be necessary, are imposed in these conditions.

The administration of effluent consents, be they as formal standards or consent conditions stipulated with reference to the particular situation, poses very real problems. If we neglect for the moment what parameters should be monitored, certain contrasts with the well managed radioactive effluents are obvious. Relatively few industries, apart from the very best, monitor the composition of their effluent on a regular basis and keep adequate records of the amount and quality of the effluent released, even though such a monitoring programme could lead to an improvement in the financial return from the plant. An overall improvement of waste control could be achieved by the maintenance of such logs and by the routine, but intermittent inspection, as is the case with the nuclear power industry. The more obvious effluent and environmental parameters, e.g. pH and temperature, could be monitored readily by means of automated equipment accessible only to the regulating body, but many chemical components may not be susceptible to this treatment; here, spot samples could and should be taken at intermittent though not fixed intervals. Such samples should be analysed chemically and subjected to bioassay. In the latter case, the acute test, as a 24, 48 or 96 hr L.C.$_{50}$, has great potential value, for, if its relationship to the effects of chronic exposure is properly understood, it offers a simple, cheap method of monitoring effluent quality. Few effluents have a constant or near constant composition, but such methods would cover all released except the grossly excessive which results from unscrupulousness or an uninformed act of carelessness. The latter can be obviated only by a programme of education for the process operatives, but the former is difficult to detect except in those circumstances where the effluent has a character readily measured in samples taken from the environment. The problem of monitoring the environment, like that of the effluent itself, may be relatively simple with respect to such parameters as temperature, oxygen concentration, pH and clarity of the water. The problem may be more difficult where trace metals and exotic chemicals are involved and may depend upon the adequate selection of stations where the concentration of these materials in the biota or the sediments may be measured. Here, as in much else of this work, much depends upon the selection of stations for the monitoring programme. One can only make what appear to be reasonable assumptions in this respect and yet, in time, find that a critical station was excluded from the original programme[143, X].

References

1. Klein, L. (1966). "River Pollution. III. Control", Butterworths, London.
2. Jeger, L. (1970). "Taken for Granted", Report of the Working Party on Sewage Disposal. H.M.S.O., Ministry of Housing and Local Government.
3. Tarzewell, C. M. (1969). In "Proc. 2nd Annual N.E. Regional Antipollution Conference, University of Rhode Island, July 22–25, 1969", pp. 1–4.
4. Preston, A. (1964). Rep. Challenger Soc. 16, 34–35.
5. Pritchard, D. W. (1959). In "Trans. 2nd Seminar on Biological Problems of Water Pollution, April 20–24, 1959", 11 pp. U.S. Public Health Service, Robert A. Taft Sanitary Engineering Center, Cincinnati 26, Ohio.
5a. International Commission on Radiological Protection. (1955). Brit. J. Radiol. N.S. (Suppl.) 6, 1–92.
6. Ancellin, J., Avargues, M. and Bovard, P. (1970). In "F.A.O. Technical Conference on Marine Pollution and its Effects on Living Resources and Fishing", F.I.R.: MP/70/E-69, 12 pp. Rome, 9–18th December, 1970.
7. Wray, E. T. (1969). H.M.S.O., U.K.A.E.A., A.H.S.B. (RP) R97.
8. Straub, C. P. (1959). In "Trans. 2nd Seminar on Biological Problems in Water pollution, April 20–24, 1959", pp. 33–41. U.S. Public Health Service, Robert A. Taft Sanitary Engineering Center, Cincinnati 26, Ohio.
9. Hawes, F. B. (ed.). (1968). "Hydrobiological Studies in the River Blackwater in Relation to the Bradwell Nuclear Power Station", Central Electricity Generating Board, London.
9a. Carritt, D. E. et al. (1959). "Radioactive Waste Disposal into Atlantic and Gulf Coast Waters", Publ. No. 655. National Acad. of Sciences. Nat. Res. Council, Washington, D.C.
10. Pritchard, D. W. (1952). John Hopkins University. Chesapeake Bay Institute, Tech. Rept. No. III. Reference 52–2, 56 pp.
11. Burt, W. V. and Marriage, L. D. (1957). Sewage ind. Wastes. 29, 1385–1389.
12. Bowles, P., Burns, R. H., Hudswell, F. and Whipple, R. T. P. (1958). "Exercise Mermaid", H.M.S.O., U.K.A.E.A., A.E.R.E. E/R2625.
13. McKinley, W. R., Brooks, D. C. and Westley, R. E. (1959). Washington Department of Fisheries. Fisheries Res. Pap. 2, 84–87.
14. Pritchard, D. W. (1959). John Hopkins University. Chesapeake Bay Institute. Tech. Rept. No. 18. Ref. No. 59–3, 72 pp.

15. Carpenter, J. H. (1960). *In* "Proc. 33rd. Conf. Maryland-Delaware Water and Sewage Assoc., 1960", pp. 62–78.
16. Pritchard, D. W. and Carpenter, J. H. (1960). *Bull int. Ass. Scient. Hydrol.* **20**, 37–50.
17. Delaware State Water Pollution Commission (1961). "Dispersion Studies on the Delaware River Estuary Model and Potential Applications Toward Stream Purification Capacity", 224 pp.
18. Bowden, K. F. (1962). *In* "Proc. 1st International Conference on Water Pollution Research", Section 3, Paper 33, 15 pp. Pergamon, Oxford.
19. Preddy, W. S. (1962). *In* "Proc. 1st International Conference on Water Pollution Research", Section 3, Paper 42, 19 pp. Pergamon, Oxford.
20. Scobie, J. and Porter, C. J. (1964). U.K.A.E.A., P.G. Memorandum No. 373 (CC). Unclassified.
21. Waldichuk, M. (1965). *J. Fish. Res. Bd. Can.* **22**, 801–822.
22. Waldichuk, M. (1965). *In* "Proc. 2nd International Water Pollution Research Conference, Tokyo, 1964", pp. 133–166. Pergamon, Oxford.
23. Neal, V. T. (1966). *In* "Proc. 10th Conference Coastal Engineering (Tokyo)", Vol. 2, 1463–1480. Pergamon, Oxford.
24. Abraham, G. and Van Dam, G. C. (1970). *In* "F. A. O. Technical Conference on Marine Pollution and Its Effect on Living Resources and Fishing", F.I.R.: MP/70/E-1 Rome. 9–18 1970.
25. Fischer, H. B. and Brooks, N. H. (1970). *In* "F.A.O. Technical Conference on Marine Pollution and its Effects on Living Resources and Fishing", F.I.R.: MP/70/R-4. 16 pp. Rome: 9–18 December, 1970.
26. Simmons, H. B. and Hermann, F. A. (1970). *In* "F.A.O. Technical Conference on Marine Pollution and its Effects on Living Resources and Fishing", F.I.R.: MP/70/E-103. 8 pp. Rome: 9–18 December 1970.
27. Talbot, J. W. (1970). *In* "F.A.O. Technical Conference on Marine Pollution and its Effects on Living Resources and Fishing", F.I.R.: MP/70/E-43. 18 pp. Rome: 9–19 December 1970.
28. Weidemann, H. and Sendner, H. (1970). *In* "F.A.O. Technical Conference on Marine Pollution and its Effects on Living Resources and Fishing", F.I.R.: MP/70/R-35. Rome: 9–18 December 1970.
29. Wolff, P. M. and Hansen, W. (1970). *In* "F.A.O. Technical Conference on Marine Pollution and its Effects on Living Resources and Fishing", F.I.R.: MP/70/E-46. 9 pp. Rome: 9–18 December 1970.
30. Zats, V. I. (1970). *In* "F.A.O. Technical Conference on Marine Pollution and its Effects on Living Resources and Fishing", F.I.R.: MP/70/E-48. 7 pp. Rome: 9–18 December 1970.
31. Werner, A. E. and Hyslop, W. F. (1967). *J. Fish. Res. Bd. Can.* **24**, 2137–2153.
31a. Pearl, I. A. and Benson, A. K. (1940). *Paper Trade J.* **111**, 35–36.
32. Ketchum, B. H. (1955). *Sewage ind. Wastes.* **27**, 1288–1296.
33. Robson, J. E. (1958). *J. appl. Bact.* **19**, 243–246.
34. Putman, J. T., Wildblood, A. M. and Robson, J. E. (1956). *Wat. sanit. Engr.* Sept./Oct., 99–101.
35. Keenan, C. J. *et al.* (1965). "Current Observations in Cordova Bay and Predictions on Sewage Dispersal", (Unpublished memo.), *Fish. Res. Bd. Can. Biol. St., Nanaimo, B.C.*
36. Waldichuk, M. (1966). *In* "Proc. 3rd International Conference on Water Pollution Research", Section 3, Paper 13: 22 pp. Pergamon, Oxford.

37. Scherz, J. P., Graff, D. R. and Boyle, W. C. (1969). *Photogramm. Engng.*, January 1969, 38–43.
38. Pritchard, D. W. (1958). *In* "Proc. 2nd U.N. Geneva Conference, 1958", pp. 410–414.
39. Pritchard, D. W. (1960). *In* "Disposal of Radioactive Wastes in the Sea", pp. 229–248. I.A.E.A., Vienna.
40. Mauchline, J. (1963). H.M.S.O., U.K.A.E.A., AHSB (RP)R 27: 70 pp.
41. Mauchline, J. and Templeton, W. L. (1963). *Nature, Lond.* **198**, 623–626.
42. Osterberg, C., Kulm, L. D. and Bryne, J. V. (1962). *Science, N.Y.* **13** 916–917.
43. Osterberg, C. (1965). *Ocean Sci. Ocean Engng.* **2**, 968–979.
44. Osterberg, C., Cutshall, N. and Cronin, J. (1965). *Science, N.Y.* **15**, 1585–1587.
45. Park, K., George, M., Miyake, Y., Saruhashi, K., Katsauragi, Y. and Kanazawa, T. (1965). *Nature, Lond.* **208**, 1084–1085.
46. Curl, H., Cutshall, N. and Osterberg, C. (1965). *Nature, Lond.* **205**, 275–276.
47. Jennings, D., Cutshall, N. and Osterberg, C. (1965). *Science, N.Y.* **148**, 948–950.
48. Cutshall, N., Johnson, V. and Osterberg, C. (1966). *Science, N.Y.* **152**, 202–203.
49. Osterberg, C., Cutshall, N., Johnson, V., Cronin, J., Jennings, D. and Frederick, L. *In* "Disposal of Radioactive Wastes in the Sea", pp. 321–335, I.A.E.A., Vienna.
50. Jennings, C. D. and Osterberg, C. (1969). *In* "Ecological Studies of Radioactivity in the Columbia River Estuary and Adjacent Pacific Ocean", Progress Report, Dept. of Oceanography, Oregon State Univ., Corvallis, 69–9, 222–228.
51. Peeters, E. and Mertens, M. (1970). *In* "F.A.O. Technical Conference on Marine Pollution and its Effect on Living Resources and Fishing", F.I.R.: MP/70/E-29: 8 pp. Rome, 9–18 December 1970.
52. Joseph, J. (1970). *In* "F.A.O. Technical Conference on Marine Pollution and, its Effects on Living Resources and Fishing", F.I.R.: MP/70/Rep. 2. Rome 9–18 December 1970.
53. Johnson, V., Cutshall, N. and Osterberg, C. (1967). *Wat. Resour. Res.* **3**, 99–102.
54. Gross, M. G. (1970). *In* "F.A.O. Technical Conference on Marine Pollution and its Effects on Living Resources and Fishing", F.I.R.: MP/70/R-5. Rome, 9–18 December 1970.
55. Olausson, E. (1970). *In* "F.A.O. Technical Conference on Marine Pollution and its Effects on Living Resources and Fishing", F.I.R.: MP/70/R-31. Rome, 9–18 December, 1970.
56. Perkins, E. J. and Williams, B. R. H. (1966). H.M.S.O., U.K.A.E.A., P.G. Report 587 (CC).
57. O'Connor, B. A. and Croft, J. E. (1967). *Effluent and Water Treatment Journal.* July 1967, 365–374.
58. Osterberg, C. L., Pearcy, W. G. and Curl, H. (1964). *J. mar. Res.* **22**, 2–12.
59. Perkins, E. J., Williams, B. R. H. and Gorman, J. (1966). H.M.S.O., U.K.A.E.A., P.G. Report 752 (CC): 17 pp.
60. Chipman, W. A., Rice, T. A. and Price, T. J. (1958). *Fishery Bull. Fish Wildl. Serv. U.S.* **58**, 279–292.

61. Chipman, W. A. (1959). *In* "Trans. 2nd Seminar on Biological Problems in Water Pollution, April 20–24, 1959", pp. 8–14. U.S. Public Health Service, Robert A. Taft Sanitary Engineering Center, Cincinnati 26, Ohio.
62. Donaldson, L. R. (1959). *In* "Trans. 2nd Seminar on Biological Problems in Water Pollution, April 20–24, 1959", pp. 1–7. U.S. Public Health Service, Robert A. Taft Sanitary Engineering Center, Cincinnati 26, Ohio.
63. Jones, R. F. (1960). *Limnol. Oceanogr.* 5, 312–325.
64. Mauchline, J. and Taylor, A. M. (1964). *Limnol. Oceanogr.* 9, 303–309.
65. Osterberg, C. (1962). *Science, N.Y.* 138, 529–530.
66. Osterberg, C. (1962). *Limnol. Oceanogr.* 7, 478–479.
67. Osterberg, C., Small, L. and Hubbard, L. (1963). *Nature, Lond.* 197, 883–884.
68. Osterberg, C., Pattullo, J. and Pearcy, W. (1964). *Limnol. Oceanogr.* 9, 249–257.
69. Osterberg, C., Pearcy, W. and Kujala, N. (1964). *Nature, Lond.* 204, 1006–1007.
70. Carey, A. G., Pearcy, W. G. and Osterberg, C. L. (1966). *In* "Disposal of Radioactive Wastes into Seas, Oceans and Surface Waters", pp. 303–319. I.A.E.A., Vienna.
71. Osterberg, C., Carey, A. G. and Pearcy, W. (1966). *In*. "Proc. 2nd International Oceanographic Congress, Moscow", No. 320—Sym. IV.
72. Wallauschek, E. and Lutzen, J. (1964). "Study of Problems Relating to Radioactive Waste Disposal into the North Sea. **II.** General Survey on Radioactivity in Sea Water and Marine Organisms", Health and Safety Office, European Nuclear Energy Agency, O.E.C.D., 1–96. Paris 1964.
73. Renfro, W. C. and Osterberg, C. (1969). *In* "Proc. 2nd National Symposium on Radioecology. Ann. Arbor, Michigan 1967", pp. 372–378 U.S.A.E.C. Div. Tech. Info. Ext., 1969.
74. Ketchum, B. H. and Bowen, V. T. (1958). *In* "Peaceful Uses of Atomic Energy. Proc. 2nd International Conference held at Geneva Sept. 1958", U.N. Publication, 402—OIC724: 11 pp.
75. Osterberg, C., Carey, A. G. and Curl, H. (1963). *Nature, Lond.* 200, 1276–1277.
76. Pearcy, W. G. and Osterberg, C. (1964). *Nature, Lond.* 204, 440–441.
77. Riley, J. P. and Segar, D. A. (1970). *J. mar. biol. Ass. U.K.* 50, 721–730.
78. Haderlie, E. C. (1970). *Proc. Challenger Soc.* 4, 81–82.
79. Halstead, B. W. (1970). *In* "F.A.O. Technical Conference on Marine Pollution and its Effects on Living Resources and Fishing", F.I.R.: MP/70/R-6. Rome, 9–18 December 1970.
80. Suess, M. J. (1970). *In* "F.A.O. Technical Conference on Marine Pollution and its Effects on Living Resources and Fishing", F.I.R.: MP/70/E-42. Rome, 9–18 December 1970.
81. Blumer, M. J. (1970). *In* "F.A.O. Technical Conference on Marine Pollution and its Effects on Living Resources and Fishing", F.I.R.: MP/70/R-1. Rome, 9–18 December 1970.
82. Nitta, T. (1970). *In* "F.A.O. Technical Conference on Marine Pollution and its Effects on Living Resources and Fishing", F.I.R.: MP/70/R-16. Rome, 9–18 December 1970.
83. Wellings, S. R., Chuinard, R. G. and Bens, M. (1965). *Ann. N.Y. Acad. Sci.* 126, 479–501.
84. Kiyoura, R. (1962). "Minimata Disease and Water Pollution", Proc. 1st International Conference on Water Pollution Research, Section 3, Paper No.

36: 13 pp. Held in London, 3–7 September 1962. Pergamon, Oxford. (N.B. This paper *must* be read in conjunction with that of Moore, 1962).

85. Moore, B. (1962). "Minimata Disease and Water Pollution", Proc. 1st International Conference on Water Pollution Research, Section 3. Discussion of Paper No. 36 (which must be read in conjunction with this discussion). Pergamon, Oxford.

86. Albert, A. (1968). "Selective Toxicity", Methuen, London.

87. Thomas, E. (1799). *Mem. Med. Soc. Lond.* **5**, 94–111.

88. Hueper, W. C. (1963). *Ann. N.Y. Acad. Sci.* **108**, 961–1038.

88a. Friend, M. and Trainer, D. O. (1970). *Science N.Y.* **170**, 1314–1316.

88b. Vos, J. G. and Driel-Groothenhuis, L. van (1972). *Sci. Total. Environ.* **1**, 289–302.

89. Alexander, W. B., Southgate, B. A. and Bassindale, R. (1931). *D.S.I.R. Water Pollution Research. Tech. Pap.* No. 2, 60 pp.

90. Alexander, W. B., Southgate, B. A. and Bassindale, R. (1935). *D.S.I.R. Water Pollution Research. Tech. Paper* No. 5, 171 pp.

91. Southgate, B. A. *et al.* (1938) (1961). *D.S.I.R. Water Pollution Research. Tech. Paper* No. 7, 337 pp.

92. Perkins, E. J., Williams, B. R. H., Bailey, M., Hinde, A. and Gorman, J. (1963–66). H.M.S.O., U.K.A.E.A., P.G. Reports 500(CC), 550(CC), 551(CC), 577(CC), 578(CC), 487(CC), 604(CC), 605(CC), 611(CC), 650(CC) 752(CC), 753(CC) and U.K.A.E.A., Chapelcross Works H.P. and S. Note No. 2(12 pts).

93. Smith, J. E. (1968). "'Torrey Canyon' Pollution and Marine Life", University Press, Cambridge.

94. Howells, G. P., Eisenbud, M. and Kneip, T. J. (1970). *In* "F.A.O. Technical Conference on Marine Pollution and its Effects on Living Resources and Fishing", F.I.R.: MP/70/E-18. 8 pp. Rome, 9–18 December 1970.

95. Kneip, T. J., Howells, G. P. and Wrenn, M. E. (1970). *In* "F.A.O. Technical Conference on Marine Pollution and its Effects on Living Resources and Fishing", F.I.R.: MP/70/E-20: 11 pp. Rome, 9–18 December 1970.

96. MacGinitie, G. E. (1935). *Am. Midl. Nat.* **16**, 629–765.

97. Cole, H. A. (1970). *In* "F.A.O. Technical Conference on Marine Pollution and its Effects on Living Resources and Fishing", F.I.R.: MP/70/R-20. 12pp. Rome, 9–18 December 1970.

98. Crapp, G. B. (1970). *In* "Symposium on the Ecological Effects of Oil Pollution on Littoral Communities. Held at the Zoological Society, London, 30th November–1st December 1970", (E. Cowell, ed.), Institute of Petroleum, London.

99. Buchanan, J. B. (1966). *J. mar. biol. Ass. U.K.* **46**, 97–111.

99a. Lewis, J. R. (1970). *In* "F.A.O. Technical Conference on Marine Pollution and its Effects on Living Resources and Fishing", F.I.R.: MP/70/E-22. Rome, 9–18 December 1970.

100. James, A. and Head, P. C. (1970). *In* "F.A.O. Technical Conference on Marine Pollution and its Effects on Living Resources and Fishing", F.I.R.: MP/70/E-19: 6 pp. Rome, 9–18 December 1970.

101. Glover, R. S., Robinson, G. A. and Colebrook, J. M. (1970). *In* "F.A.O. Technical Conference on Marine Pollution and its Effects on Living Resources and Fishing", F.I.R.: MP/70/E-55. Rome, 9–18 December 1970.

101a. Russell, F. S., Southward, A. J., Boalch, G. T. and Butler, E. I. (1971). *Nature, Lond.* **234**, 468–470.
101b. Idler, D. R. (1969). Keynote address presented at the Pollution Conference of the Atlantic Section, The Chemical Institute of Canada, St. Mary's University, Halifax, N.S., 24–26th August 1969. Chemistry in Canada (and *Albright Magazine*, February 1970: 3).
102. Fujiya, M. (1964). *In* "Conf. Water Poll. Res.", Sect. III. No. 5: 6 pp. Tokyo, Japan, Aug. 24–28, 1964.
103. Alderdice, D. F. (1967). *Canadian Fisheries Reports*, No. 9, 33–39.
104. Smith, R. I. (1959). *Mar. Biol., Biol. Colliquium* 1959, Oregon State University, Corvallis.
105. Klekowski, R. Z. and Duncan, A. (1966). *Verh. Verein theor. angew. Limnol.* **16**, 1753–1760.
106. Newell, R. C. (1964). *Proc. zool. Soc. Lond.* **142**, 85–106.
107. Spencer Davies, P. (1966). *J. exp. mar. Biol. Ecol.* **1**, 271–281.
107a. Spencer Davies, P. (1967). *J. mar. biol. Ass. U.K.* **47**, 61–74.
108. Walne, P. R. (1966). *Fish. Invest., Lond. Ser.* II, **25**(4): 53 pp.
109. Perkins, E. J. (1970). *Chemy Ind.* 1970, 14–22.
110. Erlander, S. R. (1969). *Science Journal.* **5**(5), 60–65.
111. Raymont, J. E. G. and Shields, J. (1962). *In* "Proc. 1st International Conference on Water Pollution Research" Section 3, Paper No. 43: 13 pp. Pergamon, Oxford.
112. Cairns, J., Waller, W. T. and Smrchek, J. C. (1968). *Revta Biol., Lisb.* **7**, 75–91.
113. Wilson, D. P. (1968). *J. mar. biol. Ass. U.K.* **48**, 177–182.
114. Alderdice, D. F. (1962). *In* "Trans. of 3rd Seminar on the Biological Problems of Water Pollution", pp. 320–325, U.S. Public Health Service, Robert A. Taft Sanitary Engineering Center, Cincinnati, Ohio. P.H.S. Publ No. 999-WP-25.
115. Alderdice, D. F. (1963). *J. Fish. Res. Bd. Can.* **20**, 525–550.
116. Howard, T. E. and Walden, C. C. (1969). *Pulp Pap. Can.* **72**(1): 73–77.
117. Woelke, C. E. (1968). *NW Sci.* **42**, 125–133.
118. Shirer, H. W., Cairns, J. and Waller, W. T. (1968). *Water Res. Bull.* **4**, 27–43
119. Lennon, R. E. (1964). *U. S. Dept. of the Interior, Bureau of Sport Fisheries and Wildlife, Circular* **185**, 1–15. Washington D.C.
120. Federal Water Pollution Control Adminstration. (1968). "Water Quality Criteria", Report of the National Technical Advisory Committee to the Secretary of the Interior: 66–92. Washington, D.C. April 1, 1968.
121. Oregon Administrative Rules, State Sanitary Authority (1967). "Water Pollution", Ch. 334: 1–2 g. 7.1.67. Salem, Oregon.
122. Water Pollution Control Commission State of Washington. (1967). "Water Quality Standards for Interstate and Coastal Waters of the State of Washington and a Plan for Implementation and Enforcement of Such Standards", A Regulation: 23 pp., Dec. 4 1967, Olympia, Washington.
123. Water Pollution Control Commission, State of Washington. (1968). "Water Pollution Control Laws" 15 pp., Sept. 1968, Olympia, Washington.
124. Bittel, R. and Lacourly, G. (1970). *In* "F.A.O. Technical Conference on Marine Pollution and its Effects on Living Resources and Fishing", F.I.R.: MP/70/E-51: 7 pp. Rome, 9–18 December 1970.
125. Cuperus, K. W. (1970). *In* "F.A.O. Technical Conference on Marine

Pollution and its Effects on Living Resources and Fishing", F.I.R.: MP/70/ E-54: 3 pp. Rome, 9–18 December 1970.

126. Lesaca, R. M. (1970). *In* "F.A.O. Technical Conference on Marine Pollution and its Effects on Living Resources and Fishing", F.I.R.: MP/70/R-7: 4 pp. Rome, 9–18 December 1970.

127. Lopuski, J. (1970). *In* "F.A.O. Technical Conference on Marine Pollution and its Effects on Living Resources and Fishing", F.I.R.: MP/70/R-28: 6 pp. Rome, 9–18 December 1970.

128. Moore, G. (1970). *In* "F.A.O. Technical Conference on Marine Pollution and its Effects on Living Resources and Fishing", F.I.R.: MP/70/R-15: 25 pp. Rome, 9–18 December 1970.

129. Pontavice, E. du (1970). *In* "F.A.O. Technical Conference on Marine Pollution and its Effects on Living Resources and Fishing", F.I.R.: MP/70/ E-80: 8 pp. Rome, 9–18 December 1970.

130. Ministry of Housing and Local Government (1968). "Standards of Effluents to Rivers with Particular Reference to Industrial Effluents", H.M.S.O., London.

131. Bonderson, P. (1962). "Scientific Parameters of Marine Waste Discharge", Discussion of Paper No. 37 by H. F. Ludwig and B. Onodera. *In* "Proc. 1st International Conference on Water Pollution Research", Section 3: 6 pp. Pergamon, Oxford.

132. Ludwig, H. F. and Onodera, B. (1962). *In* "Proc. 1st International Conference on Water Pollution Research", Section 3, Paper No. 37: 17 pp. Pergamon, Oxford.

133. Waldichuk, M. (1967). *Can. Fish. Rep.* **9**, 24–32.

134. Millipore Filter Corporation. (1960). "Microbiological Analysis of Water and Milk", Bedford, Mass.

135. Hoak, R. D. (1962). *In* "Proc. 1st International Conference on Water Pollution Research", Section 1, Paper No. 7: 19 pp. Pergamon, Oxford.

136. Gafford, R. D. (1970). *In* "F.A.O. Technical Conference on Marine Pollution and its Effects on Living Resources and Fishing", F.I.R.: MP/70/E-15. Rome, 9–18 December 1970.

137. Noble, V. E. (1970). *In* "F.A.O. Technical Conference on Marine Pollution and its Effects on Living Resources and Fishing", F.I.R.: MP/70/E-96. Rome, 9–18 December 1970.

138. Royal Commission on Sewage Disposal. 8th Report Vol. II. Appendix. Cmd. 694. H.M.S.O. London. 1913.

139. Ainsworth, G. (1968). *Process Biochemistry* **3** (12), 11–14.

140a. Klein, L. (1968). "River Pollution. I. Chemical Analysis", Butterworths, London.

140b. Klein, L. (1962). "River Pollution. II. Causes and Effects", Butterworths, London.

141. Eckenfelder, W. W. (1970). *In* "F.A.O. Technical Conference on Marine Pollution and its Effects on Living Resources and Fishing", F.I.R.: MP/70/ R-22. Rome, 9–18 December 1970.

142. Pearson, E. A. (1972). *In* "Coastal Zone Management" (J.F. Peel Brahtz, ed.), Wiley, New York, London, Sydney and Toronto.

143. Johnston, R. (1973). *Proc. Challenger Soc.* 1973. (In press).

X. Perkins, E. J. Previously unpublished work.

16

The Biological Effects of Waste Disposal

Clearly, an individual pollutant may influence living organisms in a number of ways, and the individual effect may not be the same for all phyla, e.g. a suspended solid may influence a fish by clogging its gills and an alga by reducing the amount of sunlight available for photosynthesis. The polluted situation may arise from mechanical effects due to suspended and settled solids, or a disturbance of the ionic ratios of sea water, or directly toxic effects, or by a change in the physical regimen of the habitat. Few individual pollutants have only one of these attributes, and most effluents are complex in effect, e.g. sewage has a high B.O.D. and therefore lowers the oxygen content of the water, its ammonical fraction is both toxic and has an oxygen demand, and it has a solid content which may settle in those areas where turbulence is poor.

In a situation where widespread mortality is overt the problem is not scientific since the effect and the need for a remedy are equally obvious. It is the possible physiological consequences in apparently non-lethal situations which are so important to fishery and other interests and which, in general, are so poorly understood at the present time. These effects include inhibition or promotion of growth; lowered reproductive potential, i.e. lowered fecundity, poor brood survival, or an inability to settle at the end of the larval phase; behavioural changes, e.g. a disturbance of the predator/prey relationship which results when a sub-lethal concentration of a pollutant affects the ability of a prey to resist a predator, or the ability of a predator to attack the prey; inhibition or other effects in the crustacean moult cycle; effects upon respiration; and effects upon the mechanisms of osmo-regulation and ionic regulation. In each of these cases, one can visualize that a long term population change may result.

Certain aspects of these problems, e.g. suspended and sedimented solids, pH, and oxygen content of the water, are susceptible to individual treatment and will be considered thus, in other cases, it is preferable to consider the situation as a whole.

Suspended Solids

The ecological effects of suspended solids may be due to one of six effects, thus:

1. Mechanical or abrasive action (e.g. clogging of gills, and irritation of tissues).
2. Blanketing action or sedimentation.
3. Reduction of light penetration.
4. Availability as a surface for growth of bacteria and fungi.
5. Adsorption and/or absorption of various chemicals.
6. Reduction of temperature fluctuations.

The relative importance of these effects is variable, some species are more sensitive than others, and in the individual species not all stages in the life history are equally susceptible[1].

There is a paucity of information relevant to the marine environment, but it is worth making some points from the considerable body of knowledge of the freshwater habitat. Fish gills are important sites of nitrogenous and chloride excretion, although the chloride regulation of sunfish is not impaired when its gills are damaged by alkyl benzene sulphonate[2]. Nevertheless, wood fibre suspensions at concentrations between 100 and 800 p.p.m. reduced the active metabolism of fathead minnows by 1·3 mg/kg per minute at 21°C for every doubling of the concentration[3]. However, the lethal action of wood fibres depends upon type and the dissolved oxygen concentration. In 96-hour bioassays, the mortality of minnows in a 2000 p.p.m. suspension of conifer groundwood, conifer kraft and aspen groundwood was 78%, 24% and 4% respectively. Mortality increased with increased temperature and decreased oxygen concentrations[4]. Trout can withstand exposure to 50–200 p.p.m. of wood fibre for nine months and evince only a progressive decrease in growth rate with increased concentration[5]; rainbow trout exposed to suspensions of kaolin and diatomaceous earth at concentrations of 30, 90 and 270 p.p.m. showed no effect, a little mortality and >50% mortality in 2–12 weeks respectively[6]. Effects of blanketing by settled sediment, photosynthetic inhibition, bacterial development and adsorption and/or absorption of chemicals upon the surfaces provided have all been observed in the freshwater environment[1].

In the sea, the feeding rate (as indicated by the pumping rate) of the oyster, *Crassostrea virginica* is high and most efficient when the number of food organisms is small, and the turbidity is low. However, an addition, of 0·1 g/litre, i.e. 100 p.p.m. of silt, causes an average reduction of 57% in pumping rate: severe cases show a 90% reduction. Prolonged exposure to excessive amounts of suspended material will cause death in both

Crassostrea and *Mytilus edulis*; even if removed to clean sea water the effect may be irreversible[7]. Bivalve larvae are also affected by the sediment load of the water (p. 94).

In studies upon the Japanese bivalves *Tapes semidecussata* (Japanese little-neck clam), *Meretrix meretrix lusoria*, *Crassostrea gigas* (oyster) and *Mytilus edulis* (mussel) exposed to suspension of organic and inorganic materials, viz. algae, diatomaceous earth, kaolin, bentonite, talc, shale, charcoal powder and soluble starch, it was found that they never stopped pumping, whatever the quality of these particles, at concentrations ranging from 10–20 mg/litre (i.e. 10–20 p.p.m.): moreover all these materials were ingested. Each of these species showed little reduction in pumping rate in concentrations of bentonite which ranged from 500 to 1000 mg/litre, but when the concentrations exceeded 10 mg/litre *Tapes semidecussata* and *Meretrix meretrix lusoria* commenced production of pseudo-faeces which increased in amount with concentration of the suspension. The amount of the true faeces did not increase at concentrations exceeding 10 p.p.m.[8].

In studies of the effect of wood fibres, produced by a paper-board mill, upon some elements of the estuarine fauna, viz. fish—*Cottus scorpius*, scorpion fish; *Pholis gunnellus*, gunnel; *Anguilla anguilla*, eel: crustacea—*Carcinus maenas*, shore crab; *Eupagurus bernhardus*, hermit crab; *Hyas araneus*, spider crab: molluscs—*Buccinum undatum*, whelk: and echinoderms; *Asterias rubens*, starfish—*Psammechinus miliaris*, sea-urchin, it was found that the echinoderms were poorly resistant and the fish and crustacea highly resistant to wood fibre suspension (Table 16.1). The failure of the

Table 16.1. The effect of chronic exposure of estuarine animals to ground-wood fibre suspensions[X]

Species	Wood fibre concentration (mg/litre)	Time to 50% mortality (days)
Buccinum undatum	2342	18
Hyas araneus juv.	1446	41
Hyas araneus (adult)	1509	> 59
Carcinus maenas (adult)	850	> 77
Eupagurus bernhardus	1252	> 59
Asterias rubens	1000	33

echinoderms was apparently related to a need to keep the external surfaces free of particulate materials and the tube feet were adversely affected; on removal to clean water, recovery and rehabilitation was poor. Failure in the whelk, *Buccinum*, apparently arose from a difficulty in cleansing fibres from the mantle cavity for protracted periods: *ca.* 50% of the dead whelks had a compacted mass of wood fibres and mucus in the mantle cavity[X].

Considerable amounts of waste from the Cornish china clay workings are discharged to Mevagissey and St. Austell Bays annually. Within a line drawn from the Dodman Point to Gribbin Head, approximately two-thirds of the sediments contain significant amounts of china clay, and one sixth of the area surveyed is covered by this material[9]. These deposits of china clay have profoundly affected the bottom fauna. Close inshore where the rate of deposition is greatest, the sediment is nearly sterile, but a little further offshore, where the rate of deposition is less, a rich *Echinocardium/filiformis* community is present: this community was characterized by *Echinocardium cordatum, Amphiura filiformis, Labidoplax digitata, Lumbrinereis impatiens, Melinna palmata, Pectinaria auricoma, Owenia fusiformis, Dosinia lupinus, Abra alba, Abra prismatica, Cultellus pellucidus, Spisula subtruncata* and *Turritella communis*. Further offshore in the fine gravel ("coral sand") or fine sand, which was little affected by china clay, the fauna was characterized by *Spatangus purpureus, Venus fasciata, Ophiothrix fragilis, Glycymeris glycymeris, Venerupis rhomboides, Gari tellinella* and *Amphioxus lanceolatus* at the Coral Sand stations, and *Echinocardium cordatum, Acrocnida brachiata, Nephtys hombergi, Dosinia lupinus, Venus striatula, Gari fervensis, Tellina fabula, Ensis ensis* and *Acanthocardia echinata* at the fine sand stations. The biomass of areas affected by china clay was greatly increased compared to those areas which were not: an increase largely due to sedentary polychaetes and small deposit feeding lamellibranchs. Flatfish, viz. plaice, *Pleuronectes platessa*; dab, *Limanda limanda*; and sole *Solea solea*, caught on the clay banks, fed actively upon the polychaetes and lamellibranchs[10]. Apparently, the clay waste deposition improved these bays as fish feeding grounds[10], but this conclusion should be viewed in the light of the relationship between soil composition and the success and fecundity of *Echinocardium cordatum* (p. 343).

In some areas, the dumping of sewage sludge has been practised for long periods. In 1904, Glasgow Corporation began such dumping off the Garroch Head, Isle of Bute, in the Firth of Clyde. Initially, 210 000 tons/annum were dumped, but this reached 1 000 000 tons by 1970. The operation is performed in water of 300 ft depth which permits the discharged sludge to become distributed over a wide area. In general, there is little evidence of obvious harm to the enviornment and no deterioration in fishing success has been observed. Nevertheless, at the centre of the area from which dispersal occurs, qualitative changes are indicated by the presence of large numbers of the polychaete *Cirratulus* spp. and of the tubificid oligochaete *Peloscolex* sp.; the latter is typical of heavily polluted areas in the Clyde estuary[11].

Dredge and sewage sludge wastes, originating in the New York Metropolitan area, are dumped at selected sites, in water of 10 m depth, in New

York Bight and Long Island Sound. These wastes were dumped at an annual rate of 8×10^6 tons and $9 \cdot 6 \times 10^6$ tons in the periods 1960–63 and 1964–68 respectively, and this material represents the largest source of sediments entering the North Atlantic Ocean from the North American continent. In those areas, *ca.* 20 sq. miles, which are particularly affected by the sewage sludge normal benthic populations are absent: in such areas, the sediments are rich in heavy metals and contain $> 10\%$ of organic matter in a dried sample (Fig. 16.1). These sediments are black and have

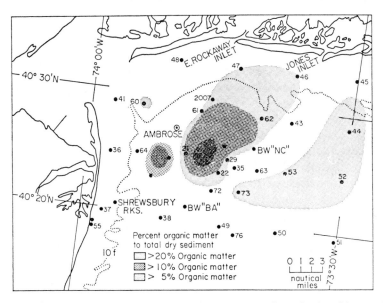

Fig. 16.1. Percentage of organic matter in dry sediments collected at benthic sampling stations in New York Bight. Biologically impoverished bottoms roughly conform to the area characterized by greater than 10% organic matter (after Pearce[12]).

an odour characteristic of sewer sludge, hydrogen sulphide and occasionally petrochemicals. During the summer months, the water above the dredge spoil and sewage disposal areas has impoverished oxygen concentrations. The sediments surrounding the impoverished areas are characteristically inhabited by the burrowing sea anemone, *Cerianthus americanus*, rhychocoeles, polychaetes, and the protobranch bivalves *Yoldia limatula* and *Nucula proxima*. To the east of the sewage sludge ground *Cerianthus*, and polychaetes from the families Ampharetidae, Cossuridae, Flabelligeridae, Maldanidae and Nereidae are lacking; instead several species of bivalve mollusc, including *Tellina agilis*, a species characteristic of clean sands, are present. The benthic fauna outside the dredge spoil area is characteristic of, and reflects the diversity of bottom types which surround it: for example, to the east, the fauna is very limited and characterized by the surf clam, *Spisula solidissima*, and *Tellina*, the polychaetes are dominated by *Prionospio malmgreni* and *Spiophanes bombyx*; the sand dollar, *Echinarachnius parma* is very common and comprises the greater proportion of the biomass in some samples. These species are all characteristic of medium to coarse sand found in high energy, wave swept, coastal environments.

The crab, *Cancer irroratus*, is widely and abundantly distributed in the New York Bight. Few living ones are taken alive in the sewage sludge area, but, here, during the summer months, when this species migrates offshore, many moribund individuals are taken. These crabs exhibit a considerable erosion of the exoskeleton and the gill chambers are filled with fine debris characteristic of these sediments. These results have been confirmed in the laboratory tests performed with *C. irroratus* and the lobster *Homarus americanus*; under these conditions, the symptoms appear within 6 weeks after the initial exposure[12].

In the Thames estuary, dumping of sewage sludge is carried on in the Sea Reach and the Black and Barrow Deeps[12a]. Here no effects comparable with those in New York Bight have been recorded, presumably due to the strong currents which prevent an accumulation on the bed at the site of dumping. Evidently, the influence of sewage sludge depends upon the amount dumped and degree of dispersion by the currents which influence the dumping ground.

Many substances which are soluble in fresh water, or in the effluent by which they are released to the sea, precipitate upon contact with the sea water, but little is known of their effect. The salts of aluminium, for example, are widely used as a flocculant, to sweep down unwanted materials, when fresh water of a high quality is required. The adsorption of many elements upon the sediments is a function of their Fe or Mn content (p. 121). Salts such as phosphates precipitate as the calcium salt when they reach the sea.

Such a loss of calcium may be important when it is recalled that one of the most important functions of calcium is to control the permeability of semipermeable membranes. In its absence these membranes lose their integrity, and become porous and leaky[13]: animals e.g. *Gunda ulvae* are unable to osmoregulate[14-17]. An increase in the relative content of Ca ions in water increases the ability of animals to survive in brackish water[18], whereas water which is deficient in Ca, inhibits volume regulation, as indeed does water which is deficient in potassium or magnesium[19, 20]; a similar effects are noted in fresh water molluscs[21]. Even minor changes in the environment may modify the volume regulating capacity of animals[18]; moreover, the coelenterates, echinoderms, polychaetes, lamellibranch and gastropod molluscs and ascidians have a poorly developed ability to regulate the ionic content of their body fluids. On the other hand, the decapod crustacea, cephalopods and fish have this ability well developed[22]. Clearly, in an ionically unbalanced situation many marine animals could find themselves in difficulty. Indeed, the relatively poor fauna of the Black Sea compared with that of the Mediterranean is probably due to such an imbalance (p. 182). Therefore, problems may arise in situations where the normal ionic balance of sea water is disrupted, and since estuarine waters may show some variation from the composition characteristic of the open sea, estuarine animals are stressed naturally to some degree. Sea water, in general, is a well balanced mixture of ions, which occur in the same relative proportions all over the world (p. 5). We are now in a position to examine a widely quoted fallacy. Frequently, those concerned with the engineering and chemical aspects of effluents state that because sea water contains salt to the value of 35‰ it is correct to release, to the environment, solutions at concentrations which are substantial in comparison with 35‰. Often such a solution is a single solute, and does not therefore represent a balanced addition to the environment; even if no precipitation occurs, the ionic balance of the receiving waters may well be changed markedly, especially when releases are made to interface currents.

Nevertheless, the composition of the dissolved salts, maintained within narrow limits, is critical to marine life[23, 24]. An increase of 1‰ in the concentration of the K^+ or Mg^{2+} ion in sea water will render the teredine *Bankia setacea* inactive[24]. Superficially, sodium sulphate seems harmless and the 24 hr L.C.$_{50}$ of *Eupagurus bernhardus* (which tends to accumulate SO_4^{2-} ions compared to sea water[22]) with respect to this salt is > 6400 p.p.m.[25]; however, a continuous exposure of this decapod to sea water containing an additional 500 p.p.m. of this salt (i.e. 162 p.p.m. Na^+ and 338 p.p.m. SO_4^{2-}) killed the test animals in 3 months whereas no death occurred in controls[X]. An addition of 500 p.p.m. of Na_2SO_4 represents an increase of 1·5% and 12·5% in the Na^+ and SO_4^{2-} ions respectively: a relatively small imbalance in the environmental ionic structure to cause death. The need to recognize this problem was highlighted at the F.A.O. Conference in Rome, 1970, where it was considered that the influence of toxicants upon the ionic balance, especially K^+ and Na^+, of animal sera should be investigated[26].

Another fallacy may be added to that which considers the composition of solutes added to sea water to have no influence upon the biota. All estuaries carry a variable sediment load, therefore it may be argued that high suspended solid loadings can be liberated in an effluent. If we disregard the water criteria in Chapter 15, the evidence given above (p. 514) indicates very clearly that estuarine animals are influenced even by moderately low sediment loadings. Looked at from the point of view of the individual estuary, the Solway Firth, turbulent and shallow though it is, only rarely, and for relatively short periods, carries, in the bottom waters, a sediment loading which exceeds 1 g/litre (i.e. 1000 p.p.m. w/v) and average loadings are very much less than this[27]. In the Blackwater estuary, Essex, which appears to carry a heavy burden of sediment, the sediment concentrations recorded in the cooling water system of Bradwell Power Station ranged from a maximum of 750 p.p.m. to a minimum of 14 p.p.m. w/v; from the estuary itself maximum and minimum values of 124 p.p.m. w/v and 6 p.p.m. w/v respectively were recorded[28].

pH Effects

Although the importance of pH is recognized widely in the literature, little is known of its direct effects, and from fresh water sources especially, a considerable amount is known of its indirect effects[29]. Of course, sea water is weakly buffered (p. 7) but it is insufficiently recognized that the buffering *is weak* and subject to natural change by respiration and photosynthesis.

The anions of the strong mineral acids, viz. HCl, H_2SO_4, HNO_3, are relatively innocuous and the essential factor is the hydrogen ion concentration. Goldfish will survive in a solution of sulphuric acid having a pH *ca.* 4·0 and in hydrochloric acid at a pH *ca.* 4·5. Sticklebacks are probably able to survive indefinitely in hydrochloric acid solutions of pH 5·2 and 5·4, and it appears that pH 5·0 is critical for most species of fresh water fish. In more acid solutions gill secretions are coagulated and asphyxia follows, or corrosion of the gills occurs[30]. Nevertheless, sticklebacks retreat from water more acid than pH 5·4 and more alkaline than pH 11·4[31]. The toxicity of the weaker acids is due to their anions or undissociated molecules; oxalic acid, for example, is effective not because of the hydrogen ions which it liberates, but because it precipitates calcium[32]. The limits of pH for the growth of most microorganisms range from pH 4·0 to pH 9·0, but most species have a fairly narrow range, favourable for growth, within these limits[33]. The foraminiferan *A. beccari* can survive more than 25 min, but less than 75 min, at pH 2·0, while it can tolerate pH 9·5 for more than 14 hr, but less than 37 hr[34]. However, solution of the foraminiferan test begins at pH 7·2 and is marked at pH 7·0[35]; fossil specimens exposed to sea water at pH 6·5 dissolve completely in 9 days[36]. The lethal limits of the shore crab, *Carcinus maenas*, were formerly considered to be pH 6·0 and pH 9·0[37]; the lower limit for a 24 hr exposure is pH 5·0[38], but pH 5·5 is fatal in 48 hr[39]. At these pH values, when the acid is hydrochloric, the exoskeleton loses its calcium and, in consequence, the animal is deprived of its power of locomotion. The eggs of herring develop normally within the pH range 6·7–8·7, but in sea water having a pH < 6·7 development is retarded[40]. Young chinook salmon subjected to a low pH in flowing sea water apparently tolerated a pH between 5·51 and 6·62 for a 72 hr exposure period, and the apparent tolerance to high pH was between 9·24 and 9·56. However, the value of 9·12 was probably critical and an increased exposure would probably have reduced the tolerance of these animals[41].

Within the lethal limits, small changes of pH can have pronounced physiological effects. The transport of a portion of an excised gill of the lamellibranch mollusc, *Ostrea circumpicta*, is slowed by as small a change as 0·4 units of pH; a 22 hr exposure at pH 5·9 (a lower limit of some pH authorizations) approximately halves the speed of crawl[42, 43] (Table 16.2).

Table 16.2. The effect on the mechanical activity of cilia of varying the pH of sea water by adding HCl: the relative mean speed of crawl after various periods in sea water at various pH (after Tomita[43])

pH of sea water, salt error not corrected	Time (hr) Relative speed of crawl (speed in normal sea water = 100)					
	$\frac{1}{2}$	1	2	4	7	22
8·2	99	96	90	87	106	120
7·8	94	89	85	83	90	101
7·3	99	92	91	88	89	90
6·8	91	87	85	86	87	90
5·9	77	63	61	63	64	65
5·5	71	55	56	58	59	55
5·4 (1)	49	27	22	19	19	21
5·5 (2)	41	21	10	0		

The pH of the body fluid of many marine animals is slightly more acid than the normal pH of sea water, ranging from 6·47–6·56 in *Ciona intestinalis* to 7·24–7·90 in *Octopus vulgaris* and 7·94 in *Scomber scombrus*. The lobster heart has a pH optimum of 7·4, but continues to beat normally over the range of pH 7·0–8·0: outside these limits pH can affect its tonus, amplitude and frequency. The cardiac and smooth muscles of other animals show similar sensitivity[22]. The blood of the shore crab, *Carcinus*, has a normal pH value of 7·53 ± 0·37[22], and upon immersion in sea water at pH 5·5, the pH of *Carcinus* blood falls likewise[39] (Fig. 16.2).

The oxygen carrying capacity of crustacean blood is affected, *in vitro*, by carbon dioxide partial pressures, pH and temperature, although the functional effect in the body is less certain. An increase in the CO_2 tension

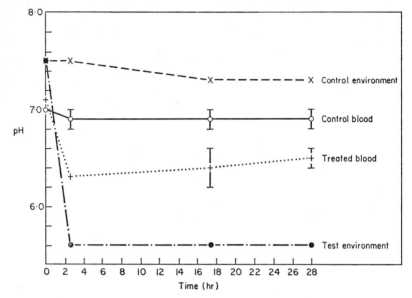

Fig. 16.2. Depression of the pH of the blood of a *Carcinus maenas* subjected to an environmental reduction of pH (after Parker[39]).

may or may not lead to an increase in the ventilation of crustacean arthropods. In a number of cases, viz. aquatic amphipods, the crabs *Trichodactylus* and *Eriocheir*, a low pH in the external medium provokes a response similar to those correlated with a high CO_2 concentration, while in *Astacus*, *Homarus gammarus*, and *Eriocheir* the effects of CO_2 seem to be directly attributable to its lowering of the pH, but simultaneous measurements of pH and pCO_2 have apparently not yet been made. It has been suggested that, in *Eriocheir*, changes in the blood pH brought about by alterations in its CO_2 and O_2 pressures are the immediate rate controlling factors for the pacemaker, but experimental proof is lacking[44].

In addition, to the direct effect of the hydrogen ion concentration, the pH of a particular body of water may have an important influence in potentiating the activity of any toxins which may be present. In fresh water, alkalies which dissociate to a high degree, appear not to be toxic below pH 9·0, but alkalis which are poorly dissociated are often toxic at a pH less than 9·0 and their toxicity is due to the cation or the undissociated molecule, e.g. NH_4OH, Carbon dioxide which is not bound as the HCO_3^- or CO_3^{2-} is highly toxic also, and to prevent adverse effects its concentration should not exceed 25 mg/litre. The acute toxicity of some heavy metals, e.g. Zn, Cd, increases as the pH is raised from 5 to 9, when the concentration of calcium and magnesium remains constant[29]. The acute toxicity of

Table 16.3. The median tolerance limits for guppies, *Lebistes*, exposed to neutralized and unneutralized whole Kraft pulp mill effluent (after Howard and Walden[45]).

	TL$_m$ as % effluent by volume, at exposures of		
	24 hr	48 hr	96 hr
Whole effluent	73·0	64·0	54·8
(neutralized)	62·0	48·3	48·3
	70·0	63·7	56·0
	52·0	52·0	52·0
Whole effluent	14·6	11·5	8·4
(unneutralized)	13·5	10·7	10·1

whole Kraft mill effluent to guppies, *Lebistes reticulatus*, is reduced about 5–6 times if the effluent is neutralized before the commencement of the test[45] (Table 16.3): similar results have been obtained in studies of whole effluent toxicity in relation to invertebrates characteristic of the Solway Firth[X].

Oxygen

The level of dissolved oxygen, in an estuary, is clearly of fundamental importance to its would be inhabitants, although response of individual species or a group of species may be highly variable. The mysids, as a group, are sensitive to reduced oxygen levels. On the Northumberland coast, *Neomysis integer* rarely occurs in those estuaries where pollution has depleted the oxygen content of the water[46, 47]. In New York Bight, the gammarid amphipods, so important elsewhere, were absent from sludge and spoil disposal areas, and on the periphery of these areas were represented only by occasional specimens of *Unciola irrorata* and *Monoculodes edwardsii*[12]. In waters with a low oxygen content, adjacent to a sulphite mill at Prince Rupert, British Columbia, the formerly rich bottom fauna of Porpoise Harbour has been replaced by the amphipod *Anisogammarus pugettensis* and the occasional individual of the isopod *Synidotea* sp. at the sea bed. The amphipods taken, at the surface, were predominantly *Allorchestes angustus*, and some *Anisogammarus pugettensis*. These amphipod species occur widely on the North American Pacific coast; although not reported previously from polluted and/or anoxic waters, *A. angustus* is common in warmer, shallow, brackish bays and estuaries among *Phyllospadix* and *Zostera*. *A. pugettensis*, an inhabitant of cold, slightly brackish waters, has been found in relatively stagnant water in mud, dead algal masses and dead eel grass in the Queen Charlotte Islands[48]. Similarly, *Jassa falcata* and corophiids were the most abundant animals in the polluted bays of Los Angeles and Long Beach[49]: amphipods may be used as indicators of enrichment[50].

Due to an increasing salinity below the halocline, which has in consequence become more stable, the bottom waters of the Baltic have become more stagnant since the early years of this century. Concurrently, an increased nutrient supply, presumably arising from pollution, increased the B.O.D. of the deeper water. As a result, the oxygen content of the deeper waters declined progressively from 1900 onwards[51] (Fig. 16.3), and from

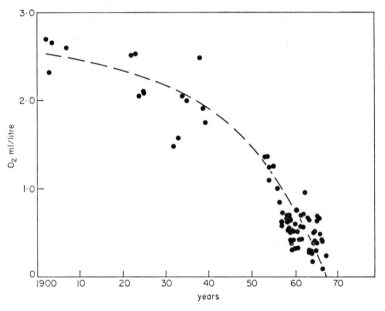

Fig. 16.3. Mean values of O_2 (ml/litre) below the halocline in Landsort Deep (F 78), 1902–67 (after Dybern[51]).

1957 onwards the deeper parts of the Gotland, Gdansk and Bornholm Basins, became anaerobic; H_2S was produced[51, 52] (Fig. 16.4) and catastrophic death of the bottom fauna resulted[52] (Fig. 16.5).

Just as low oxygen concentrations may kill the biota, super saturation of atmospheric gases, especially oxygen, in sea water, can be fatal also. In 1959, in Galveston Bay, Texas dissolved oxygen concentrations reached 250% of saturation and were associated with the death of 300 spotted seatrout, *Cynoscion nebulosus* (mean length 50 cm); in addition postlarval largescale menhaden, *Brevoortia patronus*; bay anchovies, *Anchoa mitchilli*; juvenile Atlantic croakers, *Micropogon undulatus*; small eels, probably *Myrophis punctatus*; and a longnose gar *Lepisosteus osseus* were affected[53].

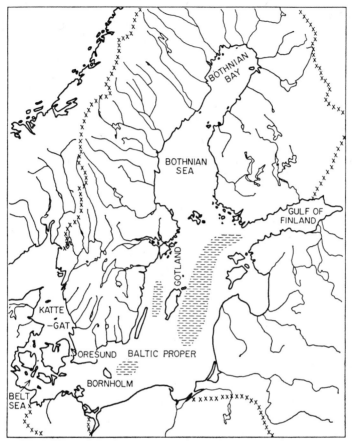

Fig. 16.4. Approximate distribution of H_2S containing bottom waters in September 1968 ≡≡≡ H_2S containing water (after Dybern[51]).

The oxygen concentrations of estuarine waters are of great importance to the diadromous fish, e.g. salmon, *Salmo salar*, Atlantic salmon; *Oncorhynchus kisutch*, coho salmon; sea lampreys, *Petromyzon*, or eel, *Anguilla*, which must pass through estuaries on their way to or from the spawning and feeding grounds. The sockeye salmon, *Oncorhynchus nerka*, becomes distressed when the oxygen concentration of the water is reduced to 3·5 p.p.m.: although some survive to a level of 1·2 p.p.m., death begins below 3 p.p.m. and is rapid below 2·5 p.p.m.; chinook, *O. tschawytscha*, are affected similarly[54]. The survival of juvenile coho salmon, *O. kisutch*, in low oxygen concentrations is related to temperature: at higher temperatures a higher concentration of oxygen is required[55] (Fig. 16.6). When maintained, during the winter, for 30 days at a temperature of 18°C and an

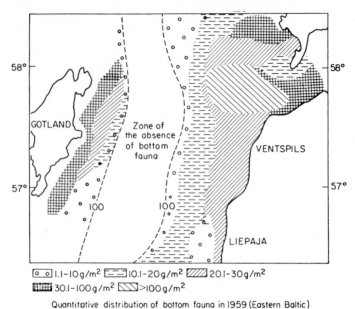

Quantitative distribution of bottom fauna in 1959 (Eastern Baltic)

Fig. 16.5. The dead zone in the Gotland area and adjacent waters in 1959 (after Segerstråle[52]).

oxygen concentration of 2 p.p.m., the juvenile coho consumed little food and lost weight. Those maintained in an oxygen concentration of 2·9 p.p.m. fed well and gained weight but were probably affected[55].

In tests performed in a swimming chamber it was found that adult coho salmon when forced to maintain a sustained swimming speed fatigued in water drawn from the Duwanish estuary, which contained 4·5 mg/litre of oxygen. Prior to fatigue, the salmon attempted to acclimate to the hypoxic environment by using anaerobic metabolism as a partial energy source[56].

In the estuarine environment, generally, minimum dissolved oxygen concentrations of 0·75–2·5 mg/litre are necessary for test species to resist death for 24 hr, and most marine species die when the dissolved oxygen content falls below 1·25 mg/litre. Reduced swimming speeds and changes in blood and serum constituents occur at dissolved oxygen (D.O.) levels of 2·5–3·0 mg/litre. D.O. levels between 5·3 and 8 mg/litre are satisfactory for survival and growth, but in excess of 17 mg/litre adverse effects are produced. Diurnal or other fluctuations in D.O. from to 8 mg/litre, produce significantly more physiological stress in fishes than fluctuations from 3 to 6 mg/litre[29].

A fall in the oxygen concentration may increase the activity of any toxicants present in the water. In fresh water the acute toxicity of zinc is

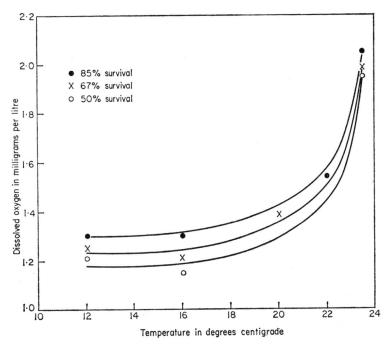

Fig. 16.6. Influence of dissolved oxygen and temperature on 24-hr survival of juvenile coho salmon in the fall season (after Davison *et al.*[55]).

increased by 50% when the dissolved oxygen falls from 6–7 mg/litre to 2 mg/litre[29, 57, 58].

Many estuaries are in a deplorable condition due largely to the B.O.D. of the sewage released to their waters.

Fiordic estuaries present an especial difficulty in this respect since the exchange of waters above and below sill depth may be very low (p. 63). In the Holy Loch, Firth of Clyde, for example, the retention time of materials which are added to, and remain in, the surface layers, is short, varying between 1 and 7 days. At depths below 20 ft however, there is a pronounced oxygen deficit which, coupled with the very slow circulation, at these depths, indicates that the loch is incapable of receiving with safety any pollutant which might settle rapidly and exert a high B.O.D. Currently, 355 lb/day of B.O.D. are being added to the loch and steps are being taken to reduce this to about 190 lb/day[59].

The city of Oslo, in 1945, had a population of 400 000 and discharged its sewage into the inner part of Oslo Fjord, which is partially isolated from the remainder by the narrows and shallows at the Drøbak sill. Even in the early 1930s, a marked oxygen deficiency at depth resulted, especially in the arm known as the Bonne Fjord[60–62], and by 1950, when the population of Oslo had grown to 480 000 the deep waters of the Bonnefjord were void of oxygen and H_2S occurred up to the 75–50 m level. In the western branch of the main fiord low oxygen values were recorded as far out as Drøbak Sound, but outside this normal values occurred[62].

The coastal plain estuaries of the Tees, Mersey, Tyne and Thames, and New York Harbour are all affected by pollution to some degree. Of these, the Thames is a notable example of the decline which accompanies the release of excessive amounts of sewage (or indeed any other organic matter if sufficient is released) to an estuary, but the Thames is also one of the major successes in the field of pollution abatement. Although minor tributaries, such as the Fleet, were heavily polluted as early as the fourteenth century, the Thames was a major salmon river until about 1810; at this time the water closet was introduced, and from then on domestic sewage, formerly removed from London as "night soil" and dumped on the fields, was passed straight into the Thames[63, 64]. By the 1850s, it was so foul that it had an offensive stench and fish had disappeared from long stretches of the river. In the early 1900s, efforts to clean the river had succeeded to the extent that migratory fishes were once more found upstream of London. This situation continued until World War I, but at the end of hostilities, a rapid expansion of home building without the appropriate treatment facilities brought about a return to the earlier evil conditions.

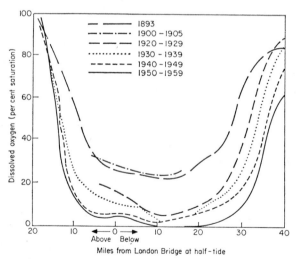

Fig. 16.7. Oxygen sag curves during successive decades for third quarters (July–September) when flow of the River Thames at Teddington was around 250 m.g.d. (Reproduced with the permission of the Controller of H.M. Stationery Office).

This decline in the condition of the Thames is illustrated by the *oxygen sag curves* (Fig. 16.7) for the period 1893–1959[65]. These deplorable conditions led to a classic study by the Water Pollution Research Laboratory which commenced in 1949[66]. The following important conclusions were derived from these studies, thus:

1. During dry weather, an anaerobic zone, lacking dissolved oxygen extended for about 20 miles in the middle reaches of the estuary. Under these conditions, the sulphate present in sea water is reduced to H_2S (p. 119) which is responsible for the foul smell.
2. The effect of a polluting discharge depends not only upon its B.O.D. load, but upon the point at which discharge is allowed to take place. In the Thames, the maximum deoxygenation was approximately halved for every 10 miles the outlet was placed nearer the seaward end of the estuary.
3. The nature of the nitrogenous matter present in sewage has an important influence upon events in the estuary. If the nitrogenous matter is ammonia then it is oxidized thus

$$NH_3 + 2O_2 \rightarrow HNO_3 + H_2O.$$

When however some nitrate is present, denitrifying bacteria utilize the oxygen present in the nitrate to oxidize organic matter thus

$$2HNO_3 \rightarrow N_2 + H_2O + 5[O].$$

This occurs when the dissolved oxygen level of the estuary has fallen to *ca.* 5–10% of saturation (i.e. $\leqslant 1$ p.p.m.); the oxygen level remains constant until all the nitrate has been utilized; then, and only then, does sulphate reduction and sulphide formation commence. Indeed on occasion, nitrate has been added to rivers to prevent or delay bacterial reduction of sulphate[63, 65–71].

The bacterial oxidation of ammonia to nitrate takes place in two steps, viz. (1) $NH_3 \rightarrow NO_2^-$, (2) $NO_2^- \rightarrow NO_3^-$, and to perform this oxidation the oxidizing bacteria require iron[72]. The production of nitrite by the oxidation of ammonia in sewage may lead to high values of this anion, which can be used to estimate the relative degree of pollution in a fiord or coastal plain estuary. The bacteria responsible for NO_2^- formation are able to carry out this activity over a wide range of oxygen concentrations[73].

4. The Thames, and by implication, other estuaries have a more limited capacity for self purification than was believed hitherto. This may be a consequence of the fact that the Thames is fairly deep, and since re-oxygenation from the atmosphere is relatively slow and has to be

distributed through a relatively large body of water compared with the surface area, it is less complete than would be the case in a shallower estuary.

5. Apparently, the sludge and the mud in the bed of the Thames has a low oxygen demand. However, the dredging performed to remove sludge, deepens the estuary and therefore inhibits re-oxygenation.

6. The heated effluents (*ca.* 4–5 °C rise) of power stations tend to accelerate de-oxygenation and, therefore, the emission of H_2S.

7. The numerous factors which affect dissolved oxygen concentration in the Thames estuary can be assessed quantitatively. Consequently, the changes in dissolved oxygen due to an additional source of pollution can be predicted[63, 65–71].

Based on these surveys, the construction of new sewage systems and treatment plants was undertaken. A situation, in which a survey performed in 1957 produced no evidence of fish living between Richmond and Gravesend, was replaced by one in which, since September 1967, 50 species (including such interesting species as the John Dory, *Zeus faber*, scad, *Trachurus trachurus* and Nilsson's pipefish, *Syngnathus rostellatus*) have been caught on the band screens of the power stations. In the Deptford Tunnel region, flat fishes are caught, flounders regularly, plaice, dab, brill and sole commonly; grey mullet and bass also occur commonly. Cod and whiting occur, and bib, pouting, tadpole fish (*Raniceps raninus*) and haddock are caught fairly frequently[64]. In an age which has tended to despair over the worsening trend in most estuaries, the relatively rapid rejuvenation of the Thames is a worthy example of what can be achieved by men who are determined.

It was widely considered, in early works on estuarine pollution, that the assimilative capacity of the estuary depended upon the amount of water which entered and passed through the estuary, and upon a high tidal range. In other words, the more the freshwater run-off and the greater the tidal turbulence, the more rapidly would polluted water be removed from the estuary[74]. However, in the Clyde estuary, where a neutrally buoyant particle released at Glasgow Bridge may take between 3 and 25 days to reach the outer Firth, this is not true. Indeed, the levels of dissolved oxygen in the estuary are much lower at the time of spring tides than at the time of neap tides[75, 76]. However, like the Thames, it is now possible to forecast the effect of a pollutant introduced into the Clyde estuary. Furthermore, like the Thames, the Clyde is improving due to the efforts of the Clyde River Purification Board, and the prospect that the Clyde will become a salmon river once more is now a real possibility rather than just a dream of a few far-sighted men.

Because estuarine waters are generally turbid the absorption of dissolved oxygen through the sea's surface is of evident importance. The exchange coefficient is somewhat reduced by salinity and synthetic detergents lower it by nearly 20%[63].

Finally, it is important to appreciate that as one major contributor to a pollutional situation is dealt with, so another, less important in the formerly more acute situation, may assume an important influence. The River Kelvin, a tributary of the Clyde, was formerly anaerobic. Now, however, this situation has been rectified, but other pollutants are preventing the survival of fish in this river[77].

By means of light and dark bottle techniques, attempts have been made to estimate the photosynthetic re-oxygenation of estuaries, e.g. the Delaware. Although re-oxygenation by this means undoubtedly occurs[78-80], its relative importance compared with that due to the algae of the microbenthos and re-oxygenation through the surface of the estuary is still uncertain.

Sewage

Oxygen deficiencies of estuarine waters induced by sewage are not the only attribute of this effluent. The receiving waters may be rendered objectionable by the presence of oil, grease, faeces, paper, food particles and smell, although such overt symptoms do permit avoidance. Pronounced biological effects may result either from directly toxic effects or by eutrophication; public health hazards can arise.

Biological Consequences of Sewage Pollution

Where aeration is satisfactory, the effect of sewage, by adding nutrients, is that of enrichment. The recycling of nutrients is essential to the proper functioning of biological systems, but the addition of excessive amounts leads to a decline in the populations of the receiving waters (Chapter 14). An increase of the nutritional standard of the environment, with respect to nitrogen and phosphorus, is termed *eutrophication*. Many consider eutrophication and pollution to be synonymous, but this is not so. A body of water may become eutrophic due to certain kinds of pollution, but all pollution is not eutrophication, e.g. arsenic and D.D.T. discharged into a body of water do not enrich the environment, but, depending upon concentration, may have powerful negative effects upon aquatic life[81].

This problem of enrichment of the fresh waters of the world is both widespread and a cause for considerable concern[81-93]. The varied sources which contribute to eutrophication in fresh water are summarized in Table 16.4, and since fresh waters flow into estuaries, the contributions

Table 16.4. Summary of estimated nitrogen and phosphorus reaching Wisconsin surface waters (after Hasler[93])

Source	N (lbs. per year)	P (lbs. per year)	N (% of total)	P (% of total)
Municipal treatment facilities	20 000 000	7 000 000	24·5	55·7
Private sewage systems	4 800 000	280 000	5·9	2·2
Industrial wastes*	1 500 000	100 000	1·8	0·8
Rural sources				
Manured lands	8 110 000	2 700 000	9·9	21·5
Other cropland	576 000	384 000	0·7	3·1
Forest land	435 000	43 500	0·5	0·3
Pasture, woodlot and other lands	540 000	360 000	0·7	2·9
Ground water	34 300 000	285 000	42·0	2·3
Urban run off	4 450 000	1 250 000	5·5	10·0
Precipitation on water areas	6 950 000	155 000	8·5	1·2
Total	81 661 000	12 557 500	100·0	100·0

* Excludes industrial wastes that discharge to municipal systems. Table does not include contributions from aquatic nitrogen fixation, waterfowl, chemical deicers and wetland drainage.

to each are similar. However, the situation in individual countries and estuaries may show a considerable departure from this pattern. There is also one notable exception from this table, viz. highly concentrated farm effluents.

Detergents have been implicated in the process of eutrophication, especially in North America, where detergents contain very much more tripolyphosphate than they do in Europe. However, an emphasis upon one source of one particular algal nutrient may distract attention from other contributory factors and thus prevent an improvement which might otherwise be possible[94]. Nevertheless, Liebig's Law of the Minimum should not be forgotten. This law states that growth is limited by the factor present in minimal quantity. Phosphorus is the limiting nutrient in fresh water, conversely the amount of nitrate in the sea is generally limiting: this situation arises because marine plants do not have the marked ability of freshwater plants to remove nitrogen from the atmosphere[95]

Eutrophication is a progressive process under the continued addition of nutrients. However, the process can be halted, and indeed reversed, if the appropriate measures are taken. For example, Lake Washington first received raw and then treated sewage from the City of Seattle. By 1959, treated sewage effluent from a population of 64000 was poured into the lake. Its condition became such that the alga *Oscillatoria rubescens* gave rise to dense blooms which awakened the public conscience and led to the construction of the Seattle Metropolitan Sewerage Disposal Scheme at a cost of $135 000 000. Sewage from Seattle is now treated at the Renton and West Point plants which pass effluent to Puget Sound, either directly or by the Duwanish River, and these receiving areas are the subject of continued supervision. Meanwhile, Lake Washington has undergone a dramatic improvement and the dense blooms of *Oscillatoria rubescens* are no longer seen[93, 96-105].

As in the freshwater environment the effects of eutrophication in inshore waters have been noted on a global scale[51, 52, 60-62, 95, 104-118]. The intense development of the green algae *Enteromorpha* and *Ulva*, especially the latter, occurs frequently where sewage is released to an estuary. In summer, where these growths are excessive they may be thrown up in large quantities at the high-water mark, there to decompose and give rise to the evil-smelling H_2S[29, 95, 106, 110, 112]; *Ascophyllum nodosum* may become abundant in enriched conditions also.

In Biscayne Bay, Florida, the flowering plants *Halophila baillonis* and *Halodule wrightii*, together with the echinoderm *Amphioplus abditus* become abundant in waters which contain mineralized effluents from sewage treatment plants. Where heavy organic enrichment occurs, the algae, *Gracilaria blodgettii* and *Agardhiella tenera*, the worm *Diopatra cupera* and the amphipods, *Erichthonius brasiliensis* and *Corophium acherusicum* become very common[29, 119].

In Oslo Fjord, where sewage pollution has been progressive throughout this century, the attached algae evince responses which fall into three natural groups[108]; thus:

Group I.   (species withstanding a very high degree of pollution): *Enteromorpha intestinalis, E. crinita, Urospora penicilliformis* and *Bangia fuscopurpurea*.

Group II.   (species tolerating a moderate degree of pollution): *Enteromorpha clathrata, E. compressa, Ulva lactuca, Ectocarpus confervoides, Elachista fucicola, Fucus serratus, F. spiralis, F. vesiculosus, Acrochaetium, Ceramium rubrum, C. strictum, Erythrotrichia carnea, Hildenbrandia prototypus, Polysiphonia violacea* and *Porphyra umbilicalis*.

Group III. (species which seem to withstand only a low degree of pollution): *Cladophora crystallina, Chorda filum, Chordaria flagelliformis, Chondrus crispus* and *Polysiphonia nigrescens*.

On the rocky coast of California, sewage discharges have had a marked adverse effect upon the beds of *Macrocystis*[120] and tumorous growths were noted on algae near the discharges[121]. In British waters, the "kelp forests" of *Laminaria hyperborea*, are adversely affected where effluents, including domestic wastes, are released[122]. Where algal changes take place, profound faunal changes may result[121].

On the Norwegian coast, the alga *Gracilaria confervoides* occurs in abundance in areas where it is liable to be inundated with water containing H_2S; indeed this alga is able to survive immersion for 40 days in sea water containing 5 ml/litre of H_2S, at a temperature of $3°C$ and in the dark. Some algae are said to be able to withstand exposure to H_2S for a whole year if the temperature is low[123].

Where sewage nutrients are added to an estuary, the increased production which results may cause huge concentrations of planktonic algae in the surface layers, and may contribute to the oxygenation of these layers. At depth however, there may be little production of oxygen by algae, since light cannot penetrate the intense blooms above. Indeed, during these periods of large plankton production in the surface layers there may be extended periods of anaerobosis in the mud of a coastal plain estuary or in the water below sill depth in a fiord. Bloom timing is a function of hydrographic conditions[61, 62, 104, 105, 107, 109, 111, 113].

Many algal species may be able to take advantage of nutrient enrichment[51]: in the inner fiord of Oslo Fjord, abundant populations of diatoms, coccolithophorids, dinoflagellates and other forms, develop at a time when the waters of the outer fiord are poor in phytoplankton. The general character of this plankton is striking in that it consists partly of regular coastal species and partly of Atlantic species typical of oceanic communities[61, 62]. Where nutrient levels, i.e. N and P, are adequate diatom blooms may be regulated, at times, by silica concentrations in the inner Oslo Fjord[109].

During the summer, the plankton regularly contains large populations of dinoflagellates. These dinoflagellates produce toxins which are accumulated by shellfish and cause fatalities when eaten: mussel collection in the inner fiord and in the waters outside the sill at Drøbak has been banned[62]. Indeed this development of toxic dinoflagellate blooms or "red tides" seems to be an inevitable consequence of enrichment and this' like eutrophication, has been reported on a global scale[112, 114, 124–126] (p. 14). Red tides are, however, not the sole adverse effect of the presence of sewage pollution. At Portage Inlet, Vancouver Island, B.C., the oyster, *Ostrea lurida*, has effectively disappeared under its pressure, the runs of cutthroat trout and coho salmon have been substantially reduced, and herring spawning, in the upper inlet, has decreased steadily during a 20 yr recording period[112]. In the Bonne Fjord, there is no zooplankton below 50 m depth, the population of the prawn *Pandalus borealis* was destroyed and a 50 yr old trawl fishery ceased as a result of the pollution of this arm of Oslo Fjord. Although the growth rate of cod taken in the inner Oslo Fjord is as good as in other areas investigated, the hatching of cod eggs is impaired[62].

Off Marseilles there is no distinct community associated with sewage pollution, but a series of "facies", each characterized by three or four species, indicative of pollution, can be found. Members of these species can show increased abundance and can live in well-defined communities or in sediments poor in dissolved oxygen. At the source of pollution, several concentric circles of population can be traced thus:

1. An azoic zone of maximum pollution which also lacks macroscopic vegetation.
2. A polluted zone characterized by *Capitella capitata* and *Scolecolepis fuliginosa* which are joined more or less progressively by *Nereis caudata*, *Staurocephalus rudolphii* and *Cirriformia tentaculata*.
3. A sub-normal zone characterized by the molluscs *Corbula gibba* and *Thyasira flexuosa* and a rich fauna of polychaetes. Here, the species indicative of pollution have almost disappeared and are replaced by those animals which would be expected in the absence of pollution.
4. A zone of pure water lying beyond a more or less well-defined transitional margin[127].

In some places, one or more of these zones may be suppressed, and in general the situation is much like that at Long Beach and Los Angeles Harbours, California[127, 128]. In Lake Mariut, a small, shallow, brackish-water lake, adjoining the Mediterranean coast of Egypt, sewage and industrial pollution has reduced the catch of fish from 9 977 815 kg in 1961 to 1 868 000 kg in 1967. This effect is not a direct result of the addition of toxic matter, but is due to a lowering of the dissolved oxygen levels by the B.O.D. loading[129]. In some areas, the increased development of tumours, in fish, has been ascribed to the presence of sewage outfalls[130, 131].

Adverse effects are not noted in every situation where sewage effluents are added to the environment and eutrophication occurs[93, 100, 102, 104, 105, 113, 132]. Nutrient levels may be raised in offshore waters by the discharge of eutrophic water from estuaries[133–137]. These nutrients may lead to an increase in primary productivity[137]; however, in the absence of data regarding the composition of the phytoplankton (which could theoretically be composed entirely of species which secrete neurotoxins), it is unwise to consider the effect as beneficial. Nevertheless, the influence of London and the Thames estuary upon the southern North Sea is well authenticated.

Here, in the 1930s, it was found that an abundance of nitrates and phosphates, derived from London's sewage, was present[133]. Subsequently, it was shown that this water had a marked effect upon the fisheries. Unlike water from the Continental rivers which is rapidly swept away to the north-east, by a current which has come from the English Channel, that from the Thames is held in a wedge by two streams of oceanic water of a higher salinity, viz. that from the Channel and the Atlantic influx from the North. About 2900 tons of phosphorus per year passes into this wedge and spreads through 171 cubic miles of English coastal waters, north to the Humber, and gives rise to an increase of 4 mg of P per m³. In this area, the catch of fish per unit area is about double the corresponding catch in the rest of the North Sea, in the English Channel, and in the Kattegat/Skagerrak region, and is about 25 times the catch reckoned for the Baltic Sea as a whole. Some two-thirds of this increased catch is ascribed to the nutrients derived from London[134, 138].

In the Baltic, despite the anaerobic basins which have resulted from eutrophication, the periodic overturn and uplift of the deeper, nutrient rich layers into the impoverished upper layers induced an unusually high and hitherto unattained level of zooplankton production in 1962. In turn, this abnormally high production of zooplankton led to a striking increase in catches of cod. A similar event occurred in the early 1930s when the annual production of fish rose from *ca.* 30 000 to *ca.* 180 000 tons; at this time the annual catch of cod rose from about 3000 tons in the early 1930s to about 70 000 tons in the early 1940s[52].

It has been suggested that a sewage effluent should be stripped of its nutrients, by biological methods, before it is released to the environment[138a]. Since sewage has been used with advantage in fresh water fish farming practice, its use in marine fish culture is now being considered[139–142].

Health Hazards Associated with Sewage Pollution

In Britain, as elsewhere in the world, it has long been the custom to discharge sewage, untreated either by comminution or chlorination, to the open sea or estuarine waters. To improve a sewage effluent before discharge to such waters it may be subjected to one of the following processes[63]:

1. Screen, to remove grosser solids.
2. Maceration by comminutor to break up solids into fine particles. This does not remove the pollution but merely disguises it. Screening and maceration do help to improve the appearance of bathing beaches to the untutored, but not to the perceptive eye.
3. Sedimentation, to remove settleable solids and so avoid putrefying sludge or mud banks.
4. Chlorination of settled sewage. This may be necessary in summer to avoid odours.
5. Aerobic treatment of the settled sewage, e.g. percolating filters, activated sludge process or oxidation ponds.
6. Sludge disposal.

Current practice may be either to do nothing or rarely to do better than stage 2 before sewage is released to the marine environment. In 1959, the Medical Research Council concluded that "with the possible exception of a "few aesthetically revolting beaches" around England and Wales, there is little risk to health in bathing in sewage contaminated sea water"[143]; a view shared by the municipality of Barcelona[144], and a British Working Party in 1970. However, the Justice of the Peace and Local Government Review[145], and others[146], showed little enthusiasm for this viewpoint.

It is generally conceded, however, that a considerable risk may arise from eating contaminated shellfish. In Europe and North America, the main risks are associated with eating raw shellfish which have absorbed pathogens from the water. The most important bacterial agents are the *Salmonella* spp. which include the organism capable of causing typhoid, the *Shigella* spp. which cause dysentery, and certain of the *Clostridia* which can produce exotoxins pathogenic to man. This source may give rise to virus infections of which infective hepatitis is the best documented[146–148]. When cooked there is little danger of infection from shellfish[148].

In the past, great use has been made of *Escherichia coli* as an indicator of pathogens present in sea water and in sea foods. In the former case, considerable efforts have been made to determine the factors which lead to the decrease of coliform bacteria in estuaries and where it is certain that dilution and biological inactivation play an important part in the removal of this bacterium, viruses, fungi and algae[149–151]. In the latter case, it has formed the basis of sanitary control both for the environmental waters, Table 16.5, and for shellfish supplied to the market, Table 16.6. The

Table 16.5. Bacterial standards for the waters in which shellfish cultivation is practiced (after Wood[148])

Class	France Condition	*E. coli*/100 ml
1	Satisfactory	0
2	Acceptable	1–60
3	Suspicious	60–120
4	Unfavourable	> 120

Class	United States Condition	*E. coli*/100 ml
1	Growing area, approved area	70
2	Moderately polluted	70–700
3	Closed area for shellfish culture	> 700

practice employed depends upon the individual country, and the individua situation, but, in either case, high standards are maintained[147, 148, 152]. However, it has become evident that *E. coli* is not a very satisfactory indicator of pollution. For example, salmonellae can survive for long periods in the sea and can be present in coastal waters of relatively low coliform count ($\times 2400/100$ ml). Entero-viruses have been isolated from coastal waters with a low coliform count[146]. National public health authorities are divided in their attitude to *E. coli,* and although it is still regarded as acceptable it is the subject of discussion once more.

Although the bacterial and virus infections which may arise from untreated sewage constitute a significant hazard, these are not the sole source of risk from this source. Faecal matter, released to estuaries and inshore waters, constitutes a rich source of food for seagulls, wading birds and other scavengers (when investigating a "new" area, the investigator automatically looks to the seagulls for an indication of the whereabouts of a sewer, since large flocks are almost permanently in residence at these situations). The eggs of the beef tapeworm *Taenia saginata* can pass through the alimentary tract of seagulls and still retain their viability and infectivity. Since such gulls "commute' between the rich source of sewer

Table 16.6. Comparison of bacterial standards for shellfish (all expressed as number per 100 g or per 100 ml of tissue) (after Wood[148])

Quality	British standard (No. of *E. coli*/100 ml by roll tube)	French standard (No. of *E. coli*/100 ml in phenol broth)		U.S. standard (No. of faecal coliforms /100 g by MPN technique)
Acceptable	90% of samples—up to 200 10% of samples—up to 200–500	Class I	Less than 100	Up to 230
		II	Between 100 and less than 500	—
Suspicious	—	III	Between 500 and less than 1500	—
Unacceptable	>500	IV	⩾1500	>230

borne food and coastal fields, these eggs may be carried inland to be deposited upon pastures and thus returned to cattle and man once more. There is circumstantial evidence to suggest that an apparent increase in bovine cysticercosis in Great Britain is related to the breakdown of a formerly reliable sewage treatment system[153, 154].

The Pulp and Paper Industry

There is relatively little information upon the effects of this industry in the marine environment, and most is known of its influence in fresh water. About 50% of the income of some countries, or states of some countries, e.g. Oregon and Washington in the U.S.A. and British Columbia in Canada, is derived from this source. In Europe, the Scandinavian countries are pre-eminent in this industry: the U.K. share is small, but growing, although it faces some difficulty from the giants, especially as trade barriers are lowered, e.g. in the E.F.T.A. group. Japan has a flourishing industry based on the bulk transport of wood chips from the western states of North America. Traditionally, soft woods, e.g. hemlock, Sitka spruce, Douglas fir, from northern latitudes have been used in the process, since in general hardwoods do not yield the longer fibre desired. However, the use of pulp is increasing, and in the United States supplies of wood for pulping are now being sought as far south as the state of Mississippi.

Whatever the method of pulping finally undertaken, the logs are first debarked either by mechanical or hydraulic debarkers, and then chipped. The waste from both these processes, i.e. hog-fuel, is passed to the boilers of the pulp-mill power station, where it is then burnt. This burning presents some technical difficulty and when supplies of hog-fuel exceed the capacity of the power station other methods of disposal have to be sought. During the period up to debarking, i.e. including the period of logging in the forests and log storage, the bark may give rise to a number of problems. The decay of bark takes so long and its oxygen demand is sufficient to create conditions of oxygen deprivation in spawning grounds, moreover fine bark particles clog gravel and cause mortality of salmon eggs. Bacteria, *Sphaerotilus* sp., can grow on bark and suffocate both eggs and alevins. Where spawning areas are blanketed by bark, they are avoided by the sockeye salmon, *Oncorhynchus nerka*[155]. The wood sugars which are released during log storage and hydraulic barking operations contribute a significant B.O.D. which can promote bacterial growth and thus have an adverse influence[156]. Bark contains a considerable quantity of tannins and lignins which can give a colour to water: the rate at which bark contributes colour-producing substances to water depends upon the type of log and the intactness of the bark. Cross-cut end sections of logs tend to expedite the release of colour and soluble organics. The rate of extraction is approximately constant in flowing waters, and is apparently the same in fresh and saline waters. In periods up to 80 days, the length of immersion does not affect the leaching rate[157].

Mechanical Pulp Production

The mechanical process in which the wood chips are ground to pulp between rotating metal plates produces a fibreboard or cardboard, brown in colour because it still contains lignin. The amount of pulp produced by this method is approximately equal to 100% of the wood input. Additives to the process include aluminium sulphate, cationic starches, china clay and titanium dioxide in amounts which depend upon the extent to which the finished product is surfaced with bleached chemical pulp. In general, this process is not regarded as being a significant source of waste in comparison with the chemical pulp process[158]. When working efficiently there should be little chemical loss in the mill effluent and wood fibres, which should also be at a minimum, may be the principal and only cause for concern.

Chemical Pulp Production

Chemical pulp is produced by two basic processes, viz. the sulphite process and the Kraft process. In the sulphite process wood chips are digested, at high temperature, with an aqueous solution of sulphurous acid (H_2SO_3) in which lime, or some other base, has been dissolved. The solution contains excess H_2SO_3 and is therefore acid. In the basic lime process no recovery from the *spent liquor* or *sulphite waste liquor* (S.W.L.) is possible and *ca.* 2500 gallons of this waste liquor is produced with every ton of fibre. For every ton of pulp produced, about one ton of soluble products passes to waste; the waste liquors also contain large amounts of pulp fibre[159]. Under pressure to improve the effluent quality and the economics of the process, the digestion may now be based upon sodium sulphite/sodium bisulphite in Britain and Scandinavia[160], or upon ammonium bisulphite, or magnesium bisulphite in North America. By these methods, the recovery of raw materials, by evaporating and burning the waste liquor, is possible and the process comes to resemble that of the Kraft mill.

Sulphite mills which release effluent directly to the aquatic environment are recognized as major polluters: a result of the high B.O.D.[161, 162] toxic constituents, and fibre loss. The toxic constituents are not so toxic as those from the Kraft process[159], nevertheless 8–16 p.p.m. of 10% S.W.L. are detrimental to the adult Olympia oyster, whereas adult Pacific oysters can withstand 50–100 p.p.m. The threshold of toxicity to Pacific oyster larvae ranges from 8–16 p.p.m. of 10% S.W.L., but the larvae of the Olympia oyster are more resistant and the threshold of toxicity is *ca.* 16 p.p.m. 10% S.W.L.[163].

Arising from this knowledge, an extensive survey, undertaken in the Puget Sound, showed that general adverse effects could be demonstrated upon the biota of the Sound[158]; this report was attacked vigorously since it could be faulted in at least some of its evidence[164]. However, although not perfect, it did attempt to see the situation as a whole in this area, and, as such, has some admirable attributes. Adverse effects due to such mills have been recorded at marine inlets at Prince Rupert, British Columbia (p. 523) and in the Penobscot estuary, Maine. At the latter site, relatively high waste liquor concentrations were concurrent with low pH values, and an absence of the living species of Foraminifera which occurred elsewhere in the estuary[165]. In Loch Eil and Loch Linnhe, on the Scottish west coast, a five year study has been performed relative to the effluent from a sulphite mill which uses the standard Swedish "Stora" two-stage process where extensive recovery of the sulphite waste liquor is undertaken[160]. Here, the effluent consists very largely of a dilute suspension of fibres, and little effect has resulted from this operation. Changes in the benthic populations are apparently within the known natural fluctuations of these populations and are probably caused by variations in the success of post larval recruitment to the populations involved. There has been no marked increase in the organic detrital deposition in this 5 yr period, but there has been a general increase in the B.O.D. of these waters, following the introduction of the effluent[166]. It has been reported from Scandinavia, that in the Hakefjord and Gota Alv, near Gothenburg, *Nucula nitida* has spread in the direction of a region of organic pollution; *Corbula gibba* has undergone no change of distribution, but has increased steadily in abundance over the forty year period during which the pollution of these areas has increased steadily[166].

In the Grays Harbour estuary, Washington State, U.S.A., the biota were subjected to pressure from sulphite pulp mill and sewage effluents. Here two anadromous species, viz. Chinook and silver salmon underwent a definite decline in productivity. Chum salmon, on the other hand, maintained a constant level overall, due possibly to the fact that the young of the chum migrate rapidly to sea and do not loiter in the estuarial area of Gray's Harbour. The effect of the sulphite waste liquor upon the oyster stocks was uncertain, but the proximity of domestic sewage sources led to a restriction in the areas from which marketable stock could be taken[167].

In the United States, where popular, state and Federal pressure is considerable, primary and secondary treatment of the sulphite waste liquor is extensively undertaken[168–173]. This process was commenced in

1925 by the formation of a Stream Improvement Committee of the American Pulp and Paper Association[173], and in 1939 by the formation of the Pulp Manufacturers Research League[174]. Effluent treatment is now taken so far that it is considered necessary to remove colour by the addition of lime or activated charcoal[173]. Work has been undertaken to improve the quality and colour of dilute effluents by use of electrodialysis[175] and reverse osmosis[174, 176–179], of which, the latter has been particularly successful.

The assimilative capacity of a stream, and the employment of various means to enhance it, were once considered to be legitimate practice in effluent disposal[180–182], but as more knowledge of the sub-lethal effects of effluents accumulates this approach is coming into disfavour; indeed in British Columbia, every effort is now made to ensure that water returned to a water course is of a high quality and the doctrine of stream assimilative capacity is considered to have no justifiable basis. However in view of the effects noted on p. 438, this may prove to be an exaggerated view of the problem.

Spent sulphite liquor is one of the greatest potential sources of organic materials still available in large volume and at low cost. From this raw material (Table 16.7) a wide variety of products can be made, these include emulsifying and dispersing agents, oil well drilling mud additives, cement and concrete additives, carbon black dispersants, ore flotation agents,

Table 16.7. Composition of spent sulphite liquor solids. A comparative approximation of calcium base liquor solids (after P.M.R.L. Bulletin[183])

	% Content			
	Softwoods		Hardwoods	
Lignosulfonates	55		42	
Hexose sugars	14		5	
Mannose		40·0		10·0
Galactose		20·0		2·5
Glucose		10·0		7·5
Pentose sugars	6		20	
Xylose		25·0		77·5
Arabinose		5·0		2·5
		100%		100%
Miscellaneous				
Sugar acids and residues	12		20	
Resins and extractives	3		3	
Ash	10		10	
	100%		100%	

18*

refractory and ceramic additives, boiler feed water treatments, sequestering agents, linoleum paste, binder for foundry cores, road binder, soil stabilizers, binder for ore pellets, briquettes and other products and dust abatement sprays. In addition, it can be used in the preparation of such high value products as Torula type food yeast, protein-lignin glue, aldonates, chrome-lignin gels, vanillin, vanillic acid and other derivatives, desugared lignosulphonates, modified phenol-formaldehyde resins and tanning agents[159, 183, 184].

In the Kraft process the wood chips are digested, at high temperature, with a mixture of caustic soda (55%) and sodium sulphide (45%)[185]. The process is alkaline, and extensive recovery from the waste liquors is undertaken. The effluent released to the environment consists of weak washings and waste from which no recovery is possible; however it contains mercaptans, sulphides, disulphides, sulphate soaps, turpentine and methyl alcohol, of which the first four are known to be highly toxic to fish and invertebrates[159, 185]. The activity of the sulphide compounds depends upon their state of oxidation, e.g. the minimum lethal concentrations of Na_2SO_4, Na_2SO_3 and Na_2S to *Daphnia pulex* are 5000, 300 and 10 p.p.m. respectively; where SH groups are present, the toxicity decreases with the number of SH groups thus the 4 hr L.C.$_{50}$ of H_2S, MeSH and Me_2S for *Daphnia pulex* is 2·82, 7·86 and 521·0 p.p.m.[186].* The effects of a Kraft mill may result from a high B.O.D.[187], but although the B.O.D. of this waste may reduce the dissoved oxygen levels of the receiving waters and thus influence toxicity indirectly, the toxicity of the Kraft mill waste is not due to B.O.D.† The toxicity of the total effluent is variable, and that of the contributing streams may be very different; effluent toxicity is markedly reduced if it is neutralized before release to the aquatic environment. Exposure of the underyearlings of sockeye salmon to low doses of Kraft mill effluent increases their tolerance to acute dosage. The toxicity of the effluent decreases if it is allowed to age before treating the fish[45, 188, 189].

The toxicity of neutralized Kraft bleach pulp waste (N.B.K.) is decreased substantially by biological treatment[190]. In a situation where it is usual to treat even neutralized effluents, e.g. in British Columbia, the need to understand sub-lethal effects becomes increasingly important. The alevins of both sockeye and pink salmon incubated in varying concentrations of N.B.K., grew only poorly in concentrations *ca.* 1/10 and 1/20 of the average TL_m for sockeye fingerlings. Moreover, this neutralized waste promotes the growth of slime in gravel and this may lead to increased mortality of incubating eggs and alevins; however, the exposure of adult sockeye to

* Oxidation in the environment is considered now to render these substances innocuous.
† B.C. Research have shown it to be due to resin acids.

N.B.K. did not affect the viability of ova and sperm[190] Tests with juvenile salmon, i.e. coho and kokanee, indicate that swimming performance respiration and haematocrit are affected by sub-lethal concentrations of neutralized Kraft mill effluent: the swimming performance is influenced by concentrations of unbleached white water as low as 10% of the 96 hr TL_m (96 hr L.C.$_{50}$). At sub-lethal concentrations of 20–30% N.B.K. the cough rate frequency increased up to 6–7 hr and then declined to the control rate of 3 coughs/min. Observations upon changes in blood lactate have proved inconclusive so far[191–195].

In situations where turbulence is reduced and there is a poor net flow from the area, considerable amounts of wood fibre may accumulate on the sea or estuarine bed. In these situations, microorganisms attack and break down the fibre, and, if the accumulations are large, significant amounts of gas may be evolved. Under anaerobic conditions liberal amounts of CH_4, CO_2, H_2 and H_2S are produced. Aeration inhibits the production of CH_4 and H_2S[196]. Sediments containing wood fibres which are decomposing with the formation of H_2S may be inimical to marine life. At Bellingham Harbour, Puget Sound, sediment concentrations of fibre which exceeded 1% were lethal to salmon smolts, and a concentration of 0·1% appeared to be the threshold of visible distress; H_2S concentrations of 2·3 and 0·3 p.p.m. were recorded in these sediments[197].

Paper machine wastes contain cellulose fibres, size, filler, e.g. china clay, $CaSO_4$, $BaSO_4$, colouring, i.e. Dyestuffs, glue and casein. Of these materials the cellulose fibres and $BaSO_4$ are known to have an adverse effect upon fish and invertebrates[149]. Glue from plywood plants by affecting the phytoplankton and fostering obnoxious slime growths, creates an unhealthy climate for bottom invertebrates, resident fish and fish spawning beds. Glues used in plywood for outside use are more toxic than those intended for inside use. These materials coagulate in sea water, they are in consequence less toxic to salmon in this medium than in fresh water[41, 198].

All pulp and paper manufacturing plants have problems in controlling microbial slimes. The slimicides used are all very toxic and many, though not all, are based on mercury compounds[199, 200]. In the pulp industry, phenylmercuric acetate has become an important means of slime control and the predominating conservant for wet pulp. It is from this source, and the run-off from fields containing seeds which have been dressed with alcyle and alcoxyalcyle mercury compounds that a problem of the mercury content of fish, normally taken for human consumption, has arisen[201].

Crude Oil and Petroleum Products

The occurrence of oil in estuaries presents a number of special problems, which arise partly from the nature of the oils and petroleum products and partly from the peculiar nature of the estuarine environment with its extensive areas of sand and mud flats, and salt marshes. Oils released to the sea, where they are not soluble, spread out to form thin films, which may become adsorbed on the surface of sediments and may even penetrate the sediments. The formation of emulsions, the leaching of water soluble fractions and the coating and tainting of rocks, and shore-dwelling plants and animals create further problems.

The principal sources of oil pollution are numerous. Listed in order of their destructiveness to ecosystems, they are[29]:

1. Sudden and uncontrolled discharge from wells towards the end of drilling operations.
2. Escape from wrecked and submerged oil tankers.
3. Spillage of oil during loading and unloading operations, leaky barges and accidents during transport.
4. Discharge of oil-contaminated ballast and bilge water into coastal areas and on the high seas.
5. Cleaning and flushing oil tanks at sea. On the average, the ships are estimated to release 2–3% of oil in 1000–2000 tons of waste.
6. Spillage from various shore installations, refineries, railroads, city dumps, garages and various industrial plants.

Natural submarine seepages make a contribution also[202], but the influence of man's activities is of greater significance[203].

The impact of oil spillages is generally complex, since the spill is almost always accompanied by clean-up measures. These normally include the use of oil emulsifier detergents which are both toxic in themselves and influence the toxicity of the oil. In such conditions, especially in a case like that of the *Torrey Canyon*, widespread mortality of marine life, particularly shore dwellers and birds, occurs and on the shore at least, marked ecological changes of a variable duration result. Much work in this field has been concerned with descriptions of the consequences of the spill plus the clean-up measures[204–213]. It is, however, more instructive to consider the component effects of oil and detergents upon marine life.

Of all groups the sea birds are most affected by an oil spill[29, 202, 214]: the feathers become fouled with oil, and most die from hypothermia and toxic effects resulting from preening. Ducks, gulls, terns and waders which haunt the shoreline are less affected than the birds of the open sea[202, 214]. Attempts to rehabilitate seriously affected birds are usually a dismal failure[214–216].

Crude oils have widely differing compositions, depending upon their source, and the components of the individual oil have widely differing toxicities, although the lighter, volatile fractions are the more toxic: the low boiling-point saturated hydrocarbons produce anaesthesia and narcosis at low concentrations, at high concentrations they cause cell damage and death[217]. The low boiling point aromatic hydrocarbons have the greater toxicity, and olefinic hydrocarbons occupy an intermediate position[218].

In March 1957, an oil tanker, the S.S. *Tampico Maru* was wrecked near Ensenada, on the coast of California. The wreck released *ca.* 60 000 barrels of heavy diesel oil. Many sub-littoral and littoral animals died; these included abalones, *Haliotis cracherodii*, *H. fulgens* and *H. rufescens*; lobster, *Panulirus interruptus*; barnacles, *Balanus glandula* and *Chthamalus fissus*; shore crabs, *Pachygrapsus crassipes*; molluscs, *Acmaea* sp., *Tegula funebralis*, *Mytilus californianus*, and *Tivela stultorum*; sea urchins, *Strongylocentrotus franciscanus*, *S. purpuratus*; sea stars, *Pisaster giganteus*, *P. ochraceus*. Among the animals which survived were an inhabitant of the splash zone, *Littorina planaxis* and a large intertidal anemone *Anthopleura xanthogrammica*. Of the plants, the principal inhabitant *Macrocystis pyrifera* was apparently unharmed, and an extensive growth of this species followed the death of herbivorous echinoids but the flora was reduced to four species. Recovery gradually took place and was extensive at the end of 2 years; however, even in 1967, recovery was not complete, and the proportions of the individual species were not the same as they were at the time of the incident[29, 202, 204, 207].

In the Black Sea, phytoplankton species are highly sensitive to the presence of oil in sea water, although the individual species differ considerably in sensitivity, e.g. *Ditylum brightwellii* is 3–4 orders more sensitive than *Melosira moniliformis*[219] (Table 16.8). Similarly, the life span of the

Table 16.8. Reaction of algae exposed to different concentrations of oil pollution (ml/litre) for 5 days (after Mironov[219]).

Algae	Death of cells 100%	No cell division or cell division delayed	Treated cells not different from controls
Diatoms			
Chaetoceros curvisetum	1·0–0·1	0·01	0·001–0·0001
Coscinodiscus granii	1·0	1·0–0·1	0·01–0·0001
Ditylum brightwellii	0·1–0·0001	—	—
Licmophora ehrenbergii	1·0	0·1–0·001	0·0001–0·00001
Melosira moniliformis	1·0	1·0–0·1	0·01
Dinoflagellates			
Glenodinium foliaceum	1·0–0·1	0·1–0·01	0·001–0·0001
Gymnodinium kovalevskii	1·0–0·001	0·001–0·0001	0·00001
G. wulfii	1·0–0·1	0·01–0·0001	—
Peridinium trochoideum	1·0	1·0	0·1–0·00001
Prorocentrum trochoideum	1·0	0·1–0·00001	—

copepods *Acartia clausi, Paracalanus parvus, Penilia avirostris, Centropages typicus* and *Oithona nana* are all significantly shortened by exposure to sea water which contains oil, mazout and diesel oil at a concentration of 0·001 ml/litre. Moreover, the nauplii of *Acartia clausi* and *Oithona nana* die on the third or fourth day of exposure to sea water containing 0·001 ml/litre of mazout: the larvae of *Balanus* sp. and *Pachygrapsus marmoratus* are affected at concentrations of 0·1–0·01 ml/litre of oil products in sea water. Benthic animals are sensitive also; the crustacean *Diogenes pugilator* dies at a concentration of 0·01 ml/litre, having first abandoned its shell. Oil at a concentration of 1·0 g/kg of bottom sediment accelerated the death of *Nereis diversicolor*, on the other hand, *Pachygrapsus marmoratus* remained active after 15 days exposure to concentrations of 1·0 and 0·1 ml/litre of mazout in sea water[219].

Fish eggs and larvae are very sensitive to oil products, and may be killed by concentrations ranging from 10^{-5} to 10^{-3} ml/litre[219]. Larvae are most sensitive at the stage when the yolk sac is resorbed and feeding commences, but, depending upon the oil used, a wide range of effect is noted. Natural oil dispersions gradually lose their toxicity[220].

In Novorossiyak Bay, Black Sea, beds of the oyster *Ostrea taurica* have been destroyed and other valuable species of mollusc, viz. *Spisula subtruncata, Tapes rugatus* and *Pecten ponticus* have declined under the pressure of a mixed daily discharge of 15 000–30 000 m³ of combined domestic sewage and petroleum refinery wastes. In bioassay tests, the copepod *Acartia clausi* died after 24 hr exposure to water taken 25 m away from the outfall; *Calanus* exposed in a sample taken 1 m away from the outfall died in 5 days, but survived a 10-day exposure to water taken 5, 10 and 25 m away from the outfall. The larvae of decapod crustacea and gastropod molluscs died in 3 and 4 days when exposed to water taken 10 and 25 m respectively, from the outfall[29].

In a series of studies based on Milford Haven, it has been shown that the short term effect of oil on a salt marsh is due to oil which adheres firmly to the plants and of which only a little is washed off by the tide. Leaves may remain green under an oil film for some days, but they eventually become yellow and die. Recovery takes place by the production of new shoots. Marshes recover from single spillages, or from successive, well-spaced spillages. However, if the oilings exceed three or four per month, the vegetation declines rapidly; recovery from this state is slow and probably takes some years. In Southampton Water, the cord grass, *Spartina X townsendii*, subjected to a refinery discharge containing 25 p.p.m. oil and 2 p.p.m. phenols died over an area of about 25 ha.

Seasonal effects are important: perennials which die back above the surface, but perennate below it, e.g. *Juncus gerardii, Spartina X townsendii*, may thus be protected from a winter spillage. On the other hand, seedlings and annuals rarely recover from oil, and are therefore most vulnerable during the summer. A spring and summer oiling may also reduce flowering and seed production in many plants, e.g. *Festuca rubra* (Red fescue), *Plantago maritima* (Sea plantain) and *Spartina X townsendii*. The variation in susceptibility of salt marsh plants is summarized in Table 16.9. The

Table 16.9. The susceptibility of different species of salt marsh plants, and the reasons for damage (after Cowell, Baker and Crapp[221])

Group 1	Very susceptible—Shallow rooting plants with no, or small, food reserves; quickly killed by oil and cannot recover; e.g. *Suaeda maritima* (Seablite); seedlings all species.
Group 2	Susceptible—Shrubby perennials with exposed branch ends which are badly damaged by oil; e.g. *Halimione portulacoides* (Sea purslane).
Group 3	‚Susceptible—Filamentous green algae. Though filaments are quickly killed, populations can recover rapidly by growth and vegetative reproduction of any unharmed fragments or spores.
Group 4	Intermediate—Perennials which usually recover from one spillage or up to four light experimental oilings, but decline rapidly if chronically polluted e.g. *Spartina X townsendii*; *Puccinellia maritima* (Sea poa).
Group 5	Resistant—Perennials, usually of rosette form, with large food reserves (e.g. tap roots). Most of them die down in winter. Some have survived twelve experimental oilings e.g. *Armeria maritima* (Thrift).
Group 6	Very resistant—Perennials of Group 5 type which have in addition a resistance to oils at the cellular level; e.g. members of the Umbelliferae.

responses of these groups of plants differ because of their different morphology, anatomy and physiology. In general, the fucoid algae are highly resistant to the consequences of oil pollution.

Most littoral animal species are resistant to highly toxic oils, even when spilt at intervals of 1 month. However, the gastropod inhabitants at high shore levels, viz. *Littorina neritoides*, *L. saxatilis* and *L. littoralis*, are affected by the thicker oils. This may be a consequence of the oil increasing the volume and weight of the animal and rendering it more liable to dislodgement. On a shore which was subjected to a long term chronic exposure of a continuous discharge of oil, a shore community dominated by *Patella* and barnacles was replaced by *Fucus vesiculosus*; a function perhaps of the high resistance to the consequences of oil pollution normally displayed by the fucoids[221, 222]. This observation regarding barnacles is curious, since results in the Black Sea indicate that *Balanus* sp. are at least moderately tolerant of the presence of oil in sea water[219]. In North America it was found that certain benthic animals, viz. the hydrozoan *Tubularia crocea*, *Balanus balanoides*, embryos of the toadfish *Opsanus tau*, the hardshell clam, *Mercenaria mercenaria*, and the oyster, *Crassostrea virginica*, were not very tolerant to oil adsorbed upon sand, but that most could withstand 1 part of oil in 1000 parts of sea water, i.e. 1000 p.p.m., for 24 hr[202]. The concentration of oil in the outfall at Milford Haven rarely exceeded 20–25 p.p.m. of oil, and 50 p.p.m. was the maximum; some dilution occurred before the effluent impinged upon the shore where the effect was observed[222]. Adult *B. balanoides* is tolerant not only to moderate amounts of oil, but it is very tolerant, perhaps it is the most tolerant animal species, to effluents which are toxic and have a low pH[X]; hence the interesting nature of the result from Milford Haven. Perhaps the explanation may lie in the sensitivity of barnacle cyprids and newly metamorphosed individuals to oil.

Little is known of the effects of oil spillages upon sandy beaches[222]. Nevertheless, *Mya arenaria* was killed by oil in the sand flats at Staten Island[223]. However, where oil emulsifier is used the oil/emulsifier mixture may penetrate the sediment and cause considerable mortality. Oil emulsifiers by changing the viscosity of the interstitial medium, bring about changes in soil hardness, and a temporary quicksand may result[213]. Despite these problems, once an oil has entered a soil, or come into contact with a surface in sufficiently thin layers, it undergoes microbial breakdown which may be of an extended duration[202, 224–227a]. Indeed, in North America, particularly with reference to the highly productive marshland areas such as Barataria Bay, attempts are being made to find a biological process which can be used to clean up petroleum spills[226].

Oil emulsifier detergents have been used widely to clean up the results of an oil spill, but it is usually found that the emulsifier and emulsifier/oil mixtures are more toxic than the oil alone[25, 209, 212, 213, 220, 222, 228–230]; indeed there is evidence to suggest that where an oil emulsifier of low toxicity is used, the mixture of oil and emulsifier is more toxic than either alone[220] (Fig. 16.8). Moreover, although natural oil dispersions obtained without chemicals gradually lose their toxicity, those stabilized with dispersants keep or increase their deleterious effect for some days[220].

Various other methods have been used for cleaning up oil spills. These include collection, sinking, mechanical removal of stranded oil, booms to prevent entry to estuaries, attempts at burning the oil from off the sea's surface and the washing of rocks with high pressure hoses[231–233]. Bearing in mind the instability of soils in offshore and outer estuarine areas, and their general importance in the economy of the sea, sinking of oil was never a very attractive concept. The influence of oil emulsifiers and other low viscosity products upon sediments to which oil products may become adsorbed make it even less attractive. In practice, however, anyone faced with an oil spill is placed on the horns of a dilemma, since what was good practice in one place is not necessarily good in another. In estuarine areas extreme care is necessary, and in Britain it is recommended that the use of

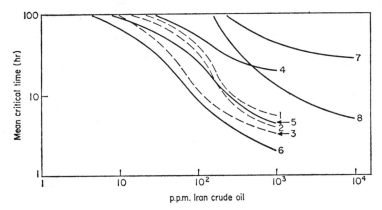

Fig. 16.8. Effect of Iran crude oil and COREXIT 7664 on one-day-old herring larvae:
1,4 : dispersions without dispersant; 2,5 : dispersions with 10 p.p.m. of COREXIT 7664; 3,6 : dispersions with 100 p.p.m. of COREXIT 7664; ----=test started 1 hr after preparing of dispersions; ——— = test started 50 hr after preparing of dispersions; 7 : dissolved compounds from oil film of plotted amount; 8 : solution of COREXIT 7664 (after Kuhnhold[220]).

dispersants be avoided, especially in those areas which are important fish nursery grounds; nevertheless these materials are frequently used in estuaries.

Obviously prevention is better than cure, and to this end, the oil industry is taking such action as it can. It was formerly the practice to wash down tanks, and flush them out once a tanker had put to sea. Now by gathering the washings in one tank, and retaining the oil, that is by adopting the "Load-on-Top" procedure, losses from this source are minimized[234]. Nevertheless, the most outstanding improvement must come from the routing of tankers to avoid collision or wreck and minimizing the loss of oil when such an event does occur.

The problem of oil pollution is not restricted to the death of marine organisms. When a pollution incident has occurred tainting of the flesh of fish and shellfish, especially molluscs, may be significant and may last for some time; a considerable financial loss to fishermen may result therefore[212, 218, 235, 236]. Disappearance of the "tainting taste" of oil may not be a good indicator of the elimination of pollutants, due to oil, from the fish body; indeed shellfish exposed to the West Falmouth spill retained fuel oil, to which they had been exposed, for several months after the accident[218, 237]. Carcinogenic hydrocarbons have been found in both plankton and benthic invertebrates[29]. Results from the West Falmouth spill suggest that shellfish and other fishery resources which have been exposed to oil pollution should be viewed with caution[237]; however, like so much else in the pollution field, much more needs to be learned with regard to this problem.

Tainting Substances

The problem of taint in fish and shellfish has been touched on above. However, petroleum is not the only source of flesh taint. The production of H_2S in association with the dumping of sewage sludge, or waters receiving sewage, or black liquor from Kraft pulp mills, may lead to a discoloration and a tainting of oyster tissues[29, 238]. Where the copper content of the water is high, oysters which live in these waters become green in colour and have a metallic taste: the normal copper content of sea water is 0·0038–0·005 mg/litre and oysters here contain from 20–80 mg/litre of wet weight; in areas contaminated with copper to give a concentration of 0·019 mg/litre, the content of the oysters rose to 124·5–392·0 mg/litre wet weight[29]. The green colour of the gills of the European oyster, recorded in rich feeding areas such as the Strood Channel in the U.K., and in France, and in the American oyster in North Carolina and Chesapeake Bay, is due to a pigment absorbed from the diatom *Navicula* which lives on these grounds[29]: such oysters, unlike those tainted with Cu, are a gourmet's delight.

Toxic Substances

The sea receives a bewildering array of effluents (p. 461) many containing materials which are directly toxic in effect; these include heavy metals, pesticides, detergents, ammonia and ammonium compounds, cyanides, sulphides, fluorides, tar, gas and coke wastes, wastes from petroleum refining and laboratory wastes. Of the many substances recognized as potentially toxic pollutants of the marine environment few have been studied in sufficient detail to allow us to define their maximum allowable concentrations[29]. The deficiencies of present day toxicity studies, with reference to marine organisms (p. 487), do not permit other than a patchy discussion of this topic. The poverty stricken approach to this work which mainly produces long, depressing and not very valuable lists of median tolerance limits do not depress this author alone[239]. In North America, it is accepted as a fundamental concept that all "effluents containing foreign materials are harmful and are not permissible until laboratory tests have shown the reverse to be true. It is the obligation of the agency producing the effluent to demonstrate that it is harmless rather than of the pollution abatement agencies to demonstrate that the effluent is causing damage[29].

Under such an attitude, a climate for the proper study of the physiological effects of toxic materials has been created: an attitude likely to become world wide if the sense of the recent F.A.O. Technical Meeting upon Marine pollution is a reliable indicator.

Heavy Metals

Marine organisms possess the ability to concentrate elements at levels greater than they occur in the environment. In connection with this ability three considerations are of significance, thus:

1. All, except six elements, are concentrated to a degree. The exceptions are fluorine, bromine, magnesium and sulphur for which the concentration factor is 1, sodium and chlorine are weakly and strongly rejected respectively.
2. Cations (including metallic elements such as iron which may exist in the sea in a colloidal form) have an affinity for living matter which depends upon valency thus:

    tetra and trivalent elements > divalent transition elements > divalent group IIA metals > univalent group I metals. Within the tetravalent and trivalent subgroup the affinity for plankton and brown algae differs:

    plankton: $Fe > Al > Tl > Cr, Si > Ga$
    brown algae: $Fe > La > Cr > Ga > Li > Al > Si.$

    These organisms show similar differences with respect to divalent metals, also:

    plankton: $Zn > Pb > Cu > Mn > Co > Ni > Cd$
    brown algae: $Pb > Mn > Zn > Cu, Cd > Co > Ni.$

    The prominence of lead is interesting since it has no known biological function.

    The greater ease with which the heavier elements are taken up in these organisms, compared with the lighter elements, may be a function of the ease with which they are polarized.
3. Anions have an affinity for living materials, thus:

    Nitrates > trivalent anions > divalent anions > univalent anions.

It seems likely that most polyvalent metallic elements are more or less chelated by organic matter. The uptake of ions is an active, metabolic process in which the ions are probably transported across the cell membrane with metabollically produced organic molecules. The trace metal concentrations in some commercial shellfish are given in Table 16.10.

Mode of Toxic Action

All elements are toxic if the concentration is high enough, and some are notorious even at low concentrations. Copper, an essential micronutrient

Table 16.10. The concentrations of trace metals in shellfish in parts per million of wet weight (after Pringle et al.[240])

Element	West Coast, U.S. Crassostrea gigas	East Coast, U.S.						
		Crassostrea virginica Range	Mean	Mya arenaria Range	Mean	Mercenaria mercenaria Range	Mean
Zinc	86–344	180–4120	1428	9·0–28	17	11·50–40·20	20·6
Copper	7·80–37·50	7·0–517	91·50	1·20–90	5·80	1·0–16·50	2·6
Manganese	0·90–16	0·14–15·0	4·30	0·10–29·90	6·70	0·7–29·70	5·8
Iron	15·30–91·40	31–238	67·00	49·70–1710	405	9·0–83·0	30
Lead	0·10–4·50	0·10–2·30	0·47	0·10–10·20	0·70	0·10–7·50	0·52
Chromium	0·10–0·30	0·04–3·40	0·10	0·10–5·0	0·10	0·19–5·80	0·20
Nickel	0·10–0·20	0·08–1·80	0·19	0·10–2·30	0·27	0·10–2·40	0·24
Cobalt	0·10–0·20	0·06–0·20	0·40	0·10–0·20	0·52	0·10–0·20	0·31
Cadmium	0·20–2·10	0·10–7·80	3·10	0·10–0·90	0·27	0·10–0·73	0·19

of all organisms, is highly toxic at low concentrations; other micro-nutrients are toxic when supplied in excess, but not all are so striking as copper. However there is an optimum range of concentration, which is sometimes quite narrow, for the supply of each element to each organism[29]. Indeed, it is a characteristic feature of these heavy metals that the dose response curve is *oligodynamic*, i.e. that are stimulatory at low doses, but at higher levels they are toxic[241]; in the presence of increased concentrations of these materials the respiration rate of *Nereis*, for example, may be raised even at toxic concentrations[242].

The most important mechanism of toxic action is thought to be poisoning of the enzyme systems. The more electronegative metals, particularly Cu, Hg and Ag, have a great affinity for amino, imino, and sulphydryl groups, and such metals are chelated readily by organic molecules. Attempts have been made to correlate the toxicity of the metals with their electro-negativeness, the insolubility of their sulphides or the order of stability of their chelated derivatives. It seems likely that all the divalent transition metals, as well as the other electronegative metals, which form insoluble sulphides, e.g. Ag, Mo, Sb, Tl and W, are poisons by virtue of their reactivity with proteins, and especially with enzymes. Since living cells contain a wide variety of enzymes, the variations in toxicity which result are considerable. Moreover, metals which give rise to similar toxic effects may have their effect through different enzymes and more atoms of metal are absorbed by an inactivated enzyme than are required to block the reactive sites.

Toxic effects may arise from causes other than the poisoning of enzyme systems. Antimetabolites, such as chlorate and arsenate, may occupy sites for nitrates and phosphates respectively: this group includes fluorides, bromates, borates, permanganates, antimonates, selenates, tellurates, tung-states and beryllium. Stable precipitates or chelates may be formed with essential metabolites, viz. by Al, Be, Sc, Ti, Y, Zr, reacting with phosphate; Ba with sulphate, or Fe with ATP. The decomposition of essential meta-bolites may be catalysed by La; other lanthanide cations decompose ATP. Au, Cd, Cu, Hg, Pb and U combine with the cell membrane either to affect its permeability, and thus the transport of Na, K, Cl and organic molecules across the membranes, or to rupture it. Electrochemically and structurally important elements in a cell may be replaced and functional failure follows, e.g. Li may replace Na, Cs replace K, or Br replace Cl. Metallo-organic compounds may be either more toxic than the metal ion, e.g. ethyl mercuric chloride, or much less so, e.g. the cupric ion and copper salicylaldoxime.

The elements can be divided into four groups with regard to their pollution potential[240], thus:

1. Very high pollution potential: Ag, Au, Cd, Cr, Cu, Hg, Pb, Sb, Sn, Te, Zn.
2. High pollution potential: Ba, Bi, Ca, Fe, Mn, Mo, Ti, U.
3. Moderate pollution potential: Al, As, B, Be, Br, Cl, Co, F, Ge, K, Li, Na, Ni, Rb, V, W.
4. Low pollution potential: Ga, I, La, Mg, Nb, Si, Sr, Ta, Zr.

The sources of heavy metals reaching the sea are very varied and include the burning of fossil fuel (> 5000 tons of Hg/yr enter the earth's atmospere from this source, compared with 3000 ton/yr added to the environment from other sources[243]); antifouling paints (a source of Cu, Hg); seed dressings and slimicides (Hg)[201]; power station corrosion products (Cu, Zn, Cr)[242]; mining either in the sea itself or as a run-out from spoil heaps and mines inland carried to the sea in rivers (these include Sb, As, Be, Bi, Cd, Cr, Co, Cu, Pb, Hg, Ni, Se, Sn, U and Zn[244]); the internal combustion engine (Pb); and, of course, industrial processes, of which the most notable case is the occurrence of Minimata disease in Japan due to an effluent containing mercury (p. 481). In the context of these problems it is worth remembering that "only in coastal areas and in estuaries has man increased the concentration of mercury in the sea by about 1%". Yet alarm with respect to this element is now global.

In the sea, the metals about which a significant amount is known are few. Silver, normally present at a concentration of 0·0003 mg/litre, is known to be highly toxic to mammals and plants. Arsenic is found in sea water at a concentration of ca. 0·003 mg/litre; it is moderately toxic to plants and highly toxic to animals, especially as AsH_3. An 8 day exposure of salmon to 5 mg/litre As_2O_3 was very harmful and mussels are killed in 3–16 days by a concentration of 16 mg As_2O_3/litre. Cadmium is moderately toxic to all organisms and is a cumulative poison in mammals; its carbonate and hydroxide are insoluble and are precipitated at a high pH. This element acts with other substances to produce a synergistic increase in toxicity. The adult American oyster, *Crassostrea virginica*, has 8 and 15-week TL_m values of 0·2 and 0·1 mg/litre of $[Cd^{2+} Cd(NO_3)_2]$ respectively; sub-lethal concentrations of this material depresses growth.

Lead is present in sea water at a concentration ca. 0·00003 mg/litre, and it is toxic to marine animals; lobsters maintained in lead-line tanks died in 20 days, *Crassostrea virginica* has 12 and 18-week TL_m values of 0·5 and 0·3 mg/litre of lead; exposure to 0·1 and 0·2 mg/litre for < 12 weeks induced changes in mantle and gonadal tissue. Lead is less toxic than Hg, Cu, hexavalent Cr, Zn or Ni to the giant kelp *Macrocystis pyrifera*. Nickel is found in sea water at a concentration of ca. 0·0054 mg/litre, it is highly phyto-toxic, but is not so harmful to animals. Long-term exposure of oysters to a concentration of 0·121 mg/litre Ni caused a considerable mortality.

Zinc is found in sea water at a concentration of *ca.* 0·01 mg/litre, small amounts are said to be dangerous to oysters. The photosynthetic rate of the giant kelp *Macrocystis pyrifera* was apparently unaffected when the plant was exposed to 1·31 mg Zn/litre as $ZnSO_4$, however 50% inactivation occurred at 10 mg/litre[29]. In the salmon parr (*Salmo salar*) from the Miramichi River, New Brunswick, the incipient lethal level (I.L.L.), which is defined as the threshold between lethal and non-lethal concentrations, was 48 μg/litre for copper and 600 μg/litre for zinc in soft water. Avoidance thresholds were found at 0·09 I.L.L. of zinc, 0·05 I.L.L. of copper and 0·02 I.L.L. of equitoxic copper–zinc mixtures[245]. In lobsters, *Homarus vulgaris*, from sea water containing Zn at a concentration of 5 μg/litre, the blood contained 10 and 50 μg/g. A long exposure to sea water containing about 6 μg/g of Zn, mostly in the serum; the soft tissues contained between 100 μg/litre failed to alter the Zn concentrations of the blood, muscle and gonads, but increased the levels in the urine, excretory organs, hepato pancreas and gills. It seems likely that the Zn is absorbed from high concentrations of Zn in sea water via the gills, and this is removed from the circulation by urinary excretion and absorption by the hepato pancreas[246]. The laminarian seaweed *Laminaria digitata* does not regulate zinc, but absorbs it as a gradual process throughout the life of the plant; this process of absorption is not accompanied by the exchange of zinc. Once zinc has been absorbed, therefore, it shows little tendency to be lost from the plant[247].

The best known effects of heavy metals upon marine organisms are those due to mercury, copper and chromium, in that order.

Mercury is found in sea water at a concentration of 0·00003 mg/litre[29], and is found always in the atmosphere as trace amounts of the metallic mercury or as dimethyl mercury. Volatization takes place from sea water (cf. NH_3), soil and from the lungs and body surfaces of mammals. Bacteria and viruses can fix mercury and transform it biochemically. The formation of methyl mercury, which was the cause of Minimata disease, is considered to be by microbial means in sediments[248], but this has been difficult to establish[249].

Phytoplankton, e.g. *Chaetoceros costatum*, *Chlamydomonas angulosa* and *Phaeodactylum tricornutum* accumulate mercury, principally by surface adsorption[248], as do fresh water plants which do not incorporate methylmercuric hydroxide, phenylmercuric acetate, methoxyethylmercuric hydroxide, mercuric chloride and mercuric nitrate in their tissues to any appreciable degree[201]. Little is known of the bioaccumulation and biochemical transformation of mercury in higher marine organisms[248].

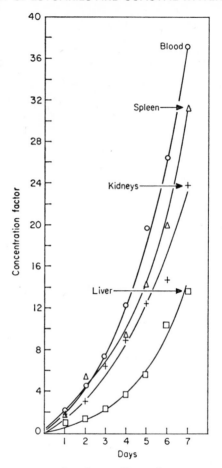

Fig. 16.9. Uptake of mercuric nitrate direct from water by the cod (after Hannerz[201]).

However, uptake is rapid (Fig. 16.9) and of the compounds listed above methyl mercury is taken up the most rapidly[201]. The concentration of the electrolytes in sea water has an important influence, since the uptake of methoxyethyl mercury is lower in brackish water than in sea water[201]. Similar marked concentration of Hg has been observed in the molluscs *Venus japonica*, *Venerupis philippinarum* and *Tapes*. The distribution of the mercury compounds within the animal depends upon its structure and the individual mercury compound[248]. In cod, of 32·5 cm length, the 7 day concentration factor of mercuric chloride in the gills, blood, spleen, kidneys, liver and muscles was 3668·20, 39·2, 31·22, 23·89, 13·69 and 3·38 respectively[201]. In *Venerupis philippinarum* the descending order of mercury accumulation, for phenyl mercury, was the mid-gut gland, Bojanus' organ, intestine and gill, and for mercuric chloride, Bojanus' organ, pericardium, gill and mid-gut gland[248].

The spider crab, *Maia squinado*, concentrates mercuric chloride from sea water, principally in the blood, antennary glands and gills: most of the $HgCl_2$ in the blood is combined with protein. When mercuric chloride is replaced by *n*-amyl mercuric chloride, the mercury again concentrates in the gills and various internal organs, but the amount detected in the blood is small, and none is detected in the urine. Prolonged exposure to $HgCl_2$ raises the total quantity of amino-N in the urine, a similar rise is noted with n-amyl mercuric chloride. Animals poisoned with $HgCl_2$ and returned to normal sea water show an abnormally high urine: blood ratio of amino-N several weeks later. In contrast, poisoning with $HgCl_2$ has no effect on the urine: blood ratio of total sulphate[250].

Table 16.11. The half time required for the elimination of mercury compounds from marine animals (after[248, 254–256])

Species	Compound	Half-time (day)
Venerupis decussata	Mercuric chloride	~100
Venerupis decussata	Methyl mercury nitrate	480
Mytilus galloprovincialis	Methyl mercury nitrate	1000
Carcinus maenas	Methyl mercury nitrate	400
Platichthys flesus (flounder)	Methyl mercury nitrate	700–1200
Anguilla anguilla (eel)	Methyl mercury nitrate	~1000
Serranus scriba (sea perch)	Methyl mercury nitrate	270

When the larvae of the barnacle *Elminius modestus* are subjected to equitoxic solutions of $HgCl_2$ and n-amyl mercuric chloride, they take up the poisons at the same rate in both treatments. The pattern of uptake of these compounds by the prawn *Palaemon serratus* is similar to that of *Maia squinado*[251]. When bipartite mixtures of copper (as copper sodium citrate) and mercury (as chloride, iodide and ethylmercuric chloride) are tested with *Artemia* synergistic effects are noted. Further, *Artemia*, which have been pretreated with a sub-lethal dose of copper, are markedly sensitized to poisoning by mercuric chloride and to a lesser extent by mercuric iodide and ethylmercuric chloride. These effects can be ameliorated if after treatment with the copper the *Artemia* are washed with solutions of either cysteine or reduced glutathione[252].

The mercury taken up is lost only slowly from seals[253], fish[201, 254], and invertebrates[248] and depends on the species and type of mercury compound[248, 255, 256] (Table 16.11).

In fish, the uptake of Hg takes place primarily from the water through the outer epithelia[201], but uptake with the food is important also[254]. However, the type of mercury compound may dictate the route by which uptake occurs, e.g. methyl mercury is more readily accumulated from the food than methoxyethyl mercury, but even here the predator/prey relationship is important for the difference is greater with cod fed upon mussels than with pike fed on fish fry. In fresh water fish, the uptake of inorganic mercury from the food is apparently negligible[201].

In laboratory culture, the efficiency of the transfer of mercuric chloride through the food chain; sea water → *Pseudomonas fluorescens* → *Artemia salina* → *Venerupis philippinarum*, was about 0·1%, while the transfer efficiency through the food chain without *Artemia* was estimated to be 1%. Once the mercury has reached the fish, these fish represent an important route for the transfer of mercury to higher trophic levels[248].

It is evident, from the above, that each mercury compound must be treated individually in toxicity studies, even though related derivatives have similar effects. Depending upon dose and chemical form, mercury poisoning can cause retarded growth, loss of weight, decreased viability and, in extreme cases, death. The precise mechanisms are not understood, but the absorbed poison may inhibit vital metabolic processes or affect the exchange capacity of the animals' external or internal surfaces[248].

Sporelings of the red algae *Plumaria elegans*, *Polysiphonia lanosa*, *Spermothamnion repens*, *Antithamnion plumula*, *Ceramium pedicellatum* and *C. flabelligerum* are all more sensitive to *n*-propyl mercuric chloride than HgCl₂; however, the sporelings of the first three species which have a high lipid content are much more sensitive to *n*-propyl mercuric chloride, which is a highly lipid soluble, than to HgCl₂ which is not, but with the last three species, which do not have such a high lipid content, the values for the two compounds are more nearly the same[257]. The red alga *Plumaria elegans* submerged for 18 hr in a range of mercury compounds was significantly more sensitive to organic than to inorganic forms[258] (Table 16.12). An essentially similar relationship was observed when *Artemia* and *Elminius* larvae were tested against these compounds; *Elminius* was always more sensitive than *Artemia*[259].

The harpacticoid copepod *Nitocra spinipes* suffer 50 and 100% mortality when exposed for 24 hr to 0·70 and 4·00 p.p.m. HgCl₂ respectively[260]. It is usual to find that juveniles are more sensitive than adults to toxic substances, however, although 1 p.p.m. HgCl₂ killed 90% of the adults of the barnacles of *Balanus balanoides* and *B. eburneus* within 48 hr, it required 16 p.p.m. to affect the metamorphosis of the cyprids[261].

The tolerance of fouling organisms to anti-fouling paints is variable, and to an anti-fouling paint containing 14–90% of mercury (dry weight of

Table 16.12. Comparative toxicities of inorganic and organic mercury compounds to spore-lings of the red algae, *Plumaria elegans* (after Boney, Corner, and Sparrow[258])

Compound	18 hr L.D.$_{50}$ (p.p.b. Hg)	Compound	18 hr L.D.$_{50}$ (p.p.b. Hg)
Methyl mercuric chloride	44	Isoamyl mercuric chloride	19
Ethyl mercuric chloride	26	Phenyl mercuric chloride	54
n-propyl mercuric chloride	13	Phenyl mercuric iodide	104
n-butyl mercuric chloride	13	Mercuric iodide	156
n-amyl mercuric chloride	13	Mercuric chloride	3120
Isopropyl mercuric chloride	28		

paints) the order of decreasing tolerance is *Polysiphonia* sp., *Balanus amphitrite*, *Bugula neritina*, *Balanus improvisus*, *Watersipora cucullata*, *Anomia* sp., *Enteromorpha* sp., *Hydroides parvus*, hydroids and tunicates; of these, only the first five species are found frequently on surfaces coated with paints containing mercury[262]. The larvae of some bryozoans, worms, molluscs and brine shrimps exposed to mercuric chloride solutions at 18–25°C reached the 50% death point in 2 hr, at the following Hg^{2+} concentrations: *Watersipora cucullata* 0·1 p.p.m.; *Spirorbis lamellosa* 0·14 p.p.m.; *Bugula neritina* 0·2 p.p.m.; *Galeolaria caespitosa* 1·2 p.p.m.; *Mytilus edulis* 13 p.p.m.; *Crassostrea commercialis* 180 p.p.m.; and *Artemia salina* 1800 p.p.m.[263].

Fish are as sensitive to mercury compounds as invertebrates. The symptoms of acute mercury poisoning are rigidity, widely-spread fins, lazy movements and "hanging" on the surface with the hind part of the body turned downwards. The euryhaline *Salmo gairdneri* and *Gasterosteus aculeatus* are more sensitive than freshwater species. They are also more sensitive to phenylmercuric acetate than to $HgCl_2$, and the survival time is decreased with increased temperature, and increased with increasing fish size. In the short term, 1 p.p.m. Hg^{2+} (as $HgCl_2$) is fatal for *Gasterosteus aculeatus*, on prolonged exposure this is reduced to 0·01–0·02 p.p.m. Hg^{2+}[248].

Copper is found in sea water at a concentration of 0·003 mg/litre: it is accumulated by plants and animals and is used as a respiratory pigment by certain annelids, crustacea and molluscs. It is highly toxic to algae, and in various formulations is used as an algicide[264–267]; it is toxic to seed plants, to invertebrates (copper compounds are used to control the snail vectors of schistosome dermatitis[268]), and is moderately toxic to mammals. However, it is not considered to be a cumulative poison like lead or mercury[29].

The toxicity of copper to aquatic organisms is related to the species, and the physical and chemical characteristics of the water. It acts synergistically with Zn, Cd and Hg, but there is little action with Ca. It is an element in anti-fouling paints: barnacles and related fouling organisms are killed in 2 hr by 20–30 mg/litre copper[29]. *Mytilus edulis* is killed in 12 hr by 0·55 mg/litre[261]. Copper is especially toxic to the bivalve molluscs *Mya arenaria*, soft-shelled clam; *Mercenaria mercenaria*, hard-shelled clam, and the oyster *Crassostrea virginica*. Copper at an experimental level of 0·02 p.p.m. is extremely toxic to *Mya*, and at 0·1 and 0·2 p.p.m. is of sufficient toxicity to reduce the duration of an experiment. At the 0·2 and 0·1 p.p.m. environmental level the uptake rate of Cu in *Mya* is 3 p.p.m. per kg per day and 5·6 p.p.m. per kg per day respectively, at 20°C; the decreased uptake at the upper level indicating the toxicity of the copper.

At a concentration of 0·05 p.p.m. the uptake rate at a temperature of 25–26°C was 20 p.p.m. per kg per day: this increased rate of uptake and concentration was a function of temperature[240]. Concentrations of Cu which exceed 0·1–0·5 mg/litre are toxic to oysters, the 96 hr TL_m is 1·9 mg/litre; oysters maintained in water at 0·13 to 0·5 mg/litre accumulate Cu and are unfit for human consumption[29]. The polychaete worm *Nereis diversicolor* dies in 2–3 days and 4 days when exposed to 1·5 and 0·05 mg/litre Cu respectively[242]. Net photosynthesis of the giant kelp *Macrocystis pyrifera* is inhibited to the 50% level in 2–5 days, and to the 70% level in 7–9 days on exposure to 0·1 mg Cu/litre: visible injury appeared in 10 days. Copper was slightly less toxic than mercury, but more so than nickel, chromium, lead or zinc[29].

Chromium is found in sea water at a concentration of 0·00005 mg/litre. Its toxicity in sea water varies with species, temperature, pH, valency, and synergistic or antagonistic effects. Under long term exposure the hexavalent form is no more toxic than the trivalent form[29], although toxicity of these forms is apparently influenced by salinity[41]. The photosynthesis of the kelp *Macrocystis pyrifera* exposed to 1 mg/litre chromium was reduced by 0, 10–20 and 20–30% after 2, 5 and 7–9 days contact respectively: at a concentration of 5 mg/litre, an exposure of 4 days was necessary to induce a 50% reduction in photosynthesis. The mortality of oysters exposed to the metals chromium, molybdenum and nickel at a concentration of 10–12 μg/litre for two years was seasonal *ca.* 63–73% of the mortalities took place in the period May to July, i.e. at a time of increased physiological activity. The mortalities also increased with temperature[29]. The threshold of toxicity for juveniles of the prawn *Palaemon serratus* is a little less than 5 p.p.m. Cr: at concentrations from 10–80 p.p.m. Cr, 100% mortality occurred in one week, at 5 p.p.m. there was no death in the first week, but a few animals died in the following 21 days. Larger individuals of *P. serratus* have a toxicity threshold of *ca.* 10 p.p.m. Cr. *Carcinus maenas*, the shore crab, exposed to 20, 40 and 60 p.p.m. of sodium chromate suffered a mortality of 8, 9 and 50% after a 12 day exposure period. The polychaete, *Nereis virens*, suffered a heavy mortality after a 2–3 week exposure to 2–10 p.p.m. Cr: the threshold of toxicity appeared to be 1·0 p.p.m. Cr[242].

Detergents and Surfactants

During the period since World War II synthetic detergents have replaced a majority of the soap products. First concern regarding these substances arose from the foaming which they caused in the rivers of Europe and North America. The term "detergent" is loosely applied to a variety of commercial products and to their surface active constituents, i.e. surfactants. For clarity it is essential to make a distinction between detergent and surfactant. The former is best reserved for commercial preparations, tailor-made for many uses, domestic and industrial, ranging from shampoos to oil emulsifiers and to bactericidal creams. Detergents contain a variety of chemical compounds to give the product its desired properties, including a relatively small proportion of a surface-active agent or surfactant: these subsidiary constituents are:

Filler which is an inorganic or organic product employed to produce the desired presentation and/or concentration; it is normally inert, e.g. sodium sulphate, water, alcohol.

Ancillary is a subsidiary constituent which imparts added properties, but not in the case of a washing detergent, related to the washing action as such. Ancillaries are usually present in small quantities, e.g. optical bleaches, corrosion inhibitors, antistatic agents, colouring matter, perfumes and bactericides.

Builder is a subsidiary constituent, usually inorganic, which, with reference to the washing action, adds its characteristics to those of the essential constituents. In other words, by synergising with the surfactant they increase the surfactants detergent capacity, e.g. tripolyphosphates, carbonates and silicates.

Booster is a subsidiary constituent of a detergent, usually organic which enhances certain properties of the essential constituents, e.g. alkanolamides and amine oxides[269].

Surfactants belong to four principal groups—anionic, non-ionic, cationic and ampholytic—and only the first two are of concern in the sea. Unlike the anionic surfactants, the non-ionic surfactants are used more in industry than in the home. Although most detergents released to the sea, usually with untreated sewage, contain anionic and non-ionic surfactants, it is the non-ionic ones, used as oil emulsifiers, that have produced the most spectacular effects. While biological effects are important, the physical effects of these substances should not be ignored. They lower the surface tension, reduce the exchange coefficient of oxygen by 20% [63, 270] (p. 6) and reduce wave heights[271, 272, X]. The influence of the detergent is not due to the surfactant portion alone, the eutrophication of many bodies of fresh water has been attributed to enrichment arising from the polyphosphates in detergents, but the extent of their contribution is uncertain.

Marine life is sensitive to detergents, but the anionic and non-ionic detergents are poisonous to a highly variable degree. The resistance of these organisms depends on species, age and habitat: shore-dwelling species are more resistant than those from below the low-water mark[25, 272]. The dose response curve is *paradoxical*, i.e. more concentrated solutions may be less toxic than those which are less concentrated[241].

The chemical nature of the surfactant is important too, highly branched anionic alkylbenzenesulphonates are not markedly biodegradable, but with a reduction in branching to a straight alkylic chain, bio-degradability increases: a long, straight unbranched chain with many C atoms is most easily biodegraded. However, in this case the toxicity of the molecule increases compared with that containing the unbranched chain[270, 273]. Where a detergent is biodegradable the properties and toxicity of the degradation products may be uncertain, however, a biodegradable non-ionic detergent, which contains ethoxylate groups, degrades to polyols which are markedly synergistic with respect to the foaming of anionic surfactants in rivers. Experiments upon *Eupagurus bernhardus* with detergents containing nonionic surfactants with differing numbers of ethoxylate groups have shown that the toxicity increases with the increasing number of EO groups and then diminishes once more[Y].

Experiments conducted upon the amphipod *Gammarus tigrinus* and the eel *Anguilla anguilla* in Germany, and mummichog *Fundulus heteroclitus*, and the eel *Anguilla rostrata* in North America, have shown that at the same concentration of anionic detergent, the toxicity may at first fall with a slight increase in salinity from the freshwater condition, but with a marked increase in salinity the toxicity is increased significantly[29, 273, 274]. Apparently, this change is related to the considerable decrease of surface tension in brackish water and in sea water[273].

The nonionic surfactants are much more toxic than anionic surfactants to marine animals[275, 276], moreover biodegradation does not diminish the toxicity of these non-ionic surfactants with time[276].

The *Torrey Canyon* incident, in which large amounts of oil-emulsifer were used, led to a considerable amount of work in this field. An oil emulsifier consists of a surfactant portion, dissolved in a liquid carrier whose function is to permit a thorough mixing of the surfactant with the oil, plus a small amount of stabilizing element. The surfactants used in the oil-emulsifiers at the time of the *Torrey Canyon* affair, were carried in a highly toxic aromatic solvent; not unnaturally a widespread mortality resulted[25, 208, 209, 213, 222, 277–279]. Since then, considerable efforts have been made to prepare a formulation which was relatively non-toxic to marine life. This has been successful to a considerable degree as far as animals are concerned, but new and old formulations are equi-toxic to salt marsh plants[221, 222]. Furthermore the discussion rather lacks meaning when it is recalled than an oil/oil-emulsifier mixture may be more toxic than either of the constituents (p. 551).

It is a characteristic of the oil-emulsifier formulation and of the surfactant constituent, that a delayed mortality and growth abnormalities and inhibition occur even at sub-lethal concentrations. This has been observed in the embryos and larvae of fish[280]. The larvae of the polchaete *Sabellaria spinulosa* showed all indications of recovery after treatment with B.P.1002, and then subsequently died after some weeks in a healthy condition[278]. Single 24 hr doses of oil emulsifiers and of non-ionic surfactants alone produced, over 10–22 weeks, delayed mortality and inhibition of growth in the winkle. *Littorina littorea*, the rough periwinkle, *L. saxatilis*, and the dog-whelk, *Nucella lapillus*; in the case of *L. saxatilis* at < 1/3000 of the 24 hr L.C.$_{50}$[281]; this halt in growth, though prolonged, is of a temporary nature[282]. In addition, *L. saxatilis* is more susceptible to predation by the shore crab, *Carcinus maenas*, at doses as low as 1/30 000 24 hr L.C.$_{50}$, and *Nucella lapillus* is significantly more susceptible to predation by the whelk, *Buccinum*, when treated at *ca.* 1/100 TL$_m$[272, 281].

Treatment of cultures of the alga *Phaeodactylum tricornutum* with a 5·6 p.p.m. solution of Gamasol and a 8 p.p.m. solution of its surfactant element, reduced cell division by 50%[283]. The filamentous green algae of the *Cladophora* type are more sensitive to these substances than the fucoids, and the young plants of *Bryopsis* are also very sensitive. The relatively high resistance of the intertidal red algae *Polysiphonia lanosa* and *Porphyra umbilicalis* may be associated with the fact that they are algae which are normally subjected to some drying during periods of tidal exposure[284]. The young, non-motile, reproductive cells of the green alga *Prasinocladus marinus* are more sensitive to B.P.1002 and its component fractions than are the "aged" cysts. The effect of these toxic agents was much reduced at a low temperature (4°C) although rapid changes in the chloroplasts and pyrenoids were observed. In contrast, to the animal work noted above, a reduction in salinity brought about a marked increase in the toxic effects[285].

It is not easy to study the effects of long term treatment upon the reproduction of marine animals. However, the freshwater gastropod *Physa fontinalis* when dosed with 1 p.p.m. Essolvene weekly, spawned earlier, layed larger egg masses and died earlier than the untreated animals. The eggs of the treated animals took longer to hatch than those of the controls, but the young were apparently not affected by the treatment which their parents received[272].

The young of the rainbow trout *Salmo gairdneri* show marked differences in their resistance to the action of the non-ionic surfactant, nonyl phenol ethoxylate. The alevins, immediately after hatching, survived concentrations up to 42 p.p.m. for 6 hr. This resistance decreased gradually with the absorption of the yolk sac, so that when it was completely absorbed, and before feeding commenced, the lethal concentration was 2·5 p.p.m.; after feeding it rose slightly, but showed no increase from 15 days onwards; a treatment of 5·2 p.p.m. for 6 hr was toxic to both 40 and 210 day old fish. It is considered that these changes are not correlated with the respiration of these fish, but with the changes in metabolic activity in the post-hatch period[286].

During development, the marine fish herring, lemon sole, pilchard, plaice, sole and haddock show a broadly similar pattern; embryos within the chorion are much more resistant than the larvae and had a 100 hr L.C.$_{50}$ > 20 p.p.m. The most sensitive stage of the embryo appeared to be before gastrulation, and exposure to dispersants at a concentration > 20 p.p.m. shortly after fertilization disrupted normal cell division. Once hatched the tolerance of the larvae fell sharply, i.e. the 100 hr. L.C.$_{50}$ fell from > 20 to < 8 p.p.m. with the shedding of the chorion. These young fish became gradually more susceptible until at the resorption of the yolk sac and during the period of first feeding the 100 hr L.C.$_{50}$ fell to ca. 2 p.p.m. after which it rose once more to > 8 p.p.m. at metamorphosis[280].

Anionic detergents may react with calcium, and thus in turn interfere with cell membrane structures, and osmoregulation. Cytoplasmic membranes, being composed of a protein/lipoid "sandwich", appeared to be affected by surfactants becoming adsorbed on to the membrane or competing for adsorption with an existing membrane component. The former leads to fixing, and the latter to lysis[217]. Surfactants alter the permeability of the cell membrane, and hence their use with herbicides and insecticides to ensure greater penetration and effect. This ability of a surfactant to become firmly fixed in a cytoplasmic membrane leads one to question the effect and breakdown of biodegradable surfactants in the sea. First, they do not undergo a marked decrease in toxicity with time[276] and the rate of breakdown is very slow[273]. Furthermore, studies of biodegradability in the laboratory are made with special reference to sewage works and under very special conditions[287]. In the sea, there are few bacteria in the main body of the water and most are associated either with surfaces or with sediments, therefore the opportunities for contact between bacteria and detergent leading to biodegradation are rather small. Will a surfactant which has become incorporated in a cytoplasmic membrane undergo biodegradation? In addition, the behaviour of surfactants in sea water is very much an open question, for micelle formation offers a poor explanation of their behaviour[25, 272, 281]; elucidation of this interesting physicochemical problem could provide an explanation for the paradoxical survival and growth response curves shown by organisms treated with these substances[241, 281].

Although there may be some doubt about the physico-chemical behaviour of these interesting substances, and concerning their mode of action, they do have some important effects at the tissue level. A concentration of 1 p.p.m. causes foaming in rivers. The gills of *Lepomis* (sunfish) are damaged by 3 p.p.m. A. B. S. and <3 p.p.m. L.A.S. Bullhead maintained in 0·5 p.p.m. experience erosion of taste buds and an impairment of receptor functions related to swimming and feeding behaviour. Fish so affected did not recover fully after 6 weeks in detergent free water[288]. In studies on *Fundulus heteroclitus*, mummichog, exposed to alkyl benzene sulphonate, no effects on red blood cell number, gonadosomatic index or liver condition were recorded[274]. However, "oil spill' detergents and the constituent surfactants are biochemically powerful agents[289].

Pesticides

Pesticides may be described as natural and synthetic materials used to control unwanted or noxious animals and plants[29]. "The effects of the vast majority are relatively non-selective and usage therefore may result in undesirable, even unanticipated side effects"[290]. For convenience, they are classified as herbicides, insecticides, fungicides. fumigants and rodenticides, of which the first two only have a known impact upon the marine environment. Both of these kinds of substances have been used in ever increasing amounts since the end of the 1939–45 war. It is difficult to obtain data of their application, but in 1965 the United States alone produced more than 875×10^6 lb, which represented a 10% increase in production over 1964. The use, and abuse, of these substances lead to a widespread and continuing public concern, following the publication of "Silent Spring"[291].

Herbicides

Herbicides, including algicides, are widely used in the terrestrial and fresh water aquatic environments to "control" unwanted plant growth. Such substances include a wide range of organic synthetics the fate of which, having reached a water-course, is largely unknown. In fresh water, however, they have a toxicity to animal life which is comparable with the B.P.1002 generation of oil-emulsifier detergents[290, 292–301]. Much less is known regarding the marine environment, though presumably some estuaries must receive significant amounts of these substances. It is, however, known that the herbicide and defoliant 3-animo-1,2,4-triazole (3AT) inhibits cell production in sporelings of the red algae *Antithamnion plumula, Plumaria elegans, Callithamnion tetricum, Nemalion multifidum* and *Brongniartella byssoides*. Moreover, short term immersions in a culture medium containing 3AT have a lasting effect upon the sporelings; long term exposures induce chlorosis. This inhibition of growth by 3AT is less marked with the sporelings of sub-littoral red algae than with those of the intertidal zone[302].

Insecticides

As with so many other pollution problems much more is known of the effects of these substances upon freshwater organisms than upon marine organisms. Massive fish kills, particularly in the water ways of the United States, have resulted from the use of these substances[290]. These are overt symptoms for the insecticides are extremely toxic to aquatic vertebrates and invertebrates alike: biological magnification occurs along the food chain[290, 303–312].

The diatom *Navicula seminulum* var. *Hustedtii* is relatively resistant to dieldrin (TL_m 12·8 p.p.m.) compared to fish and aquatic invertebrates[313]. Fish, viz. mosquito fish, *Gambusia affinis*, and the Atlantic salmon, *Salmo salar*, can metabolize D.D.T. to D.D.E. and T.D.E., but whether other metabolites are produced is uncertain[314–315]. Some species of fish, viz. *Gambusia affinis*, *Notemigonus crysoleucas*, *Ictalurus natalis* and amphibia, viz. *Acris crepitans* and *A. gryllus*, have developed resistant strains to D.D.T., Endrin and Aldrin[316–324]. However, an unfortunate consequence of this resistance is that such resistant strains are likely to be toxic to predators, including man[325, 326].

Most species have not developed resistant strains, and it is these with which the succeeding paragraphs are concerned. Streams to which D.D.T. has been applied show a marked reduction in the quality and quantity of the invertebrate fauna[327]. Fish exposed to chlorinated hydrocarbons evince symptoms of hyperactivity and hypersensitivity to disturbance; however, those exposed to Dursban, an organo-phosphorus insecticide, show partial paralysis and loss of orientation before death, but are not hyperactive[328]. Fish, *Ophicephalus punctatus*, *Heteropneustes fossilis*, *Trichogaster fasciatus* and *Barbus stigma* exposed to Dieldrin, B.H.C. and Lindane undergo histopathological changes of the liver[329–331]. Guppies, *Poecilia reticulata*, exposed for 14 months to treatments with Dieldrin near the estimated "biologically safe concentration", i.e. 0·0018, 0·0056 and 0·01 p.p.m., first increased in total numbers compared with the controls, and then towards the end of the test period, evinced deleterious symptoms at the highest concentration tested. The initial increase in numbers, in the test solutions compared with the controls, was considered to be due to a slight change in the feeding habits of the adults, i.e. an inhibition of feeding upon the fry, induced by the dieldrin[332]. The median tolerance limits of the sunfish, *Lepomis gibbosus*, with respect to Dieldrin are 24 hr L.D.$_{50}$ 0·0155 p.p.m., 48 hr L.D.$_{50}$ 0·012 p.p.m., 72 hr L.D.$_{50}$ 0·0075 p.p.m. and 96 hr L.D.$_{50}$ 0·0067 p.p.m. Chronic exposure, for 12 weeks, to a concentration of 0·00168 p.p.m. of dieldrin affected the oxygen consumption and swimming ability, i.e. cruising speed, of the fish[333]. In those areas where the use of D.D.T. is high, the progeny of adults which have lived in these waters is liable to a high mortality: in Lake George, New York State, the lake trout fishery has declined steadily since 1950 from this cause[334, 335].

The essential difficulty created by the chlorinated hydrocarbons, viz. D.D.T., Endrin and Dieldrin, is the cumulative nature of their toxicity[336] and although they are metabolized to some degree, no one seems sure of the extent of this metabolism. The much publicized Coho salmon scare of the Great Lakes in 1969, is an example of this problem. The Coho were introduced to take advantage of the large stocks of food provided by the alewife, *Alosa pseudoharengus*, and thus provide sport and food. In 1969, tests on the salmon caught showed them to have a mean D.D.T. residue of 15 p.p.m., whereas the maximum limit accepted by the Food and Drugs Administration was 5 p.p.m. Consequently, such fish were condemned. This problem was exacerbated by the fact that at least 50% of the gas chromatogram could be ascribed to P.C.B.s, or even non-toxic fractions. Current opinion has it that many D.D.T. determinations are either in error to a similar degree or are of a non-toxic mimic of D.D.T. With such difficulties facing investigators and regulatory agencies alike, it seems probable that the problem of D.D.T. will remain diffuse for some time.

The chief source of pesticides in the estuarine environment and biota is agricultural practice and malpractice[29, 337]; it is the dispersal in drainage systems and the possible accumulation in estuaries which renders coastal fisheries especially vulnerable to their toxic effects[29].

The acute toxicities of many pesticides are known[338-341]; insecticides are much more toxic to marine animals than are herbicides[342] (Table 16.13). These laboratory tests have shown that economically important animals are very sensitive to insecticides[343]. Indeed, concentrations which are insufficient to control many pestiferous insects, including several species of salt-marsh mosquito, may cause considerable harm to marine organisms. These harmful effects include inhibition of phytoplankton productivity, death or immobilization of fish, molluscs and crustaceans, death of eggs and larvae of bivalve molluscs and deleterious changes in molluscan and fish tissues[344]. The larvae of the oyster *Crassostrea virginica* will die after 4 days exposure to 1 p.p.m. of D.D.T. in sea water, growth is almost stopped by 0·05 p.p.m. and is poor at 0·025 p.p.m.[343]. The adult will live in the presence of D.D.T. at levels as high as 0·1 p.p.m. in the

Table 16.13. Relative toxicity of common types of pesticides to estuarine fauna compared to herbicides rated as unity (after Butler[842])

Pesticide type	Zooplankters	Shrimp	Crab	Oyster	Fish
Herbicide	1	1	1	1	1
Insecticide					
Organophosphorus compound	$\frac{1}{2}$	1000	800	1	1
Polychlorinated hydrocarbon compounds	3	300	100	100	500

environment, but at levels 1000 times less, i.e. 0·1 μg/litre, its growth or production is reduced by 80%, i.e. to 20% of normal growth: in similar conditions, a shrimp population would suffer a 20% mortality and menhaden a disastrous mortality[29]. Other substances are not as toxic as D.D.T., e.g. Dipterex and Parathion do not kill more larvae than die in the controls at 1 p.p.m., although growth is somewhat retarded; at this concentration Gunthion causes neither mortality nor retarded growth[341]. Conversely other insecticides will kill 50% of shrimps after 48 hr exposure to concentrations of only 30–50 nanograms per litre[29].

Biological accumulation and trophic magnification is a characteristic of these substances in the marine, just as it is in the freshwater environment[337, 345, 346]. Off California, some organisms are such effective concentrators that they may eventually build up pesticide residues to levels 100 000 greater than those in their food[345]. In the estuaries of Texas, a similar trophic magnification of D.D.T. resulted in the reproductive failure of sea trout populations in the lower Laguna Madre in 1969; the sea trout in other Texas estuaries remained unharmed because of their different food chain interactions. In these estuaries, there is an accumulation of D.D.T. upon the sediments which built up to a plateau over the years. These residues did not reflect the seasonal levels of water borne pesticide pollution of the environment, however, the sedimentary concentrations of D.D.T. may be resuspended physically by storms and recycled in the biota. With the introduction of restrictions upon the agricultural use of D.D.T., decreased residues in the biota of adjacent estuarine areas were reflected within three years[337]. In Britain, organochlorine residues in waterways and fish arise from the careless discharge of chemical concentrates and sheep dips, field spraying or long-term leaching, and are due only to dieldrin, B.H.C. and D.D.T. or its breakdown products[308]. However, it should not be forgotten that these substances are volatile and that atmospheric translocation and deposition by rainfall occurs on a global scale[345].

Industrial toxicologists, physicans and others concerned with the health of persons, who routinely work with dangerous or hazardous materials, have developed a stress profile technique based upon the hypothesis that metabolic disturbances have a far greater probability of detection if a number of parameters are measured simultaneously. This interesting technique has been applied to the influence of chlorinated hydrocarbons upon the fish *Sphaeroides maculatus*[344, 347] (Table 16.14). Results obtained by such a technique may be plotted up on a series of bar graphs arranged radially (Fig. 16.10) and the method may be applied to invertebrates also[344]. The quahog, *Mercenaria mercenaria*, when subjected to chronic treatment of sub-lethal doses of D.D.T. and lindane, experiences metabolic

Table 16.14. Effect of endrin on various blood and tissue constituents of northern puffer *Sphaeroides maculatus*, surviving 96 hr exposure to 1·0 μg/litre (after Eisler[344], Eisler and Edmunds[347]).

Constituent	Controls	1·0 μg/litre endrin	% deviation from controls
Liver Ca—g/kg ash	127·5	10·0	−92
Liver Zn—g/kg ash	37·5	4·5	−88
Gill Zn—g/kg ash	4·4	3·0	−32
Liver K—g/kg ash	165·0	130·0	−21
Liver Na—g/kg ash	137·0	120·0	−12
Haemoglobin—g/100 ml	7·86	8·39	+7
Serum Ca—mg/100 ml	13·8	14·8	+7
Serum Na—mg/100 ml	225·4	246·1	+9
Serum K—mg/100 ml	66·8	91·4	+37
Serum cholesterol—mg/100 ml	269·4	440·8	+64

alterations indicative of an increase in glucose degradation and a suppression gluconeogenesis[348].

As with freshwater organisms, the problem of most concern to man is the contamination of his food supplies by biological magnification[349–351]. This problem is not uniform throughout the world. For example, analyses of seal blubber have shown that the levels of contamination in the Baltic, North Sea and Irish Sea are very much higher than those of the Arctic, Atlantic and Pacific oceans, when judged by the presence of pesticide residues and P.C.B. residues. Dieldrin and P.C.B. levels are highest around Great Britain and D.D.T. levels highest in the Baltic and Gulf of St. Lawrence; there is no evidence that seals in good condition are affected by these residues[349]. In the processing of natural products, e.g. vegetable oils and fish oils, attempts have been made to eliminate the chlorinated hydrocarbon residues present by vacuum steam deodorization, vacuum steam distillation and refining, and hydrogenation and deodorization. The degree of elimination is variable and depends upon the nature of the residue, e.g. hydrogenation and deodorization will remove 90% of D.D.T., but only *ca.* 67–79% of D.D.E. was removed by this means[351].

Associated with, although not used as, pesticides two groups of organochlorine compounds have come into prominence in recent years. The P.C.B.s or polychlorinated biphenyls are widely used in industry as plasticizers, lubricants, heat transfer media and insulators. They are compatible with paints, synthetic resins, varnishes and waxes, and improve such properties of these products as their chemical resistance, water resistance and flexibility. Their presence in the biosphere was first recognized in 1966[352], and they are now known to be of global occurrence[349, 350, 352]. Some toxicological effects have been produced

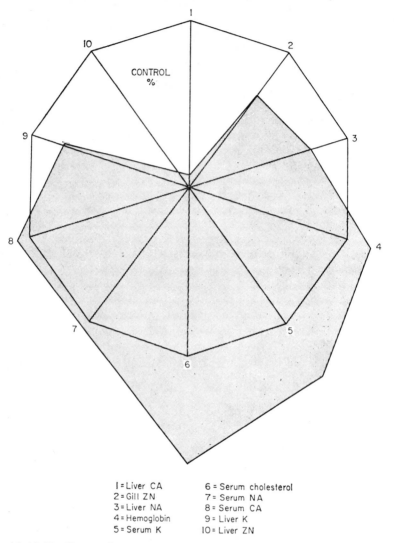

I = Liver CA   6 = Serum cholesterol
2 = Gill ZN   7 = Serum NA
3 = Liver NA   8 = Serum CA
4 = Hemoglobin   9 = Liver K
5 = Serum K   10 = Liver ZN

Fig. 16.10. Profile on effect of 1·0 μg/litre of endrin on selected blood and tissue constituents of northern puffer, *Sphaeroides maculatus*, surviving exposure for 96 hr (after Eisler[344]).

experimentally in birds, and it has been inferred that these substances are a hazard to marine life[352].

The other group of chemicals are the unwanted chlorinated aliphatic hydrocarbons formed as a by-product in vinyl chloride production. These substances are dumped at sea, and dead plankton and fish have been associated with these dumpings. It has been shown experimentally that the 24 hr L.C.$_{50}$ of *Gadus morhua, Pleuronectes platessa, Ophiura texturata* and *Leander adspersus* with respect to these wastes are *ca.* 6, 15, 38 and 12 p.p.m. respectively; *ca.* 13 p.p.m. causes a 50% reduction in the photosynthesis of *Dunaliella* sp. These substances are distributed widely in the sea and their breakdown rate is apparently slow[353].

Other Toxicants

Ammonia and Ammonium Compounds

Many industrial waste discharges contain ammonia. Depending upon the pH and the oxygen concentration, ammonia is very toxic to marine life within the range 1·0–25 mg/litre. Its toxicity depends primarily upon undissociated NH_4OH and nonionic ammonia: in fresh water, it is considered that when the pH exceeds 8·0 the total ammonia expressed as N should not exceed 1·5 mg/litre. In sea water, a concentration of 1·0 mg/litre NH_3 impairs the ability of haemoglobin to combine with oxygen and fish may suffocate[29]; however, the photosynthesis of diatoms is not inhibited by this concentration[354].

Cyanides

Hydrocyanic acid, or hydrogen cyanide, and the cyanides are extremely poisonous and of a considerable industrial importance. At a pH > 8·2, hydrogen cyanide is dissociated, but with a decrease in pH its toxicity rises since the toxicity is due to the undissociated molecule[29]: in fresh water, a fall in pH from 8·0 to 6·5 increases the toxicity of nickelo-cyanide a thousand fold[355]. The toxic action of cyanides rises rapidly with temperature[29]. When removed to water free from cyanides, fish can recover from a short exposure to concentrations < 1·0 mg/litre, which apparently act as an anaesthetic: at these concentrations, it seems that the fish convert cyanide to thiocyanate, an ion which does not inhibit respiratory enzymes. Complex cyanides may be more or less toxic than HCN: complexes of zinc and cadmium are more toxic; that between CN and nickel is less toxic than the CN itself at high pH levels[29].

Both cyanide, as HCN, and ammonia, as NH_3, are present, in an undissociated state, in blast furnace gas washing[356]. The 96 hr L.C.$_{50}$ of these washings with respect to the hermit crab, *Eupagurus bernhardus*, is > 10% of effluent in sea water of salinity 30‰. Chronic exposure to low concentrations produced a 50% mortality after 17 weeks at a 200 times dilution i.e. 0·5% effluent[356a].

Tar, Gas and Coke Wastes

Coal is distilled to yield gas, coke and tarry materials from which a variety of organic chemicals are produced. This process gives rise to a watery ammoniacal gas liquor which contains free ammonia, ammonium salts, cyanide, sulphide, thiocyanate, pyridine, phenols, cresols, xylenols, and aromatic acids. Ammonia may be removed and "spent gas liquor" produced. The materials in this effluent are all highly toxic, e.g. freshwater fish and lower aquatic life are affected by 2–75 mg/litre of cresols and 0·1–50 mg/litre of phenols[29].

A coke-oven effluent from a steel works was considerably more toxic than the effluent from the blast furnace gas washings: the 48 hr L.C.$_{50}$ with respect to *Eupagurus bernhardus* was $\sim$1·0% of the effluent in sea water at a salinity of 30‰; animals maintained chronically in a 200 times dilution suffered a 100% mortality in 29 to >114 days[356a].

Sulphides and Fluorides

There is apparently no information regarding the toxicity of these substances to marine biota. However, in fresh water, sulphide concentrations in the range 1·0–25·0 mg/litre are lethal in less than 3 days, and fluorides are toxic to fish at concentrations which exceed 1·5 mg/litre[29].

Phosphorus

Yellow phosphorus is remarkably stable when released to the sea, and especially when absorbed on to a sediment or other solid support. It is extremely toxic to herring, salmon, lobster and gammarid amphipods to which the incipient lethal levels (I.L.L.) are <2·5, 18, 40 and 3–4 mg/litre respectively. These toxic effects are irreversible and probably cumulative. An accidental pollution by yellow phosphorus may therefore have biological consequences over a much wider area than that adjacent to the spill. Death in fish and invertebrates is apparently due to asphixia resulting from massive haemolysis and coagulation of the blood respectively[357].

Large Power Stations

It has become customary to refer to the effects of large power stations as being those of thermal pollution. However, this seems a poor term to use. It can be demonstrated in a coastal plain estuary, which has a large expanse of mud and sand flats exposed by the receding tide, that the heating effects due to isolation of these flats, and a subsequent exchange of heat between the flats and the incoming tide, may be of the same order or dwarf that due to a large power station sited on the estuary (p. 34). Moreover, the possible effects of such a power station are not confined to heating of water alone.

In the past two decades, large power stations have been sited on estuaries, or on the open coast, in increasing numbers, because there is insufficient water in many rivers, especially in Europe, for cooling purposes. For example, the new Fawley Power Station of the Central Electricity Generating Board has an output of 2000 megawatts, and requires 50×10^6 gal/hr for the purposes of condenser cooling: this water experiences a rise of 8°C, over the ambient, on passing through the condensers. The potential consequences, which arise from this operation, with respect to living organisms are:

1. the effects of chlorination;
2. the effects of warming the water;
3. the effects whereby the power station acts as a fishing machine;
4. where the station generates its heat by means of a nuclear reactor, there will generally be some rather small amount of radioactive waste from the spent fuel "cooling pond". This is a relatively minor issue in power station operation, although it has considerable political overtones. However, it is both more convenient and more logical to consider it under the heading of radioactivity below.

The Effects of Chlorination

The need for chlorination to clear fouling organisms has already been discussed (Chapter 12), it is sufficient to say there that the best current practice is to inject enough chlorine into the system to ensure a residual concentration of *ca.* 0·5 mg/litre. This is sufficient to prevent fouling problems and at the outlet gives a residual concentration of less than 0·1 mg/litre, which is quickly absorbed by suspended solids present in sea water[358]. Oyster larvae can withstand 2·5 p.p.m. of chlorine for periods greater than 10 min, even at temperatures as high as 30°C; indeed, larvae, in oyster culture, are exposed to 3 p.p.m. of chlorine for 5 min, before transfer to fresh, sterile seawater. This method ensures that bacteria do not build up in the larval cultures. In British waters, at least, it seems unlikely that this source will kill oyster larvae which either pass through a cooling water system or are present in the receiving waters. The nauplii of *Elminius* are more sensitive than oyster larvae, free chlorine concentrations in excess of 0·5 p.p.m. cause a heavy mortality and a reduced rate of growth in the survivors, nothing is known of effects upon *Elminius* at concentrations <0·5 p.p.m. of free chlorine. The larvae of *Crepidula*, slipper limpet, and *Littorina*, winkle, can probably withstand much heavier doses than can barnacle nauplii, copepods and other crustacean and worm larvae[359]. In recent years, the development of fish farming has led to the use of power station outfalls to grow fish: here both *Pleuronectes platessa*, plaice, and *Solea solea*, sole, survive the low concentrations of residual chlorine[360].

At Hunterston Nuclear Generating Station average residual chlorine levels of 0·02–0·35 p.p.m. are recorded; levels of 0·5 and 0·7 p.p.m. are reached occasionally. These high levels quickly kill plaice larvae, and the higher levels of the more average figure would also kill. Plaice eggs exposed for 3 days, at 7·5°C, to a range of chlorine concentrations experienced 36% mortality at 0·36 p.p.m. Cl_2; at 0·01 p.p.m. mortality was 9%. Although some mortality, of eggs and larvae, may result from this cause within the power station, it seems unlikely to occur in the sea itself, but this remains to be determined[361].

A scare raised in the late 1950s concerned the possible formation of chloramines in sea water and their likely effects upon marine life. In fresh water, chlorine first forms hypochlorite which reacts thus in the presence of ammonia:

$$OCl' + NH_3 \xrightarrow[\text{H}_2\text{O}]{\text{in}} NH_2Cl + OH'$$
monochloramine

$$NH_2Cl + OCl' \longrightarrow NHCl_2 + OH'$$
dichloramine

It was presumed that this reaction would occur at low concentrations of chlorine in sea water. However, there is no evidence to suggest that it occurs at concentrations compatible with economic power station operation. Indeed, as with so many other substances, a knowledge of the chemistry of chlorine in sea water, rather than the simple system of fresh water, would be invaluable.

Effects of the Warm Water Effleunt

The effects of the warm water flow, which in a large power station is equivalent to that of a sizeable river, may be due to direct and indirect effects thus:

Direct effects:
1. death through the direct effects of heat, particularly upon μ-flagellates, plankton and planktonic larvae;
2. induction of physiological aberations, i.e. in growth, respiration and feeding;
3. interference with a spawning and other critical activities;
4. competitive replacement by more tolerant species;
5. encouragement of exotics and unwanted pest species (exotics such as *Balanus amphitrite* are known to occur near some power stations[362]);
6. enhancement of toxicity of substances dissolved in sea water;
7. effects upon the external metabolites.

Indirect effects include:

1. loss of a food supply, e.g. death of μ-flagellates on passage through the cooling water system, could adversely affect survival of larvae such as the oyster;
2. changes in a sedimentation regime could influence the biota;
3. changes in the external metabolite regime could influence the populations and production of the estuary.

For most of these possibilities relatively little information is available. Moreover, current knowledge would suggest that in the higher latitudes, at least, only species near to their geographical limit are likely to be affected adversely. However, in tropical and sub-tropical waters the situation seems to be quite different.

The invertebrates *Sphaeroma hookeri, Gammarus locusta, G. zaddachi salinus, Palaemonetes varians, Potamopyrgus jenkinsi* and the fish *Gasterosteus aculeatus* and *Gobius minutus*, all inhabitants of brackish waters, can pass through cooling water (C.W.) systems apparently without harm, although the larger individuals may be mutilated and die. Both *Palaemonetes varians* and *Gasterosteus aculeatus* lived for a few days in the laboratory after passage through this system[363, 364]. Nothing is known of the influence of a passage through the C.W. system upon the μ-flagellates.

With regard to the immediate effects of the warmed water upon the life adjacent to the outfall: at Cavendish Dock, Barrow-in-Furness, the specific composition of invertebrates was the same at the intake and at the outfall. However, algae, *Enteromorpha intestinalis* and *E. ahlneriana* grew abundantly in the outfall, but not in the intake; benthic invertebrates living in the outfall appeared earlier than those in the intake[365].

At Chalk Point, Maryland, also, the species composition in the intake and outfall channel did not differ, further the set of attached organisms took place earlier and they were more abundant in the outfall channel than in the intake. However, during the summer when temperatures were high there was an apparent reduction in number or disappearance of flatworms, and colonial hydroids, and an increased barnacle growth in the outfall area[366].

Oysters collected in the immediate vicinity of the discharge at Bradwell Nuclear Generating Station were in poor condition, but, those collected from the barrier wall just 80 m from the point of discharge were in better condition than those taken from layings on the adjacent south shore of the Blackwater estuary[367]. At Smithtown Bay, Long Island, sediments, overlain by a heated effluent from an electric power station, had *Elphidium clavatum* as the only foraminiferal inhabitant[368].

With respect to large bodies of water to which such warmed effluents are released in temperate waters, a variety of effects have been noted. At Bradwell in Essex, where it was thought that the warm effluent would have such an impact upon the estuary, marginal effects only have been observed. No mortality of oysters has been attributable to the power station, which had no impact upon the fatally cold conditions of the 1962–63 winter when 80–90% of the Blackwater oyster stocks died. The condition of the oysters on the south side of the estuary is poorer than on the north side and in the adjacent Roach and Crouch estuaries. However, this is probably a reflection of the poor nature of the oyster grounds on this side of the Blackwater[369].

Elsewhere, however, subtle biological changes have been reported in waters associated with these discharges. In Southampton Water, the annual occurrence of species of the copepod *Acartia* has been modified, so that *Acartia tonsa* has assumed a greater importance than it held formerly; a change which may be due to a temperature change in these waters[370]. Further north at Cavendish Dock, Barrow-in-Furness, Roosecote Power Station was commissioned in September 1953. The copepods *Calanus finmarchicus* and *Acartia clausi*, recorded in the period 1955–59, disappeared by 1957–58 and were replaced by *Acartia tonsa*[364]. This could have been due to salinity changes, but on the evidence available the case is not strong, neither is it for a direct effect due to temperature. *Acartia tonsa* is a more southerly species than *A. clausi* and *A. longiremis*[371], and its range extends into sub-tropical waters[372]. Is there in fact a more subtle mechanism at work? Nothing is known of the effect of passage through the condensers, upon those important substances the external metabolites. Yet the condensers, inducing as they do a rise of temperature and pressure are in effect chemical reactors, theoretically capable of inducing chemical changes in relatively unstable organic materials; furthermore chlorine by its ability to react with the double bonds in an unsaturated carbon chain may not be without influence. Are these power stations producing an external metabolite situation, of a sub-tropical character, in bodies of water which do not otherwise posses these properties? If it were so a relatively tough animal like *Acartia tonsa* and having its geographical range, would be admirably suited to take advantage of the situation provided. An interesting speculation and one which would surely repay further investigation.

At Hunterston, in the area affected by the outfall, a population of the gastropod mollusc *Nassarius reticulatus* experienced an increase in shell height in proportion to the cube root of the shell weight. These shells were lighter and thinner, for a given height, than those taken from control stations at Millport. In the population on Hunterston Sands, the laying of egg capsules commenced in January, reached a peak in May, and finished by the end of July; at Millport, egg laying began in mid-April, reached a peak in July, and finished in early September. Spawning in this animal commences when the temperature maximum exceeds 8°C; at Hunterston the winter maximum did not fall below 12°C, but at Millport the maximum winter temperatures did not exceed 8·5°C[373].

In studies at Morro Bay, California, it was found that within 150 m of the outfall the temperatures remained 6°C above normal, but within 230 m of the outfall the communities remained essentially unchanged[374].

In a series of experimental studies, related to the possible effects of a warmed water discharge to the Cape Cod Canal, it was shown that the mussel, *Mytilus edulis*, is adversely affected by temperatures >77°F (>25°C) and temperatures >80·5°F kill in a few hours: sub-lethal effects were noted in the range 77–80·5°F. Curiously, prolonged intermittent exposures of mussels to normally lethal temperatures had no effect. Similarly such intermittent exposure had no effect upon the winkle *Littorina littorea*. The feeding rates of *Thais* (*Nucella*) *lapillus*, dog-whelk, *Asterias*, starfish, and *Carcinus*, shore crab, upon mussels were dramatically reduced by temperatures ⩾25°C[375].

The distribution of benthic animals is dependent upon temperature. For example, *Carcinus maenas* thrives, but does not breed, at temperatures from 14 to 28°C[362]: here temperature limits the population, but migration of organisms can occur from outside the heated area[376]. The American oyster, *Crassostrea virginica*, is present in waters with a temperature range from 4 to 34°C; but *Ostrea edulis*, the European oyster, is restricted to waters of 0–20°C. The former ceases feeding below 7°C, and above 32°C ciliary activity is diminished until at 42°C almost all bodily functions cease or are reduced to a minimum. *Ostrea lurida* remains predominantly closed from 4 to 6°C, at 6–8°C it opens *ca.* 6 hr per day, and at 15°C stays open 23 hr/day. Little is known of the effects of exposing oysters to temperatures above 32–34°C for prolonged periods[376].

Individual fish species have preferred temperatures, they have an ability to perceive temperature differences as low as 0·5°C and when temperatures rise above these preferred temperatures they move away[376, 377]. The young and eggs, however, may be at some risk, and the temperature range for successful hatching of the eggs of the Striped Bass, California Killifish and California grunion are 12·8–23·9, 16·6–28·5 and 14·8–26·8°C respectively[376].

Just as some adult fish move away when the water temperature exceeds the preferred temperature, so other fish may move in. Mullet characteristically occur near to outfalls in British estuaries and are sought there by sea anglers. Curiously, the valve rubber from a pneumatic tyre is used as a bait by some who are successful. Apparently a similar use of warm water outfalls is made by American sea anglers[378].

It has been worked out for freshwater streams that, while diatoms, green and blue-green algae may grow together, if the temperature is raised from 68 to 104°F say, the number of diatom species declines continuously, the species of green algae first increase in number to a maximum at 86°F when they too begin to decline, and the blue-green species increase in number continuously from 68 to 104°F and particularly from 86 to 104°F[376, 379]: it is not known whether such changes are likely to occur in the marine environment. However, the turtle grass, *Thalassia*, deteriorates under temperature stress and may have a use as an indicator of thermal pollution[380].

Of all communities which are exposed to warm water effluents, those in sub-tropical and tropical estuaries are most at risk. In the United States, particular concern is now felt relative to a proposed power station at Biscayne Bay, Florida. Here extensive field and laboratory studies have shown that temperatures sustained above 33°C can cause extensive mortality of some of the most important macroalgae and sea-grasses. Loss of these plants would have undesirable effects upon the many herbivores and detritus feeders which use these plants as a source of shelter and food: these include the juveniles of commercial species such as shrimp, lobster and fishes. Many of these commercially important animals, viz. shrimp and crabs, and other invertebrate adults, have upper thermal limits of 33–37°C. Indeed, tropical estuarine organisms live so close to their upper lethal limits in summer, that in "developing" these estuaries, careful planning is necessary to prevent a loss of food production and recreational properties[381].

Like the benthos, the plankton is at greater risk during the summer. The copepod, *Acartia tonsa*, is least abundant in Biscayne Bay during the summer; when exposed for 1½ hr to 34°C a large percentage of this animal dies. However, these upper thermal limit experiments are very sensitive to changes in water quality and "in experimental procedures involving biological systems at ambient temperatures in the upper twenties and thirties (°C), we are dealing with situations highly sensitive to even the slightest (and probably unwitting) modification, including removal and isolation from the original system"[372]. In this connection too, it is well to remember that certain temperature ranges, especially 30–33°C and 44–46°C may exert a profound and sometimes dominating effect on biological activity through changes in the structure of water. Furthermore, it is possible that general limits on thermal pollution may be derived on a purely physico-chemical basis[382].

Finally, in cold freshwater sources, e.g. the Columbia River, scrap fishes may carry low virulence strains of *Chondrococcus columnaris*, the agent of columnaris disease. When these fish enter warm waters as at Hanford these strains may develop and cause active infection of the fish[383].

Large Power Stations as Fishing Machines

It is sufficient to recall that a large power station may circulate from 20 to 50×10^6 gal/hr, through its culvert system, to appreciate that such a problem must exist. This flow, the equivalent of a river, passes through a variety of screens designed to prevent rubbish from reaching that part of the system which is liable to be damaged or blocked by such materials, viz. pumps and condenser tubes. Of these screens, the rotating band screens, in particular, are a potential catching mechanism and indeed are to some extent in all power stations. There is, however, no literature of this aspect of large power station operation.

Radioactivity

Excluding the military sources of fission products, artificial radio-nuclides in the sea arise either from plants which reprocess the fuel elements which have been "burnt-out" in nuclear reactors, or from the "cooling ponds" into which the latter are placed in the early period after removal from a reactor. Materials from the reprocessing plants constitute the more important of these two sources. However, some $99 \cdot 9\%$ of all radioactive material from this process is far too active for release to the environment and is therefore stored. The remaining $0 \cdot 1\%$ is made up of plant washings so dilute that no other method is possible. A measure of the problem can be appreciated when it is considered that most normal polluting sources are of the order $\geqslant 1$ p.p.m. whereas radioactive sources are of the order of 10^{-12}–10^{-16} p.p.m.

The radioactivity which arises from fuel element "cooling ponds" (not to be confused with that associated with cooling towers) is due either to corrosion of the irradiated fuel can, or due to a leakage of fission products from a can which has burst. An effluent from a fuel processing plant contains a mixture of radionuclides which includes ^{106}Ru, ^{103}Ru, ^{95}Zr, ^{95}Nb, ^{144}Ce, ^{137}Cs, ^{89}Sr and ^{90}Sr, and of these ^{106}Ru is predominant[384], (Table 16.15). From a cooling pond on the other hand, although these materials may be present, the corrosion products such as ^{65}Zn are much more in evidence. Only some of these nuclides have a metabolic function, e.g. Sr and Zn, others, e.g. Ru and Nb, appear to have none. It is those nuclides which have a metabolic function and which are therefore actively concentrated by living organisms which apparently present the greatest hazard.

Table 16.15. Mean monthly discharge of
radioactive effluent from Windscale 1959 and
1960 (after Mauchline[384])

	Curies per month 1959	1960
Total α activity	5·5	6·4
Total β activity	7659	6461
106Ru	2955	3302
90Sr	130	43

Ruthenium and niobium, which form complexes in sea water, are not entirely excluded from the living tissue. Niobium is concentrated by ascidians[385]. During 1963, ruthenium (as 106Ru) was present in the muscle of Solway Firth flounder, *Platichthys flesus*, at a concentration of 5–10 pCi/g ash[386]; it was also present at a concentration of $0·94 \pm 0·17$ pCi/g wet weight of *Raia clavata* taken from the N.E. Irish Sea and the Solway Firth[387]. Nitrosyl 106Ru, from Windscale, accumulates upon the surface of the diatom *Phaeodactylum* and the attached algae *Porphyra umbilicalis*, *Ulva lactuca* and *Laminaria digitata* upon which it complexes with the extracellular polysaccharide material. The adsorption takes place with equal ease upon living or dead samples of these large algae[388]; *Ulva lactuca* takes up more 65Zn when dead than alive[389]. Algae are characterized by a "luxury consumption" of nutrients[390], a process which bears some similarity with the mode of uptake of these nuclides.

Once liberated to the sea water radioactive nuclides may be classed either as ionic and fully-soluble, or as particulate or radiocolloidal when they become adsorbed upon silt and organic matter (p. 122). The ionic forms diffuse through and with the solvent water mass and the general principles of estuarine hydrography apply to them. The radiocolloids on the other hand move as the sediments move and are distributed similarly (p. 149).

The nuclides, which are metabolized, are distributed in the tissues of the fauna depending upon fluctuations in the supply of the individual nuclide[391] and upon the condition of the animal. In the Solway Firth, 90Sr levels in the muscle of the flounder, *Platichthys flesus*, showed a marked peak in July 1962 and 1963, which was not directly correlated with the concentration of 90Sr in the waters of the Firth[386]. This peak coincided with the time at which the flounder were feeding most actively, were putting on flesh and were in their best condition[392].

As one might expect, since the colloidal radio nuclides become attached to sedimentary and organic material, particularly the silt/clay fraction, the distribution of such activity is closely related to the distribution of this type of sediment[393–396]. In the Solway Firth, 106Ru was most abundant at the accretion edge of a marsh which had a specific activity about 10 times that of the rest of the shore and *ca.* 3 times that of the grass marsh[395]. In general, the radioactivity in the flora and fauna was related to the distribution and movement of the sediments, so that, for example, the specific activity in *Pelvetia canaliculata* and *Fucus spiralis* was much higher than that in *Ascophyllum nodosum* and *F. vesiculosus*[397]. Although the amounts of such activity were high in mussels living adjacent to or upon silty areas, these animals showed a discrimination in favour of ^{106}Ru compared with ^{95}Zr/^{95}Nb[386] (Table 16.16). In this area individuals of

Table 16.16. Discrimination factors of mussels, *Mytilus edulis* in ^{106}Ru: ^{95}Zr/^{95}Nb during the period 1962–63 (after Perkins, Williams and Gorman[386])

Site	No. of samples	Discrimination Factor $\left(\dfrac{\text{Mussel } ^{106}\text{Ru}: \ ^{95}\text{Zr}/^{95}\text{Nb}}{\text{Soil } ^{106}\text{Ru}: \ ^{95}\text{Zr}/^{95}\text{Nb}}\right)$ Mean	Range
Balcary Bay	11	1·9	0·4–3·5
Kippford Ford	12	3·4	1·4–5·7
Southerness	4	4	1·8–5·7

certain species, viz. *Arenicola marina*, *Corophium volutator*, *Mytilus edulis*, *Cardium edule*, *Scrobicularia plana*, *Macoma balthica* and *Mya arenaria*, which inhabited the finer sediments were subjected to a higher dose of radiation (max $\geqslant 0.04$ mR/hr) than the inhabitants of coarser sediments, viz. *Scoloplos armiger*, *Owenia fusiformis*, *Lanice conchilega*, *Haustorius arenarius* and *Tellina tenuis* (maximum radiation dose $\leqslant 0.02$ mR/hr)[386].

Fission products generally emit more β than γ radiation, so that experimental studies which consider the effects of the latter only upon aquatic organisms, give only a limited amount of information upon the effects of the radiation from fission products upon these organisms[398]. Nevertheless, a body of useful information has been obtained by exposing these organisms to X-rays and γ radiation.

The ability of a living organism to tolerate radiation is related to its phylogeny and ontogeny. Primitive organisms are more tolerant than the highly organized vertebrates, and adults are more tolerant than juveniles: protozoa may tolerate hundreds of thousands of roentgens, but mammals can only tolerate between 200 and 1000 r[399] (Table 16.17). Relatively little is known of the resistance of marine organisms to radiation; however, those tested exhibit a range of resistance comparable with that shown between phyla, and one species, *Palaemonetes* sp. is slightly less resistant than most mammals[400, 401] (Table 16.18). Salinity is an important factor

Table 16.17. Radiation resistance of different animals
and micro-organisms (after Alexander[399])

	Radiation dose necessary to kill (L.D.$_{50}$)
Mammals	200–1000 r
Goldfish	700 r
Frog	700 r
Tortoise	1 500 r
Newt	3 000 r
Snail	10 000 r
Yeast	30 000 r
Escherichia coli	10 000 r
Fruit-fly (adult) and other insects	60 000 r
Amoeba	100 000 r
Paramecia	300 000 r

in resistance to radiation, and the effects of a radiation dose and the temperature-salinity interaction upon the mummichog, *Fundulus heteroclitus*, was significant at the 5% level of probability. The temperature-salinity interaction caused an increase in mortality with a decrease in salinity at temperatures below 20°C, and an increase in mortality with an increase in salinity at temperatures above 20°C[400].

Chronic exposure of marine animals to very low doses of radiation apparently have little observable effect. The eggs of plaice, *Pleuronectes platessa*, which received 0·6–500R from a ^{137}Cs source, at rates of 10 mR/hr–1R/hr, from fertilization to hatching, showed no significant difference in survival or in the number of abnormal larvae produced. Exposure of *P. platessa* eggs from fertilization to hatching to ^{90}Sr–^{90}Y in the concentration range 10^{-10} Ci/litre–10^{-4} Ci/litre had no effect upon mortality or abnormal larval production[401, 402]. Similarly, *Fundulus heteroclitus* eggs and larvae were apparently unaffected by ^{137}Cs at concentrations of 3×10^{-7}, 3×10^{-6}, 3×10^{-5} Ci/litre. However, the dividing cells of the Black Sea scorpionfish, *Scorpaena porans*, was affected progressively as the concentration increased 10^{-10} to 10^{-5} Ci ^{90}Sr–^{90}Y/litre, and was statistically significant at concentrations which exceeded 10^{-9}Ci/litre. When exposed to sea water containing either ^{65}Zn, ^{51}Cr or ^{90}Sr–^{90}Y in the concentration range 10^{-2} Ci/litre to 10^{-8} Ci/litre, the larvae of the Pacific oyster, *Crassostrea gigas*, showed a significant increase in larval abnormalities at 10^{-4}, 10^{-4} and 10^{-3} Ci/litre of ^{65}Zn, ^{51}Cr and ^{90}Sr–^{90}Y respectively[401].

Once ingested the nature of the effect produced by a radio nuclide depends upon the particular nuclide[401] (Table 16.19). In the rainbow trout, *Salmo gairdneri*, the radiation syndrome including leukopenia, anorexia, loss of scales, lethargy, and growth depression was very

Table 16.18.   Radiation resistance of different species of marine animals (after Templeton et al.(401), Rice et al.(400))

Organism	Experimental Conditions		Median tolerance limit to radiation	
	Salinity ‰	Temperature °C	Period-days	L.D.$_{50}$ (rads)
Invertebrates				
Callinectes sapidus "Blue crab"	30	—	30	2000–110 000
Uca pugnax "Fiddler crab"	30	20	40	42 000
U. pugilator "Fiddler crab"	30	20	40	} 9600–18 000
U. minax "Fiddler crab"	30	20	40	
Palaemonetes sp. "Grass shrimp"	30	20	40	215
Adult fish	—	—	30	1050–5550

Table 16.19.   Effect of chronic ingestion of ^{32}P, ^{90}Sr–^{90}Y, or ^{65}Zn on yearling rainbow trout, Salmo gairdneri (after Templeton, Nakatani and Held(401))

Treatment μCi/g fish/day	Duration of feeding (weeks)	Growth depression	Significant mortality	Leukopenia	Gut damaged	Concentration at end of feeding (μCi/g wet)	
						Bone	Muscle
^{32}P							
0·006	25	no	no	no	no	—	—
0·06	25	wk 17	no	4 mos.	no	1·8	0·23
0·60	25	wk 17	yes	17 days	yes	—	—
^{90}Sr–^{90}Y							
0·005	21	no	no	no	no	2·1	0·0022
0·05	21	no	no	wk 15	no	28	0·078
0·50	21	wk 12	wk 15	wk 15	yes	248	0·27
^{65}Zn							
0·01	17	no	no	no	no	—	—
0·10	17	no	no	no	no	—	—
1·0	17	no	no	no	no	4·0	0·35
10·0	10	no	no	wk 10	no	—	—

pronounced for fish fed on ^{39}P at the higher levels, i.e. 0·60 μCi/g fish/day, and to a lesser extent for fish fed ^{90}Sr–^{90}Y, i.e. 0·50 μCi/g fish/day. The fish were able to absorb much higher levels of ^{65}Zn on a μCi/g/fish basis than of ^{32}P or ^{90}Sr–^{90}Y, because much of the energy from the γ emitting ^{65}Zn was not absorbed by the trout[401].

Radiation at low levels has been found to induce hyperactivity in fish; growth in periphyton, marine invertebrates (including *Callinectes sapidus*, blue crab) and rainbow trout; and metabolic activity in the eggs and fry of Salmon, *Salmo salar*. However, little is known of the underlying mechanism or of the significance of these observations[401].

References

1. Cairns, J. (1968). *Engng Bull. Purdue Univ.* **129**, 16–27.
2. Cairns, J. and Scheifer, A. (1966). *Progve Fish Cult.* **28**, 128–132.
3. McLeod, J. C. and Smith, L. L. (1966). *Trans. Am. Fish. Soc.* **95**, 71–84.
4. Smith, L. L., Kramer, R. H. and McLeod, J. C. (1965). *J. Wat. Pollut. Control Fed.* **37**, 130–140.
5. Herbert, D. W. M. and Richards, J. M. (1963). *Air Wat. Pollut* **7**, 297–302.
6. Herbert, D. W. M. and Merkins, J. C. (1961). *Air Wat. Pollut* **5**, 46–55.
7. Loosanoff, V. L. (1961). *Proc. Gulf Caribb. Fish. Inst.* 14th Annual Session, 80–94.
8. Chiba, K. and Ohsima, Y. (1957). *Bull. Jap. Soc. scient. Fish.* **23**, 348–354. Abstract in: *Records of Research in the Faculty of Agriculture, Univ. Tokyo* No. 8 (1957–58): 54–55.
9. Portmann, J. E. (1970). *J. mar. biol. Ass. U.K.* **50**, 577–591.
10. Howell, B. R. and Shelton, R. G. J. (1970). *J. mar. biol. Ass. U.K.* **50**, 593–607.
11. Mackay, D. W. and Topping, G. (1970). *Effluent and Water Treatment Journal*, Nov. 1970, 7 pp.
12. Pearce, J. B. (1970). *In* "F.A.O. Technical Conference on Marine Pollution and its Effects on Living Resources and Fishing", F.I.R.: MP/70/E-99. Rome, 9–18 December 1970.
12a. Shelton, R. G. J. (1971). *Mar. Pollut. Bull.* **2**, 24–27.
13. Albert, A. (1968). "Selective Toxicity", Methuen, London.
14. Pantin, C. F. A. (1931). *J. exp. Biol.* **8**, 63–72.
15. Pantin, C. F. A. (1931). *J. exp. Biol.* **8**, 82–94.
16. Pantin, C. F. A. and Weil, E. (1931). *J. exp. Biol.* **8**, 73–81.
17. Beadle, L. C. (1934). *J. exp. Biol.* **11**, 382–396.
18. Schlieper, C. (1958). *Kieler Meeresforsch.* **11**, 22–33.
19. Beadle, L. C. (1937). *J. exp. Biol.* **14**, 56–70.
20. Ellis, W. G. (1937). *J. exp. Biol.* **14**, 340–350.
21. Todd, M. E. (1964). *J. exp. Biol.* **41**, 665–677.
22. Nicol, J. A. C. (1968). "The Biology of Marine Animals", Pitman, London.
23. Prosser, C. L. and Brown, F. A. (1962). "Comparative Animal Physiology", Saunders, Philadelphia.
24. Trussell, P. C. (1967). *Sea Front.* **13**, 234–243.
25. Perkins, E. J. (1968). *Fld Stud.* **2** (Suppl.), 81–90.
26. Halsband, E. (1970). *In* "F.A.O. Technical Conference on Marine Pollution

and its Effects on Living Resources and Fishing", F.I.R.: MP/70/E-17. 13 pp. Rome, 9–18 December, 1970.

27. Perkins, E. J., Williams, B. R. H., Bailey, M., Hinde, A. and Gorman, J. (1964–66). U.K.A.E.A., Chapelcross Works, Health Physics Note No. 2.

28. Spencer, J. F. (1968). *C.E.R.L. Laboratory Note* No. RD/L/N 53/68.

29. Federal Water Pollution Control Administration (1968). "Water Quality Criteria", U.S. Department of the Interior: 234 pp.

30. Doudoroff, P. and Katz, M. (1950). *Sewage ind. Wastes*, **22**, 432–458.

31. Jones, J. Erichsen. (1964). "Fish and Pollution", Butterworths, London.

32. Ellis, M. M. (1937). *Bull. U.S. Bur. Fish.* **48**, 365–437.

33. Rose, A. H. (1968). "Chemical Microbiology", Butterworths, London.

34. Bradshaw, J. S. (1968). *Limnol. Oceanogr.* **13**, 26–38.

35. Arnal, R. (1961). *Bull. geol. Soc. Am.* **72**, 427–478.

36. Schafer, C. T. (1970). In "F.A.O. Technical Conference on Marine Pollution and its Effects on Living Resources and Fishing", F.I.R.: MP/70/E-36. Rome, 9–18 December 1970.

37. Bateman, J. B. (1933). *J. exp. Biol.* **10**, 355–368.

38. Perkins, E. J., Gilchrist, J. R. S. and Logan, J. (1970). *Trans. J. Proc. Dumfries. Galloway Nat. hist. Antiq. Soc.* Ser. 3, **47**, 13–14.

39. Parker, J. (1971). "A Toxicological Investigation of Phosphoric Acid and its Derivatives", University of Strathclyde. B.Sc. Thesis.

40. Kelley, A. M. (1946). *J. Fish. Res. Bd Can.* **6**, 435.

41. Holland, G. E., Laseter, J. E., Neumann, E. D. and Eldridge, W. E. (1964). *State of Washington, Department of Fisheries, Research Bulletin*, **5**, 264 pp.

42. Nomura, S. and Tomita, G. (1933). *J. Shanghai Sci. Inst.* Sect. IV. **1**, 29–39.

43. Tomita, G. (1934). *J. Shanghai Sci. Inst.* Sect. IV. **1**, 77–84.

44. Wolverkamp, H. P. and Waterman, T. H. (1960). *In* "The Physiology of Crustacea", (T. H. Waterman, ed.), Vol. 1. pp. 35–100, Academic Press, New York.

45. Howard, T. E. and Walden, C. C. (1965). *Tappi* **48**, 136–141.

46. Jørgensen, O. M. (1929). *Proc. Univ. Durham phil. Soc.* **8**, 41–54.

47. Green, J. (1968). "The Biology of Estuarine Animals", Sidgwick and Jackson, London.

48. Waldichuk, M. and Bousfield, E. L. (1962). *J. Fish. Res. Bd. Can.* **19**, 1163–1165.

49. Barnard, J. L. (1958). *Calif. Fish Game.* **44**, 161–170.

50. Olson, T. A. and Burgess, F. J. (Eds) (1967). "Pollution and Marine Ecology", Wiley–Interscience, New York.

51. Dybern, B. I. (1970) In "F.A.O. Technical Conference on Marine Pollution and its Effects on Living Resources and Fishing", F.I.R.: MP/70/R-3. Rome, 9–18 December 1970.

52. Segerstråle, S. G. (1969). *Progress in Oceanography*, **5**, 169–183.

53. Renfro, W. C. (1963). *Trans. Am. Fish. Soc.* **92**, 320–322.

54. Chapman, W. McL. (1939). *Trans. Am. Fish. Soc.* **69**, 197–204.

55. Davison, R. C., Cardwell, R. S., Mearns, A. J., Newcomb, T. W. and Watters, K. W. (1959). *Sewage ind. Wastes* **31**, 950–966.

56. Smith, L. S., Cardwell, R. D., Mearns, A. J., Newcomb, T. W. and Watters, K. W. (1970). "F.A.O. Technical Conference on Marine Pollution and its Effects on Living Resources and Fishing", F.I.R.: MP/70/E-40: 7 pp. Rome, 9–18 December 1970.

57. Cairns, J. and Scheier, A. (1958). *In* "12th Ind. Waste Conf. Proc. 1957", Purdue Univ., Lafayette, Ind.
58. Lloyd, R. (1961). *J. exp. Biol.* **38**. 447.
59. Mackay, D. W. (1969). *J. inst. Wat. Pollut. Control*, 1969, 3–11.
60. Braarud, T. and Ruud, J. T. (1937). *Hvalråd. Skr.* No. 15, 1–56.
61. Braarud, T. (1945). *Hvalråd. Skr.* No. 28, 1–142.
62. Braarud, T. (1955). *Proc. Internat. Assoc. of theoretical and applied Limnol.* **12**, 811–813.
63. Klein, L. (1966). "River pollution. **III**. Control", Butterworths, London.
64. Wheeler, A. (1970). *Science Journal.* **6** (11), 28–32.
65. Ministry of Housing and Local Government (1961). "Pollution of the Tidal Thames", H.M.S.O., London.
66. Water Pollution Research Laboratory (1964). *D.S.I.R. Wat. Pollut. Res.* Tech. Paper No. 11. H.M.S.O., London.
67. Preddy, W. S. (1954). *J. mar. biol. Ass. U.K.* **33**, 645–662.
68. Gameson, A. L. H. (1959). *In* "The Effects of Pollution on Living Material" (W. B. Yapp, ed.), Symposia of the Institute of Biology, London. No. 8, 51–59.
69. Wheatland, A. B. (1959). *In* "The Effects of Pollution on Living Material" (W. B. Yapp, ed.), Symposia of the Institute of Biology, London. No. 8, 33–50.
70. Wheatland, A. B., Barrett, M. J. and Bruce, A. M. (1959). *J. Proc. Inst. Sew. Purif.* 1959 (2), 149–159.
71. Water Pollution Research Laboratory (1965). *Notes on Water Pollution.* No. 28. H.M.S.O., London.
72. Spencer, C. P. (1956). *J. mar. biol. Ass. U.K.* **35**, 621–630.
73. Braarud, T. and Føyn, E. (1951). *Avh. norske Vidensk Akad. Oslo, I. Mat.-Naturv. Klasse.* 1951. No. 3, 24 pp.
74. O'Connor, D. J. (1960). *J. sanit. Engng Div. Am. Soc. civ. Engrs.* **2472**, SA3, 35–55.
75. Mackay, D. W. and Fleming, G. (1969). *Water Res.* **3**, 121–128.
76. Mackay, D. W. and Gilligan, J. (1972). *Water Res.* **6**, 183–190.
77. Mackay, D. W. Personal communication.
78. Goldman, C. J. (1962). "Photosynthetic Oxygenation of a Polluted Estuary". A discussion of Paper No. 35, Section 3 by C. J. Hull. *In* "Proc. 1st. International Conference on Water Pollution Research", Pergamon, Oxford.
79. Hull, C. J. (1962). *In* "Proc. 1st International Conference on Water Poll. Research", Section 3, Paper 35. Pergamon, Oxford.
80. Kaplovsky, A. J. (1962). "Photosynthetic Oxidation of a Polluted Estuary", Proc. 1st Internat. Conf. Water Poll. Res. A discussion of Paper No. 35, Section 3 by C. J. Hull. State of Delaware, Water Pollution Commission, 20 pp.
81. Stewart, K. M. and Rohlich, G. A. (1967). "Eutrophication—A Review", *State of California, State Water Quality Control Board*, Publication No. 34, 190p.
82. Caspers, H. (1964). *Verh. int. Verein. theor. angew. Limnol.* **15**, 631–638.
83. Fruh, E. G., Stewart, K. M., Lee, G. F. and Rohlich, G. A. (1966). *J. Wat. Pollut. Control. Fed.* **38**, 1237–1258.
84. Corey, R. B., Hasler, A. D., Lee, G. F., Schraufuagel, F. H. and Wirth, T. L. (1967). "Excessive Water Fertilization", Report to the Water Subcommittee, Natural Resources Committee of State Agencies. Madison, Wisconsin, 31 January 1967.

85. Hasler, A. D. and Swenson, M. E. (1967). *Science N.Y.* **158**, 278–282.
86. Koidsumi, K., *et al.* (1967). "Standing Crop of Higher Aquatic Plants in Lake Suwa" (Materials for the limnology of Lake Suwa 1).
87. Hasler, A. D. and Ingersoll, B. (1968). *Nat. Hist.* **77** (9), 8–31.
88. Koidsumi, K., Sakurai, Y., Kawashima, S. and Nagasawa, T. (1968). *Jap. J. Ecol.* **18** (4), 167–171.
89. Mackenthum, K. M., Keup, L. E. and Stewart, R. K. (1968). *J. Wat. Pollut. Control. Fed.* **40**, No. 2, Part 2, R72-R81.
90. Rohlich, G. A. (1968). *In* "Proc. 41st Annual Convention of the Soap and Detergent Association", New York City, January 25th, 1968, 3–11.
91. Skulberg, O. M. (1968). *Mitt. int. Verein. theor. angew. Limnol.* **14**, 187–200.
92. Thomas, E. A. (1968). *Mitt. int. Verein theor. angew. Limnol.* **14**, 231–242.
93. Hasler, A. D. (1969). *In* "Symposium on Ecosystems—Evolution and Revolution", Paper No. 7, 425–443.
94. Hudson, E. J. and Marson, H. W. (1970). *Chemy Ind.* 1970, 1449–1458.
95. Waldichuk, M. (1968). "Phosphorus—a problem in eutrophication". Meeting of the Vancouver Island, Section of the Chemical Institute of Canada, 27th November, 1968, Nanaimo, B.C.; 47 pp. (Mimeo).
96. Sylvester, R. O. (1952). *Washington Pollution Control Commission, Tech. Bull.* No. 13, 28 pp. Sept. 1952.
97. Edmonson, W. T., Anderson, G. C. and Peterson, D. R. (1956). *Limnol. Oceanogr.* **1**, 47.
98. Anderson, G. C. (1961). *In* "Algae and Metropolitan Wastes", Robert A. Taft Sanitary Engineering Center, Cincinnati, Ohio, TRW61-3 (1961).
99. Peterson, D. R. (1955). *Washington Pollution Control Commission, Tech. Bull.* No. 18.
100. Issac, G. W., Farris, G. D. and Gibbs, C. V. (1964). *Municipality of Metropolitan Seattle, Seattle, Washington. Water Quality Series.* No. 1, 37 pp.
101. Stockner, J. G. and Benson, W. W. (1967). *Limnol. Oceanogr.* **12**, 513–532.
102. Domenowske, R. S. and Matsuda, R. I. (1968). *In* "P.N.P.C.A. Conference, Penticton, B.C., October 1968". 27 pp.
103. Edmondson, W. T. (1968). *In* "Water Resources Management and Public Policy" (T. H. Campbell and R. O. Sylvester, ed), pp. 139–178, University of Washington, Seattle.
104. Welch, E. B. (1968). *J. Wat. Pollut. Control Fed.* **40**, 1711–1727.
105. Welch, E. B. (1969). "Factors Initiating Phytoplankton Blooms and Resulting Effects on Dissolved Oxygen in Duwanish River Estuary, Seattle, Washington", Geological Survey Water-Supply Paper 1873A: 62 pp.
106. Royal Commission on Sewage Disposal. (1911). Seventh Report Vol. 1 Cmd. 5542. H.M.S.O., London.
107. Braarud, T. and Hope, B. (1952). *FiskDir. Skr Serie Havunders.* **9**(16), 26 pp.
108. Grenager, B. (1957). *Nytt. Mag. Bot.* **5**, 41–60.
109. Hasle, G. R. and Smayda, T. J. (1960). *Nytt. Mag. Bot.* **8**, 53–75.
110. Wilkinson, L. (1962). "Nitrogen transformations in a polluted estuary", *In* "1st International Conference on Water Pollution Research", Section 3, Paper 47, 20 pp. Pergamon, Oxford.
111. Barlow, J. P., Lorenzen, C. J. and Myren, R. T. (1963). *Limnol Oceanogr.* **8**, 251–262.
112. Waldichuk, M. (1968). *Pacific Search.* 1968, 16.

113. Berge, G., Ljøen, R. and Palmork, K. H. (1970). *In* "F.A.O. Technical Conference on Marine Pollution and its Effects on Living Resources and Fishing", F.I.R.: MP/70/E-72: 8 pp. Rome, 9–18 December 1970.
114. Clutter, R. I. (1970). *In* "F.A.O. Technical Conference on Marine Pollution and its Effects on Living Resources and Fishing", F.I.R.: MP/70/E-52. Rome, 9–18 December 1970.
115. Cory, R. L. (1970). *In* "F.A.O. Technical Conference on Marine Pollution and its Effects on Living Resources and Fishing", F.I.R.: MP/70/E-53. Rome, 9–18 December 1970.
116. Gupta, R. S. (1970). *In* "F.A.O. Technical Conference on Marine Pollution and its Effects on Living Resources and Fishing", F.I.R.: MP/70/E-37: 8 pp. Rome, 9–18 December 1970.
117. Haahtela, I. (1970). *In* "F.A.O. Technical Conference on Marine Pollution and its Effects on Living Resources and Fishing", F.I.R.: MP/70/E-85: 11 pp. Rome, 9–18 December 1970.
118. Platt, T., Conover, R. J., Loucks, R., Mann, K. H., Peer, D. L., Prakash, A. and Sameoto, D. D. (1970). *In* "F.A.O. Technical Conference on Marine Pollution and its Effects on Living Resources and Fishing", F.I.R.: MP/70/E-30 10 pp. Rome, 9–18 December 1970.
119. McNulty, J. K. (1955). *Bull. mar. Sci. Gulf Caribb.* **11**, 394–447.
120. North, W. J. (1964). *Proc. int. Conf. Wat. Pollut. Res.* **1**, 247–437.
121. North, W. J., Stephens, G. C. and North B. B. (1970). *In* "F.A.O. Technical Conference on Marine Pollution and its Effects on Living Resources and Fishing", F.I.R.: MP/70/R-8: 22 pp. Rome, 9–18 December 1970.
122. Bellamy, D. J., John, D. M., Jones, D. J., Starkie, A. and Whittick, A. (1970). *In* "F.A.O. Technical Conference on Marine Pollution and its Effects on Living Resources and Fishing", F.I.R.: MP/70/E-65. Rome, 9–18 December 1970.
123. Stokke, K. (1956). *In* "Proc. 2nd International Seaweed Symposium" (T. Braarud and N. A. Sorensen, eds), pp. 310–214. Pergamon, Oxford.
124. Cole, H. A. (1970). *In* "F.A.O. Technical Conference on Marine Pollution and its Effects on Living Resources and Fishing", F.I.R.: MP/70/R-20. Rome, 9–18 December 1970.
125. Iwasaki, H., *et al.* (1968). *J. Fac. Fish. Anim. Husb. Hiroshima Univ.* **7**, 259–267.
126. Pincemin, J-M. (1970). *In* "F.A.O. Technical Conference on Marine Pollution and its Effects on Living Resources and Fishing", F.I.R.: MP/70/E-100: 3 pp. Rome, 9–18 December 1970.
127. Bellan, G. (1970). *Mar. Pollut. Bull.* **1**, 59–60.
128. Reish, D. J. (1959). *Allan Hancock Found. Occ. Pap.* **22**, 117.
129. Saad, M. S. H. (1970). *In* "F.A.O. Technical Conference on Marine Pollution and its Effects on Living Resources and Fishing", F.I.R.: MP/70/E-35. Rome, 9–18 December 1970.
130. Young, P. H. (1964). *Calif. Fish Game* **50**, 33–41.
131. Halstead, B. W. (1970). *In* "F.A.O. Technical Conference on Marine Pollution and its Effects on Living Resources and Fishing", F.I.R.: MP/70/R-6: 21 pp. Rome, 9–18 December, 1970.
132. Matsuda, R. I., Isaac, G. W. and Dalseg, R. D. (1968). "Fishes of the Green-Duwanish River." *Municipality of Metropolitan Seattle, Seattle, Washington, Water Quality Series* No. 4: 38 pp.

133. Graham, M. (1938). *Fish. Invest., Lond.* Ser. II, **16**, No. 3.
134. Carruthers, J. N. (1954). *Intelligence Digest Supplement,* October, 1954.
135. Blanc, F., Leveau, M. and Szekielda, K. H. (1969). *Marine Biology.* **3**, 233–242.
136. Dugdale, R. C., Kelley, J. C. and Becacos-Kontos, T. (1970). *In* "F.A.O. Technical Conference on Marine Pollution and its Effects on Living Resources and Fishing", F.I.R.: MP/70/E-79: 12 pp. Rome, 9–18 December 1970.
137. James, A. and Head, P. C. (1970). *In* "F.A.O. Technical Conference on Marine Pollution and its Effects on Living Resources and Fishing", F.I.R.: MP/70/E-18: 6 pp. Rome, 9–18 December 1970.
138. Hardy, A. C. (1970). "The Open Sea", Part One: The World of Plankton, Collins, London. The Fontana New Naturalist.
138a. Fitzgerald, G. P. (1960). *In* "Trans. Seminar on Algae and Metropolitan Wastes, April 27–29, 1960", U.S. Public Health Service, Robert A. Taft Engineering Center, Cincannati, Ohio.
139. Allen, G. H. (1970). *In* "F.A.O. Technical Conference on Marine Pollution and its Effects on Living Resources and Fishing", F.I.R.: MP/70/R-13: 26 pp. Rome, 9–18 December 1970.
140. Chan, G. L. (1970). *In* "F.A.O. Technical Conference on Marine Pollution and its Effects on Living Resources and Fishing", F.I.R.: MP/70/E-10: 4 pp. Rome 9–18 December, 1970.
141. Stirn, J. (1970). *In* "F.A.O. Technical Conference on Marine Pollution and its Effects on Living Resources and Fishing", F.I.R.: MP/70/E-105: 7 pp. Rome, 9–18 December 1970.
142. Parsons, T. R., McAllister, C. D., Le Brasseur, R. J. and Barraclough, W. E. (1970). "Technical Conference on Marine Pollution and its Effects on Living Resources and Fishing", F.I.R.: MP/70/E-58: 14 pp. Rome, 9–18 December 1970.
143. Medical Research Council. (1959). "Sewage Contamination of Bathing Beaches in England and Wales", Memo No. 37. H.M.S.O., London.
144. Josa, F. (1962). *In* "Proc. 1st International Conference Water Pollution Research", Section 3, 1–15. Pergamon, Oxford.
145. Brooks, P. F. (1970). *Justice of the Peace and Local Government Review.* October 17, 1970. 134, 772–773.
146. Yoshpe-Purer, Y. and Shuval, H. I. (1970). *In* "F.A.O. Technical Conference on Marine Pollution and its Effects on Living Resources and Fishing", F.I.R.: MP/70/E-47: 15 pp. Rome, 9–18 December 1970.
147. Metcalf, T. G., Vaugh, J. M. and Stiles, W. C. (1970) "F.A.O. Technical Conference on Marine Pollution and its Effects on Living Resources and Fishing", F.I.R.: MP/70/E-24: 9 pp. Rome, 9–18 December 1970.
148. Wood, P. C. (1970). *In* "F.A.O. Technical Conference on Marine Pollution and its Effects on Living Resources and Fishing", F.I.R.: MP/70/R-12: 14 pp. Rome, 9–18 December 1970.
149. Ketchum, B. H., Ayers, J. C. and Vaccaro, F. R. (1952). *Ecology* **33**, 247–258.
150. Mitchell, R. (1970). *In* "F.A.O. Technical Conference on Marine Pollution and its Effects on Living Resources and Fishing", F.I.R.: MP/70/E-26: 7 pp. Rome, 9–18 December 1970.
151. Shuval, H. I., Thompson, A., Fattal, B., Cymbalista, S. and Wiener, Y. (1970). *In* "F.A.O. Technical Conference on Marine Pollution and its Effects on Living Resources and Fishing", F.I.R.: MP/70/E-38: 12 pp. Rome, 9–18 December 1970.

152. Hunt, D. A. (1970). *In* "F.A.O. Technical Conference on Marine Pollution and its Effects on Living Resources and Fishing", F.I.R.: MP/70/E-87: 6 pp. Rome, 9–18 December 1970.
153. Silverman, P. H. (1956). *Sanitarian, Lond.* **65**, 77.
154. Silverman, P. H. and Griffiths, R. B. (1955). *Ann. trop. Med. Parasit.* **49**, 436–450.
155. Servisi, J. A., Martens, D. W. and Gordon, R. W. (1970). *International Pacific Salmon Fisheries Commission. Prog. Rept.* No. 24: 28 pp.
156. Ziebell, C. D. and Mills, A. D. (1963). *Washington Pollution Control Commission, Research Bulletin.* **63**-1, 33 pp. May 1963.
157. Graham, J. L. and Schaumburg, F. D. (1969). *In* "Proc. 24th Purdue Industrial Waste Conference, Lafayette, Indiana, May 6th, 1969", 2 pp.
158. U.S. Department of the Interior and Washington State Pollution Control Commission (1967). "Pollutional effects of pulp and paper mill wastes in Puget Sound". A report on studies conducted by the Washington State Enforcement Project: 474 pp.
159. Van Horn, W. M. (1949). *Trans. Wis. Acad. Sci. Arts Lett.* **39**, 105–114.
160. Editor. (1967). *Steam and Heating Engineer*, February and March 1967: 12 pp.
161. Tully, J. P. (1949). *Paper Trade J.* July 28, 1949, 1–6.
162. Tully, J. P. (1949). Forest Products Research Society. 3rd Annual National Meeting, Grand Rapids, Michigan, May 2–4, 1949: 1–12.
163. Gunter, G. and McKee, J. (1960). "On Oysters and Sulfite Waste Liquor", Report to Washington Pollution Control Commission, February 1960: 93 pp.
164. Federal Water Pollution Control Administration. (1967). "Conference on the Matter of Pollution of Navigable Waters of Puget Sound, the Strait of Juan de Fuca and their Tributaries and Estuaries (Washington)", held in Seattle, Washington, 6–7 September 1967, 6 October 1967. F.W.P.C.A., U.S. Dept. of the Interior. 3 volumes: 618 pp.
165. Schafer, C. T. and Gupta, B. K. S. (1969). "Foraminiferal Ecology in Polluted Estuaries of New Brunswick and Maine", Report A.O.L. 69–1: 24 pp. Unpublished manuscript.
166. Pearson, T. H. (1970). *In* "F.A.O. Technical Conference on Marine Pollution and its Effects on Living Resources and Fishing", F.I.R.: MP/70/E-62. Rome, 9–18 December 1970.
167. Peterson, D. R., Wagner, R. A. and Livingston, A. (1957). *Washington Pollution Control Commission, Tech. Bull.* No. 21, 52 pp.
168. Holderby, J. M. and Wiley, A. J. (1950). *Sewage ind. Wastes.* **22**, 61–70.
169. Wisniewski, T. F., Wiley, A. J. and Lueck, B. F. (1956). *Tappi* **39**, 65–71.
170. Fisher, D. R. (1968). "Biological Treatment of Sulphite Pulping Effluents at Weyerhauser Company, Gray's Harbour", 13 pp. Mimeo. Weyerhauser Co., Cosmopolis, Washington.
171. Miller, A. M. (1968). "Review of Clarifier Solids Handling and Disposal Experience, Scott Paper Company, Everett, Washington", National Council for Air and Stream Improvement, meeting held at Portland, Oregon, October 2–3, 1968: 4 pp.
172. Rozycki, Z. F. (1968). "Effluent Control Program at Publishers Paper Co., Oregon City, Oregon", National Council for Air and Stream Improvement, meeting held at Portland, Oregon, October 2–3 1968: 13 pp.
173. N.C.A.S.I. (1969). Report to Members, 1968. National Council of the Paper Industry for Air and Stream Improvement Inc.: 31 pp.

174. Pulp Manufactures Research League (1968). Annual Report: 11 pp.
175. Dubey, G. A., McElhinney, T. R. and Wiley, A. J. (1965). *Tappi* **48**, 95–98.
176. Wiley, A. J., Ammerlaan, A. C. F. and Dubey, G. A. (1967). *Tappi* **50**, 455–460.
177. Nelson, W. R. and Walraven, G. O. (1968). *Pulp Pap.* August 18, 1968: 30–31, 48.
178. Ammerlaan, A. C. F., Lueck, B. F. and Wiley, A. J. (1969). *Tappi* **52**, 118–122.
179. Ammerlaan, A. C. F. and Wiley, A. J. (1969). *In* "Proc. New Orleans meeting of the A.I. Ch.E., March 17–20, 1969": 18 pp.
180. Wagner, H. (1956). *Special Communication to the German Hydrological Yearbook* No. 15. Federal Office for Hydrology, Coblenz. Translated at Kresge-Hooker Science Library Services: 90 pp.
181. Lueck, B. F., Wiley, A. J., Scott, R. H. and Wisniewski, T. F. (1957). *Sewage ind. Wastes*, **29**, 1054–1065.
182. Scott, R. H., Wisniewski, T. F., Leuck, B. F. and Wiley, A. J. (1958). *Sewage ind. Wastes*, **30**, 1496–1505.
183. Pulp Manufacturers Research League. "Spent Sulphite Liquor". General information bulletin: 12 pp.
184. Van Horn, W. M. (1953). *The Paper Industry*, March–June, 1953: 8 pp.
185. Van Horn, W. M. (1949). "Limnological Aspects of Water Supply and Waste Disposal", *Publs. Am. Ass. Advmt.* 1949, 49–55.
186. Werner, A. E. (1963). *Can. Pulp Pap. Ind.* **16**(3), 35–43.
187. Greer, A., Gillespie, R. E. and Trussell, P. C. (1956). *Tappi* **39**, 599–603.
188. Howard, T. E. and Walden, C. C. (1971). *Pulp Pap. Can.* **72**, 73–79.
189. Walden, C. C. (1965). *West. Fish.* **71**(1), 4 pp.
190. Servisi, J. A., Stone, E. T. and Gordon, R. W. (1966). *International Pacific Salmon Fisheries Commission, Prog. Rept.* No. 13: 34 pp.
191. Howard, T. E. and Walden, C. C. (1967). *In* "Proc. 97th Annual Conference of the American Fisheries Society, Toronto, Ontario, September 15th, 1967", 15 pp.
192. Schaumburg, F. D., Howard, T. E. and Walden, C. C. (1967). *Water Res.* **1**, 731–737.
193. Howard, T. E., Bowen, D. L. and Walden, C. C. (1968). *Pulp Pap. Can.* May 17, 1968, 4 pp.
194. Howard, T. E., Walden, C. C. and Eade, B. L. (1970). "Effects of Kraft Mill Effluent upon Swimming Performance of Juvenile Coho Salmon", *In* "57th Annual Technical Meeting", Pulp and Paper Association of Canada, Montreal, January 1971, pp. 1–16.
195. Walden, C. C. and Howard, T. E. (1967). *Pulp Pap. Can.* February 16, 1968, 7 pp.
196. Werner, A. E. (1968). *Pulp Pap. Can.* **69**(5), 3–12.
197. Servisi, J. A., Gordon, R. W. and Martens, D. W. (1969). *International Pacific Salmon Fisheries Commission. Prog. Rept.* No. 23, 38 pp.
198. Bodien, D. G. (1969). "Plywood Plant Glue Wastes Disposal. Final Report". A Technical Projects Report No. FR-5: 84 pp. U.S. Dept. of the Interior, Federal Water Pollution Control Administration, Pacific Northwest Laboratory, Corvallis, Oregon.
199. Van Horn, W. M. (1943). *Paper Trade Jl.* **117**(24), 33–35.
200. Van Horn, W. M. and Balch, R. (1955). *Tappi* **38**, 151–153.
201. Hannerz, L. (1968). *Institute of Freshwater Research, Drottningholm, Rept.* No. 48: 176 pp.

202. ZoBell, C. E. (1962). *In* "Proc. 1st International Conference on Water Pollution Research", Section 3, Paper 48, 27 pp. Pergamon, Oxford.
203. Ludwig, H. F. and Rich, L. G. (1962). *In* "Proc. 1st International Conference on Water Pollution Research", Discussion of paper by C. E. ZoBell. Pergamon Oxford.
204. North, W. J., Neuschul, M. and Clendenning, K. A. (1965). *In* "Pollution Marins Parles Micro-organisms et Produits Petroliere. Symposium de Monaco, Avril 1964". pp. 335–354.
205. Bellamy, D. J., Clarke, P. H., John, D. M., Jones, D. J., Whittick, A. and Darke, T. (1967). *Nature, Lond.* **216**, 1170–1173.
206. Mercer, I. D. (Editor) (1967). *J. Devon Trust for Nature*, July 1967 (Suppl.), 72 pp.
207. North, W. J. (1967). *Sea Front.* **13**, 212–217.
208. O'Sullivan, A. J. and Richardson, A. J. (1967). *Nature, Lond.* **214**, 448, 451, 452.
209. Carthy, J. D. and Arthur, D. R. (editors) (1968). *Fld Stud.* 2(Suppl.) 198 pp.
210. Nelson-Smith, A. (1968). *J. appl. Ecol.* **5**, 97–107.
211. Ranwell, D. S. (1968). *The Lichenologist* **4**, 55–56.
212. Simpson, A. C. (1968). *M.A.F.F. Laboratory Leaflet* (*New Series*) No. 18: 43 pp.
213. Smith, J. E. (1968). "'Torrey Canyon' Pollution and Marine Life", University Press, Cambridge.
214. Bourne, W. R. P. (1968). *Fld Stud.* 2(Suppl.), 99–121.
215. Beer, J. V. (1968). *Fld Stud.* 2(Suppl.) 123–129.
216. Beer, J. V. (1968). *Wildfowler*, **19**, 120–124.
217. Goldacre, R. J. (1968). *Fld Stud.* 2(Suppl.), 131–137.
218. Blumer, M. (1970). *In* "F.A.O. Technical Conference on Marine Pollution and its Effects on Living Resources and Fishing", F.I.R.: MP/70/R-1. Rome, 9–18 December 1970.
219. Mironov, O. G. (1970). *In* "F.A.O. Technical Conference on Marine Pollution and its Effects on Living Resources and Fishing", F.I.R.: MP/70/E-92: 4 pp. Rome, 9–18 December 1970.
220. Kuhnhold, W. W. (1970). *In* "F.A.O. Technical Conference on Marine Pollution and its Effects on Living Resources and Fishing", F.I.R.: MP/70/E-64: 10 pp. Rome, 9–18 December 1970.
221. Cowell, E. B., Baker, J. M. and Crapp, G. B. (1970). *In* "F.A.O. Technical Conference on Marine Pollution and its Effects on Living Resources and Fishing", F.I.R.: MP/70/E-11; 11 pp. Rome, 9–18 December 1970.
222. Cowell, E. B., Crapp, G. B., Baker, J. M., Dudley, G., Withers, R. G., Sullivan, G. E. and Ottway, S. (1970). "Ecological Effects of Oil Pollution on Littoral Communities", Symposium held at the Zoological Society of London, 30 November and 1 December, 1970.
223. Nelson, T. C. (1925). *Rept. to U.S. Commissioner of Fisheries for* 1925. Appendix V, 171–181.
224. Gunkel, W. and Trekel, H.-H. (1967). *Helgoländer wiss. Meeresunters.* **16**, 336–348.
225. Wallhauser, K. H. (1967). *Helgoländer wiss. Meeresunters.* **16**, 328–335.
226. Meyers, S. P. and Ahearn, D. G. (1970). *In* "F.A.O. Technical Conference on Marine Pollution and its Effects on Living Resources and Fishing", F.I.R.: MP/70/E-25: 10 pp. Rome, 9–18 December 1970.

227. Bloom, S. (1970). *J. mar. biol. Ass. U.K.* **50**, 919–923.
227a. Johnston, R. (1970). *J. mar. biol. Ass. U.K.* **50**, 925–937.
228. Kuhl, H. and Mann, H. (1967). *Helgoländer wiss. Meeresunters.* **16**, 321–327.
229. Rosenthal, H. and Gunkel, W. (1967). *Helgoländer wiss. Meeresunters.* **16**, 315–320.
230. Griffith, D. de G. (1970). *In* "F.A.O. Technical Conference on Marine Pollution and its Effects on Living Resources and Fishing", F.I.R.: MP/70/E-16. Rome, 9–18 December 1970.
231. Beynon, L. P. (1968). *Hydrospace.* **1**, 17–27.
232. Hellmann, H. and Marcinowski, H.-J. (1970). *In* "F.A.O. Technical Conference on Marine Pollution and its Effects on Living Resources and Fishing", F.I.R.: MP/70/E-86: 4 pp. 9–18 December 1970.
233. Tay Estuary Oil Pollution Scheme (1970). Conference of Local Authorities and other bodies held in Marryat Hall, Dundee, 15 to 16th April, 1970.
234. Brockis, G. J. (1967). *Helogländer wiss. Meeresunters.* **16**, 296–305.
235. Simpson, A. C. (1962). I.C.E.S. Shellfish Committee, 1962, No. 84: 4 pp. (Mimeo).
236. Sidhu, G. S., Vale, G. L., Shipten, J. and Murray, K. E. (1970). *In* "F.A.O. Technical Conference on Marine Pollution and its Effects on Living Resources and Fishing", F.I.R.: MP/70/E-39: 9 pp. Rome, 9–18 December 1970.
237. Blumer, M., Souza, G. and Sass, J. (1970). *W.H.O.I.* 70–1, January 1970: 14 pp.
238. Galtsoff, P. S. (1947). *Fish and Wildlife Service Fish. Bull.* **43**, 59–186.
239. Mawdesley-Thomas, L. E. (1971). *New Scient.* **49**, 74–75.
240. Pringle, B. H., Hissong, D. E., Katz, E. L. and Mulawka, S. T. (1968). *J. sanit. Engng. Div. Proc. Am. Soc. civ. Engrs.* 94(SA3), Paper 5970, 455–475.
241. Schatz, A., Schalscha, E. B. and Schatz, V. (1964). *Compost Science* Spring 1964, 26–30.
242. Raymont, J. E. G. and Shields, J. (1962). *In* "Proc. 1st International Conference on Water Pollution Research", Section 3, Paper 43, 1–27. Pergamon, Oxford.
243. Hammond, A. L. (1970). *Science, N.Y.* **171**, 788–789.
244. Portmann, J. E. (1970). *In* "F.A.O. Technical Conference on Marine Pollution and its Effects on Living Resources and Fishing", F.I.R.: MP/70/E-32: 8 pp. Rome, 9–18 December 1970.
245. Sprague, J. B., Elson, P. F. and Saunders, R. L. (1965). *Int. J. Air. Wat. Pollut.* **9**, 531–543.
246. Bryan, G. W. (1964). *J. mar. biol. Ass. U.K.* **44**, 549–563.
247. Bryan, G. W. (1969). *J. mar. biol. Ass. U.K.* **49**, 225–243.
248. Keckes, S. and Miettinen, J. K. (1970). *In* "F.A.O. Technical Conference on Marine Pollution and its Effects on Living Resources and Fishing", F.I.R.: MP/70/R-26: 34 pp. Rome, 9–18 December 1970.
249. Rissanen, K., Ericama, J. and Miettinen, J. K. (1970). *In* "F.A.O. Technical Conference on Marine Pollution and its Effects on Living Resources and Fishing", F.I.R.: MP/70/E-61: 4 pp. Rome, 9–18 December 1970.
250. Corner, E. D. S. (1959). *Biochem. Pharmac.* **2**, 121–132. *J. mar. biol. Ass. U.K.* **39**, 414 (1960).
251. Corner, E. D. S. and Rigler, F. H. (1958). *J. mar. biol. Ass. U.K.* **37**, 85–96.
252. Corner, E. D. S. and Sparrow, B. W. (1956). *J. mar. biol. Ass. U.K.* **35**, 531–548.

253. Tillander, M. and Miettinen, J. K. (1970). *In* "F.A.O. Technical Conference on Marine Pollution and its Effects on Living Resources and Fishing", F.I.R.: MP/70/E-67: 4 pp. Rome, 9–18 December 1970.
254. Järvenpää, T., Tillander, M. and Miettinen, J. K. (1970). *In* "F.A.O. Technical Conference on Marine Pollution and its Effects on Living Resources and Fishing", F.I.R.: MP/70/E-66: 6 pp. Rome, 9–18 December 1970.
255. Miettinen, J. K., Heyraud, M. and Keckes, S. (1970). *In* "F.A.O. Technical Conference on Marine Pollution and its Effects on Living Resources and Fishing", F.I.R.: MP/70/E-90: 8 pp. Rome, 9–18 December 1970.
256. Ünlü, M. Y., Heyraud, M. and Keckes, S. (1970). *In* "F.A.O. Technical Conference on Marine Pollution and its Effects on Living Resources and Fishing", F.I.R.: MP/70/E-68: 6 pp. Rome, 9–18 December 1970.
257. Boney, A. D. and Corner, E. D. S. (1959). *J. mar. biol. Ass. U.K.* **38**, 267–275.
258. Boney, A. D., Corner, E. D. S. and Sparrow, B. W. (1959). *Biochem. Pharmac.* **2**, 37–49.
259. Corner, E. D. S. and Sparrow, B. W. (1957). *J. mar. biol. Ass. U.K.* **36**, 459–472.
260. Barnes, H. and Stanbury, F. A. (1948). *J. exp. Biol.* **25**, 270–275.
261. Clarke, G. L. (1947). *Biol. Bull. mar. biol. Lab., Woods Hole.* **92**, 73–91.
262. Weiss, C. M. (1947). *Biol. Bull. mar. biol. Lab., Woods Hole.* **93**, 56–63.
263. Wisely, B. and Blick, R. A. P. (1967). *Aust. J. mar. freshwat. Res.* **18**, 63–72.
264. Fitzgerald, G. P. and Faust, S. L. (1963). *Wat. Sewage Wks.* **110**, 296–298.
265. Fitzgerald, G. P. and Faust, S. L. (1963). *Appl. Microbiol.* **11**, 345–351.
266. Fitzgerald, G. P. (1964). *Appl. Microbiol.* **12**, 247–253.
267. Fitzgerald, G. P. and Faust, S. L. (1965). *Wat. Sewage Wks* July 1965: 5 pp.
268. Howard, T. E., Halverson, H. N. and Walden, C. C. (1964). *Am. J. Hyg* **79**, 33–44.
269. Tarring, R. C. (1967). *In* "Proc. Institute of Water Pollution Control, Annual Conference held at Torquay, 20–23 June, 1967", 2A, 2–8.
270. Marchetti, R. (1965). *Stud. Rev. gen. Fish. Counc. Medit.* **26**, 32 pp. Rome: F.A.O.
271. van Dorn, W. G. (1953). *J. mar. Res.* **12**, 249–276.
272. Perkins, E. J. (1970). *Br. Sci. News. Spectrum.* No. 73(6), 2 pp.
273. Mann. H. G. W. (1970). *In* "F.A.O. Technical Conference on Marine Pollution and its Effects on Living Resources and Fishing", F.I.R.: MP/70/E-23: 3 pp. Rome, 9–18 December 1970.
274. Eisler, R. (1965). *Trans. Am. Fish. Soc.* **94**, 26–31.
275. Bellan, G., Caruelle, F., Foret-Montardo, P., Kaim-Malka, R. A. and Leung-Tack, K. (1969). *Tethys* **1**, 367–374.
276. Bellan, G., Foret, J.-P., Foret-Montardo, P. and Kaim-Malka, R. A. (1970). *In* "F.A.O. Technical Conference on Marine Pollution and its Effects on Living Resources and Fishing", F.I.R.: MP/70/E-7: 5 pp. Rome, 9–18 December 1970.
277. Corner, E. D. S., Southward, A. J. and Southward, E. C. (1968). *J. mar. biol. Ass. U.K.* **48**, 29–47.
278. Wilson, D. P. (1968). *J. mar. biol. Ass. U.K.* **48**, 177–186.
279. Wilson, D. P. (1968). *J. mar. biol. Ass. U.K.* **48**, 183–196.
280. Wilson, K. W. (1970). *In* "F.A.O. Technical Conference on Marine Pollution and its Effects on Living Resources and Fishing", F.I.R.: MP/70/E-45: 7 pp. Rome, 9–18 December 1970.

281. Perkins, E. J. (1970). *Chemy Ind.* **1970**, 14–22.
282. Bryan, G. W. (1970). *J. mar. biol. Ass. U.K.* **49**, 1067–1092.
283. Davis, C. C. (1970). *In* "F.A.O. Technical Conference on Marine Pollution and its Effects on Living Resources and Fishing", F.I.R.: MP/70/R-2: 13 pp. Rome, 9–18 December 1970.
284. Boney, A. D. (1968). *Fld Stud.* **2**(Suppl.), 55–72.
285. Boney, A. D. (1970). *J. mar. biol. Ass. U.K.* **50**, 461–473.
286. Marchetti, R. (1965). *Ann. appl. Biol.* **55**, 425–430.
287. Water Pollution Research Laboratory. Biodegradation Aeration Screening Test.
288. Bardach, J. E., Fujiya, M. and Hall, A. (1965). *Science, N.Y.* **148**, 1605.
289. Manwell, C. and Baker, C. A. M. (1967). *J. mar. biol. Ass. U.K.* **47**, 659–675.
290. Johnson, D. W. (1968). *Trans. Am. Fish. Soc.* **97**, 398–424.
291. Carson, R. (1962). "Silent Spring", Houghton-Mifflin, Boston, Mass.
292. Hughes, J. S. and Davis, J. T. (1962). *Proc. La. Acad. Sci.* **25**, 86–93.
293. Lachlan, J. (1963). *Can. J. Bot.* **41**, 35–40.
294. Funderburk, H. H. and Lawrence, J. M. (1963). *Weeds* **11**, 217–219.
295. Hughes, J. S. and Davis, J. T. (1963). *Weeds* **11**, 50–53.
296. Walker, C. R. (1963). *Weeds* **11**, 226–232.
297. Harp, G. L. and Campbell, R. S. (1964). *J. Wildl. Mgmt* **28**, 308–317.
298. Sanders, H. O. (1970). *J. Water Pollut. Control Fed.* **42**, 1544–1550.
299. Walker, C. R. (1964). *Weeds* **12**, 134–139.
300. Walker, C. R. (1964). *Weeds* **12**, 267–268.
301. Walker, C. R. (1964). *Wat. Sewage Wks.* **111**(3), pp. 7.
302. Boney, A. D. (1963). *J. mar. biol. Ass. U.K.* **43**, 643–652.
303. Lüdemann, D. and Kayser, H. (1962). *Z. angew. Zool.* **49**, 447–463.
304. Lüdemann, D. and Neumann, H. (1962). *Sonderabdruck Anzeiger für Schädlingskunde* **35**, 5–9.
305. Mathur, D. S. (1963). *Zoologica Pol.* **13**, 99–108.
306. Mulla, M. S. (1963). *Mosquito News.* **23**, 299–303.
307. Muncy, R. J. and Oliver, A. D. (1963). *Trans. Am. Fish. Soc.* **92**, 428–431.
308. Holden, A. V. (1965). *Rep. Challenger Soc.* **3**(17), 37.
309. Velsen, F. P. J. and Alderdice, D. F. (1967). *J. Fish. Res. Bd. Can.* **24**, 1173–1175.
310. Fletcher, W. W. (1968). *Philos. J.* **5**, 151–159.
311. Sanders, H. O. and Cope, O. B. (1968). *Limnol. Oceanogr.* **13**, 112–117.
312. Saunders, J. W. (1969). *J. Fish. Res. Bd. Can.* **26**, 695–699.
313. Cairns, J. (1968). *Mosquito News.* **28**, 177–179.
314. Prather, J. W. and Ferguson, D. E. (1966). *J. Miss. Acad. Sci.* **72**, 317.
315. Greer, G. L. and Paim, U. (1968). *J. Fish. Res. Bd. Can.* **25**, 2321–2326.
316. Ferguson, D. E., Tisdale, J. L., Callahan, R. L. and Catching, G. (1962). *In* "Proc. 16th Annual Conf. Southeastern Assoc. of Game and Fish Commissioners: October 14–17, 1962, Charleston, S. Carolina", pp. 142–145.
317. Vinson, S. B., Boyd, C. E. and Ferguson, D. E. (1963). *Science, N.Y.* **139**, 217–218.
318. Vinson, E. B., Boyd, C. E. and Ferguson, D. E. (1963). *Herpetologica* **19**, 77–80.
319. Ferguson, D. E., Ludke, J. L. and Murphy, G. G. (1966). *Trans. Am. Fish. Soc.* **95**, 335–344.
320. Ferguson, D. E. (1967). *In* "Trans. 39th N. American Wildl. and Nat. Res. Conf." 103–107.

321. Ferguson, D. E. (1968). *In* "Reservoir Fishery Resources Symposium, Athens, Georgia, April 5–7, 1957", pp. 531–536.
322. Ludke, J. L., Ferguson, D. E. and Burke, W. D. (1968). *Trans. Am. Fish. Soc.* **97**, 260–263.
323. Ferguson, D. E. and Bingham, C. R. (1966). *Trans. Am. Fish. Soc.* **95**, 325–326.
324. Ferguson, D. E. and Bingham, C. R. (1966). *Bull. Environ. Toxicol.* **1**, 97–103.
325. Ferguson, D. E., Ludke, J. L., Finley, M. T. and Murphy, G. G. (1967). *J. Miss. Acad. Sci.* **13**, 138–140.
326. Rosato, P. and Ferguson, D. E. (1968). *Bio. Sci.* **18**, 783–784.
327. Dimond, J. (1967). *Maine Forest Service and Conversation Foundation (Washington, D.C.), Bull.* No. 23, 21 pp.
328. Ferguson, D. E., Gardner, D. T. and Lindley, A. L. (1966). *Mosquito News* **26**, 80–82.
329. Mathur, D. S. (1965). *Sci. Cult.* **31**, 258–259.
330. Mathur, D. S. (1964). *Proc. natn. Acad. Sci. India.* B, **34**, 337–338.
331. Mathur, D. S. (1962). *Proc. natn. Acad. Sci. India.* B, **32**, 429–434.
332. Cairns, J., Foster, N. R. and Loose, J. J. (1967). *Proc. Acad. nat. Sci. Philad.* **119**, 75–91.
333. Cairns, J. and Scheier, A. (1964). *Noctulae Naturae* No. 370, 10 pp.
334. Dean, H. J. (1963–64). *N.Y. St. Conserv.* (1963–64), 33–35.
335. Burdick, G. E. (1964). *Trans. Am. Fish. Soc.* **93**, 127–136.
336. Dugan, P. R., Pfister, R. M. and Sprague, M. L. (1963). *N.Y. St. Dep. Hlth, Res. Rep.* No. 10, Pts. I and II, 80 pp. and 130 pp.
337. Butler, P. A., Childress, R. and Wilson, A. J. (1970). *In* "F.A.O. Technical Conference on Marine Pollution and its Effects on Living Resources and Fishing", F.I.R.: MP/70/E-76. Rome, 9–18 December 1970.
338. Butler, P. A. (1962). *U.S. Dept. of the Interior, Fish and Wildlife Service, Circ.* No. 143, 20–24.
339. Butler, P. A. (1963). *U.S. Dept. of the Interior, Fish and Wildlife Service, Circ.* No. 167, 11–25.
340. Butler, P. A. (1964). *U.S. Dept. of the Interior, Fish and Wildlife Service, Circ.* No. 199, 9–15.
341. Butler, P. A. (1965). *U.S. Dept. of the Interior, Fish and Wildlife Service. Circ.* No. 226, 65–71.
342. Butler, P. A. (1966). *Am. Fish. Soc. Spec. Publ.* No. 3, 110–115.
343. Loosanoff, V. L. (1959). *In* "Trans. 2nd Seminar on Biological Problems in Water Pollution. April 20–24, 1959", U.S. Public Health Service, Robert A. Taft Sanitary Engineering Center, Cincinnati, Ohio: 5 pp.
344. Eisler, R. (1970). *In* "F.A.O. Technical Conference on Marine Pollution and its Effects on Living Resources and Fishing", F.I.R.: MP/70/E-12: 9 pp. Rome, 9–18 December 1970.
345. Ernst, W. (1970). *In* "F.A.O. Technical Conference on Marine Pollution and its Effects on Living Resources and Fishing", F.I.R.: MP/79/E-14: 3 pp. Rome, 9–18 December 1970.
346. Haderlie, E. C. (1970). *Proc. Challenger Soc.* **4**(2): 81–82.
347. Eisler, R. and Edmunds, P. H. (1966). *Trans. Am. Fish. Soc.* **95**, 153–159.
348. Engel, R. H., Neat, M. J. and Hillman, R. E. (1970). *In* "F.A.O. Technical Conference on Marine Pollution and its Effects on Living Resources and Fishing", F.I.R.: MP/70/E-13: 7 pp. Rome, 9–18 December 1970.

349. Holden, A. V. (1970). *In* "F.A.O. Technical Conference on Marine Pollution and its Effects on Living Resources and Fishing", F.I.R.: MP/70/E-63: 14 pp. Rome, 9–18 December 1970.

350. Koeman, J. H. and van Genderen, H. (1970). *In* "F.A.O. Technical Conference on Marine Pollution and its Effects on Living Resources and Fishing", F.I.R.: MP/70/E-21: 12 pp. Rome, 9–18 December 1970.

351. Stout, V. F., Beezhold, F. L. and Houle, C. E. (1970). *In* "F.A.O. Technical Conference on Marine Pollution and its Effects on Living Resources and Fishing", F.I.R.: MP/70/E-106: 8 pp. Rome, 9–18 December 1970.

352. Koeman, J. H., Brauw, Ten Noever De, Brauw, M. C. and de Vos, R. H. (1969). *Nature, Lond.* **221**, 1126–1128.

353. Jensen, S., Lange, R., Jernelov, A. and Palmork, K. H. (1970). *In* "F.A.O. Technical Conference on Marine Pollution and its Effects on Living Resources and Fishing", F.I.R.: MP/70/E-88: 8 pp. Rome, 9–18 December 1970.

354. Natarajan, K. V. (1970). *J. Wat. Pollut. Control Fed. Res. Suppl.* **42**, R 184–R 190.

355. Doudoroff, P. (1956). *Sewage ind. Wastes* **28**, 1020.

356. Buchanan, D. (1967). *Wat. Pollut. Control.* **66**, 570–582.

356a. Perkins, E. J. (1972). *Mar. Pollut. Bull.*, **3**, 86–88.

357. Zitko, V. *et al.* (1970). *J. Fish. Res. Bd. Can.* **27**, 21–29.

358. Holmes, N. (1970). *Mar. Poll. Bull.* **1**, 105–106.

359. Waugh, G. D. (1964). *Ann. appl. Biol.* **54**, 423–440.

360. Nash, C. E. (1970). *Mar. Pollut. Bull.* **1**, 28–30.

361. Alderson, R. (1970). *In* "F.A.O. Technical Conference on Marine Pollution and its Effects on Living Resources and Fishing", F.I.R.: MP/70/E-3: 8 pp. Rome, 9–18 December 1970.

362. Naylor, E. (1965). *Adv. Mar. Biol.* 1965, 63–103.

363. Markowski, S. (1959). *J. Anim. Ecol.* **28**, 243–258.

364. Markowski, S. (1962). *J. Anim. Ecol.* **31**, 43–51.

365. Markowski, S. (1960). *J. Anim. Ecol.* **29**, 349–357.

366. Naumann, J. W. and Cory, R. L. (1969). *Chesapeake Sci.* **10**, 218–226.

367. Coughlan, J. (1969). *C.E.R.L. Laboratory Note* No. RD/L/N7/69: 12 pp.

368. Schafer, C. T. (1970). *In* "F.A.O. Technical Conference on Marine Pollution and its Effects on Living Resources and Fishing", F.I.R.: MP/70/E-36: 14 pp. Rome, 9–18 December 1970.

369. Hawes, F. B. (ed.). (1968). "Hydrobiological Studies in the River Blackwater in Relation to the Bradwell Nuclear Power Station", C.E.G.B., London. 65 pp.

370. Raymont, J. E. G. and Carrie, B. G. A. (1964). *Int. Revue ges. Hydrobiol. Hydrogr.* **49**, 185–232.

371. Bowman, T. E. (1961). *Chesapeake Sci.* **2**, 206–207.

372. Reeve, M. R. and Cosper, E. (1970). *In* "F.A.O. Technical Conference on Marine Pollution and its Effects on Living Resources and Fishing", F.I.R.: MP/70/E-59: 4 pp. Rome, 9–18 December 1970.

373. Barnett, P. R. O. (1970). *Proc. Challenger Soc.* **4**(2), 77–78.

374. Adams, J. R., Gormly, H. L. and Doyle, M. J. (1970). *Mar. Pollut. Bull.* **1**, 140–142.

375. Pearce, J. B. (1969). *Chesapeake Sci.* **10**, 227–233.

376. Christianson, A. G. (1968). "Industrial Waste Guide on Thermal Pollution", U.S. Dept. of the Interior, Federal Water Pollution Control Administration, Corvallis, Oregon: 112 pp.

377. Alasbaster, J. (1962). *In* "Proc. 1st International Conference Water Pollution Research", Section 1, Paper 1, 1–28. Pergamon, Oxford.
378. Zeller, R. W., Simison, H. E., Weathersbee, E. J., Patterson, H., Hansen, G. and Hildebrandt, P. (1969). "A survey of thermal power plant cooling facilities". Pollution Control Council, Pacific Northwest Area, 54 pp.
379. Cairns, J. (1968). *Scientist and Citizen.* **10**, 187–198.
380. Wood, E. J. F. (1969). *Chesapeake Sci.* **10**, 172–174.
381. Bader, R. G., Roessler, M. A. and Thorhaug, A. (1970). *In* "F.A.O. Technical Conference on Marine Pollution and its Effects on Living Resources and Fishing", F.I.R.: MP/70/E-4: 6 pp. Rome, 9–18 December 1970.
382. Drost-Hansen, W. (1969). *Chesapeake Sci.* **10**, 281–288.
383. Ordal, E. J. and Pacha, R. E. (1967). *In* "Proc. 12th Pacific Northwest Symposium on Water Pollution Research, Corvallis, Oregon, 7 November, 1963", U.S. Dept. of the Interior, Federal Water Pollution Control Administration, Northwest Region, pp. 39–53.
384. Mauchline, J. (1963). H.M.S.O., U.K.A.E.A., A.H.S.B. (RP)R27.
385. Carlisle, D. B. (1958). *Nature, Lond.* **181**, 933.
386. Perkins, E. J., Williams, B. R. H. and Gorman, J. (1966). H.M.S.O., U.K.A.E.A., P.G. Report 752(CC): 17 pp.
387. Mauchline, J. and Taylor, A. M. (1964). *Limnol. Oceanogr.* **9**, 303–309.
388. Jones, R. F. (1960). *Limnol. Oceanogr.* **5**, 312–325.
389. Gutknecht, J. (1961). *Limnol. Oceanogr.* **6**, 426–431.
390. Droop, M. R. (1968). *J. mar. biol. Ass. U.K.* **48**, 689–733.
391. Renfro, W. C. and Osterberg, C. (1969). *In* "Proc. 2nd National Symposium on Radioecology, Ann Arbor, Michigan, 1967" (D. J. Nelson and F. C. Evans, eds), pp. 372–379, U.S.A.E.C. Conference Number 67053.
392. Williams, B. R. H., Perkins, E. J. and Hinde, A. (1965). H.M.S.O., U.K.A.E.A., P.G. Report 611 (CC).
393. Mauchline, J., Taylor, A. M. and Ritson, E. B. (1964). *Limnol Oceanogr.* **9**, 187–194.
394. Peeters, E. and Mertens, M. (1970). *In* "F.A.O. Technical Conference on Marine Pollution and its Effects on Living Resources and Fishing", F.I.R.: MP/70/E-29: 8 pp. Rome, 9–18 December 1970.
395. Perkins, E. J. and Williams, B. R. H. (1966). H.M.S.O., U.K.A.E.A., P.G. Report 587 (CC).
396. Perkins, E. J. and Williams, B. R. H. (1966). H.M.S.O., U.K.A.E.A., P.G. Report 753 (CC).
397. Williams, B. R. H. and Perkins, E. J. (1965). H.M.S.O., U.K.A.E.A., P.G. Report 650 (CC).
398. Fontaine, M. (1955). *J. Cons. perm. int. Explor. Mer* **21**, 241–249.
399. Alexander, P. (1959). "Atomic Radiation and Life", Penguin (Pelican), Harmondsworth, Middlesex.
400. Rice, T. R., Baptist, J. P. and Cross, F. A. (1970). *In* "F.A.O. Technical Conference on Marine Pollution and its Effects on Living Resources and Fishing", F.I.R.: MP/70/E-4: 9 pp. Rome, 9–18 December 1970.
401. Templeton, W. L., Nakatani, R. E. and Held, E. (1970). *In* "F.A.O. Technical Conference on Marine Pollution and its Effects on Living Resources and Fishing", F.I.R.: MP/70/R-10: 19 pp. Rome, 9–18 December 1970.
402. Brown, V. M. and Templeton, W. L. (1964). *Nature, Lond.* **203**, 1257–1259.
 X. Perkins, E. J. Previously unpublished work.
 Y. Gribbon, E. and Perkins, E. J. Previously unpublished work.

17

The Management of Estuaries and Coastal Waters

"Try killed can't long ago."
Anon.

This, the final chapter, must clearly be something of an examination of the problems facing those responsible for estuaries and coastal waters and of the choices which must be made if they are to be managed adequately in the future. The irritating parental admonition of old Herefordshire that "Try killed can't long ago" may be considered apposite, for the saying evidently recognized that can't should be interpreted as a lack of will, but that the whole history of mankind has been the accomplishment of the apparently impossible.

To recapitulate, the continental shelf comprises some 3% of the earth's surface and is the site of almost all the commercial fisheries and most other exploitation of the sea. The seaboard represents a proportion only of this 3%, and thus forms an insignificant proportion of the world as a whole, but it is here that human maritime activity is most concentrated and it is here too that the nursery and fishing grounds of so many commercially important species are found. Inevitably, estuaries and coastal waters, like the rest of the earth, will be subjected to increasing pressure and exploitation in the future.

For reasons, not always sound, many interests, which conflict mutually and with the fisheries, are seeking either a foothold or total domination of substantial areas of this critically important part of the marine environment. Such interests include military installations, housing (particularly the second house), construction of new deep water harbours, construction of new industrial complexes (including power stations, oil terminals and refineries, steel works and miscellaneous chemical works), and leisure activities such as climbing, sand-yachting, dune buggying, marinas (with the concommitant problems of waste and sea beds fouled by moorings), and the general wear and tear produced in the continual use by a multitude of people with nothing better to do. In the absence of adequate management these pressures can hardly do other than adversely affect the water quality. Yet a high water quality is essential if fish farming in our waters is ever to be a success. Even without a deterioration in water quality the fisheries are under pressure from sea angling and the problem of over-fishing is in no way helped by the widespread poaching which occurs. Moreover, it is often forgotten or ignored by these other interests that the fisheries are a resource that, with good management, i.e. taking care neither to overfish nor to produce a poor environmental quality, is there for man's taking forever. Surely, this gift is not to be despised in a world already very aware of a large, expanding human population and inadequate means of sustenance.

Although one customarily receives the assurance that this or that "development" of the seaboard is necessary, it is perhaps instructive to consider some of these sources of conflict from the point of view of optimum management of global resources, and whether indeed they are as necessary as the protagonists would have everyone believe.

In the national interest, much of the coastline of Britain is dominated by military activity. In each case, no other activity is tolerated on land or on the seashore and the adjacent sea fishing grounds cannot be used (and thus pressure on those available is increased). True some fishing and nursery grounds may gain a respite from fishing activity, but is a country like Britain, already some 200 yr past its optimal condition, really managing its assets to its best advantage when such a significant proportion of its coastal resources are devoted to such an essentially wasteful end?

Salt marshes and dune lands are viewed with hungry eyes by "developers" who characteristically employ the euphemism "reclamation" to describe their activities, and thus presuppose that such areas make no contribution to the welfare of man either through the fisheries, or the grazing of farm animals, or just as a place of peace. Given these attributes, can the spoilation by toxic spoil heaps, and the creation of industrial sites, schemes of second houses, and marinas really be justified? For surely the active promotion of such schemes on a world wide scale carries the clear implication that man may work and play, but eat an adequate diet he must not. Moreover, since these lands are all low lying and subject to floods, storm surges and tsunami, their "development" brings an inevitable, but real, risk to human life and well-being.

Apart from the schemes noted in the foregoing paragraph, the oil industry bears a considerable responsibility for the destruction and alteration of salt marsh and shallow bay in North America. Nevertheless, despite these experiences, it seems evident that precisely the same mistakes will be made in Britain. Is it really justifiable to destroy or permanently change areas of natural beauty, of scientific interest and of some consequence in filling the belly of man for the sake of a relatively short term gain? By their nature, oil rigs will undergo considerable movement during their working life. Where then is the environmental management in a situation of world over-capacity in ship-building, and where existing shipyards can construct the rigs necessary, and yet it is proposed to erect temporary construction yards relatively so little nearer the site of use, that in a lifetime of movement of the rigs the gain must be negligible. It is on the debit side, however, that the real disadvantages appear: since the construction of these rigs requires a particular skill, unlikely to be available at these prospective sites, an immigrant labour force will be required, surely not an advantage when the performance is not strictly necessary. It may seem good sense to the companies involved to suggest that once the job is done that the wrecked lands can be restored, but can they? Where a large blow out of dunes has occurred naturally the erosion train, so easy to start, has proved very difficult to halt. Moreover, by even a "temporary" interference with sand dune and marshland for this purpose a diminished supply of nutrients and external metabolites to the inshore fishing grounds must result[1] and the grounds themselves may be modified physically. Similarly, the provision of houses which exceeds that normally required by the host community must lead to an inevitable loss of agricultural land. Yet by using a migrant work force members are lost to the parent community which in consequence undergoes adverse changes in social structure. Indeed, can it ever be justifiable to undertake new developments, with the concommitant migration of a work force, when older industrial areas are either in desperate need of improvement or are perfectly capable of handling the work required? No one who suggested, nor is it suggested, that we should abandon the use of petroleum products, could expect a hearing, for this would be both impractical and gain little support. However, the environmental record of the oil industry as a whole, as opposed to that of specific companies, is poor and is likely to remain so until such time as the best principles of environmental management are taken into its calculations.

More difficult to understand, and probably the least defensible of the indignities imposed by man upon the shores and coastal waters, is dumping. Why if a material is too toxic to release to a river should it be considered more acceptable in coastal waters where marked reconcentration and return to the estuaries and marshlands frequently occurs? Why, too, should a hole or deep in the coastal seabed be regarded as a suitable repository for these materials? Certainly around Britain, the seabed has such mobility that for a hole or deep to persist, it must be maintained by active hydrographical forces, and thus what goes in must come out! The urge to fill the marine deeps may be irresistible, but at best is often a waste of valuable materials and at worst is a confession of failure with results that seriously conflict with fisheries[2, 3, 4] (p. 449), leisure use and other more adequately managed industrial uses of the sea. Gravel extraction, the converse of dumping, is often practised since this represents an easily accessible and cheap source of this essential material. However, such extraction may have far reaching consequences, for despite all protests and with the consent of the Board of Trade, the contractors for the extension to Devonport Dockyard removed in the period 1897–1902 some 650 000 tons of intertidal shingle at Hallsand, Devon. In the winter of 1903–04, 12 houses were lost to the sea and despite the construction of sea wall, 24 of the 25 remaining habitations, robbed of the protecting shingle bank, were demolished or badly damaged by a north-easterly gale which coincided with spring tides in January, 1917[5]. Such effects and the observed effects of coal mine and iron works waste on the coast of Cumberland (p. 451) suggest that very great care should be taken either to avoid or to minimize the consequences of such actions. Moreover, they do lead one to question the often glib statements made with respect to oil and natural gas terminals and the temporary shipyards noted above. In an adequately managed seaboard these questions must be resolved.

Because modern power stations require increasingly large amounts of cooling water more are being sited on estuaries and open coasts. Surely no one who has endured the illumination by paraffin lamp and candle light, the dreadful old kitchen ranges, the slightly rancid food which accompanied a lack of refrigeration in summer (a condition which was found in Britain even as late as the 1950s) would seriously advocate that power stations are unnecessary. But, do large power stations really need to waste such large quantities of commercially valuable SO_2, which in the wrong place, i.e. the atmosphere, produce industrial hazes, scorched plants and a reduction in the quality of life of every living organism; the latter day philisophy of the increasingly tall stack merely begs the question since this material must come to earth somewhere. In this field, though, the use of the warm water effluent for farming fish is a welcome advance towards the retrieval of at least some of the 70% of the total heat input lost in these effluents.

Similarly, no one who has experienced the small house at the bottom of the garden, especially in the depths of winter, would seriously advocate that this is the happiest substitute for the water closet. Indeed, the water closet, with the many accompanying advances in hygiene, has done much to improve the lot of mankind. A measure of this improvement may be judged by the manner in which diseases such as cholera have been brought under control in Western Europe. Even as late as 1832, the town of Dumfries, in southern Scotland, with a population of ca. 12 000, suffered an epidemic of this disease in which 837 fell sick, and of these 422 died; in the period during which the disease was rampant the life and habits of everyone in this southern Scottish town and the surrounding areas was totally disrupted[6].

With these advantages gained, can there really be any justification for pouring large quantities of untreated sewage into estuary or sea, bringing eutrophication and infective hepatitis in its train? Given the will, surely there must be a better way to employ these valuable materials, not least perhaps to return nutrients and trace elements to the producing area. No one, but a madman, would suggest that some should not return to the sea since a recycling of nutrients and trace elements is necessary to the maintenance of healthy biological systems. It is merely the amount, scale and manner of this operation which is in question.

Again, the standard of housing which the developed countries enjoy compared with the hovels suffered by their grandparents or great-grandparents[7], indicates clearly what man can accomplish when he really makes up his mind: although even here reality is far short of the attainable.

Nevertheless, it seems to be regarded as axiomatic that achievement must be accompanied by a progressive degeneration of the environment, but must it? There seems to be little point in arguing that some change in the environment must result not only from man's increasing numbers, but also from the artifacts of his cilivilization. However, it is becoming daily more necessary to make conscious choices of what development is acceptable and where it should be allowed to take place. For example, would it be wise to ban totally all development on some estuaries and confine it to those for which it is impossible to see any future other than as a receptacle for effluents, and even in this case, since there may be positive hazards to health and fisheries in the adjacent sea, to what extent must such situations be cleaned up. Alternatively, would it be wiser to "spread the load" over all such situations, but insist that all must be performed and operated to a very high standard. Such decisions can be made only if adequate information is available[8]. However, even in the contemporary situation improvements can be made, and the author has been privileged to work with industries who dissent from the view that to gain industrial benefit man must live in a sty. These industries have taken active steps to determine what is the state of the environment and them, if all is not well, to rectify the situation[9].

Economics have often been used to saddle the environment or self-sufficient communities with unwelcome and destructive developments. No one has really attempted to put a price on the value of a fishery in terms other than that of the catch landed, but, if we are to make balanced decisions, economic assessments of productive value and estimated long term yield must be available for comparison with the total yield from an industrial plant of finite life. Similarly, real efforts must be made to define the cost in ill-health which accrues from pollution of all kinds: it is insensate to make ever increasing investments in a National Health Service if the cause of illness flourishes unabated, and if by removal of this cause the money could be saved or diverted to more worthy causes. These issues remain untouched and though this is clearly not an easy field of study in which to attain precision, economists have a trule vital role to play here if the environment is to be managed rationally.

Experience has shown that the biological disciplines have the great responsibility, towards mankind as a whole, of providing valid information concerning the effect of human activities. A woolly attitude, which either postulates that all such activity is necessarily bad or provides incomplete and ill-considered evidence regarding the human influence, is merely doing a disservice to mankind.

However, the responsibility does not rest with the biologist alone; too often, the answer to a problem in oceanic or coastal biology, pure or applied, lies in a knowledge of the chemistry of sea water and the effect of sea water upon added substances, and of the relationship of both with the sediments. It is to be hoped that increasing numbers of chemists will become interested in these problems in the near future. If they do not, it is no exaggeration to state that many vital aspects of estuarine and coastal biology will remain beyond our grasp.

In the physical environment, much valuable insight has been gained by elegant definitions of the average situation, but in the future such studies must be tempered by the need for a much greater knowledge of "cell" formation and maintenance, and of that portion of the water mass which borders on the shore. These phenomena are evident to all who live with inshore waters, but such knowledge must be formalized. As with chemistry, such information is obviously necessary to an understanding of the biology of an estuary and of inshore waters.

Finally, it is often forgotten, although it was recognized in 1921[10], that it is not a vast mass of information accumulated over a small time scale and for a very small area which carries the mark of value in this field. Rather, it is the very long run of relatively simple measurements carried out over a wide area which is likely to produce the greatest advances in estuarine and coastal biology. Sophisticated methods clearly have their place, but perhaps it is worth remembering that sophistication once had another meaning, and that weariness obscures issues and leads to a lack of judgement. Now that man is so abundant and has the powerful tools of technology at his disposal, he is perhaps the single most important factor in the neritic environment. This being so, it is implicit that he will have to understand these waters more thoroughly then he does now, and, given that understanding, he must manage this attractive environment for the benefit of all. Since the stakes are so high, there is no room for consideration of short term gain; man must begin to think of mankind, instead of man. After all, try really did kill can't a very long time ago.

References

1. Perkins, E. J. (1974). *In* "The Biology of Plant Litter Decomposition" (C. H. Dickinson and G. T. F. Pugh, eds), Academic Press, London.
2. Berge, G., Ljøen, R. and Palmork, K. H. (1970). *In* "F.A.O. Technical Conference on Marine Pollution and its Effects on Living Resources and Fishing", F.I.R.: MP/70/E-73. Rome, 9–18 December 1970.
3. Jensen, S., Jernelov, A., Lange, R. and Palmork, K. H. (1970). *In* "F.A.O. Technical Conference on Marine Pollution and its Effects on Living Resources and Fishing", F.I.R.: MP/70/E-88. Rome, 9–18 December 1970.
4. Saila, S. B. (1971). *Mar. Pollut. Bull.* **2**, 125–127.
5. Wilson, A. (1970). *New Scient.* **45**, 311.
6. Wallace, E. and Duncan, T. T. (1845). "The New Statistical Account of Scotland", Vol. 4 (Dumfries), 6–10, Blackwood, London and Edinburgh.
7. Woodforde, J. (1969). "The Truth about Cottages", Routledge and Kegan Paul, London.
8. Brahtz, J. F. P. (ed.). (1972). "Coastal Zone Management", Wiley, New York.
9. Perkins, E. J. (1972). *Br. Sci. News. Spectrum.* **97**, 12–13.
10. Allen, W. E. (1921). *Ecology*, **2**, 2, 215–219.

Subject Index

distribution and turbulence, 253–254, 256, 325, 329–330
fishery, 294, 407–411
food, 327, 329, 380–381
growth, 327, 331, 378–379
growth, factors controlling, 271, 378–379
hatcheries, 412
influence of sediment movement, 308–310, 441–442
influence of sewage, 535
influence of suspended solids, 94, 310, 369, 514–515
larvae, 94, 181, 273, 339–341, 378, 408
larval settlement, 182, 246, 255, 327, 329–331, 340, 356, 359, 369, 378, 408, 418
larval settlement, critical velocity, 358–360, 369
longevity, 328–330
movement, 329
parasites, 413–415
penetration of estuaries, 174–175, 177, 294
predators and parasites, 245–247, 329, 332, 409, 411, 413–415, 416, 486
radiation resistance, 586
resistance to desiccation, 250
spawning, 272–273
salinity tolerance, 181, 329, 331, 369, 379, 489
temperature tolerance, 329, 369, 380, 581
tolerance of industrial pollution, 542
tolerance of reducing conditions, 327, 330
tolerance of turbulence, 309–310
toxic effects upon, 374, 521, 541, 548–550, 556, 558–560, 562–563, 570–572, 577
uptake of radioactivity, 584–585
Laminarin, 192, 265, 267
Lead, 121, 443, 448, 476–477, 480, 553, 554–556

Lichens
distribution and salinity, 162
penetration of estuaries, 162
tolerance to wave exposure, 251, 256, 295
zonation, 240, 256, 281, 295
Liebigs Law of the Minimum, 533
Light, 15, 18–20, 34–36, 112
albedo, 18, 112
absorption by nitrate, 20
absorption by phytoplankton, 19–20
absorption by sea water, 18–20
absorption by sediment, 19, 34–36, 112
absorption coefficient, see extinction coefficient
attenuation coefficient, 35–36
extinction coefficient, 19
Secchi Disc, 19, 35, 472
transmission exponent, see extinction coefficient
Limiting nutrient, 124, 533
Lunar periodicity in breeding, 337

M

Macrofauna
breeding, 167, 181–182, 332–345, 487
commercial importance, 294, 331, 399–412
communities, 296–301
distribution and anoxic waters, 526
distribution and latitude, 244–246
distribution and salinity, 165–168, 171–182, 294, 318, 321, 324, 326–327, 329–330, 332
distribution and sediment type, 291–293, 296–301, 308–311, 320, 321, 324, 332
distribution and temperature, 259, 268
distribution and water depth, 292–293, 323–324, 329
growth, 267–273, 329, 332–345
influence of desiccation, 171, 295, 326

Systematic Index

A

Geographic Index